Matrix Methods
in Finite
Mathematics

Matrix Methods in Finite Mathematics

An Introduction with Applications to Business and Industry

Steven C. Althoen
University of Michigan—Flint

Robert J. Bumcrot
Hofstra University

W • W • Norton & Company • Inc • New York

Copyright © 1976 by W. W. Norton & Company, Inc.

First Edition

Library of Congress Cataloging in Publication Data

Althoen, Steven C
    Matrix methods in finite mathematics.

    Includes index.
    1. Matrices. 2. Linear programming. I. Bumcrot,
Robert J., 1936–   joint author. II. Title.
QA188.A38   519.7′2   75-38841
ISBN 0-393-09192-9

Published simultaneously in Canada
by George J. McLeod Limited, Toronto

Printed in the United States of America

1 2 3 4 5 6 7 8 9

Contents

Preface   ix

**How to Use
This Book**   xi

Chapter 1   Matrix Arithmetic   1

   1-1   What Is a Matrix?   2
   1-2   Special Types of Matrices   6
   1-3   Multiplication of a Matrix by a Number   10
   1-4   Addition and Subtraction of Matrices   16
   1-5   Matrix Multiplication   25
   1-6   Some Facts about Matrix Arithmetic   45
         Review Exercises   52
         Supplementary Exercises   55
         Summary   57

Chapter 2   Linear Equations   59

   2-1   Coordinates and Lines in the Plane   60
   2-2   Systems of Two Linear Equations in Two Variables   70
   2-3   Linear Equations at Appalachian Creations   79
   2-4   Gauss-Jordan Reduction   84
   2-5   Shortcuts in Gauss-Jordan Reduction   97
   2-6   The Geometry of Linear Equations in Three Variables 106
         Review Exercises   111
         Supplementary Exercises   113
         Summary   118

Chapter 3 Inverse Matrices   119

3-1   Matrix Notation for a System of Linear Equations   120
3-2   Identity Matrices and the Inverse of a Matrix   129
3-3   Calculating the Inverse of a Matrix   136
3-4   Applications of Inverse Matrices   143
3-5   Notation and Properties of Matrix Inversion   149
3-6   A Formula for the 2×2 Inverse   156
      Review Exercises   **160**
      Supplementary Exercises   **161**
      Summary   **165**

Chapter 4 Transportation Problems   167

4-1   Introduction   168
4-2   The Northwest Corner Method   170
4-3   Patterns of Change   177
4-4   Patterns of Change (continued)   190
4-5   Unequal Supply and Demand   199
4-6   The Assignment Problem   210
      Review Exercises   **216**
      Supplementary Exercises   **217**
      Summary   **221**

Chapter 5 The Simplex Method   223

5-1   Introduction   224
5-2   The Geometry of Linear Programming in Two
      Variables   228
5-3   The Simplex Method for Maximization Problems   238
5-4   The Algebra of the Simplex Method   251
5-5   The Simplex Method for Minimization Problems   258
5-6   Linear Programming at Appalachian Creations   267
      Review Exercises   **273**
      Supplementary Exercises   **274**
      Summary   **276**

Chapter 6   Additional Topics   **277**

6-1   Game Theory   278
6-2   Introduction to Markov Chains   288
6-3   Regular and Absorbing Chains   302
6-4   More on the Simplex Method   319
6-5   Determinants and their Applications   346
6-6   Linear Programming and Computers   362
      Supplementary Exercises   **385**

Appendix A   Review of Arithmetic and Algebra   **389**

Appendix B   Solutions and Answers to
Selected Exercises   **399**

Index   **447**

To Jane

Preface

In the last three decades some powerful mathematical tools have been developed to aid in the process of decision making in modern business. This book is an introduction to some of the more important of these techniques. Its principal audience is the student who is planning a career in business, though it is equally suitable for the liberal arts student who wants to understand the uses of mathematics. Linear programming, the principal topic of this book, provides a very clear example of the application of mathematics to practical problems. Every student, even with a limited mathematical background, can learn to use the simplex method to solve complex and obviously important problems.

In this book, each concept and method is illustrated with examples based on the hypothetical problems of imaginary business organizations. While a few of these examples are quite complicated, none are as involved as those generally encountered in real life, so that readers are not overwhelmed with data. We hope that the occasionally lighthearted nature of some of these examples will help make the learning process a pleasant one.

To make the text as accessible as possible to beginning students, mathematical abstraction is held to a minimum. Formal statements and superfluous terminology are avoided. The instructor may be surprised to note the omission of such terms as *set*, *vector*, and *vector space*. Double subscripts are also avoided. In more traditional treatments, the student is forced to surmount difficult barriers of notation and terminology in order to master concepts that can just as easily be expressed in plain English. Certainly, rather sophisticated mathematical machinery would be needed to provide rigorous proofs of some of our assertions. Yet a student can understand these assertions, even see why they are true, without such proofs.

The following diagram indicates the interdependence of the various sections of the book. It also gives our assessment of their relative difficulty, the bottom level being the most difficult.

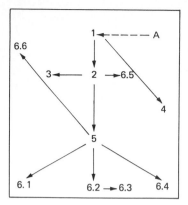

The practical computational core of the book is contained in Sections 1-1–1-5, 2-3–2-4, 3-1–3-4, 4-1–4-6, 5-3–5-6, and 6-1–6-4. With advanced students, such as well-prepared MBA students, it would be possible to cover this core in a semester or perhaps even in a quarter. But for most courses, it will be better to include some of the introductory, motivational, and ancillary material in the other sections, even if some of the practical computations must be omitted for reasons of time. A course for liberal arts students fulfilling a general university requirement might concentrate, for example, on Chapters 1, 2, and 5 (omitting Section 5-4), taking up Chapter 4 or selected topics from Chapter 6 as time allowed.

Two different techniques for minimization with the simplex method are presented. The first uses duality (Section 5-5), the second proceeds by maximizing the negative of the objective variable (Section 6-4). These two sections are interchangeable, except that the technique of Section 5-5 is used in Section 5-6. Students may find Section 6-4 more interesting in that it also includes the technique called Phase I, which expands the domain of applicability of the simplex methods for maximization problems and is the method generally used by computers.

A preliminary version of this book was used in a number of classes of business and liberal arts students at Hofstra University. We are grateful for their patience and constructive criticism, as well as for that of their instructors: Ronald Baslaw, Alfred Vassalotti, and John Weidner. Detailed reviews of the revised manuscript were provided by Karen Collins, Daniel Cullen, Leonard Gillman, Paul Knopp, Stephen Maurer, Lawrence Verner, and John Weidner. These were of great value in improving both the style and the mathematical content of the work. Thanks are due to Joseph B. Janson II and Katherine Hyde of W. W. Norton & Company for arranging for the reviews and preparing the final manuscript for publication. Harold Kuhn was kind enough to let us draw on his definitive knowledge of the field. The Hofstra University computer center was most helpful in the production of the computer printouts in Section 6-6. Finally, special thanks are due to Marcia Althoen, whose cheerful industry throughout the typing and editing of each draft was most appreciated.

Flint, Michigan   S.A.
New York City   R.B.

How to Use This Book

You have only to follow our directions to learn all of the computational techniques described in this book. No mathematical background beyond elementary algebra is required. The only requirement is that you be willing to practice each technique by working an appropriate number of exercises. It would be helpful if you were proficient in working with fractions and negative numbers. If you feel unsure in these areas, try the exercises in Appendix A, all of which are answered in Appendix B. If you have trouble, work through this appendix before starting the book. Otherwise, merely refer to it as needed.

While most of the ideas presented are fully explained in the text, a few facts are needed that are not obvious and that cannot be fully justified without using mathematics beyond the scope of this book. We have consistently introduced such facts with the phrase "it can be shown that." You should have no trouble understanding and accepting these facts, even without proper justification.

The exercises following each section are presented in increasing order of difficulty. If you can work several of the hardest numerical exercises in a set, you probably don't need to work all of the easier ones. If a section contains some word problems (called practical problems in the text), be sure to work some of them, too. This will help develop your ability to apply the techniques you are learning.

Exercises are coded as follows: [A] Before the exercise number means that the answer is given in Appendix B; [C] means that the answer and/or a comment is given in Appendix B; * means that we think the exercise is more difficult than most (all these exercises are answered in Appendix B).

Additional practice may be obtained by working some of the supplementary exercises following each chapter. These exercises are not arranged by section; you must decide which technique to use in solving each one.

Besides all of this, it's a good idea to create and solve your own exercises. One reason the instructor is so far ahead of the class is that he or she is frequently forced to make up good problems for the tests. Often you can learn more from making up one good problem than from solving many.

The first five chapters end with summaries and review exercises. (Since Chapter 6 is a collection of separate topics, it has no summary or review exercises.) In reviewing for an examination on one of these chapters, begin by going through the summary and the flow charts to which it refers. Then give yourself a "closed book" examination consisting of the review exercises. All of these exercises are answered in Appendix B.

Sometimes students waste hours in confusion because of a typographical error in the text. We've tried hard to eliminate these errors, but if you find yourself stuck on what you think should be an easy point, please allow for the possibility that you're right and we're wrong!

Chapter 1
Matrix
Arithmetic

When the managers of Appalachian Creations, Inc., the famous (fictitious) maker and retailer of camping and sporting goods, decided to computerize their operation, they found it necessary to tabulate their production data to make them suitable for data-processing machines. Table 1 is a small part of AC's complete production table for a day in July. AC makes many items at many factories, but throughout our discussions we will concentrate on a few selected factories and a few selected items, so as to shorten our calculations. Of course, a computer that is capable of performing many computations per second would not be bothered by the tedium of considering the complete table.

Table 1
Production Table for July 17

Item	Factory		
	Danbury, CT	Springfield, MA	Rutland, VT
Light sleeping bags	11	13	7
Heavy sleeping bags	3	4	0
Down jackets	5	7	2
Two-person tents	8	15	0
Four-person tents	0	14	4

Without the words in the table, we would not be able to tell what the various numbers represent. For example, we would not know that the "5" represents 5 down jackets produced at Danbury on July 17. A machine, on the other hand, is only capable of dealing with the numbers in the table. Thus, in solving a problem concerning this table, we would feed the machine the data

$$\begin{bmatrix} 11 & 13 & 7 \\ 3 & 4 & 0 \\ 5 & 7 & 2 \\ 8 & 15 & 0 \\ 0 & 14 & 4 \end{bmatrix} \qquad \mathbf{1}$$

and then resupply the headings after the particular numerical problem had been solved.

A rectangular array of numbers such as that shown above is called a **matrix** (plural: **matrices**). This book is about matrices and their application to various types of problems, some of which arise from business situations like the one described above.

We will also want to view a matrix as a listing of rows and columns of numbers. For example, the matrix

$$\begin{bmatrix} 2 & 7 & 1 & 4 \\ 5 & 2 & 3 & 8 \\ 0 & 9 & 1 & 6 \end{bmatrix} \qquad \qquad \mathbf{2}$$

has three rows: 2 7 1 4; 5 2 3 8; 0 9 1 6; and four
columns: $\begin{smallmatrix} 2 \\ 5 \\ 0 \end{smallmatrix}$; $\begin{smallmatrix} 7 \\ 2 \\ 9 \end{smallmatrix}$; $\begin{smallmatrix} 1 \\ 3 \\ 1 \end{smallmatrix}$; $\begin{smallmatrix} 4 \\ 8 \\ 6 \end{smallmatrix}$. In nearly all of our applications the rows
are used more frequently than the columns. In order to emphasize
this, when we mention facts about rows and facts about columns
together, we will always give the row information first.

To illustrate the previous remark, we define the **size** of a matrix: A 3 × 4 (read "3 by 4") matrix is a matrix with 3 rows and 4
columns. For example, matrix **2** above is a 3 × 4 matrix. The AC
production table for July 17 involves the 5 × 3 matrix **1**.

Problem 1   Determine the size of the matrix $\begin{bmatrix} -1 & 7 \\ 0 & 1 \\ 3 & 7 \end{bmatrix}$

Solution   Since the matrix has 3 rows and 2 columns, it is a 3 × 2 matrix.

In practical problems, matrices of great size are encountered.
For instance, the shipping-cost matrix of a large oil company
might easily involve 3000 rows (service stations) and 25 columns
(refineries).

In changing tables into matrices we delete the table headings
and let the position of each number in the matrix tell us what the
number represents. The location of a particular entry becomes
very important. For example, in matrix **1** the two 7's represent
different items (light sleeping bags produced at Rutland and down
jackets produced at Springfield). It is often necessary to refer to a
particular entry of a given matrix. We do this by giving first its
row and then its column. Thus, in the 3 × 5 matrix

$$\begin{bmatrix} 2 & 7 & 8 & 4 & 0 \\ 5 & 2 & 3 & 8 & 5 \\ 0 & 9 & 1 & 6 & 4 \end{bmatrix} \qquad \qquad \mathbf{3}$$

we call 7 the (1,2) **entry**, since 7 is in the first row and the second
column. We also say that (1,2) is the **address** of 7. Similarly, 6 is
the (3,4) entry, and (3,3) is the address of 1. Observe that the
order in which the entry numbers are given matters considerably.
Thus, the (3,2) entry, 9, and the (2,3) entry, 3, are different

numbers in distinctly different positions. To emphasize that the order is important, a notation such as (2,3) is often referred to as an ordered pair of numbers, or more simply as an **ordered pair.**

Problem 2　In the 2×3 matrix $\begin{bmatrix} 4 & -1 & 0 \\ 3 & 1 & 4 \end{bmatrix}$ what are the (1,1) entry, the (2,3) entry, and the (3,2) entry?

Solution　The (1,1) entry is 4, since 4 is in the first row and first column. The (2,3) entry is also 4, since there is a 4 in the second row, third column. There is no (3,2) entry, since the matrix only has 2 rows.

Problem 3　In the 3×2 matrix $\begin{bmatrix} 1 & 2 \\ 7 & 0 \\ -1 & 0 \end{bmatrix}$ what are the addresses of 2, -1, and 0?

Solution　The number 2 is the (1,2) entry, since it is in the first row, second column; -1 is in the third row, first column, so it is the (3,1) entry; and 0 is both the (2,2) entry and the (3,2) entry.

Sometimes we may wish to refer to a matrix without writing it out. For this purpose we denote a matrix by a single capital letter. For example, we might denote production matrix **1** by P so that in later discussions we could speak of "the matrix P" rather than "the matrix $\begin{bmatrix} 11 & 13 & 7 \\ 3 & 4 & 0 \\ 5 & 7 & 2 \\ 8 & 15 & 0 \\ 0 & 14 & 4 \end{bmatrix}$." For another example, suppose we denote matrix **3** by A. Then we may say "the (1,2) entry of A is 7" and "the address of the entry 6 of A is (3,4)" and "the second row of A is 5　2　3　8　5."

In this book we will not make heavy use of this sort of notation. It is of some importance, however, that the reader be familiar with the use of capital letters to represent particular matrices, since this frequently occurs in other books as well as in some technical business reports. For these reasons we will employ this notation from time to time.

Exercises 1.1　Throughout this book, exercises are coded as follows:
A before the exercise number means that the answer is given in Appendix B; C means that the answer or a comment are given in Appendix B; * means that the authors feel that the exercise is more difficult than most (answers to all such exercises are given in Appendix B).

1. Determine the size of each of the following matrices:

^{A}(a) $\begin{bmatrix} 2 & 3 & 1 \\ 4 & 8 & 6 \\ 1 & 5 & 7 \end{bmatrix}$

(b) $\begin{bmatrix} 8 & 4 \\ 2 & 3 \\ 6 & 7 \end{bmatrix}$

(c) $\begin{bmatrix} 2 & 7 & 3 & 6 \\ 8 & 5 & 9 & 4 \end{bmatrix}$

^{A}(d) $\begin{bmatrix} 2 \\ 8 \\ 7 \\ 6 \\ 3 \end{bmatrix}$

(e) $[3 \quad 6 \quad 7 \quad 8 \quad 2]$

(f) $\begin{bmatrix} 3.84 & 7.92 \\ 0.00 & -2.15 \end{bmatrix}$

^{A}(g) $\begin{bmatrix} -2/3 & 0 & 1/5 \\ 2/5 & -6/7 & 2^3/8 \\ 0 & 0 & 0 \\ -1 & 8 & 37 \end{bmatrix}$

2. In the matrix $\begin{bmatrix} 0 & 7 & 3 \\ 5 & 4 & 2 \\ 6 & 2 & 1 \end{bmatrix}$ what number is at each of the following addresses?

   (a) (1,3) ^{A}(b) (2,4)   (c) (1,1) ^{A}(d) (3,2)   (e) (2,3)

3. In the matrix $\begin{bmatrix} 2 & 1 & 3 & 4 \\ 8 & 6 & 1 & 0 \\ 5 & 7 & 0 & 5 \end{bmatrix}$ what is the address of each of the following numbers?

   ^{A}(a) 2   (b) 7 ^{A}(c) 3   (d) 6   (e) 8   (f) 0

4. ^{A}(a) Write the 2×3 matrix that has $3/2$ at address (2,2), 64 at addresses (1,3) and (2,1), 0 at address (2,3), and $7/4$ everywhere else.

   (b) Write the 3×2 matrix that has $-2/3$ at address (2,2), 46 at addresses (3,1) and (1,2), 0 at address (3,2), and $4/7$ everywhere else.

5. Write the 4×5 matrices in which
   ^{A}(a) each entry is equal to the larger of the numbers in its address.
   (b) each entry is equal to the sum of the numbers in its address.
   ^{A}(c) each entry is equal to the product of the numbers in its address.
   (d) each entry is equal to the first number in its address minus the second number in its address.
   ^{A}(e) each entry is equal to the number of its row.
   (f) each entry is equal to the number of its column.
   *(g) each entry is equal to the average of the numbers in its address.

*6. A 2 × 2 matrix A has −3 at address (1,1); the numbers in each row add up to 1; the numbers in each column add up to the same (unknown) number. Write out A.

A7. Tabulate the following data and write the relevant matrix:

> Springfield Plastics Company produces wastebaskets, hard hats, and vacuum cleaner cases, among other things. The company has four factories. The Springfield, Missouri factory produces 2520 wastebaskets, 3240 hard hats, and 760 vacuum cleaner cases a day. In Springfield, Massachusetts, 2000 wastebaskets, 1780 hard hats, and 370 vacuum cleaner cases are produced daily. Springfield, Illinois, runs slightly behind the others, producing 1950 wastebaskets, 1530 hard hats, and 250 vacuum cleaner cases per day. In Springfield, Ohio, the daily runs are 800 wastebaskets, 1360 hard hats, and 125 vacuum cleaner cases.

C8. The Petaluma Hamburger Emporium Corporation (PHE) has grown tremendously since it started with one roadside stand in a small California town in 1936. Originally the stand sold only hamburgers, cheeseburgers, french fries, and three types of soft drinks. By dint of careful scientific planning and management, PHE now controls a chain of 5724 "mini-restaurants" located in every state, every Canadian province, and many countries in Europe. The mini-restaurants, called "stores" by the data-processing department of PHE world headquarters (in San Francisco), are numbered from 1, the store in Petaluma, to 5724, the store just opened in Fairbanks, Alaska. Every store has exactly the same menu, which currently lists 35 separate food and beverage items, although prices vary somewhat from region to region. One of the most important outputs of the data-processing department is the *production matrix,* which gives the quantity of each item sold in one week by each store. This matrix is prepared each Friday from data gathered from the preceding week.

(a) What is the size of the PHE production matrix? (There are two possible answers to this question. Give the answer that is the more practical and explain why.)

(b) How many entries does the PHE production matrix have?

Section 1-2
Special Types of Matrices

Three types of matrices occur often enough to warrant giving them names.

A **square matrix** is a matrix with the same number of rows as columns. For example,

$$\begin{bmatrix} 0 & -14 & 6 & 8 \\ -1 & 2 & -1 & -4 \\ 3 & 8 & 0 & 2 \\ 4 & 7 & 3 & -1 \end{bmatrix}$$

is a 4 × 4 square matrix.

A **row matrix** is a matrix with only one row. For example,

$$[2 \quad 7 \quad 1 \quad 4 \quad 5 \quad 0]$$

is a 1×6 row matrix. Similarly, a **column matrix** contains only one column. For example,

$$\begin{bmatrix} 2 \\ -1 \\ 7 \\ 4 \end{bmatrix}$$

is a 4×1 column matrix.

Problem 1   Which of the matrices of Exercise 1 in Section 1-1 are square, which are row matrices, and which are column matrices?

Answer   Matrices a and f are square, matrix e is a row matrix, and matrix d is a column matrix. Matrices b, c, and g are none of these special types.

In Section 1-1 we saw how to refer to a particular entry in a matrix. We will also need to be able to refer to a particular row or column of a given matrix. Consider for example the matrix

$$\begin{bmatrix} 3 & 1 & 4 & 1 & 3 \\ -1 & 2 & 2 & 2 & 8 \\ 2 & 5 & 1 & 0 & 1 \\ 4 & 3 & 7 & 6 & 4 \end{bmatrix}$$

The symbol R_1 (read "R one") denotes the row matrix whose entries are of the first row of this matrix:

$$R_1 = [3 \quad 1 \quad 4 \quad 1 \quad 3]$$

Similarly,

$$R_4 = [4 \quad 3 \quad 7 \quad 6 \quad 4]$$

In general, the letter R indicates that we are talking about a row and the subscript tells us which particular row we are talking about.

The idea is exactly the same for columns except now we will use a subscripted C. In the previous matrix, the first and fifth column matrices are:

$$C_1 = \begin{bmatrix} 3 \\ -1 \\ 2 \\ 4 \end{bmatrix}, \qquad C_5 = \begin{bmatrix} 3 \\ 8 \\ 1 \\ 4 \end{bmatrix}$$

Problem 2   A matrix A has $R_1 = R_2 = [-3 \quad 0 \quad {}^2\!/_3]$ and $C_1 = \begin{bmatrix} -3 \\ -3 \end{bmatrix}$. Write the matrix.

Solution   Since C_1 is a 2×1 matrix, A must have 2 rows. Since R_1 is a 1×3 matrix, A must have 3 columns. Thus, A is the 2×3 matrix

$$A = \begin{bmatrix} -3 & 0 & {}^2\!/_3 \\ -3 & 0 & {}^2\!/_3 \end{bmatrix}$$

Problem 3   A matrix B has $R_1 = R_2 = [-3 \quad 0 \quad {}^2\!/_3]$ and $C_1 = \begin{bmatrix} -3 \\ 0 \end{bmatrix}$. Write the matrix.

Solution   Since $R_2 = [-3 \quad 0 \quad {}^2\!/_3]$, the (2,1) entry of B must be -3; but since $C_1 = \begin{bmatrix} -3 \\ 0 \end{bmatrix}$, the (2,1) entry must be 0. Thus, there is no such matrix B. In other words, this problem, like many problems that arise when we try to formulate plans for achieving suggested goals, has no soluton. The planners must try to modify the suggested goals.

Problem 4   Write a matrix K that has $R_1 = [2 \quad -4 \quad 7]$, $C_1 = \begin{bmatrix} 2 \\ -4 \end{bmatrix}$, and $C_3 = \begin{bmatrix} 7.0 \\ 3.4 \end{bmatrix}$.

Solution   The data given specify that K be a 2×3 matrix, and they give every entry except the one at address (2,2). Therefore, we may write

$$K = \begin{bmatrix} 2 & -4 & 7.0 \\ -4 & x & 3.4 \end{bmatrix}$$

where x may be any number whatsoever. For example, if we choose x to be -3, we obtain a correct answer:

$$K = \begin{bmatrix} 2 & -4 & 7.0 \\ -4 & -3 & 3.4 \end{bmatrix}$$

However, we could just as well pick x to be 0 and obtain another equally correct answer:

$$K = \begin{bmatrix} 2 & -4 & 7.0 \\ -4 & 0 & 3.4 \end{bmatrix}$$

In fact, since there are infinitely many numbers to choose from, there are infinitely many correct solutions to this problem. We indicate that we cannot list all the specific solutions by using a letter such as x. This is a situation quite diferent from that of Problem 3: here there are an infinite number of solutions, whereas Problem 3 had none. Had this problem arisen in a planning situation, the planners would have found themselves in the happy position of having a free choice.

One final point: a 1×1 matrix such as [3] will always be treated as a number. A 1×1 matrix is a square matrix, a row matrix, and a column matrix. No other matrix can make that claim.

Exercises 1.2   1. Given the matrix $\begin{bmatrix} 3 & 0 & -1 & 4 \\ 2 & 2 & 3 & 1 \\ -1 & 8 & 2 & 6 \end{bmatrix}$, write out the following associated row and column matrices:

A(a) R_2 A(b) C_3 A(c) R_4 (d) C_4 (e) R_1

2. For each case below, write a matrix that contains the given rows and columns. What is the size of the matrix in each of your answers?

C(a) $R_1 = [2 \quad 3 \quad 7]$ (b) $R_2 = [\ 2 \quad -1 \quad 3 \quad 4 \quad 6]$
       $R_2 = [4 \quad 6 \quad 8]$ $R_4 = [-1 \quad 2 \quad 4 \quad 6 \quad 9]$

$C_1 = \begin{bmatrix} 2 \\ 4 \\ 6 \end{bmatrix}$ $C_1 = \begin{bmatrix} 2 \\ 2 \\ 0 \\ -1 \end{bmatrix}$

$C_3 = \begin{bmatrix} 7 \\ 8 \\ -1 \end{bmatrix}$ $C_4 = \begin{bmatrix} 0 \\ 4 \\ 2 \\ 6 \end{bmatrix}$

A 3. In matrix **1** of Section 1-1 what do the numbers in the matrix R_2 represent? What do the numbers in the matrix C_3 represent?

4. Denote matrix **3** of Section 1-1 by B. Write out C_4 and R_2 of B.

*5. A 6×5 matrix B has $C_1 = \begin{bmatrix} 1 \\ 2 \\ 3 \\ 4 \\ 3 \\ 6 \end{bmatrix}$, $C_3 = C_5$, $R_4 = [4 \quad 3 \quad 2 \quad 1 \quad 2]$,

and $R_3 = R_6$. How many more entries do you have to know before you can know B completely?

A6. (a) Give an example of a 3×3 matrix in which the columns are all the same but the rows are all different.

    (b) What can be said about a matrix in which all the rows are the same and all the columns are the same?

*7. If we *switch* R_1 and R_2 in the matrix $\begin{bmatrix} 1 & 2 \\ 3 & 4 \end{bmatrix}$, we get $\begin{bmatrix} 3 & 4 \\ 1 & 2 \end{bmatrix}$. Similarly, we can switch any two columns of a matrix. By switching rows and columns in the matrix $\begin{bmatrix} 1 & 2 & 3 \\ 4 & 5 & 6 \\ 7 & 8 & 9 \end{bmatrix}$, show how to change it to the matrix $\begin{bmatrix} 4 & 6 & 5 \\ 7 & 9 & 8 \\ 1 & 3 & 2 \end{bmatrix}$.

See Exercise 8 in Section 1-1.
8. Consider the PHE production matrix for the week of April 21. Menu item number 14 is "foot-long hot dog on toasted roll." Store number 3227 is the only PHE store in Tuscalosa, Alabama.

    (a) What is the address of the entry giving the total production of foot-long hot dogs on toasted rolls at the Tuscaloosa store during the week of April 21?

    (b) What information is given by C_{14}?

    (c) What information is given by R_{3227}?

    *(d) Suppose we add up all the entries in C_{14}. What does the result tell us?

    *(e) Suppose we add up all the entries in R_{3227}. What does the result tell us?

Section 1-3
Multiplication of a Matrix
by a Number

While the production schedule for Appalachian Creations changes from time to time, it is the same for each workday of any given month. Thus Table 1 in Section 1-1 serves for any day in July. If we suppose there are 22 workdays in July, then to compute the total month's production, we simply multiply each entry in the production matrix **1** of Section 1-1 by 22. The result is the matrix

$$\begin{bmatrix} 242 & 286 & 154 \\ 66 & 88 & 0 \\ 110 & 154 & 44 \\ 176 & 330 & 0 \\ 0 & 308 & 88 \end{bmatrix}$$

which we may present as Table 2.

Table 2
Production Table for July

Item	Factory		
	Danbury	Springfield	Rutland
Light sleeping bags	242	286	154
Heavy sleeping bags	66	88	0
Down jackets	110	154	44
Two-person tents	176	330	0
Four-person tents	0	308	88

Now suppose that increased demand prompts the company
to switch, on July 17, from a 7-hour day to a 9-hour day. Instead
of making 8 two-person tents daily at Danbury, the company will
now produce $9/7 \times 8 = 10\,2/7$ tents. Here the fraction $2/7$ refers to
the fact that production on the last tent of the day is only partially
completed. To derive the new production figures, we multiply
each entry of Table 1, Section 1-1, by the fraction $9/7$, to obtain
Table 3.

For a review of fraction arithmetic,
see Appendix A, Section 4-7

Table 3
Production Table for July 17
(increased production
schedule)

Item	Factory		
	Danbury	Springfield	Rutland
Light sleeping bags	$14\tfrac{1}{7}$	$16\tfrac{5}{7}$	9
Heavy sleeping bags	$3\tfrac{6}{7}$	$5\tfrac{1}{7}$	0
Down jackets	$6\tfrac{3}{7}$	9	$2\tfrac{4}{7}$
Two-person tents	$10\tfrac{2}{7}$	$19\tfrac{2}{7}$	0
Four-person tents	0	18	$5\tfrac{1}{7}$

We say that matrix **1** above was obtained from matrix **1** in
Section 1-1 by *multiplying the latter by* 22:

$$22\begin{bmatrix} 11 & 13 & 7 \\ 3 & 4 & 0 \\ 5 & 7 & 2 \\ 8 & 15 & 0 \\ 0 & 14 & 4 \end{bmatrix} = \begin{bmatrix} 242 & 286 & 154 \\ 66 & 88 & 0 \\ 110 & 154 & 44 \\ 176 & 330 & 0 \\ 0 & 308 & 88 \end{bmatrix}$$

Similarly, we obtain the matrix of Table 3 by multiplying matrix
1 in Section 1-1 by $9/7$:

$$\tfrac{9}{7}\begin{bmatrix} 11 & 13 & 7 \\ 3 & 4 & 0 \\ 5 & 7 & 2 \\ 8 & 15 & 0 \\ 0 & 14 & 4 \end{bmatrix} = \begin{bmatrix} 14\tfrac{1}{7} & 16\tfrac{5}{7} & 9 \\ 3\tfrac{6}{7} & 5\tfrac{1}{7} & 0 \\ 6\tfrac{3}{7} & 9 & 2\tfrac{4}{7} \\ 10\tfrac{2}{7} & 19\tfrac{2}{7} & 0 \\ 0 & 18 & 5\tfrac{1}{7} \end{bmatrix}$$

Obviously this operation, multiplication of a matrix by a number, can be performed with any matrix and any number, and will result in a matrix of the same size as the original matrix.

Problem 1   Multiply the matrix $\begin{bmatrix} -1 & 4 & 7 & 2 \\ 3 & 8 & 6 & 1 \\ -9 & 2 & 4 & 0 \end{bmatrix}$ by $\frac{1}{2}$.

Solution

$$\frac{1}{2}\begin{bmatrix} -1 & 4 & 7 & 2 \\ 3 & 8 & 6 & 1 \\ -9 & 2 & 4 & 0 \end{bmatrix} = \begin{bmatrix} -\frac{1}{2} & 2 & \frac{7}{2} & 1 \\ \frac{3}{2} & 4 & 3 & \frac{1}{2} \\ -\frac{9}{2} & 1 & 2 & 0 \end{bmatrix}$$

It will often be convenient to perform the reverse of this operation, that is, to factor a number out of a matrix. This is particularly useful if the matrix contains fractions. In this case it is usually best to find a common denominator for the entries in the matrix. Here are two examples.

See Appendix A, Section 4-5, for a review of common denominators.

Problem 2   Factor 2 out of the matrix $\begin{bmatrix} -2 & 0 & 4 \\ 6 & 2 & 8 \\ 14 & 8 & 4 \end{bmatrix}$

Solution   We simply divide each entry by 2:

See Appendix A, Section 5, for a review of signed numbers.

$$\begin{bmatrix} -2 & 0 & 4 \\ 6 & 2 & 8 \\ 14 & 8 & 4 \end{bmatrix} = 2\begin{bmatrix} -1 & 0 & 2 \\ 3 & 1 & 4 \\ 7 & 4 & 2 \end{bmatrix}$$

Problem 3   Write the matrix $\begin{bmatrix} \frac{2}{3} & -\frac{2}{5} \\ \frac{4}{15} & \frac{2}{15} \\ 0 & 2 \end{bmatrix}$ as a fraction times a matrix of whole numbers.

Solution   We first write all the entries in the matrix as fractions with the common denominator 15:

$$\begin{bmatrix} \frac{2}{3} & -\frac{2}{5} \\ \frac{4}{15} & \frac{2}{15} \\ 0 & 2 \end{bmatrix} = \begin{bmatrix} \frac{10}{15} & -\frac{6}{15} \\ \frac{4}{15} & \frac{2}{15} \\ 0 & \frac{30}{15} \end{bmatrix}$$

We then factor out $\frac{1}{15}$:

$$\begin{bmatrix} \frac{2}{3} & -\frac{2}{5} \\ \frac{4}{15} & \frac{2}{15} \\ 0 & 2 \end{bmatrix} = \frac{1}{15}\begin{bmatrix} 10 & -6 \\ 4 & 2 \\ 0 & 30 \end{bmatrix}$$

Note that although this is a correct answer to the problem, we can reduce the size of the numbers by factoring out a 2:

$$\begin{bmatrix} {}^2\!/_3 & -{}^2\!/_5 \\ {}^4\!/_{15} & {}^2\!/_{15} \\ 0 & 2 \end{bmatrix} = {}^2\!/_{15} \begin{bmatrix} 5 & -3 \\ 2 & 1 \\ 0 & 15 \end{bmatrix}$$

Problem 4   What number should the matrix $\begin{bmatrix} 2 & -3 & 0 \\ 0 & 4 & -2 \\ 3 & -2 & -4 \end{bmatrix}$ be multiplied by

in order to obtain the matrix $\begin{bmatrix} -6 & 9 & 0 \\ 0 & -12 & 6 \\ -9 & 6 & 12 \end{bmatrix}$

Solution   Suppose we ask the following question: What number must the (1,1) entry of the first matrix be multiplied by in order to obtain the (1,1) entry of the second matrix? Looking at these entries, we see that we are asking, "What number should 2 be multiplied by in order to obtain -6?" Elementary algebra gives us the anwer: -3. Check: $-3 \cdot 2 = -6$. Do we now know that the answer to this problem is -3? No! We know that either -3 is the answer or there is no answer. It is still necessary to check:

See Appendix A, Section 11, if needed.

$$-3 \begin{bmatrix} 2 & -3 & 0 \\ 0 & 4 & -2 \\ 3 & -2 & -4 \end{bmatrix} = \begin{bmatrix} 6 & 9 & 0 \\ 0 & -12 & 6 \\ -9 & 6 & 12 \end{bmatrix} \ \checkmark$$

So the answer really is -3. We remark that instead of working with the (1,1) entries of these matrices, we could have worked with other entries. For example, if we had considered the (1,2) entries, our question would have been, "What number must -3 be multiplied by so as to obtain 9?" Again, we obtain the answer -3. We could not have worked with the (1,3) entries of these matrices, however. Why?

Suppose we denote the first matrix of Problem 4 by A, and the second by B:

$$A = \begin{bmatrix} 2 & -3 & 0 \\ 0 & 4 & -2 \\ 3 & -2 & -4 \end{bmatrix}, \ B = \begin{bmatrix} -6 & 9 & 0 \\ 0 & 12 & 6 \\ -9 & 6 & 12 \end{bmatrix}$$

Then Problem 4 could be formulated thus: "What number must the matrix A be multiplied by to obtain the matrix B?" The result of the solution of the problem is written simply $-3A = B$.

Exercises 1.3   1. Calculate the following products:

A(a) $2 \begin{bmatrix} -1 & 0 & 3 \\ 2 & 4 & 7 \\ 3 & 8 & 5 \end{bmatrix}$

(g) $5 \begin{bmatrix} 2/5 & 3/5 & 0 \\ 7/8 & 5 & 2 \\ -2 & 1 & 2/3 \end{bmatrix}$

(b) $-3 \begin{bmatrix} 2 & -1 \\ 0 & 3 \\ -4 & 1 \end{bmatrix}$

(h) $-3 \begin{bmatrix} -1/4 & 2/3 & 2 \\ 3/8 & -3/4 & 5/6 \end{bmatrix}$

(c) $1/2 \begin{bmatrix} 2 & 0 & -4 & 6 \\ 8 & 4 & -6 & 2 \\ 2 & -2 & 0 & 4 \end{bmatrix}$

(i) $1/3 \begin{bmatrix} 9 & -2 & 4 \\ 7 & 2 & 3 \\ 6 & 9 & 0 \end{bmatrix}$

A(d) $-2/3 \begin{bmatrix} -3 \\ 6 \\ 0 \end{bmatrix}$

A(j) $-3/2 \begin{bmatrix} 2 & 3 & -5 \\ 6 & 7 & 4 \end{bmatrix}$

(k) $1/5 \begin{bmatrix} 3/8 & 1/4 & 5/6 \\ 2/3 & 1/2 & 5 \end{bmatrix}$

(e) $6 \begin{bmatrix} 2/3 & -1 & 5/6 \\ 0 & 1/2 & -4/3 \end{bmatrix}$

A(f) $-12 \begin{bmatrix} 1/3 & -1/2 & -5/6 & -5/3 \end{bmatrix}$

(l) $-3/4 \begin{bmatrix} -2 & 3/8 & -4/3 \\ 0 & 4 & 7 \\ -1/4 & -2/3 & 3/4 \end{bmatrix}$

2. Write the following matrices as the product of a fraction and a matrix of whole numbers:

A(a) $\begin{bmatrix} 3/2 \\ 5/2 \\ -1/2 \end{bmatrix}$

(f) $\begin{bmatrix} 3/4 & -5/12 \\ 1/2 & 2/3 \\ 0 & 5/6 \end{bmatrix}$

(b) $\begin{bmatrix} 2/3 & 3 & -1/3 \\ 2 & 0 & 1/3 \end{bmatrix}$

(g) $\begin{bmatrix} -9/2 & 3/2 & 9/4 & -3 \end{bmatrix}$

A(c) $\begin{bmatrix} 1/2 & -3/4 \\ 1/4 & 0 \\ 1 & 3/4 \end{bmatrix}$

A(h) $\begin{bmatrix} -6 & -9/2 & -3/8 \\ 3/2 & -15/8 & 0 \\ -15/2 & 9/4 & -3 \end{bmatrix}$

(d) $\begin{bmatrix} 1/3 & 2/9 \\ -5/9 & 2/3 \end{bmatrix}$

*(i) $\begin{bmatrix} -7/5 & 4/3 & -2 \\ 5/12 & -3/20 & 7/10 \\ 0 & 9/8 & 2/15 \end{bmatrix}$

A(e) $\begin{bmatrix} 1/4 & -1/3 \\ 2/3 & 3/4 \end{bmatrix}$

3. Which number must the first matrix be multiplied by to obtain the second?

A(a) $\begin{bmatrix} 1 & -1 & 2 \\ 3 & 0 & 1 \\ 2 & 1 & -1 \end{bmatrix}, \begin{bmatrix} 2 & -2 & 4 \\ 6 & 0 & 2 \\ 4 & 2 & -2 \end{bmatrix}$

(c) $\begin{bmatrix} 4 & -2 \\ 6 & 0 \\ -2 & 8 \end{bmatrix}, \begin{bmatrix} 2 & -1 \\ 3 & 0 \\ -1 & 4 \end{bmatrix}$

C(b) $\begin{bmatrix} 2 & 1 & 0 \\ -1 & 2 & 3 \end{bmatrix}, \begin{bmatrix} 6 & 3 \\ -3 & 6 \end{bmatrix}$

^{C}(d) $\begin{bmatrix} 1 & 0 \\ 2 & 1 \end{bmatrix}$, $\begin{bmatrix} 2 & 0 \\ 4 & 1 \end{bmatrix}$

(e) $\begin{bmatrix} 6 & 9 & 0 \\ 3 & 12 & -6 \end{bmatrix}$, $\begin{bmatrix} 4 & 6 & 0 \\ 2 & 8 & -4 \end{bmatrix}$

*(f) $\begin{bmatrix} 2/3 & -3/4 & 7/8 \\ 7/9 & 5/6 & -7/6 \\ 0 & 14/3 & -14/9 \end{bmatrix}$, $\begin{bmatrix} -8/7 & 9/7 & -3/2 \\ -4/3 & -10/7 & 2 \\ 0 & -8 & 8/3 \end{bmatrix}$

4. ^{A}(a) Factor 3 out of the matrix $\begin{bmatrix} 9 & -12 \\ 3 & 0 \end{bmatrix}$

(b) Factor -4 out of $\begin{bmatrix} -8 & 6 \\ -12 & 16 \end{bmatrix}$

^{A}(c) Factor 6 out of $\begin{bmatrix} -12 & 6 \\ 3 & -3 \end{bmatrix}$

(d) Factor -6 out of $\begin{bmatrix} -3 & 4 \\ 4 & 3 \end{bmatrix}$

(e) Factor $2/3$ out of $\begin{bmatrix} 4/3 & -2/3 \\ 4/3 & 8/3 \end{bmatrix}$

*(f) Factor $-3/4$ out of $\begin{bmatrix} -9/4 & 3/5 \\ -4/7 & 5 \end{bmatrix}$

5. What would the total July production matrix for AC be if its workers had been on the increased work schedule (Table 3) during the whole month?

6. What would be the appropriate daily and total July production matrices if AC instituted a 6-hour day?

^{A}7. Suppose that each of 1000 workers at a factory does the same amount of work. If the company reduces its staff by 200 but increases its workday from 10 to 15 hours, by what fraction should it multiply its production matrix to obtain the new daily production matrix? (Assume that the reduction in staff does not affect individual productivity.)

8. Since 1958 the vaccination clinic in Addis Ababa, Ethiopia, has administered shots for typhus, typhoid, yellow fever, and cholera to men, women, and children every day, seven days a week, as shown in Table 4.

Table 4
Number of Shots Given per
Day

	Men	Women	Children
Typhus	80	60	201
Typhoid	60	70	184
Yellow fever	42	24	97
Cholera	145	163	88

Denote by V (for vaccination) the matrix obtained from this table. Write, in terms of V, the matrices that give:

(a) the number of shots given in 1973.

A (b) the number of shots given in 1972.

* (c) the number of shots given in the 1960s.

See Exercise 8 in Section 1-2.

9. Let P be the PHE production matrix for the week of April 21. The board of directors of PHE hopes that production of every item in every store will increase by 20% in one year. Let Q be the production matrix for the week of April 21 a year later.

(a) Assuming that the board's hopes are realized, express Q in terms of P.

(b) The Tuscaloosa store sold 985 foot-long hot dogs on toasted rolls in the week of April 21 this year. Assuming that the board's hopes are realized, what is the (3227,14) entry of matrix Q?

Section 1-4
Addition and Subtraction of Matrices

Appalachian Creations has only three different daily production schedules: the one given in Section 1-1 for the summer months of June, July, and August; one for the winter months of December, January, and February (Table 5); and a third table for the rest of the year (Table 6).

Table 5
Daily Production Table for
Dec., Jan., and Feb.

Item	Factory		
	Danbury	Springfield	Rutland
Light sleeping bags	2	5	0
Heavy sleeping bags	7	9	4
Down jackets	11	15	5
Two-person tents	12	18	0
Four-person tents	0	22	5

Table 6
Daily Production Table for
Mar., Apr., May, Sept., Oct.,
and Nov.

Item	Factory		
	Danbury	Springfield	Rutland
Light sleeping bags	7	8	4
Heavy sleeping bags	4	6	3
Down jackets	8	9	0
Two-person tents	10	16	0
Four-person tents	0	19	6

Since each day's production is the same throughout the en-

JUNE						
S	M	T	W	T	F	S
1	2	3	4	5	6	7
8	9	10	11	12	13	14
15	16	17	18	19	20	21
22	23	24	25	26	27	28
29	30					

JULY						
S	M	T	W	T	F	S
	1	2	3	④	5	
6	7	8	9	10	11	12
13	14	15	16	17	18	19
20	21	22	23	24	25	26
27	28	29	30	31		

AUGUST						
S	M	T	W	T	F	S
					1	2
3	4	5	6	7	8	9
10	11	12	13	14	15	16
17	18	19	20	21	22	23
24 31	25	26	27	28	29	30

Figure 1

tire summer production period, we can use the operation introduced in the previous section to obtain the total summer
production matrix. We simply multiply the daily summer matrix
by 64, the total number of summer workdays in this particular
year (see Figure 1):

TOTAL SUMMER PRODUCTION MATRIX

$$64 \begin{bmatrix} 11 & 13 & 7 \\ 3 & 4 & 0 \\ 5 & 7 & 2 \\ 8 & 15 & 0 \\ 0 & 14 & 4 \end{bmatrix} = \begin{bmatrix} 704 & 832 & 448 \\ 192 & 256 & 0 \\ 320 & 448 & 128 \\ 512 & 960 & 0 \\ 0 & 896 & 256 \end{bmatrix} \qquad \textbf{1}$$

Similarly, we obtain the total winter production matrix by
multiplying by 63 and the total off-season production matrix by
multiplying by 128:

TOTAL WINTER PRODUTION MATRIX

$$63 \begin{bmatrix} 2 & 5 & 0 \\ 7 & 9 & 4 \\ 11 & 15 & 5 \\ 12 & 18 & 0 \\ 0 & 22 & 5 \end{bmatrix} = \begin{bmatrix} 126 & 315 & 0 \\ 441 & 567 & 252 \\ 693 & 945 & 315 \\ 756 & 1134 & 0 \\ 0 & 1386 & 315 \end{bmatrix} \qquad \textbf{2}$$

TOTAL OFF-SEASON PRODUCTION MATRIX

$$128 \begin{bmatrix} 7 & 8 & 4 \\ 4 & 6 & 3 \\ 8 & 9 & 0 \\ 10 & 16 & 0 \\ 0 & 19 & 6 \end{bmatrix} = \begin{bmatrix} 896 & 1024 & 512 \\ 512 & 768 & 384 \\ 1024 & 1152 & 0 \\ 1280 & 2048 & 0 \\ 0 & 2432 & 768 \end{bmatrix} \qquad \textbf{3}$$

Of interest to the company is their total *yearly* production.
This is obtained by adding the corresponding entries in matrices

1, 2, and **3.** For example, according to the matrices, the Springfield factory produced 960 two-person tents in the summer, 1134 in the winter, and 2048 in the off-season. Accordingly, the factory's total production of tents for the year was 960 + 1134 + 2048 = 4142. We proceed in the same fashion to obtain each of the other entries in the total yearly production table (Table 7).

Table 7
Total Yearly Production

Item	Factory		
	Danbury	Springfield	Rutland
Light sleeping bags	1726	2171	960
Heavy sleeping bags	1145	1591	636
Down jackets	2037	2545	443
Two-person tents	2548	4142	0
Four-person tents	0	4714	1339

Another problem arises. Each of the factories consigns a certain part of its daily output to a small factory outlet shop. The number of items sent to the store changes every day depending upon sales. The rest of the day's production is sent to the main store in Portsmouth, New Hampshire. Every day each factory fills out a form indicating how much of that day's production is to be retained for local sale. This information is tabulated by the company in Table 8.

Table 8
Factory Outlet Consignment,
July 17

Item	Factory		
	Danbury	Springfield	Rutland
Light sleeping bags	3	5	2
Heavy sleeping bags	1	2	0
Down jackets	0	2	2
Two-person tents	6	3	0
Four-person tents	0	3	0

Refer to Table 1, p. 2.   For example, of the 7 jackets produced on July 17 by the Springfield factory, 2 are sent to the factory outlet store in Springfield and the remaining 5 are sent to Portsmouth.

In order for the company president to determine how many of the items produced on July 17 will be available for sale at the

Portsmouth store, she has merely to subtract each entry of the daily consignment matrix from the corresponding entry of the daily production matrix. Table 9 results.

Table 9
Available Production, July 17

Item	Factory		
	Danbury	Springfield	Rutland
Light sleeping bags	8	8	5
Heavy sleeping bags	2	2	0
Down jackets	5	5	0
Two-person tents	2	12	0
Four-person tents	0	11	4

This type of addition and subtraction of matrices can be done without regard to the tables from which they originate. To add two matrices, simply add corresponding entries.

Problem 1   Add the matrices $\begin{bmatrix} 1 & 7 \\ 3 & 5 \end{bmatrix}$ and $\begin{bmatrix} 4 & 0 \\ 6 & 2 \end{bmatrix}$

Solution

$$\begin{bmatrix} 1 & 7 \\ 3 & 5 \end{bmatrix} + \begin{bmatrix} 4 & 0 \\ 6 & 2 \end{bmatrix} = \begin{bmatrix} 1+4 & 7+0 \\ 3+6 & 5+2 \end{bmatrix} = \begin{bmatrix} 5 & 7 \\ 9 & 7 \end{bmatrix}$$

To subtract matrices, simply subtract corresponding entries.

Problem 2   Perform the indicated subtraction:

$$\begin{bmatrix} 9 & 7 & 5 \\ 10 & 6 & 3 \end{bmatrix} - \begin{bmatrix} 5 & 4 & 1 \\ 7 & 2 & 3 \end{bmatrix}$$

Solution

$$\begin{bmatrix} 9 & 7 & 5 \\ 10 & 6 & 3 \end{bmatrix} - \begin{bmatrix} 5 & 4 & 1 \\ 7 & 2 & 3 \end{bmatrix} = \begin{bmatrix} 9-5 & 7-4 & 5-1 \\ 10-7 & 6-2 & 3-3 \end{bmatrix}$$

$$= \begin{bmatrix} 4 & 3 & 4 \\ 3 & 4 & 0 \end{bmatrix}$$

Notice that since addition and subtraction of matrices involves the addition and subtraction of corresponding entries, *two matrices can be added or subtracted only if they are the same size*. To help keep this in mind we will occasionally provide exercises in which matrices of different sizes are to be added.

Problem 3　Add:

$$\begin{bmatrix} 1 & 3 \\ 3 & 4 \end{bmatrix} + \begin{bmatrix} 5 \\ 6 \end{bmatrix}$$

Solution　They cannot be added.

Addition and subtraction can just as easily be performed with negative or fractional entries.

Problem 4　Perform the indicated operations:

(a) $\begin{bmatrix} -2 & 3 & -1 \\ 4 & -5 & 2 \end{bmatrix} + \begin{bmatrix} 3 & -2 & -4 \\ -2 & 3 & -1 \end{bmatrix}$

(b) $\begin{bmatrix} 7 & 3 \\ -2 & 5 \\ -4 & -1 \end{bmatrix} - \begin{bmatrix} 2 & 5 \\ 4 & -2 \\ 2 & 3 \end{bmatrix}$

(c) $\begin{bmatrix} 1/2 & 3/4 \\ 1 & 1/4 \end{bmatrix} + \begin{bmatrix} 1/4 & 1/2 \\ 3/4 & 1 \end{bmatrix}$

(d) $[\begin{matrix} 2/3 & -1/5 & 1/2 \end{matrix}] - [\begin{matrix} 1/5 & 3/4 & 1/3 \end{matrix}]$

Solution

For help with signed numbers and fraction arithmetic, see Appendix A.

(a) $\begin{bmatrix} -2 & 3 & -1 \\ 4 & -5 & 2 \end{bmatrix} + \begin{bmatrix} 3 & -2 & -4 \\ -2 & 3 & -1 \end{bmatrix} = \begin{bmatrix} 1 & 1 & -5 \\ 2 & -2 & 1 \end{bmatrix}$

(b) $\begin{bmatrix} 7 & 3 \\ -2 & 5 \\ -4 & -1 \end{bmatrix} - \begin{bmatrix} 2 & 5 \\ 4 & -2 \\ 2 & 3 \end{bmatrix} = \begin{bmatrix} 5 & -2 \\ -6 & 7 \\ -6 & -4 \end{bmatrix}$

(c) $\begin{bmatrix} 1/2 & 3/4 \\ 1 & 1/4 \end{bmatrix} + \begin{bmatrix} 1/4 & 1/2 \\ 3/4 & 1 \end{bmatrix} = \begin{bmatrix} 3/4 & 5/4 \\ 7/4 & 5/4 \end{bmatrix}$

(d) $[\begin{matrix} 2/3 & -1/5 & 1/2 \end{matrix}] - [\begin{matrix} 1/5 & 3/4 & 1/3 \end{matrix}] = [\begin{matrix} 7/15 & -19/20 & 1/6 \end{matrix}]$

We can even combine the three operations of addition, subtraction, and multiplication by a number:

Problem 5　Perform the indicated operations:

$$3\begin{bmatrix} 2 & 1 & 4 \\ -1 & 3 & 5 \end{bmatrix} + 2\begin{bmatrix} -1 & 2 & -3 \\ 2 & 1 & 2 \end{bmatrix} - 5\begin{bmatrix} 1 & 3 & 2 \\ 2 & 1 & -1 \end{bmatrix}$$

Solution

$$3\begin{bmatrix} 2 & 1 & 4 \\ -1 & 3 & 5 \end{bmatrix} + 2\begin{bmatrix} -1 & 2 & -3 \\ 2 & 1 & 2 \end{bmatrix} - 5\begin{bmatrix} 1 & 3 & 2 \\ 2 & 1 & -1 \end{bmatrix}$$

If we let $A = \begin{bmatrix} 2 & 1 & 4 \\ -1 & 3 & 5 \end{bmatrix}$,

$B = \begin{bmatrix} -1 & 2 & -3 \\ 2 & 1 & 2 \end{bmatrix}$,

$C = \begin{bmatrix} 1 & 3 & 2 \\ 2 & 1 & -1 \end{bmatrix}$,

then the result of the solution of the previous problem can be written as follows:

$3A + 2B - 5C = \begin{bmatrix} -1 & -8 & -4 \\ -9 & 6 & 24 \end{bmatrix}$

$= \begin{bmatrix} 6 & 3 & 12 \\ -3 & 9 & 15 \end{bmatrix} + \begin{bmatrix} -2 & 4 & -6 \\ 4 & 2 & 4 \end{bmatrix} - \begin{bmatrix} 5 & 15 & 10 \\ 10 & 5 & -5 \end{bmatrix}$

$= \begin{bmatrix} 6-2-5 & 3+4-15 & 12-6-10 \\ -3+4-10 & 9+2-5 & 15+4+5 \end{bmatrix}$

$= \begin{bmatrix} -1 & -8 & -4 \\ -9 & 6 & 24 \end{bmatrix}$

Exercises 1.4

1. What is the total production of bags, jackets, and tents by Appalachian Creations for the months of February, July, and September?
A2. It happened that in November the daily consignment table was the same for each one of the 18 workdays (Table 10).

Table 10
November Daily Consignment
Table

Item	Factory		
	Danbury	Springfield	Rutland
Light sleeping bags	2	3	0
Heavy sleeping bags	3	5	3
Down jackets	6	7	0
Two-person tents	3	5	0
Four-person tents	0	3	2

Use the figures in Table 10 to obtain the avaiable-output table for the entire month of November.

3. Add or subtract as indicated:

A(a) $\begin{bmatrix} 1 & 2 \\ 3 & 1 \end{bmatrix} + \begin{bmatrix} 2 & 2 \\ 1 & 4 \end{bmatrix}$

(b) $\begin{bmatrix} -1 & 2 \\ 3 & -4 \end{bmatrix} + \begin{bmatrix} 2 & 5 \\ -1 & 9 \end{bmatrix}$

A(c) $\begin{bmatrix} 3 & 1 & -1 \\ 2 & -1 & 0 \end{bmatrix} + \begin{bmatrix} 1 & 2 \\ -1 & 3 \end{bmatrix}$

(d) $\begin{bmatrix} -1 \\ 3 \\ -2 \end{bmatrix} + \begin{bmatrix} 3 \\ -2 \\ 4 \end{bmatrix}$

(e) $\begin{bmatrix} -1 & 2 & -3 \end{bmatrix} + \begin{bmatrix} 3 \\ -1 \\ 4 \end{bmatrix}$

^{A}(f) $\begin{bmatrix} -1 & -2 & 4 \\ -4 & -1 & -3 \end{bmatrix} + \begin{bmatrix} -2 & 2 & -4 \\ 3 & -1 & 2 \end{bmatrix}$

(g) $\begin{bmatrix} 2 & -2 \\ -4 & -1 \\ 2 & 4 \end{bmatrix} + \begin{bmatrix} -3 & 0 \\ 2 & -3 \\ -3 & -1 \end{bmatrix}$

^{A}(h) $\begin{bmatrix} 3 & 2 \\ 4 & 5 \end{bmatrix} - \begin{bmatrix} 2 & 1 \\ 3 & 2 \end{bmatrix}$

(i) $\begin{bmatrix} 2 & 4 \\ 3 & 1 \\ 4 & 2 \end{bmatrix} - \begin{bmatrix} 3 & 5 \\ 2 & 3 \\ 7 & 4 \end{bmatrix}$

(j) $\begin{bmatrix} -1 & -2 & 3 \\ 2 & -1 & -2 \end{bmatrix} - \begin{bmatrix} -1 & 2 & -3 & 4 \\ 0 & -2 & 3 & -1 \end{bmatrix}$

(k) $\begin{bmatrix} -1 & 2 & -3 \\ 4 & -1 & 0 \\ -2 & 0 & -3 \end{bmatrix} - \begin{bmatrix} -2 & 1 & -2 \\ -3 & -1 & 2 \\ 3 & -3 & 5 \end{bmatrix}$

^{A}(l) $\begin{bmatrix} 1/5 \\ 2/5 \\ 1/5 \end{bmatrix} + \begin{bmatrix} 3/5 \\ 7/5 \\ 4/5 \end{bmatrix}$

(m) $\begin{bmatrix} 3/8 \\ 5/8 \\ 1/8 \end{bmatrix} - \begin{bmatrix} 5/8 \\ 3/8 \\ 1/8 \end{bmatrix}$

(n) $\begin{bmatrix} 1/2 & 3/4 \\ -1/4 & -1/2 \end{bmatrix} + \begin{bmatrix} 3/4 & 1/2 \\ -3/4 & 1 \end{bmatrix}$

(o) $\begin{bmatrix} 1/5 & -2/3 & 3/5 \end{bmatrix} + \begin{bmatrix} 2/3 & 4/5 & 1/3 \end{bmatrix}$

^{A}(p) $\begin{bmatrix} 1/2 \\ 1/3 \end{bmatrix} - \begin{bmatrix} -1/3 \\ 1/2 \end{bmatrix}$

(q) $\begin{bmatrix} -1/3 & 1/5 & -1/7 \\ 2/3 & 1/7 & 3/8 \end{bmatrix} - \begin{bmatrix} 2/3 & -1/5 & 3/2 & 1 \\ -1/5 & 1/9 & 2/3 & -1/2 \end{bmatrix}$

^{A}(r) $\begin{bmatrix} 2/3 & 1/8 \\ 3/4 & 1/2 \end{bmatrix} + \begin{bmatrix} 0 & 0 \\ 0 & 0 \end{bmatrix}$

^{A}(s) $\begin{bmatrix} 3/8 \\ 2/3 \end{bmatrix} - \begin{bmatrix} 3/8 \\ 2/3 \end{bmatrix}$

^{A}(t) $\begin{bmatrix} 1 & -2 & 2/3 & -1/7 \\ 2 & -1/5 & -1/4 & 1/8 \\ 1/3 & 3/8 & -2 & 1/4 \\ -2/9 & 1/4 & -1/5 & -1/3 \end{bmatrix} - \begin{bmatrix} 1/2 & -1/4 & -1/3 & 3/7 \\ -1/3 & -3 & 1 & 3/8 \\ 2 & -1/4 & 1/2 & -2 \\ -4/9 & 1/8 & -2/3 & 1/2 \end{bmatrix}$

4. Perform the indicated operations:

^{A}(a) $3\begin{bmatrix} 2 & -1 & 3 \\ 1 & 0 & 2 \\ -1 & 4 & 1 \end{bmatrix} + 5\begin{bmatrix} 1 & -2 & 3 \\ 2 & 1 & 4 \\ 3 & 2 & 1 \end{bmatrix}$

(b) $2\begin{bmatrix} 1 & 2 \\ 4 & 2 \end{bmatrix} - \begin{bmatrix} 3 & 2 \\ 1 & 0 \end{bmatrix}$

^{A}(c) $\frac{1}{2}\begin{bmatrix} 2 \\ -4 \\ 6 \end{bmatrix} - \frac{1}{3}\begin{bmatrix} 12 \\ -15 \\ 6 \end{bmatrix}$

^{A}(d) $\frac{2}{3}\begin{bmatrix} 2 & -3 & 4 \\ 1 & 2 & 5 \end{bmatrix} - \frac{1}{3}\begin{bmatrix} 3 & 7 & 0 & -1 \\ 2 & 1 & 4 & 3 \end{bmatrix}$

(e) $\frac{1}{2}\begin{bmatrix} 3 & 4 \\ -1 & 2 \end{bmatrix} - \frac{1}{3}\begin{bmatrix} 7 & -6 \\ 2 & -4 \end{bmatrix}$

^{A}(f) $\frac{1}{5}\begin{bmatrix} \frac{1}{3} \\ -\frac{2}{5} \end{bmatrix} - \frac{2}{3}\begin{bmatrix} -\frac{1}{2} \\ \frac{3}{8} \end{bmatrix}$

^{A}(g) $2\begin{bmatrix} 2 & 1 & 3 \\ 0 & 2 & 3 \\ 1 & 1 & 1 \end{bmatrix} + \begin{bmatrix} 1 & 1 & 3 \\ 3 & 1 & 0 \\ 7 & 4 & 6 \end{bmatrix} - 3\begin{bmatrix} 1 & 1 & 3 \\ 1 & 1 & 2 \\ 3 & 2 & 2 \end{bmatrix}$

(h) $\frac{1}{2}\begin{bmatrix} 1 & 2 \\ 0 & 3 \\ -1 & 2 \end{bmatrix} + \frac{1}{3}\begin{bmatrix} 2 & 1 \\ 3 & 2 \\ -1 & 4 \end{bmatrix} - \frac{1}{5}\begin{bmatrix} 1 & 2 & -1 \\ 0 & 3 & 4 \\ 1 & 2 & 3 \end{bmatrix}$

^{C}(i) $\frac{1}{2}\begin{bmatrix} 2 & 4 & 0 \\ 4 & 6 & -2 \end{bmatrix} - \frac{1}{3}\begin{bmatrix} 3 & 6 & 0 \\ 9 & -3 & 3 \end{bmatrix} + \frac{1}{5}\begin{bmatrix} 5 & 10 \\ -10 & 0 \end{bmatrix}$

*(j) $\frac{3}{4}\begin{bmatrix} 12 & 0 \\ 5 & 3 \end{bmatrix} - \frac{7}{8}\begin{bmatrix} -1 & 2 \\ 3 & -4 \end{bmatrix} - \frac{2}{5}\begin{bmatrix} 2 & 3 \\ -1 & 0 \end{bmatrix}$

5. For each of the following pairs of matrices, determine what matrix should be added to the first to obtain the second:

^{A}(a) $\begin{bmatrix} 2 & 1 & 3 \\ 1 & 2 & 3 \end{bmatrix}, \begin{bmatrix} 7 & 0 & 4 \\ 5 & 5 & 1 \end{bmatrix}$

(b) $\begin{bmatrix} 2 & -3 \\ -1 & 4 \\ 2 & -1 \end{bmatrix}, \begin{bmatrix} 5 & -2 \\ -3 & 0 \\ -3 & 6 \end{bmatrix}$

^{C}(c) $\begin{bmatrix} 1 & 2 & 0 \\ 3 & 5 & 4 \end{bmatrix}, \begin{bmatrix} 1 & 2 & 0 \\ 3 & 5 & 4 \end{bmatrix}$

^{C}(d) $\begin{bmatrix} 2 & 0 \\ 1 & 3 \\ 4 & 5 \end{bmatrix}, \begin{bmatrix} 1 & -1 & 2 \\ 3 & 1 & 4 \\ 0 & 2 & 5 \end{bmatrix}$

^{A}(e) $[\frac{1}{3} \quad \frac{2}{3} \quad \frac{1}{2}], [\frac{5}{6} \quad \frac{1}{4} \quad -2]$

6. For each of the following pairs of matrices, determine what matrix should be subtracted from the first to obtain the second:

^{A}(a) $\begin{bmatrix} 7 \\ 3 \\ 2 \end{bmatrix}, \begin{bmatrix} 3 \\ 3 \\ 1 \end{bmatrix}$

(b) $\begin{bmatrix} 9 & -2 \\ 0 & 5 \end{bmatrix}, \begin{bmatrix} 5 & 1 \\ -2 & -4 \end{bmatrix}$

^{A}(c) $\begin{bmatrix} 6 & -3 \\ 5 & {}^1\!/_2 \\ 2 & -8 \end{bmatrix}, \begin{bmatrix} 6 & -3 & 2 \\ 1 & 4 & 0 \\ 0 & 3 & -8 \end{bmatrix}$

*(d) $\begin{bmatrix} -{}^7\!/_8 & {}^8\!/_7 & -{}^4\!/_3 \\ {}^3\!/_4 & -{}^{15}\!/_4 & {}^6\!/_7 \end{bmatrix}, \begin{bmatrix} -{}^{13}\!/_{24} & {}^{11}\!/_{28} & -{}^{32}\!/_{15} \\ {}^{43}\!/_{30} & -{}^{11}\!/_2 & 1 \end{bmatrix}$

See Exercise 8, Section 1-3.

7. Recently a second clinic opened in Addis Ababa. This one is closed on Saturday and Sunday of each week, but on the other days it gives shots as shown in Table 11.

Table 11
Number of Shots Given per Day

	Men	Women	Children
Typhus	30	30	40
Typhoid	20	30	96
Yellow fever	8	6	0
Cholera	205	107	212

(a) Write the matrix obtained from a table that shows the number of shots given on any Thursday by the two clinics combined.

(b) Let U denote the matrix obtained from Table 11. Let W be the matrix obtained from a table that shows the number of shots per week given by the combined clinics. Express W in terms of U and V.

V is obtained from Table 4 (Exercise 8, Section 1-3).

*(c) For some time the World Health Organization was using a matrix F for weekly shots for the combined clinics that was based on the false assumption that both clinics were operating every day. Show how to obtain W from F and U.

8. Like many sophisticated modern corporations, Petaluma Hamburger Emporium has created, on paper, a small "model" company for use in planning. The data related to the Petaluma Hamburger Emporium Model (PHEM) company are supplied by the PHE market-research department based on surveys of MacDougal's, PHE's arch competitor, as well as PHE itself. PHEM has three stores, called simply store 1, store 2, and store 3, each selling four items: item 1, hamburger; item 2, cheeseburger; item 3, french fries; item 4, milkshake. Production matrices for PHEM are monthly, not weekly. The production matrices for July (J), August (A), September (S), and October (O), are respectively,

$$J = \begin{bmatrix} 40 & 30 & 60 & 60 \\ 50 & 20 & 60 & 70 \\ 30 & 40 & 50 & 60 \end{bmatrix} \qquad S = \begin{bmatrix} 30 & 30 & 70 & 75 \\ 60 & 25 & 80 & 50 \\ 35 & 55 & 65 & 45 \end{bmatrix}$$

$$A = \begin{bmatrix} 45 & 35 & 65 & 70 \\ 45 & 30 & 40 & 65 \\ 40 & 60 & 70 & 35 \end{bmatrix} \qquad O = \begin{bmatrix} 40 & 35 & 70 & 60 \\ 50 & 30 & 80 & 70 \\ 30 & 60 & 65 & 60 \end{bmatrix}$$

(a) Denote by Q the PHEM production matrix for the third quarter. Express Q in terms of J, A, S, and O.

(b) Write out Q.

(c) The PHEM production matrix for the fourth quarter was

$$F = \begin{bmatrix} 120 & 100 & 190 & 185 \\ 145 & 65 & 170 & 160 \\ 110 & 145 & 195 & 155 \end{bmatrix}$$

Denote by T the production matrix for the last two months of the year. Express T in terms of the matrices above.

(d) Write out T.

Section 1-5
Matrix Multiplication

The people at Appalachian Creations were quite happy with their computer setup. Their machine was programmed to multiply a matrix by a number and to add or subtract two matrices. But then it happened. In November there was a strike at the Danbury plant that closed it down for 10 days; in Rutland, a steam pipe broke and the factory there was closed for 5 days. The Springfield factory was the only one of the three that was open for the entire month. Before this trouble, when the AC managers wanted to know the total November production of light sleeping bags, for example, they simply had the computer multiply R_1 of the November daily production matrix (see Table 12) by 18 and then they added the entries in the row:

$$18 \cdot 7 + 18 \cdot 8 + 18 \cdot 4 = 342$$

Table 12
Daily Production Table for
Mar.–May, Sept.–Nov.[a]

Item	Factory		
	Danbury	Springfield	Rutland
Light sleeping bags	7	8	4
Heavy sleeping bags	4	6	3
Down jackets	8	9	0
Two-person tents	10	16	0
Four-person tents	0	19	6

[a]This is the same as Table 6 in Section 1-4.

But now to get the total production they had to multiply the Danbury daily production by 8, the Rutland daily production by 13, and the Springfield daily production by 18. The total November light sleeping bag production was

$$8 \cdot 7 + 13 \cdot 8 + 18 \cdot 4 = 232$$

This difficulty was more distressing to them than to us because we are considering only a small part of their entire production table.

When they explained their difficulty to the resident computer expert, he calmly stated that their machine should be programmed to perform *matrix multiplication*. This made no sense at all until he explained how to multiply a row matrix and a column matrix.

How to Multiply a Row Matrix and a Column Matrix

We begin by presenting three examples:

Problem 1   Multiply $\begin{bmatrix} 1 & 2 & 3 \end{bmatrix}$ by $\begin{bmatrix} 4 \\ 5 \\ 6 \end{bmatrix}$

Solution

$$\begin{bmatrix} 1 & 2 & 3 \end{bmatrix} \begin{bmatrix} 4 \\ 5 \\ 6 \end{bmatrix} = 1 \cdot 4 + 2 \cdot 5 + 3 \cdot 6$$
$$= 4 + 10 + 18$$
$$= 32$$

Problem 2   Multiply $\begin{bmatrix} 2 & 1 & 3 & 4 \end{bmatrix}$ by $\begin{bmatrix} 5 \\ 0 \\ -1 \\ 6 \end{bmatrix}$

Solution

$$\begin{bmatrix} 2 & 1 & 3 & 4 \end{bmatrix} \begin{bmatrix} 5 \\ 0 \\ -1 \\ 6 \end{bmatrix} = 2 \cdot 5 + 1 \cdot 0 + 3 \cdot -1 + 4 \cdot 6$$
$$= 10 + 0 - 3 + 24$$
$$= 31$$

Problem 3   Multiply $\begin{bmatrix} 3 & -1 \end{bmatrix}$ by $\begin{bmatrix} -2 \\ -1 \end{bmatrix}$

Solution

$$\begin{bmatrix} 3 & -1 \end{bmatrix} \begin{bmatrix} -2 \\ -1 \end{bmatrix} = 3(-2) + (-1)(-1) = -6 + 1 = -5$$

Page 27 shows a flow chart for this operation. The idea is simply to multiply the corresponding entries and then add the results. AC's problem was easily solved by the single matrix multiplication

$$[7 \quad 8 \quad 4] \begin{bmatrix} 8 \\ 13 \\ 18 \end{bmatrix} = 232 \qquad \qquad \mathbf{1}$$

Notice that the result of multiplying a row matrix and a column matrix is a number (which may be considered to be a 1×1 matrix). Also note that the number of entries in the row matrix must be the same as the number of entries in the column matrix (or else the matrices cannot be multiplied). Here are some examples:

Problem 4   Multiply $[2 \quad 1 \quad 4 \quad 6]$ by $\begin{bmatrix} 3 \\ 1 \\ 2 \\ 4 \\ 5 \end{bmatrix}$.

Solution   Since the first matrix has fewer entries than the second, they cannot be multiplied.

**Multiplication of a Row Matrix
by a Column Matrix**

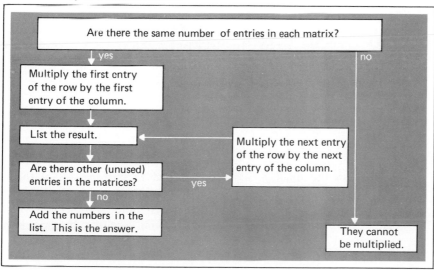

Problem 5   Multiply $\begin{bmatrix} -\frac{1}{3} & \frac{3}{4} & -\frac{1}{8} & 0 \end{bmatrix}$ by $\begin{bmatrix} 3 \\ \frac{1}{3} \\ \frac{5}{8} \\ -1 \end{bmatrix}$

Solution

$$\begin{bmatrix} -\frac{1}{3} & \frac{3}{4} & -\frac{1}{8} & 0 \end{bmatrix} \begin{bmatrix} 3 \\ \frac{1}{3} \\ \frac{5}{8} \\ -1 \end{bmatrix}$$

$$= -\frac{1}{3} \times 3 + \frac{3}{4} \times \frac{1}{3} - \frac{1}{8} \times \frac{5}{8} + 0 \times -1$$
$$= -1 + \frac{1}{4} - \frac{5}{64} + 0$$
$$= -\frac{64}{64} + \frac{16}{64} - \frac{5}{64}$$
$$= -\frac{53}{64}$$

Of course, AC makes several hundred items, and to calculate the total November production of each item requires multiplication of many row matrices by the same column matrix. To complete our table there are four multiplications left to do:

HEAVY SLEEPING BAG PRODUCTION

$$\begin{bmatrix} 4 & 6 & 3 \end{bmatrix} \begin{bmatrix} 8 \\ 13 \\ 18 \end{bmatrix} = 164 \qquad \qquad \textbf{2}$$

DOWN JACKET PRODUCTION

$$\begin{bmatrix} 8 & 9 & 0 \end{bmatrix} \begin{bmatrix} 8 \\ 13 \\ 18 \end{bmatrix} = 181 \qquad \qquad \textbf{3}$$

TWO-PERSON TENT PRODUCTION

$$\begin{bmatrix} 10 & 16 & 0 \end{bmatrix} \begin{bmatrix} 8 \\ 13 \\ 18 \end{bmatrix} = 288 \qquad \qquad \textbf{4}$$

FOUR-PERSON TENT PRODUCTION

$$\begin{bmatrix} 0 & 19 & 6 \end{bmatrix} \begin{bmatrix} 8 \\ 13 \\ 18 \end{bmatrix} = 355 \qquad \qquad \textbf{5}$$

Can the machine be programmed to perform all of these operations upon receiving a single command, thus avoiding the necessity of resubmitting the data each time? Yes, it can. The necessary procedure involves multiplying a matrix by a column matrix.

How to Multiply a Matrix by a Column Matrix

The result of this operation is always a column matrix with the same number of entries as the first matrix. Its first entry is obtained by multiplying R_1 of the first matrix by the column matrix. The second entry is obtained by multiplying R_2 of the first matrix by the column matrix, and so on. A flow chart for this type of multiplication appears on this page. We also present a few examples.

Problem 6   Multiply the matrix $\begin{bmatrix} 1 & 2 & 4 \\ 3 & 1 & 2 \\ 0 & 1 & 4 \\ 1 & 2 & 1 \end{bmatrix}$ by the matrix $\begin{bmatrix} 2 \\ 1 \\ 2 \end{bmatrix}$.

Solution   The answer will be a column matrix. We compute the product:

$$[1 \quad 2 \quad 4] \begin{bmatrix} 2 \\ 1 \\ 2 \end{bmatrix} = 1 \cdot 2 + 2 \cdot 1 + 4 \cdot 2 = 12; \text{ so the first entry in}$$

Multiplication of a Matrix by a Column Matrix

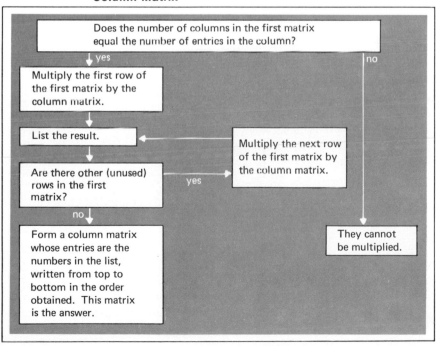

the column matrix is 12. Continuing, we compute the product:

$$[3 \quad 1 \quad 2]\begin{bmatrix} 2 \\ 1 \\ 2 \end{bmatrix} = 11, \quad [0 \quad 1 \quad 4]\begin{bmatrix} 2 \\ 1 \\ 2 \end{bmatrix} = 9, \text{ and } [1 \quad 2 \quad 1]\begin{bmatrix} 2 \\ 1 \\ 2 \end{bmatrix} = 6.$$

Thus, the answer is

$$\begin{bmatrix} 1 & 2 & 4 \\ 3 & 1 & 2 \\ 0 & 1 & 4 \\ 1 & 2 & 1 \end{bmatrix}\begin{bmatrix} 2 \\ 1 \\ 2 \end{bmatrix} = \begin{bmatrix} 12 \\ 11 \\ 9 \\ 6 \end{bmatrix}$$

Problem 7 Multiply $\begin{bmatrix} 3 & 1 & 4 \\ 2 & 8 & 1 \\ 0 & 1 & 2 \\ 3 & 1 & 5 \end{bmatrix}$ by $\begin{bmatrix} 1 \\ 3 \\ 1 \\ 4 \end{bmatrix}$.

Solution Since the first matrix has only 3 columns and the column matrix has 4 entries, they cannot be multiplied.

Problem 8

Multiply $\begin{bmatrix} {}^3\!/_2 & -{}^1\!/_3 & {}^7\!/_8 \\ {}^2\!/_5 & -{}^4\!/_5 & -{}^4\!/_{15} \end{bmatrix}$ by $\begin{bmatrix} {}^2\!/_3 \\ {}^3\!/_8 \\ -{}^1\!/_4 \end{bmatrix}$.

Solution

$$\begin{bmatrix} {}^3\!/_2 & -{}^1\!/_3 & {}^7\!/_8 \\ {}^2\!/_5 & -{}^4\!/_5 & -{}^4\!/_{15} \end{bmatrix}\begin{bmatrix} {}^2\!/_3 \\ {}^3\!/_8 \\ -{}^1\!/_4 \end{bmatrix}$$

$$= \begin{bmatrix} {}^3\!/_2 \times {}^2\!/_3 + -{}^1\!/_3 \times {}^3\!/_8 + {}^7\!/_8 \times -{}^1\!/_4 \\ {}^2\!/_5 \times {}^2\!/_3 + -{}^4\!/_5 \times {}^3\!/_8 + -{}^4\!/_{15} \times -{}^1\!/_4 \end{bmatrix}$$

$$= \begin{bmatrix} 1 - {}^1\!/_8 - {}^7\!/_{32} \\ {}^4\!/_{15} - {}^3\!/_{10} + {}^1\!/_{15} \end{bmatrix}$$

$$= \begin{bmatrix} {}^{21}\!/_{32} \\ {}^1\!/_{30} \end{bmatrix}$$

In the AC example, we have already computed each entry in the answer (equations **1–5**):

$$\begin{bmatrix} 7 & 8 & 4 \\ 4 & 6 & 3 \\ 8 & 9 & 0 \\ 10 & 16 & 0 \\ 0 & 19 & 6 \end{bmatrix}\begin{bmatrix} 8 \\ 13 \\ 18 \end{bmatrix} = \begin{bmatrix} 232 \\ 164 \\ 181 \\ 288 \\ 355 \end{bmatrix}$$

We interpret this result in Table 13.

Table 13
Total November Production

Item	Production
Light sleeping bags	232
Heavy sleeping bags	164
Down jackets	181
Two-person tents	288
Four-person tents	355

That takes care of November. In December, the Springfield factory was only open for 2 days because of a truckers' strike. To help meet the orders, Rutland stayed open for 25 days and Danbury for 30 days. To obtain the total December production, we multiply the winter production schedule matrix, obtained from

Table 14, by the column matrix $\begin{bmatrix} 30 \\ 2 \\ 25 \end{bmatrix}$:

Table 14
Daily Production Table for
Dec.–Feb.[a]

Item	Factory		
	Danbury	Springfield	Rutland
Light sleeping bags	2	5	0
Heavy sleeping bags	7	9	4
Down jackets	11	15	5
Two-person tents	12	18	0
Four-person tents	0	22	5

[a]This is the same as Table 5 in Section 1-4.

$$\begin{bmatrix} 2 & 5 & 0 \\ 7 & 9 & 4 \\ 11 & 15 & 5 \\ 12 & 18 & 0 \\ 0 & 22 & 5 \end{bmatrix} \begin{bmatrix} 30 \\ 2 \\ 25 \end{bmatrix} = \begin{bmatrix} 70 \\ 328 \\ 485 \\ 396 \\ 169 \end{bmatrix}$$

In table 15 we restore the headings to this array of numbers.

Table 15
Total December Production

Item	Production
Light sleeping bags	70
Heavy sleeping bags	328
Down jackets	485
Two-person tents	396
Four-person tents	169

As it turned out, each month that winter each factory was open a different number of days, as Table 16 shows.

Table 16
Days of Factory Operation per
Month

	Dec.	Jan.	Feb.
Danbury	30	5	20
Springfield	2	28	24
Rutland	25	15	0

To get a table listing total production for each month, it appeared at first that the accounting department would have to feed three separate problems into their machine:

DECEMBER PRODUCTION

$$\begin{bmatrix} 2 & 5 & 0 \\ 7 & 9 & 4 \\ 11 & 15 & 5 \\ 12 & 18 & 0 \\ 0 & 22 & 5 \end{bmatrix} \begin{bmatrix} 30 \\ 2 \\ 25 \end{bmatrix} = \begin{bmatrix} 70 \\ 328 \\ 485 \\ 396 \\ 169 \end{bmatrix} \qquad \textbf{6}$$

JANUARY PRODUCTION

$$\begin{bmatrix} 2 & 5 & 0 \\ 7 & 9 & 4 \\ 11 & 15 & 5 \\ 12 & 18 & 0 \\ 0 & 22 & 5 \end{bmatrix} \begin{bmatrix} 5 \\ 28 \\ 15 \end{bmatrix} = \begin{bmatrix} 150 \\ 347 \\ 550 \\ 564 \\ 691 \end{bmatrix} \qquad \textbf{7}$$

FEBRUARY PRODUCTION

$$\begin{bmatrix} 2 & 5 & 0 \\ 7 & 9 & 4 \\ 11 & 15 & 5 \\ 12 & 18 & 0 \\ 0 & 22 & 5 \end{bmatrix} \begin{bmatrix} 20 \\ 24 \\ 0 \end{bmatrix} = \begin{bmatrix} 160 \\ 356 \\ 580 \\ 672 \\ 528 \end{bmatrix} \qquad \textbf{8}$$

But this was not the case. They simply fed in the winter production matrix:

$$\begin{bmatrix} 2 & 5 & 0 \\ 7 & 9 & 4 \\ 11 & 15 & 5 \\ 12 & 18 & 0 \\ 0 & 22 & 5 \end{bmatrix} \qquad \textbf{9}$$

and the matrix from Table 4:

$$\begin{bmatrix} 30 & 5 & 20 \\ 2 & 28 & 24 \\ 25 & 15 & 0 \end{bmatrix} \qquad \textbf{10}$$

and pressed the "multiply" button. Of course, the result was a matrix. To obtain the first column of the answer, the machine multiplied the production matrix by the first (December) column of the factory day matrix, matrix **10**. The second column involved the production matrix and the January column. The third column was the product of the production matrix and the February column. The result was the winter production matrix, consisting of the three column matrices obtained above in equations **6, 7**, and **8**:

$$\begin{bmatrix} 2 & 5 & 0 \\ 7 & 9 & 4 \\ 11 & 15 & 5 \\ 12 & 18 & 0 \\ 0 & 22 & 5 \end{bmatrix} \begin{bmatrix} 30 & 5 & 20 \\ 2 & 28 & 24 \\ 25 & 15 & 0 \end{bmatrix} = \begin{bmatrix} 70 & 150 & 160 \\ 328 & 347 & 356 \\ 485 & 550 & 580 \\ 396 & 564 & 672 \\ 169 & 691 & 528 \end{bmatrix}$$

This yielded Table 17.

Table 17 Total Winter Production			
Item	**Dec.**	**Jan.**	**Feb.**
Light sleeping bags	70	150	160
Heavy sleeping bags	328	347	356
Down jackets	485	550	580
Two-person tents	369	564	672
Four-person tents	169	691	528

A flow chart for the multiplication of two matrices appears on page 34. We also provide a few examples:

Problem 9   Multiply $\begin{bmatrix} 1 & 2 & 3 \\ 1 & 1 & 2 \\ 2 & 1 & 1 \end{bmatrix}$ by $\begin{bmatrix} 3 & 1 \\ 1 & 2 \\ 2 & 3 \end{bmatrix}$.

Solution   This problem will involve six multiplications of rows by columns. Generally one does these multiplications on scratch paper and records the result in the answer matrix. However, this time we will show all the steps involved in the flow-chart procedure.
The first column is

$$\begin{bmatrix} 1 & 2 & 3 \\ 1 & 1 & 2 \\ 2 & 1 & 1 \end{bmatrix} \begin{bmatrix} 3 \\ 1 \\ 2 \end{bmatrix} = \begin{bmatrix} 1 \cdot 3 + 2 \cdot 1 + 3 \cdot 2 \\ 1 \cdot 3 + 1 \cdot 1 + 2 \cdot 2 \\ 2 \cdot 3 + 1 \cdot 1 + 1 \cdot 2 \end{bmatrix} = \begin{bmatrix} 11 \\ 8 \\ 9 \end{bmatrix}$$

The second column is

$$\begin{bmatrix} 1 & 2 & 3 \\ 1 & 1 & 2 \\ 2 & 1 & 1 \end{bmatrix} \begin{bmatrix} 1 \\ 2 \\ 3 \end{bmatrix} = \begin{bmatrix} 1 \cdot 1 + 2 \cdot 2 + 3 \cdot 3 \\ 1 \cdot 1 + 1 \cdot 2 + 2 \cdot 3 \\ 2 \cdot 1 + 1 \cdot 2 + 1 \cdot 3 \end{bmatrix} = \begin{bmatrix} 14 \\ 9 \\ 7 \end{bmatrix}$$

Hence, the answer is

$$\begin{bmatrix} 1 & 2 & 3 \\ 1 & 1 & 2 \\ 2 & 1 & 1 \end{bmatrix} \begin{bmatrix} 3 & 1 \\ 1 & 2 \\ 2 & 3 \end{bmatrix} = \begin{bmatrix} 11 & 14 \\ 8 & 9 \\ 9 & 7 \end{bmatrix}$$

Problem 10   Multiply the matrix $\begin{bmatrix} 1 & 3 & 4 & 1 \\ 2 & 1 & 8 & 0 \\ 1 & 0 & 2 & 5 \end{bmatrix}$ by the matrix $\begin{bmatrix} 2 & 0 & 3 \\ 3 & 1 & -1 \\ 1 & 2 & 2 \end{bmatrix}$.

Solution   Since the first matrix has 4 columns and the second matrix has 3 rows, they cannot be multiplied.

Problem 11   Multiply $\begin{bmatrix} 1/2 & -1/3 \\ 2/3 & 3/4 \\ 5/6 & 1/4 \end{bmatrix}$ by $\begin{bmatrix} 12 & -12 \\ 24 & 36 \end{bmatrix}$.

Multiplication of One Matrix by another Matrix

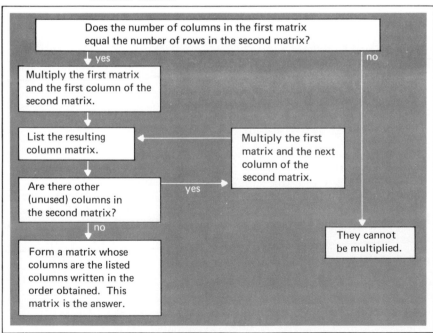

Solution

$$\begin{bmatrix} \frac{1}{2} & -\frac{1}{3} \\ \frac{2}{3} & \frac{3}{4} \\ \frac{5}{6} & \frac{1}{4} \end{bmatrix} \begin{bmatrix} 12 & -12 \\ 24 & 36 \end{bmatrix}$$

$$= \begin{bmatrix} \frac{1}{2} \times 12 + (-\frac{1}{3}) \times 24 & \frac{1}{2} \times -12 + (-\frac{1}{3}) \times 36 \\ \frac{2}{3} \times 12 + \frac{3}{4} \times 24 & \frac{2}{3} \times -12 + \frac{3}{4} \times 36 \\ \frac{5}{6} \times 12 + \frac{1}{4} \times 24 & \frac{5}{6} \times -12 + \frac{1}{4} \times 36 \end{bmatrix}$$

$$= \begin{bmatrix} 6 - 8 & -6 - 12 \\ 8 + 18 & -8 + 27 \\ 10 + 6 & -10 + 9 \end{bmatrix}$$

If we denote the first and second matrices in this problem by A and B, respectively, we may write

$$AB = \begin{bmatrix} -2 & -18 \\ 26 & 19 \\ 16 & -1 \end{bmatrix}$$

$$= \begin{bmatrix} -2 & -18 \\ 26 & 19 \\ 16 & -1 \end{bmatrix}$$

Matrix multiplication is not hard to remember once you have learned how to multiply a row by a column. For example, to obtain the (3,2) entry of the answer, multiply R_3 of the first matrix by C_2 of the second matrix:

$$R_3 \text{ (first matrix)} \times C_2 \text{ (second matrix)} =$$

the (3,2) entry of the answer

In Problem 11 the first matrix has

$$R_3 = \begin{bmatrix} \frac{5}{6} & \frac{1}{4} \end{bmatrix}$$

and the second matrix has

$$C_2 = \begin{bmatrix} -12 \\ 36 \end{bmatrix}$$

so their product,

$$R_3 C_2 = \begin{bmatrix} \frac{5}{6} & \frac{1}{4} \end{bmatrix} \begin{bmatrix} -12 \\ 36 \end{bmatrix} = -1$$

is the (3,2) entry of the answer.

In general, to obtain the entry in a particular row and column of the answer, multiply the row of the first matrix corresponding to the row of the entry by the column of the second matrix corresponding to the column of the entry.

Problem 12   What is the (4,5) entry of the product

$$\begin{bmatrix} 2 & 0 & 3 & 4 \\ 4 & 0 & 8 & 2 \\ 3 & 2 & 1 & 4 \\ 2 & 8 & 6 & 4 \end{bmatrix} \begin{bmatrix} 1 & 3 & 7 & 2 & 3 \\ 4 & 2 & 1 & 1 & 2 \\ 2 & 1 & 4 & 9 & 1 \\ 5 & 6 & 8 & 6 & 5 \end{bmatrix}$$

Solution   In the first matrix

$$R_4 = [2 \quad 8 \quad 6 \quad 4]$$

In the second matrix

$$C_5 = \begin{bmatrix} 3 \\ 2 \\ 1 \\ 5 \end{bmatrix}$$

The (4,5) entry of the answer is

$$R_4 C_5 = [2 \quad 8 \quad 6 \quad 4] \begin{bmatrix} 3 \\ 2 \\ 1 \\ 5 \end{bmatrix} = 48$$

Another helpful hint: before multiplying two matrices, write their sizes side by side. If you are asked to multiply a 3 × 4 matrix and a 4 × 2 matrix, write:

$$3 \times 4 \qquad 4 \times 2$$

First look at the inside numbers (the 4's). If they are the same, the matrices can be multiplied; otherwise, they cannot. And if the inside numbers are the same, the outside numbers (the 3 and the 2) indicate the size of the answer: in this case, a 3 × 2 matrix.

Problem 13   What size matrix results when a 3 × 5 and a 5 × 4 matrix are multiplied?

Solution

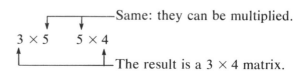

Same: they can be multiplied.

$$3 \times 5 \qquad 5 \times 4$$

The result is a 3 × 4 matrix.

Problem 14   What size matrix results when a 5 × 6 and a 7 × 4 matrix are multiplied?

Solution

not the same

$$5 \times 6 \qquad 7 \times 4$$

They cannot be multiplied.

Problem 15   What size matrix results when a 1 × 3 matrix and a 3 × 1 matrix are multiplied?

Solution

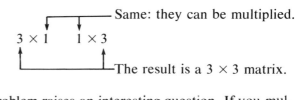

Same: they can be multiplied.

$1 \times 3 \qquad 3 \times 1$

The result is a 1×1 matrix; that is, a single number as expected from multiplying a row matrix by a column matrix.

Problem 16

What size matrix results when a 3×1 matrix and a 1×3 matrix are multiplied?

Solution

Same: they can be multiplied.

$3 \times 1 \qquad 1 \times 3$

The result is a 3×3 matrix.

This last problem raises an interesting question. If you multiply the 3×1 column matrix $\begin{bmatrix} 1 \\ 3 \\ 2 \end{bmatrix}$ and the 1×3 row matrix $[4 \quad 5 \quad 6]$, the rule says that you will get a 3×3 matrix as a result. Just how does this happen? The $(1,1)$ entry of the answer is obtained by multiplying R_1 of $\begin{bmatrix} 1 \\ 3 \\ 2 \end{bmatrix}$ by C_1 of $[4 \quad 5 \quad 6]$. R_1 is just the number $[1]$; C_1 is just the number $[4]$. Thus, the $(1,1)$ entry of the answer is $1 \cdot 4 = 4$. To obtain the $(2,3)$ entry, we multiply R_2 of $\begin{bmatrix} 1 \\ 3 \\ 2 \end{bmatrix}$ by C_3 of $[4 \quad 5 \quad 6]$. Since $R_2 = [3]$ and $C_3 = [6]$, the $(2,3)$ entry of the answer matrix is $3 \cdot 6 = 18$. The other seven entries are found in exactly the same way:

$$\begin{bmatrix} 1 \\ 3 \\ 2 \end{bmatrix} [4 \quad 5 \quad 6] = \begin{bmatrix} 1 \cdot 4 & 1 \cdot 5 & 1 \cdot 6 \\ 3 \cdot 4 & 3 \cdot 5 & 3 \cdot 6 \\ 2 \cdot 4 & 2 \cdot 5 & 2 \cdot 6 \end{bmatrix} = \begin{bmatrix} 4 & 5 & 6 \\ 12 & 15 & 18 \\ 8 & 10 & 12 \end{bmatrix}$$

Notice that the product of these two matrices in the other order is

$$[4 \quad 5 \quad 6] \begin{bmatrix} 1 \\ 3 \\ 2 \end{bmatrix} = [4 \cdot 1 + 5 \cdot 3 + 6 \cdot 2]$$
$$= [4 + 15 + 12]$$
$$= [31]$$

The product in one order is a 3 × 3 matrix and in the other order a 1 × 1 matrix, that is, a number. With numbers, the order of multiplication is immaterial: for instance, $3 \cdot 6 = 18 = 6 \cdot 3$. With matrices, the order matters quite a lot. Sometimes two matrices can be multiplied only in one order. For example, a 2 × 3 matrix can be multiplied by a 3 × 4 matrix:

$2 \times 3 \qquad 3 \times 4$; but not the other way around: $3 \times 4 \qquad 2 \times 3$.

We will explore this facet of matrix multiplication in some of the problems at the end of this section, and again in the next section. In our symbolic notation, we may say that if A and B are matrices, the products **AB** and **BA** are not necessarily equal. In fact, they are probably different even if both multiplications can be performed.

Appalachian Creations found many uses for matrix multiplication. Here is another example. One summer the managers decided to start selling their goods at three different department stores. Each contract was negotiated separately, the result being that each store paid a different price for each item purchased. Table 18 resulted.

Table 18
Price Paid by Stores (in dollars)

Store	Item				
	Light bags	Heavy bags	Jackets	Two-person tents	Four-person tents
Hartford, CT	100	120	80	100	250
Providence, RI	120	150	90	80	200
Barre, VT	150	200	100	90	200

Each store bought exactly the same number of items from each factory, as shown in Table 19.

Table 19
Items Purchased by Each Store from AC Factories

Item	Factory		
	Danbury	Springfield	Rutland
Light sleeping bags	50	50	50
Heavy sleeping bags	30	40	30
Down jackets	60	80	60
Two-person tents	40	50	0
Four-person tents	0	40	30

To obtain the total amount that each store paid each factory, the associated matrices were multiplied,

$$\begin{bmatrix} 100 & 120 & 80 & 100 & 250 \\ 120 & 150 & 90 & 80 & 200 \\ 150 & 200 & 100 & 90 & 200 \end{bmatrix} \begin{bmatrix} 50 & 50 & 50 \\ 30 & 40 & 30 \\ 60 & 80 & 60 \\ 40 & 50 & 0 \\ 0 & 40 & 30 \end{bmatrix}$$

$$= \begin{bmatrix} 17{,}400 & 31{,}200 & 20{,}900 \\ 19{,}100 & 31{,}200 & 21{,}900 \\ 23{,}100 & 36{,}000 & 25{,}500 \end{bmatrix}$$

and Table 20 resulted.

Table 20
Amount Each Store Paid Each Factory

Store	Factory		
	Danbury	Springfield	Rutland
Hartford	$17,400	$31,200	$20,000
Providence	$19,100	$31,200	$21,900
Barre	$23,100	$36,000	$25,500

In a problem arising from a practical situation, there is another condition which must be met before the matrices can be multiplied: the column headings on the first matrix must be the same as the row headings on the second matrix. The resulting table inherits its row headings from the first matrix and column headings from the second matrix. The repeated headings disappear.

We will find other uses for matrix multiplication in connection with our study of systems of linear equations (Sections 3-1 and 3-2) and Markov chains (Sections 6-2 and 6-3).

Note: While it would be possible to multiply two matrices of the same size simply by multiplying the corresponding entries, as in matrix addition, the examples above suggest that this is not as useful a technique as the one presented here.

Exercises 1.5   1. In the product

$$\begin{bmatrix} 2 & -1 & 3 & 0 & 2 \\ 1 & 1 & 2 & 1 & 4 \\ 3 & 2 & 8 & 0 & -9 \\ 1 & -4 & 6 & -2 & 1 \\ -1 & 2 & -1 & 4 & 9 \end{bmatrix} \begin{bmatrix} 1 & 3 & -1 \\ 4 & -8 & 2 \\ 2 & 1 & -3 \\ 6 & -7 & 4 \\ 8 & 6 & 6 \end{bmatrix}$$

what are each of the following entries?

A(a) (2,1)

(b) (6,3)

A(c) (5,2)

(d) (3,3)

2. Do not multiply the following pairs of matrices; simply give the size of the answer.

C(a) $\begin{bmatrix} 2 & 1 & 0 \\ 3 & 1 & 4 \\ 6 & 7 & 8 \end{bmatrix} \begin{bmatrix} 2 & 1 & 7 & 8 & 4 \\ 1 & -1 & 3 & 8 & 6 \\ 2 & 4 & 8 & 9 & 5 \end{bmatrix}$

(b) $\begin{bmatrix} 2 & 0 \\ 1 & -1 \\ 4 & 6 \\ 3 & 1 \\ 5 & 8 \end{bmatrix} \begin{bmatrix} 2 & 4 & 7 & 3 & 8 \\ 1 & 5 & 0 & 7 & 4 \end{bmatrix}$

A(c) $\begin{bmatrix} 3 & 1 & 2 & 4 \\ 0 & 1 & 1 & 2 \\ 2 & -1 & 3 & 6 \\ 4 & 2 & 1 & 8 \\ 7 & 6 & 0 & 5 \end{bmatrix} \begin{bmatrix} 3 & 2 & 7 & 6 & 1 \\ 2 & 1 & 0 & 2 & 1 \\ 4 & 5 & 7 & 8 & -1 \\ 6 & 1 & 9 & 3 & -1 \\ 3 & 2 & 0 & 2 & 5 \end{bmatrix}$

A(d) $\begin{bmatrix} 3 \\ 1 \\ 2 \\ 4 \\ 6 \\ 8 \end{bmatrix} [2 \quad 1 \quad 7 \quad 8 \quad 9]$

A3. Multiplication of two matrices results in a 5×7 matrix. What can be said about the sizes of the two matrices that were multiplied to produce this result?

4. Perform the following matrix multiplications:

A (a) $[3 \quad 2] \begin{bmatrix} 1 \\ 4 \end{bmatrix}$

(b) $\begin{bmatrix} 3 & 2 \\ 1 & 2 \end{bmatrix} \begin{bmatrix} 1 \\ 4 \end{bmatrix}$

(c) $[1 \quad 4] \begin{bmatrix} 3 & 2 \\ 1 & 2 \end{bmatrix}$

(d) $\begin{bmatrix} 1 & 4 \\ 3 & 1 \end{bmatrix} \begin{bmatrix} 3 & 2 \\ 1 & 2 \end{bmatrix}$

A(e) $\begin{bmatrix} 1 & 4 \\ 3 & 1 \\ 0 & 2 \end{bmatrix} \begin{bmatrix} 3 & 2 & 1 \\ 1 & 2 & 2 \end{bmatrix}$

C(f) $\begin{bmatrix} 1 & 1 \\ 0 & 1 \end{bmatrix} \begin{bmatrix} 1 & 0 \\ 1 & 1 \end{bmatrix}$

C(g) $\begin{bmatrix} 1 & 0 \\ 1 & 1 \end{bmatrix} \begin{bmatrix} 1 & 1 \\ 0 & 1 \end{bmatrix}$

C(h) $\begin{bmatrix} 3 & 2 \\ 4 & -1 \end{bmatrix} \begin{bmatrix} 3 & 1 \\ 2 & 1 \end{bmatrix}$

C(i) $\begin{bmatrix} 3 & 1 \\ 2 & 1 \end{bmatrix} \begin{bmatrix} 3 & 2 \\ 4 & -1 \end{bmatrix}$

A(j) $\begin{bmatrix} 1 & 0 \\ 0 & 1 \end{bmatrix} \begin{bmatrix} 3 & -1 & 5 \\ 2 & 4 & -2 \end{bmatrix}$

A(k) $\begin{bmatrix} 2 & -1 & 5 \\ 2 & 4 & -2 \end{bmatrix} \begin{bmatrix} 1 & 0 \\ 0 & 1 \end{bmatrix}$

(l) $\begin{bmatrix} 1 & -1 & -2 \\ -1 & 2 & 3 \end{bmatrix} \begin{bmatrix} 1 & -1 \\ 2 & 3 \\ -1 & -2 \end{bmatrix}$

(m) $\begin{bmatrix} 1 & -1 \\ 2 & 3 \\ -1 & -2 \end{bmatrix} \begin{bmatrix} 1 & -1 & -2 \\ -1 & 2 & 3 \end{bmatrix}$

C(n) $\begin{bmatrix} -1 & 1 & 2 \\ 2 & -2 & 1 \end{bmatrix} \begin{bmatrix} 3 & 4 & 0 & -2 \\ 1 & -2 & 2 & -1 \\ 1 & -1 & 3 & 2 \end{bmatrix}$

C(o) $\begin{bmatrix} 0 & -8 & 8 & 5 \\ 5 & 11 & -1 & 0 \end{bmatrix} \begin{bmatrix} 2 & 3 \\ 1 & 2 \\ -1 & 1 \\ -2 & 1 \end{bmatrix}$

C(p) $\begin{bmatrix} 3 & 4 & 0 & -2 \\ 1 & -2 & 2 & -1 \\ 1 & -1 & 3 & 2 \end{bmatrix} \begin{bmatrix} 2 & 3 \\ 1 & 2 \\ -1 & 1 \\ -2 & 1 \end{bmatrix}$

C(q) $\begin{bmatrix} -1 & 1 & 2 \\ 2 & -2 & 1 \end{bmatrix} \begin{bmatrix} 14 & 15 \\ 0 & 0 \\ -6 & 6 \end{bmatrix}$

A(r) $\begin{bmatrix} 2 & 1 & -1 & 3 \end{bmatrix} \begin{bmatrix} 3 \\ -1 \\ 2 \\ -1 \end{bmatrix}$

A(s) $\begin{bmatrix} 3 \\ -1 \\ 2 \\ -1 \end{bmatrix} \begin{bmatrix} 2 & 1 & -1 & 3 \end{bmatrix}$

5. Multiply the following pairs of matrices in the order given:

(a) $\begin{bmatrix} 1 & 1 & 2 \\ 3 & 2 & 3 \\ 2 & 1 & 2 \end{bmatrix}, \begin{bmatrix} 1 & -1 & 0 \\ -3 & 0 & 2 \\ 1 & 1 & -1 \end{bmatrix}$ (b) $\begin{bmatrix} 1 & 1 & 2 \\ 3 & 2 & 3 \\ 2 & 1 & 2 \end{bmatrix}, \begin{bmatrix} 0 & 0 & 1 \\ -3 & 2 & -3 \\ 2 & -1 & 1 \end{bmatrix}$

(c) $\begin{bmatrix} 1 & 1 & 2 \\ 3 & 2 & 3 \\ 2 & 1 & 2 \end{bmatrix}, \begin{bmatrix} -1 & 0 & 1 \\ 0 & -2 & -3 \\ 1 & 1 & 1 \end{bmatrix}$

A(d) $\begin{bmatrix} 1 & 1 & 2 \\ 3 & 2 & 3 \\ 2 & 1 & 2 \end{bmatrix}, \begin{bmatrix} -1 & 0 & 1 \\ 0 & 2 & -3 \\ 1 & -1 & 1 \end{bmatrix}$

A(e) $\begin{bmatrix} -8 & 3 & 1 \\ -2 & 1 & 0 \\ 13 & -5 & -1 \end{bmatrix}, \begin{bmatrix} 9 & 7 & 8 \\ 21 & 16 & 17 \\ 11 & 9 & 16 \end{bmatrix}$

A(f) $\begin{bmatrix} 1 & 2 & 1 \\ 2 & 5 & 2 \\ 3 & 1 & 2 \end{bmatrix}, \begin{bmatrix} -6 & 0 & -18 \\ -1 & 0 & -5 \\ 10 & 1 & 31 \end{bmatrix}$

A(g) $\begin{bmatrix} 1 & 1 & 2 \\ 3 & 2 & 3 \\ 2 & 1 & 2 \end{bmatrix}, \begin{bmatrix} -1 & 1 & 0 \\ 3 & -2 & -7 \\ 0 & 1 & 5 \end{bmatrix}$

A(h) $\begin{bmatrix} -1 & 0 & 1 \\ 0 & 2 & -3 \\ 1 & -1 & 1 \end{bmatrix}, \begin{bmatrix} 7 & 7 & 10 \\ 15 & 13 & 20 \\ 9 & 8 & 13 \end{bmatrix}$

A(i) $\begin{bmatrix} 1 & -2 & 1 \\ 2 & -4 & 2 \\ -1 & 2 & -1 \end{bmatrix}, \begin{bmatrix} 3 & 2 & -2 \\ 2 & 1 & -3 \\ 1 & 0 & -4 \end{bmatrix}$

(j) $\begin{bmatrix} -1 & 0 & 1 \\ 0 & 2 & -3 \\ 1 & -1 & 1 \end{bmatrix}, \begin{bmatrix} 4 & 4 & 4 \\ 8 & 8 & 8 \\ 5 & 5 & 5 \end{bmatrix}$

(k) $\begin{bmatrix} 4 & 4 & 4 \\ 8 & 8 & 8 \\ 5 & 5 & 5 \end{bmatrix}, \begin{bmatrix} -1 & 0 & 1 \\ 0 & 2 & -3 \\ 1 & -1 & 1 \end{bmatrix}$

A(l) $\begin{bmatrix} 15 & 17 & 19 \\ 25 & 37 & 46 \\ 15 & 34 & 96 \end{bmatrix}, \begin{bmatrix} 0 & 0 & 0 \\ 0 & 0 & 0 \\ 0 & 0 & 0 \end{bmatrix}$

A(m) $\begin{bmatrix} 0 & 0 & 0 \\ 0 & 0 & 0 \\ 0 & 0 & 0 \end{bmatrix}, \begin{bmatrix} 138 & 74 & -28 \\ 96 & 27 & 43 \\ 25 & 38 & 49 \end{bmatrix}$

C(n) $\begin{bmatrix} 1 & 0 & 0 \\ 0 & 1 & 0 \\ 0 & 0 & 1 \end{bmatrix}, \begin{bmatrix} 71 & 89 & 43 \\ 27 & 96 & 185 \\ 34 & 86 & 57 \end{bmatrix}$

C(o) $\begin{bmatrix} 32 & 41 & 76 \\ 194 & 38 & 49 \\ 25 & 86 & 97 \end{bmatrix}, \begin{bmatrix} 1 & 0 & 0 \\ 0 & 1 & 0 \\ 0 & 0 & 1 \end{bmatrix}$

6. Multiply the following pairs of matrices:

(a) $\begin{bmatrix} 2 & 1 & 1 \\ 1 & 1 & 2 \\ 3 & 1 & 2 \end{bmatrix}, \begin{bmatrix} 0 & -1/2 & 1/2 \\ 2 & 1/2 & -3/2 \\ -1 & 1/2 & 1/2 \end{bmatrix}$

A(b) $\begin{bmatrix} 1/2 & 1/3 \\ 2/3 & 1/4 \end{bmatrix}, \begin{bmatrix} 12 & 18 \\ 36 & -24 \end{bmatrix}$

(c) $\begin{bmatrix} 7 & -3 \\ 2 & 4 \end{bmatrix}, \begin{bmatrix} 4/22 & 3/22 \\ -2/22 & 7/22 \end{bmatrix}$

A(d) $\begin{bmatrix} 1/2 & 3/4 \\ 8/9 & 1/3 \end{bmatrix}, \begin{bmatrix} 1/3 & -3/8 \\ 2/3 & 1/4 \end{bmatrix}$

7. Table 21 is the Springfield Plastics Company's production table.

Table 21

Item	Factory			
	MO	MA	IL	OH
Wastebasket	2520	2000	1950	800
Hard hat	3240	1780	1530	1360
Vacuum cleaner case	760	370	250	125

Table 22 gives the price for each item as paid by each of 4 stores for 10 of each item:

Table 22

	Wastebasket	Hard hat	Vacuum cleaner case
Store 1	$2	$1	$3
Store 2	4	3	2
Store 3	5	2	1
Store 4	3	2	4

Assuming each store buys one fourth of each factory's output, use matrix operations to construct a table showing how much each store pays to each factory.

*8. What are the missing numbers in the following calculation?

$$\begin{bmatrix} 1 & -1 & 2 \\ 3 & 1 & 4 \\ 2 & & 1 \end{bmatrix} \begin{bmatrix} 1 & 2 & 3 \\ 4 & & 2 \\ & 3 & 6 \end{bmatrix} = \begin{bmatrix} -1 & & \\ & 19 & \\ & & 12 \end{bmatrix}$$

9. (a) If A is a 3 × 5 matrix, B is a matrix, and AB is a 3 × 7 matrix, what size is B?

   (b) If C is a 12 × 15 matrix, D is a matrix, and DC can be formed and has 9 rows, what size is D and what size is DC?

   *(c) If P is a 1 × 7 matrix, Q is a matrix with only one column, and PQ can be formed, how many entries are in the matrix QP?

C10. (a) Find a 2 × 2 matrix $\begin{bmatrix} & \\ & \end{bmatrix}$

   such that $\begin{bmatrix} 1 & 2 \\ 3 & 4 \end{bmatrix} \begin{bmatrix} & \\ & \end{bmatrix} = \begin{bmatrix} 0 & 0 \\ 0 & 0 \end{bmatrix}$.

   (b) Find a 2 × 2 matrix A such that $A \begin{bmatrix} 1 & 2 \\ 3 & 4 \end{bmatrix} = \begin{bmatrix} 1 & 2 \\ 3 & 4 \end{bmatrix}$.

   (c) Find a matrix A such that $\begin{bmatrix} 1 & 2 \\ 3 & 4 \end{bmatrix} A = \begin{bmatrix} 1 & 2 \\ 3 & 4 \\ 3 & 7 \end{bmatrix}$.

   (d) Find a matrix $\begin{bmatrix} & \\ & \end{bmatrix}$ such that $\begin{bmatrix} 1 & 2 \\ 3 & 4 \end{bmatrix} \begin{bmatrix} & \\ & \end{bmatrix} = \begin{bmatrix} 1 & 2 & 3 \\ 3 & 4 & 7 \end{bmatrix}$.

See Exercise 8 in Section 1-4.

11. PHE wishes to use its model company, PHEM, to help in its consideration of price structures. Here are two price structures, **b** and **c**, now under consideration:

	b	c
Hamburger	$.50	$.45
Cheeseburger	.60	.65
Fries	.45	.50
Shake	.50	.45

   (a) Let $B = \begin{bmatrix} .50 \\ .60 \\ .45 \\ .50 \end{bmatrix}$. Using the symbolism of Exercise 8 in Section 1-4, write the product that gives the total proceeds from each store in July, using the price structure **b**. Carry out the multiplication.

   (b) Let $C = \begin{bmatrix} .45 \\ .65 \\ .50 \\ .45 \end{bmatrix}$. Write the operation, using one of the four matrices of Exercise 8 in Section 1-4, that gives the total proceeds from each store in the third quarter, using price structure **c**.

Section 1-6
Some Facts about Matrix
Arithmetic

Now that we have seen how to add, subtract, and multiply matrices, we wish to compare these matrix operations with addition, subtraction, and multiplication of numbers. Division of matrices will be discussed in Chapter 3. One other sort of matrix operation, transposing, will be introduced in Section 5-5. None of the other matrix operations, such as taking the square root, are very useful in applications to business, though they are of use in natural science. Multiplication of matrices, on the other hand, is extremely important, and has much in common with the multiplication of numbers.

The first property of matrices we will discuss is equality. Two matrices are **equal** if they have the same size and their corresponding entries are equal. Thus, the matrices

$$\begin{bmatrix} 1 & 2 & 3 \\ 4 & 5 & 6 \\ 7 & 8 & 9 \end{bmatrix} \quad \text{and} \quad \begin{bmatrix} 1 & 2 & 7 \\ 4 & 5 & 6 \\ 3 & 8 & 9 \end{bmatrix}$$

are *not* equal, since although they have the same size and contain the same entries, the corresponding entries are not all the same. For example, the (1,3) entries are different. One use of matrix equality will be demonstrated in Chapter 3, where a single matrix equation will be used to replace many equations involving numbers. A simple illustration is provided by the matrix equation

$$\begin{bmatrix} w & x \\ y & z \end{bmatrix} = \begin{bmatrix} 3 & 1 \\ 5 & -2 \end{bmatrix}$$

which replaces the four equations $w = 3$, $x = 1$, $y = 5$, and $z = -2$.

One important property of the addition of numbers is that whenever three numbers are added together, the placement of parentheses is unimportant. For example,

$$4 + (5 + 7) = (4 + 5) + 7$$

One usually writes simply $4 + 5 + 7$, since it is immaterial which addition is performed first. The same is true for multiplication of numbers. For example,

$$4 \cdot (5 \cdot 7) = (4 \cdot 5) \cdot 7$$

Matrix multiplication and addition behave the same way.

Problem 1   Verify that

$$\begin{bmatrix} 2 & 1 \\ 1 & 3 \\ 0 & 1 \end{bmatrix} \left(\begin{bmatrix} 3 & 1 & 2 & 1 \\ 2 & 5 & 3 & 4 \end{bmatrix} \begin{bmatrix} 2 & 0 & 3 \\ 1 & 1 & 2 \\ 3 & 2 & 1 \\ 1 & 3 & 2 \end{bmatrix} \right)$$

$$= \left(\begin{bmatrix} 2 & 1 \\ 1 & 3 \\ 0 & 1 \end{bmatrix} \begin{bmatrix} 3 & 1 & 2 & 1 \\ 2 & 5 & 3 & 4 \end{bmatrix} \right) \begin{bmatrix} 2 & 0 & 3 \\ 1 & 1 & 2 \\ 3 & 2 & 1 \\ 1 & 3 & 2 \end{bmatrix}$$

Solution

$$\begin{bmatrix} 2 & 1 \\ 1 & 3 \\ 0 & 1 \end{bmatrix} \left(\begin{bmatrix} 3 & 1 & 2 & 1 \\ 2 & 5 & 3 & 4 \end{bmatrix} \begin{bmatrix} 2 & 0 & 3 \\ 1 & 1 & 2 \\ 3 & 2 & 1 \\ 1 & 3 & 2 \end{bmatrix} \right) = \begin{bmatrix} 2 & 1 \\ 1 & 3 \\ 0 & 1 \end{bmatrix} \begin{bmatrix} 14 & 8 & 15 \\ 22 & 23 & 27 \end{bmatrix}$$

$$= \begin{bmatrix} 50 & 39 & 57 \\ 80 & 77 & 96 \\ 22 & 23 & 27 \end{bmatrix} \text{ and}$$

$$\left(\begin{bmatrix} 2 & 1 \\ 1 & 3 \\ 0 & 1 \end{bmatrix} \begin{bmatrix} 3 & 1 & 2 & 1 \\ 2 & 5 & 3 & 4 \end{bmatrix} \right) \begin{bmatrix} 2 & 0 & 3 \\ 1 & 1 & 2 \\ 3 & 2 & 1 \\ 1 & 3 & 2 \end{bmatrix}$$

$$= \begin{bmatrix} 8 & 7 & 7 & 6 \\ 9 & 16 & 11 & 13 \\ 2 & 5 & 3 & 4 \end{bmatrix} \begin{bmatrix} 2 & 0 & 3 \\ 1 & 1 & 2 \\ 3 & 2 & 1 \\ 1 & 3 & 2 \end{bmatrix} = \begin{bmatrix} 50 & 39 & 57 \\ 80 & 77 & 96 \\ 22 & 23 & 27 \end{bmatrix} \text{ also.}$$

Thus, the resulting matrices are equal. This verifies this one instance of the *associative law* for multiplication of matrices.

Problem 2   Verify that

$$\begin{bmatrix} 3 & 1 & 2 \\ 1 & 0 & 1 \end{bmatrix} + \left(\begin{bmatrix} 2 & 1 & 3 \\ 4 & 2 & 2 \end{bmatrix} + \begin{bmatrix} 3 & 2 & 2 \\ 1 & 3 & 1 \end{bmatrix} \right)$$

$$= \left(\begin{bmatrix} 3 & 1 & 2 \\ 1 & 0 & 1 \end{bmatrix} + \begin{bmatrix} 2 & 1 & 3 \\ 4 & 2 & 2 \end{bmatrix} \right) + \begin{bmatrix} 3 & 2 & 2 \\ 1 & 3 & 1 \end{bmatrix}$$

Solution

$$\begin{bmatrix} 3 & 1 & 2 \\ 1 & 0 & 1 \end{bmatrix} + \left(\begin{bmatrix} 2 & 1 & 3 \\ 4 & 2 & 2 \end{bmatrix} + \begin{bmatrix} 3 & 2 & 2 \\ 1 & 3 & 1 \end{bmatrix} \right)$$

$$= \begin{bmatrix} 3 & 1 & 2 \\ 1 & 0 & 1 \end{bmatrix} + \begin{bmatrix} 5 & 3 & 5 \\ 5 & 5 & 3 \end{bmatrix} = \begin{bmatrix} 8 & 4 & 7 \\ 6 & 5 & 4 \end{bmatrix} \text{ and}$$

$$\left(\begin{bmatrix} 3 & 1 & 2 \\ 1 & 0 & 1 \end{bmatrix} + \begin{bmatrix} 2 & 1 & 3 \\ 4 & 2 & 2 \end{bmatrix} \right) + \begin{bmatrix} 3 & 2 & 2 \\ 1 & 3 & 1 \end{bmatrix}$$

$$= \begin{bmatrix} 5 & 2 & 5 \\ 5 & 2 & 3 \end{bmatrix} + \begin{bmatrix} 3 & 2 & 2 \\ 1 & 3 & 1 \end{bmatrix} = \begin{bmatrix} 8 & 4 & 7 \\ 6 & 5 & 4 \end{bmatrix} \text{ also.}$$

It should be noted that we have not shown that given *any* three matrices to add or multiply, the insertion of parentheses is irrelevant. We have merely verified that this is true for the three particular matrices above. The more general fact is not hard to verify for addition, but it is rather complicated to verify for multiplication. The interested reader is referred to *Basic Concepts of Linear Algebra,* by S. Isaak and M. N. Manougian (W. W. Norton, 1976), Theorem 1 6, p. 31. These properties of addition and multiplication are usually called, respectively, the associativity of addition and the associativity of multiplication.

Another aspect of number arithmetic is the relation between multiplication and addition. For example,

$$3 \cdot (2 + 4) = 3 \cdot 2 + 3 \cdot 4$$

Matrix arithmetic shares this property as well.

Problem 3   Verify that

$$\begin{bmatrix} 1 & 2 & 1 \\ 3 & 1 & 2 \end{bmatrix} \left(\begin{bmatrix} 2 & 1 \\ 1 & 3 \\ 2 & 1 \end{bmatrix} + \begin{bmatrix} 2 & 2 \\ 3 & 1 \\ 3 & 2 \end{bmatrix} \right)$$

$$= \begin{bmatrix} 1 & 2 & 1 \\ 3 & 1 & 2 \end{bmatrix} \begin{bmatrix} 2 & 1 \\ 1 & 3 \\ 2 & 1 \end{bmatrix} + \begin{bmatrix} 1 & 2 & 1 \\ 3 & 1 & 2 \end{bmatrix} \begin{bmatrix} 2 & 2 \\ 3 & 1 \\ 3 & 2 \end{bmatrix}$$

Solution

$$\begin{bmatrix} 1 & 2 & 1 \\ 3 & 1 & 2 \end{bmatrix} \left(\begin{bmatrix} 2 & 1 \\ 1 & 3 \\ 2 & 1 \end{bmatrix} + \begin{bmatrix} 2 & 2 \\ 3 & 1 \\ 3 & 2 \end{bmatrix} \right) = \begin{bmatrix} 1 & 2 & 1 \\ 3 & 1 & 2 \end{bmatrix} \begin{bmatrix} 4 & 3 \\ 4 & 4 \\ 5 & 3 \end{bmatrix} = \begin{bmatrix} 17 & 14 \\ 26 & 19 \end{bmatrix}$$

$$\text{and } \begin{bmatrix} 1 & 2 & 1 \\ 3 & 1 & 2 \end{bmatrix} \begin{bmatrix} 2 & 1 \\ 1 & 3 \\ 2 & 1 \end{bmatrix} + \begin{bmatrix} 1 & 2 & 1 \\ 3 & 1 & 2 \end{bmatrix} \begin{bmatrix} 2 & 2 \\ 3 & 1 \\ 3 & 2 \end{bmatrix}$$

$$= \begin{bmatrix} 6 & 8 \\ 11 & 8 \end{bmatrix} + \begin{bmatrix} 11 & 6 \\ 15 & 11 \end{bmatrix} = \begin{bmatrix} 17 & 14 \\ 26 & 29 \end{bmatrix} \text{ as well.}$$

Again, it can be shown that this relationship is always true for any three matrices, whenever the operations are **defined**—that is, whenever the sizes of the matrices are such that the indicated multiplications and additions can actually be performed. This relationship between matrix multiplication and addition is usually called the distributive law.

We have previously noted that when multiplying two matrices the order of the two matrices matters. Thus, for example,

$$\begin{bmatrix} 1 & 1 \\ 0 & 1 \end{bmatrix} \begin{bmatrix} 2 & 0 \\ 1 & 1 \end{bmatrix} = \begin{bmatrix} 3 & 1 \\ 1 & 1 \end{bmatrix}$$

but

$$\begin{bmatrix} 2 & 0 \\ 1 & 1 \end{bmatrix} \begin{bmatrix} 1 & 1 \\ 0 & 1 \end{bmatrix} = \begin{bmatrix} 2 & 2 \\ 1 & 2 \end{bmatrix}$$

Switching the order can even change the size of the answer. For example,

$$\begin{bmatrix} 2 & 1 \end{bmatrix} \begin{bmatrix} 1 \\ 3 \end{bmatrix} = \begin{bmatrix} 5 \end{bmatrix}$$

but

$$\begin{bmatrix} 1 \\ 3 \end{bmatrix} \begin{bmatrix} 2 & 1 \end{bmatrix} = \begin{bmatrix} 2 & 1 \\ 6 & 3 \end{bmatrix}$$

Finally, switching the order can result in an unworkable problem. For example, $\begin{bmatrix} 3 & 1 \\ 1 & 4 \end{bmatrix} \begin{bmatrix} 2 \\ 3 \end{bmatrix} = \begin{bmatrix} 9 \\ 14 \end{bmatrix}$ but $\begin{bmatrix} 2 \\ 3 \end{bmatrix} \begin{bmatrix} 3 & 1 \\ 1 & 4 \end{bmatrix}$ is undefined. The fact that the order of multiplication of numbers does not matter is expressed by saying that multiplication of numbers is commutative. For matrices, the proper statement is that *multiplication of matrices is not commutative.*

While matrix multiplication is not commutative, matrix addition is. For example,

$$\begin{bmatrix} 1 & 1 \\ 0 & 1 \end{bmatrix} + \begin{bmatrix} 2 & 0 \\ 1 & 1 \end{bmatrix} = \begin{bmatrix} 3 & 1 \\ 1 & 2 \end{bmatrix}$$

and

$$\begin{bmatrix} 2 & 0 \\ 1 & 1 \end{bmatrix} + \begin{bmatrix} 1 & 1 \\ 0 & 1 \end{bmatrix} = \begin{bmatrix} 3 & 1 \\ 1 & 2 \end{bmatrix} \text{ as well.}$$

There are a number of important relationships between multiplication of a matrix by a number, matrix addition, and matrix multiplication. We list them below. Note again that we are only illustrating what can be shown to be general principles.

Problem 4   Verify the following:

(a) $3\left(\begin{bmatrix} 1 & 2 \\ 4 & 2 \end{bmatrix} + \begin{bmatrix} 3 & 5 \\ 1 & 2 \end{bmatrix}\right) = 3\begin{bmatrix} 1 & 2 \\ 4 & 2 \end{bmatrix} + 3\begin{bmatrix} 3 & 5 \\ 1 & 2 \end{bmatrix}$

(b) $(3+4)\begin{bmatrix} 1 & 2 \\ 3 & 4 \end{bmatrix} = 3\begin{bmatrix} 1 & 2 \\ 3 & 4 \end{bmatrix} + 4\begin{bmatrix} 1 & 2 \\ 3 & 4 \end{bmatrix}$

(c) $5\left(\begin{bmatrix} 1 & 2 \\ 3 & 4 \end{bmatrix}\begin{bmatrix} 2 & 1 \\ 3 & 4 \end{bmatrix}\right) = \left(5\begin{bmatrix} 1 & 2 \\ 3 & 4 \end{bmatrix}\right)\begin{bmatrix} 2 & 1 \\ 3 & 4 \end{bmatrix}$

(d) $5\left(\begin{bmatrix} 1 & 2 \\ 3 & 4 \end{bmatrix}\begin{bmatrix} 2 & 1 \\ 3 & 4 \end{bmatrix}\right) = \begin{bmatrix} 1 & 2 \\ 3 & 4 \end{bmatrix}\left(5\begin{bmatrix} 2 & 1 \\ 3 & 4 \end{bmatrix}\right)$

(e) $(3 \times 4)\begin{bmatrix} 1 & 2 \\ 3 & 4 \end{bmatrix} = 3\left(4\begin{bmatrix} 1 & 2 \\ 3 & 4 \end{bmatrix}\right)$

Solution   (a) $3\left(\begin{bmatrix} 1 & 2 \\ 4 & 2 \end{bmatrix} + \begin{bmatrix} 3 & 5 \\ 1 & 2 \end{bmatrix}\right) = 3\begin{bmatrix} 4 & 7 \\ 5 & 4 \end{bmatrix} = \begin{bmatrix} 12 & 21 \\ 15 & 12 \end{bmatrix}$ and

$3\begin{bmatrix} 1 & 2 \\ 4 & 2 \end{bmatrix} + 3\begin{bmatrix} 3 & 5 \\ 1 & 2 \end{bmatrix} = \begin{bmatrix} 3 & 6 \\ 12 & 6 \end{bmatrix} + \begin{bmatrix} 9 & 15 \\ 3 & 6 \end{bmatrix} = \begin{bmatrix} 12 & 21 \\ 15 & 12 \end{bmatrix}$ also.

(b) $(3+4)\begin{bmatrix} 1 & 2 \\ 3 & 4 \end{bmatrix} = 7\begin{bmatrix} 1 & 2 \\ 3 & 4 \end{bmatrix} = \begin{bmatrix} 7 & 14 \\ 21 & 28 \end{bmatrix}$ and $3\begin{bmatrix} 1 & 2 \\ 3 & 4 \end{bmatrix}$

$+ 4\begin{bmatrix} 1 & 2 \\ 3 & 4 \end{bmatrix} = \begin{bmatrix} 3 & 6 \\ 9 & 12 \end{bmatrix} + \begin{bmatrix} 4 & 8 \\ 12 & 16 \end{bmatrix} = \begin{bmatrix} 7 & 14 \\ 21 & 28 \end{bmatrix}$ also.

(c) $5\left(\begin{bmatrix} 1 & 2 \\ 3 & 4 \end{bmatrix}\begin{bmatrix} 2 & 1 \\ 3 & 4 \end{bmatrix}\right) = 5\begin{bmatrix} 8 & 9 \\ 18 & 19 \end{bmatrix} = \begin{bmatrix} 40 & 45 \\ 90 & 95 \end{bmatrix}$ and

$\left(5\begin{bmatrix} 1 & 2 \\ 3 & 4 \end{bmatrix}\right)\begin{bmatrix} 2 & 1 \\ 3 & 4 \end{bmatrix} = \begin{bmatrix} 5 & 10 \\ 15 & 20 \end{bmatrix}\begin{bmatrix} 2 & 1 \\ 3 & 4 \end{bmatrix} = \begin{bmatrix} 40 & 45 \\ 90 & 95 \end{bmatrix}$ also.

(d) $5\left(\begin{bmatrix} 1 & 2 \\ 3 & 4 \end{bmatrix}\begin{bmatrix} 2 & 1 \\ 3 & 4 \end{bmatrix}\right) = 5\begin{bmatrix} 8 & 9 \\ 18 & 19 \end{bmatrix} = \begin{bmatrix} 40 & 45 \\ 90 & 95 \end{bmatrix}$ and

$\begin{bmatrix} 1 & 2 \\ 3 & 4 \end{bmatrix}\left(5\begin{bmatrix} 2 & 1 \\ 3 & 4 \end{bmatrix}\right) = \begin{bmatrix} 1 & 2 \\ 3 & 4 \end{bmatrix}\begin{bmatrix} 10 & 5 \\ 15 & 20 \end{bmatrix} = \begin{bmatrix} 40 & 45 \\ 90 & 95 \end{bmatrix}$ also.

(e) $(3 \times 4)\begin{bmatrix} 1 & 2 \\ 3 & 4 \end{bmatrix} = 12\begin{bmatrix} 1 & 2 \\ 3 & 4 \end{bmatrix} = \begin{bmatrix} 12 & 24 \\ 36 & 48 \end{bmatrix}$ and

$3\left(4\begin{bmatrix} 1 & 2 \\ 3 & 4 \end{bmatrix}\right) = 3\begin{bmatrix} 4 & 8 \\ 12 & 16 \end{bmatrix} = \begin{bmatrix} 12 & 24 \\ 36 & 48 \end{bmatrix}$ also.

Properties c and d above are the most useful in calculations with matrices. We now state these properties in words. When multiplying a product of two matrices by a number, one can first multiply either matrix by the number and then multiply the matrices. Similarly, one can postpone multiplying by a number until after the matrices have been multiplied. For example:

$$\begin{bmatrix} 5 & 2 \\ 6 & 3 \end{bmatrix} \left(\tfrac{1}{3} \begin{bmatrix} 3 & -2 \\ -6 & 5 \end{bmatrix} \right) = \tfrac{1}{3} \left(\begin{bmatrix} 5 & 2 \\ 6 & 3 \end{bmatrix} \begin{bmatrix} 3 & -2 \\ -6 & 5 \end{bmatrix} \right)$$

$$= \tfrac{1}{3} \begin{bmatrix} 3 & 0 \\ 0 & 3 \end{bmatrix}$$

$$= \begin{bmatrix} 1 & 0 \\ 0 & 1 \end{bmatrix}$$

All the properties in this section can be expressed concisely using capital-letter symbolism, as follows: Let A, B, and C be matrices and let r and s be numbers.

See example on page 48. If $A + B$ is defined, then

$$A + B = B + A \qquad \qquad \textbf{1}$$

See Problem 1. If $A + B$ and $B + C$ are defined, then

$$A + (B + C) = (A + B) + C \qquad \qquad \textbf{2}$$

See Problem 2. If AB and BC are defined, then

$$A(BC) = (AB)C \qquad \qquad \textbf{3}$$

See Problem 3. If AB, AC, and $B + C$ are defined, then

$$A(B + C) = AB + AC \qquad \qquad \textbf{4}$$

See example on page 48. Even if AB is defined, BA might not be defined. \qquad **5**

See examples on page 48. Even if AB and BA are both defined, AB might not equal BA. \qquad **6**

See Problem 4a. If $A + B$ is defined, then

$$r(A + B) = rA + rB \qquad \qquad \textbf{7}$$

See Problem 4b. No matter what size A is,

$$(r + s)A = rA + sA \qquad \qquad \textbf{8}$$

See Problem 4c. If AB is defined, then

$$r(AB) = (rA)B \qquad \qquad \textbf{9}$$

See Problem 4d. If AB is defined, then

$$r(AB) = A(rB) \qquad \qquad \textbf{10}$$

See Problem 4e. No matter what size A is,

$$(rs)A = r(sA) \qquad \qquad \textbf{11}$$

The requirements in some of these statements that certain products or sums be defined are essential. Thus, for example, if

$$A = \begin{bmatrix} 1 & 2 \\ 3 & 4 \end{bmatrix}, B = \begin{bmatrix} 3 & 4 \\ 2 & 1 \end{bmatrix}, \text{ and } C = \begin{bmatrix} 1 & 3 & 0 \\ 0 & 2 & -1 \end{bmatrix}$$

we cannot assert that property 4 holds. For in this case, although AB and AC are defined, B + C is not defined and hence equation **4** is meaningless.

Exercises 1.6 ^1. Multiply:

$$\begin{bmatrix} 3 & 3 & 0 \\ 1 & 5 & 3 \\ 2 & 4 & 2 \end{bmatrix} \left(\frac{1}{6} \begin{bmatrix} -2 & -6 & 9 \\ 4 & 6 & -9 \\ -6 & -6 & 12 \end{bmatrix} \right)$$

^2. Combine:

$$3 \begin{bmatrix} 1 & 2 & 4 \\ 7 & 1 & 5 \end{bmatrix} - 2 \begin{bmatrix} 1 & 2 & 4 \\ 7 & 1 & 5 \end{bmatrix}$$

3. Verify:

$$\begin{bmatrix} 1 & 3 \\ 4 & 2 \\ 0 & 1 \end{bmatrix} \left(\begin{bmatrix} 1 & 3 & 1 & 2 \\ 4 & 2 & 0 & 1 \end{bmatrix} \begin{bmatrix} 1 & 3 & 4 \\ 1 & 2 & 3 \\ 2 & 4 & 1 \\ 2 & 1 & 5 \end{bmatrix} \right)$$

$$= \left(\begin{bmatrix} 1 & 3 \\ 4 & 2 \\ 0 & 1 \end{bmatrix} \begin{bmatrix} 1 & 3 & 1 & 2 \\ 4 & 2 & 0 & 1 \end{bmatrix} \right) \begin{bmatrix} 1 & 3 & 4 \\ 1 & 2 & 3 \\ 2 & 4 & 1 \\ 2 & 1 & 5 \end{bmatrix}$$

4. Verify:

$$\begin{bmatrix} 3 & 1 & 2 \\ 4 & 2 & 5 \\ 1 & 3 & 2 \end{bmatrix} + \left(\begin{bmatrix} 1 & 0 & 2 \\ 3 & 1 & 4 \\ 2 & 1 & 2 \end{bmatrix} + \begin{bmatrix} 5 & 1 & 6 \\ 2 & 1 & 2 \\ 2 & 0 & 1 \end{bmatrix} \right)$$

$$= \left(\begin{bmatrix} 3 & 1 & 2 \\ 4 & 2 & 5 \\ 1 & 3 & 2 \end{bmatrix} + \begin{bmatrix} 1 & 0 & 2 \\ 3 & 1 & 4 \\ 2 & 1 & 2 \end{bmatrix} \right) + \begin{bmatrix} 5 & 1 & 6 \\ 2 & 1 & 2 \\ 2 & 0 & 1 \end{bmatrix}$$

5. Verify:

$$\begin{bmatrix} 1 & 2 & 1 & 4 \\ 2 & 0 & 4 & 3 \end{bmatrix} \left(\begin{bmatrix} 3 & 2 & 1 \\ 2 & 1 & 2 \\ 4 & 1 & 5 \\ 1 & 2 & 3 \end{bmatrix} + \begin{bmatrix} 1 & 0 & 1 \\ 2 & 1 & 4 \\ 1 & 3 & 1 \\ 2 & 1 & 3 \end{bmatrix} \right)$$

$$= \begin{bmatrix} 1 & 2 & 1 & 4 \\ 2 & 0 & 4 & 3 \end{bmatrix} \begin{bmatrix} 3 & 2 & 1 \\ 2 & 1 & 2 \\ 4 & 1 & 5 \\ 1 & 2 & 3 \end{bmatrix} + \begin{bmatrix} 1 & 2 & 1 & 4 \\ 2 & 0 & 4 & 3 \end{bmatrix} \begin{bmatrix} 1 & 0 & 1 \\ 2 & 1 & 4 \\ 1 & 3 & 1 \\ 2 & 1 & 3 \end{bmatrix}$$

6. Verify:

$$\left(\begin{bmatrix} 2 & 3 & -2 \\ 4 & -1 & 3 \end{bmatrix} + \begin{bmatrix} -3 & -2 & 4 \\ -1 & 4 & 3 \end{bmatrix} \right) \begin{bmatrix} -1 & 3 \\ 2 & -4 \\ 0 & -2 \end{bmatrix}$$

$$= \begin{bmatrix} 2 & 3 & -2 \\ 4 & -1 & 3 \end{bmatrix} \begin{bmatrix} -1 & 3 \\ 2 & -4 \\ 0 & -2 \end{bmatrix} + \begin{bmatrix} -3 & -2 & 4 \\ -1 & 4 & 3 \end{bmatrix} \begin{bmatrix} -1 & 3 \\ 2 & -4 \\ 0 & -2 \end{bmatrix}$$

This illustrates the "right-hand distributive law" of matrix arithmetic. Statement **4** above is the general statement of the "left-hand" law. Give the general statement of the right-hand law.

7. Use the laws of matrix arithmetic to perform *easily* the following computations:

(a) $\begin{bmatrix} -1/10 & 1/20 \\ 1/10 & -3/20 \\ 1/20 & 0 \end{bmatrix} \begin{bmatrix} -8/3 & -4/3 \\ 4 & 4/3 \end{bmatrix}$

(b) $\begin{bmatrix} -4\frac{1}{2} & -1\frac{1}{3} \\ 7\frac{1}{4} & -9\frac{1}{5} \end{bmatrix} \begin{bmatrix} 19 & -20 \\ -21 & 30 \end{bmatrix} + \begin{bmatrix} 5\frac{1}{2} & 2\frac{1}{3} \\ -6\frac{1}{4} & 10\frac{1}{5} \end{bmatrix} \begin{bmatrix} 19 & -20 \\ -21 & 30 \end{bmatrix}$

*8. The powers of a square matrix A are defined as follows: $A^1 = A$, $A^2 = AA$, $A^3 = AAA$, $A^4 = AAAA$, etc. Suppose $A = \begin{bmatrix} 1 & -2 \\ 2 & 1 \end{bmatrix}$.

Calculate A^{10} by doing *only four* matrix multiplications.

*9. Let A, B, C, D be matrices such that AB is a matrix all of whose entries are zero and $AC = D$. Find (in terms of D) $\frac{1}{3}A(\frac{2}{3}B - 6C)$.

10. "I'm doomed," said PHE's chief of data processing to his friend. "I was given what I thought was the full November production matrix N and a chart of 9 possible price structures for all our food items. I was asked to find the November proceeds from each store, assuming each different structure. So I assembled the price structures, column by column, into a 35 × 9 matrix M and multiplied to get NM, which contains all the answers. "Now," he groaned, "I find out that the November production matrix had left out a week, so that each entry in N should be 25% larger. Not only that," he sighed, "the board has decided that all of the possible price structures were too high and that each price must be reduced 20%. I have this sheaf of papers containing the old product NM, but now I have to recalculate everything, and the board expets the full results in ten minutes. What will I tell them?"

Review Exercises Reread Chapter 1, then do the following exercises without looking back. Answers to all of these exercises are in Appendix B.

▲1. (a) Present the following information in matrix form: E-Z Publishing Company publishes three monthly comic books—Little Wiggly, Super Hog, and Sam Rifle. The books are printed in Bayonne, New Jersey, and Eureka, California. The output of these factories in June, in thousands of books, was:

	Little Wiggly	Super Hog	Sam Rifle
Bayonne	40	20	30
Eureka	10	50	30

(b) What part of the matrix found in (a) gives the production of each book in Eureka? What part gives the production of Sam Rifle in both factories?

(c) What is the size of the matrix of (a)?

(d) What is the address of the entry "10" in the matrix found in (a)? Of entry "20"?

A2. (a) July production at E-Z (Exercise 1) was up 50% for each book at each factory. What was the July production matrix?

(b) How can the July production matrix be obtained from the June production matrix by using matrix operations?

A3. (a) The August production matrix at E-Z was the sum of the June and July matrices. How many copies of Little Wiggly were produced at Eureka in August?

(b) What is the August production matrix for E-Z?

A4. September production at E-Z would have been the same as June production (Exercise 1a) except that labor problems caused E-Z to run short of planned production by the following amounts:

	Little Wiggly	Super Hog	Sam Rifle
Bayonne	5	5	0
Eureka	0	15	5

(a) Write this information as a matrix.

(b) What was the actual production matrix for September?

(c) How can the production matrix for September be obtained from the matrix of (a) and the matrix of Exercise 1a by using matrix operations?

A5. (a) The June retail prices of the E-Z books are:

	Per Book (cents)	Per Thousand Books (dollars)
Little Wiggly	20	200
Super Hog	30	300
Sam Rifle	15	150

Write this information as two column matrices.

(b) Assuming all the books were sold, what was the gross June income from the Eureka operation?

(c) Given the matrix R_2 of Exercise 1b and the second matrix of (a), how can we use matrix operations to obtain the answer to (b)?

(d) Proceed as in (c) to obtain the gross June income from Bayonne.

(e) Show how to use matrix operations to obtain the answers to (c) and (d) in the form of a single column matrix by using the matrix of exercise 1a and the second matrix of (a).

^{A}6. Calculate each of the following or, when appropriate, write "not defined."

(a) $[-\frac{2}{3} \quad \frac{3}{4} \quad -1] - [\frac{4}{3} \quad -\frac{5}{6} \quad -\frac{4}{5}]$

(b) $\begin{bmatrix} 1 & 2 & 3 & 4 \\ 5 & 6 & 7 & 8 \end{bmatrix} + \begin{bmatrix} 9 & 1 & 0 \\ 3 & 2 & 1 \end{bmatrix}$

(c) $\frac{2}{3} \begin{bmatrix} -3 & 2 \\ 6 & -\frac{1}{4} \end{bmatrix}$

(d) $\frac{5}{6} \begin{bmatrix} -\frac{1}{2} \\ \frac{3}{2} \end{bmatrix} - \frac{6}{5} \begin{bmatrix} \frac{2}{3} \\ -\frac{4}{9} \end{bmatrix}$

(e) $10 \begin{bmatrix} 0.1 & -0.2 & 0.3 \\ -0.4 & 0.5 & -0.6 \\ 0.7 & -0.8 & 0.9 \end{bmatrix}$

^{A}7. Calculate each of the following or, when appropriate, write "not defined."

(a) $\begin{bmatrix} 1 & 2 \\ 3 & 4 \end{bmatrix} \begin{bmatrix} 5 & 6 \\ 7 & 8 \end{bmatrix} - \begin{bmatrix} 1 & 2 \\ 3 & 4 \end{bmatrix} \begin{bmatrix} 5 & 6 \\ 7 & 8 \end{bmatrix}$

(b) $\begin{bmatrix} 1 & 2 \\ 3 & 4 \end{bmatrix} \begin{bmatrix} 5 & 6 \\ 7 & 8 \end{bmatrix} - \begin{bmatrix} 5 & 6 \\ 7 & 8 \end{bmatrix} \begin{bmatrix} 1 & 2 \\ 3 & 4 \end{bmatrix}$

(c) $\begin{bmatrix} 1 & 2 \\ 3 & 4 \end{bmatrix} \begin{bmatrix} 5 & 6 \\ 7 & 8 \end{bmatrix} - \begin{bmatrix} 1 & 2 \\ 3 & 4 \end{bmatrix} \begin{bmatrix} 4 & 6 \\ 7 & 7 \end{bmatrix}$

(d) $\begin{bmatrix} 1 & 2 & 3 \\ 4 & 5 & 6 \end{bmatrix} \begin{bmatrix} 7 & 8 & 9 \\ 10 & 11 & 12 \end{bmatrix}$

(e) $\begin{bmatrix} 3 & 0 & -1 & 2 \\ 2 & -1 & 0 & -3 \\ 4 & -2 & 3 & -1 \\ -1 & 3 & -2 & 1 \end{bmatrix} \begin{bmatrix} 0 & -1 \\ -3 & 3 \\ 4 & 0 \\ 2 & -4 \end{bmatrix}$

^{A}8. Write each of the following as a fraction times a matrix all of whose entries are integers:

(a) $\begin{bmatrix} \frac{1}{2} & \frac{3}{2} \\ -\frac{5}{2} & \frac{7}{2} \end{bmatrix}$

(b) $\begin{bmatrix} \frac{1}{2} & \frac{3}{2} \\ -\frac{3}{2} & \frac{4}{3} \end{bmatrix}$

(c) $\begin{bmatrix} -2/3 & -3/4 & 4/5 \\ 5/6 & -6/7 & 7/8 \end{bmatrix}$

(d) $\begin{bmatrix} 9 & -6/5 \\ -3 & 12 \end{bmatrix}$

Supplementary Exercises A1. If A is a 7×4 matrix, B is a matrix, and AB is a square matrix, what size is B?

2. Find two 2×2 matrices, none of whose entries is zero, whose product is a matrix all of whose entries are zero.

A3. What number x makes the equation $\begin{bmatrix} 4 & 3 \\ -2 & 1 \end{bmatrix} - x \begin{bmatrix} 9 & -3 \\ 6 & -18 \end{bmatrix} = \begin{bmatrix} -2 & 5 \\ -6 & 13 \end{bmatrix}$ true?

4. Verify that the matrix $\begin{bmatrix} 2 & 3 & -1 \\ 4 & 1 & 6 \\ 0 & 2 & 5 \end{bmatrix}$ is a square root of the matrix

$\begin{bmatrix} 16 & 7 & 11 \\ 12 & 25 & 32 \\ 8 & 12 & 37 \end{bmatrix}$.

5. Find 2×2 matrices $\begin{bmatrix} & \\ & \end{bmatrix}$ such that:

A(a) $\begin{bmatrix} & \\ & \end{bmatrix} \begin{bmatrix} 1 & 2 \\ 3 & 4 \end{bmatrix} = \begin{bmatrix} 3 & 6 \\ 3 & 4 \end{bmatrix}$

(b) $\begin{bmatrix} & \\ & \end{bmatrix} \begin{bmatrix} 1 & 2 \\ 3 & 4 \end{bmatrix} = \begin{bmatrix} 1 & 2 \\ 9 & 12 \end{bmatrix}$

A(c) $\begin{bmatrix} 1 & 2 \\ 3 & 4 \end{bmatrix} \begin{bmatrix} & \\ & \end{bmatrix} = \begin{bmatrix} 1 & 1 \\ 3 & 2 \end{bmatrix}$

(d) $\begin{bmatrix} 1 & 2 \\ 3 & 4 \end{bmatrix} \begin{bmatrix} & \\ & \end{bmatrix} = \begin{bmatrix} 1/3 & -1/2 \\ 1 & -2 \end{bmatrix}$

A(e) $\begin{bmatrix} & \\ & \end{bmatrix} \begin{bmatrix} 1 & 2 \\ 3 & 4 \end{bmatrix} = \begin{bmatrix} 7 & 14 \\ 3/4 & 1 \end{bmatrix}$

6. If $\begin{bmatrix} 4x & 3 \\ 2 & -1 \end{bmatrix} \begin{bmatrix} 2 & -3 \\ 6x & 5 \end{bmatrix} = \begin{bmatrix} 13 & 9 \\ -1 & -11 \end{bmatrix}$, find x.

C7. If $\begin{bmatrix} 3x + 2y & 4 \\ 5x & y \end{bmatrix} = \begin{bmatrix} 2x - y & 4 \\ 10 & y \end{bmatrix}$, find x and y.

8. (a) How is a 3×4 matrix changed if the matrix $\begin{bmatrix} 7 & 0 & 0 \\ 0 & 7 & 0 \\ 0 & 0 & 7 \end{bmatrix}$ is multiplied on the left by it?

(b) How is a 3 × 4 matrix changed if it is multiplied on the right by

the matrix $\begin{bmatrix} 7 & 0 & 0 & 0 \\ 0 & 7 & 0 & 0 \\ 0 & 0 & 7 & 0 \\ 0 & 0 & 0 & 7 \end{bmatrix}$?

(c) How is a 3 × 4 matrix changed if the matrix $\begin{bmatrix} 7 & 0 & 0 \\ 0 & 1 & 0 \\ 0 & 0 & 1 \end{bmatrix}$ is multiplied

on the left by it?

(d) How is a 3 × 4 matrix changed if it is multiplied on the right by

the matrix $\begin{bmatrix} 7 & 0 & 0 & 0 \\ 0 & 1 & 0 & 0 \\ 0 & 0 & 1 & 0 \\ 0 & 0 & 0 & 1 \end{bmatrix}$?

*(e) How is a 3 × 4 matrix changed if $\begin{bmatrix} 0 & 1 & 0 \\ 1 & 0 & 0 \\ 0 & 0 & 1 \end{bmatrix}$ is multiplied on the

right by it?

*(f) How is a 3 × 4 matrix changed if it is multiplied on the left by
$\begin{bmatrix} 0 & 0 & 0 & 2 \\ 1 & 0 & 0 & 0 \\ 0 & 1 & 0 & 0 \\ 0 & 0 & 1 & 0 \end{bmatrix}$?

ᴬ 9. A major-league baseball team has 25 members and plays a schedule of 162 games. What size matrix is required to record the number of hits by each player in each game?

10. (See Exercise 9) If M is the hitting matrix for the Mets in 1975 and M′ is the hitting matrix for the Mets in 1976, what can we say about the matrix M + M′?

*11. Consider the matrix M of Exercise 10.

(a) Find a matrix B such that MB gives the total hits for the season by each team member.

(b) Find a matrix A such that AM gives the total number of hits by the team in each game.

12. The Happy Nite Motel Corporation operates a chain of 12 motels. Each motel has 30 rooms and each room can hold from 1 to 4 people. The rate matrix A is a 12 × 30 matrix giving the nightly charge *per person* for each room in each motel. The occupancy matrix for June 15, is a 30 × 12 matrix B giving the number of persons in each room of each motel on the night of June 15.

(a) What information is contained at address (5,5) of AB?

ᴬ (b) What information is contained at address (5,4) of AB?

Summary   Matrix   p. 2
   Terms   Size of a matrix   p. 3
           Address of an entry   p. 3
           Row matrix   p. 7
           Column matrix   p. 7

Computational Techniques   (1) To multiply a matrix by a number, multiply each entry by that
                               number.
                           (2) To add or subtract two matrices, add or subtract the corre-
                               sponding entries. (The matrices must be the same size.)
                           (3) To multiply one matrix by another:
                               (a) first learn how to multiply a row by a column (flow chart,
                                   p. 27);
                               (b) then learn how to multiply a matrix by a column (flow
                                   chart, p. 29);
                               (c) then use the flow chart on p. 34. (See also the two
                                   paragraphs on p. 35.)

   Note   The usual laws of arithmetic hold for matrix computations, *ex-
          cept* the commutative law of multiplication: (AB probably does
          not equal BA).

Chapter 2
Linear
Equations

In this chapter we continue to study applications of matrices. It is the function of applicable mathematics to isolate a few of the important properties of a situation under investigation. The aim is to study these properties abstractly, derive conclusions, and relate the results to the original problem. Real-life problems are too complex to admit any other kind of analysis. We describe a mathematical model that reflects as much detail as we can reasonably hope to deal with.

The most convenient, most useful, and therefore most important model is the so-called **linear model**. Virtually all business mathematics is concerned with linear models. In fact, most of mathematics is involved with linearity. This book is about nothing else. As you read the rest of this text you will come to a better understanding of exactly what a linear model entails and why so many economic situations are best studied in this way. For the moment let us just remark that the word "linear" is derived from the word "line", which denotes the simplest kind of curve. A linear model may involve more than just lines; it can also involve planes in 3-dimensional space or the appropriate analog in higher dimensions. In this chapter we begin our study of linearity by first considering lines and intersections of lines in the plane. We then move on to planes and intersections of planes in higher dimensions.

Note that while a real business problem will not involve the words "line" or "plane," business problems are often most easily handled by translating them into geometric problems where we may use our spatial intuition.

Section 2-1
Coordinates and Lines in the Plane

Figure 1
Insulation Requirements

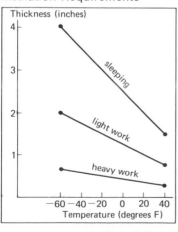

In advertising as well as at sales meetings and other presentations, the management at Appalachian Creations often finds a graphical presentation of data to be the most effective means of communication. For example, consider Figure 1, which appeared in AC's most recent catalog. This chart is typical in that it involves only straight lines. Except for a few technical reports on topics such as wind-chill at extreme altitudes, AC for the most part prepares and uses only straight-line graphs. Since improperly prepared graphs can ruin the reputation of any company, AC requires all its managers to understand the preparation and interpretation of such graphs.

The kinds of data that can be studied by coordinates and graphs are many and diverse. They range from fiscal information—for example, retirement income as a function of longevi-

Figure 2

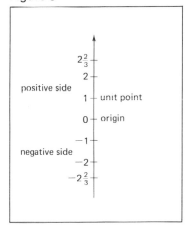

ty—to knowledge about manufacturing—for example, tent life as a function of fabric quality. While some of these data may not be linear, understanding the following material is a necessary first step to understanding the presentation of any of these kinds of data.

In this section we present a general geometric description of lines and planes, without referring to specific business applications.

Coordinates on a Line

Figure 3

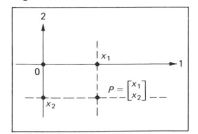

Our first geometric application involves lines and their points. It will be convenient to be able to refer to a point on a line and we do this by assigning numbers to the points in a natural way. The number to be associated with a point is called its **coordinate**. We set up a system of coordinates on a line by assigning to some point on the line, called the **origin**, the coordinate 0 and assigning to some other point, called the unit point, the coordinate 1. It is traditional to choose the unit point to the right of the origin if the line is horizontal (Figure 2), and to choose the unit point above the origin if the line is vertical (Figure 3).

The arrow on the lines in Figures 2 and 3, and similar arrows in later figures, indicate the "positive directions" established by these choices. The coordinate of a point on the positive side of the origin is simply the distance from the point to the origin in terms of the unit distance; the coordinate of a point on the negative side of the origin is the negative of the distance from the point to the origin.

Problem 1

(a) Set up a coordinate system on a horizontal line and indicate the points with coordinates $-2\frac{2}{3}$, -2, -1, 0, 1, 2 and $2\frac{2}{3}$. (b) Repeat (a) for a vertical line.

Answer

(a) See Figure 2.   (b) See Figure 3.

Coordinates on a Plane

Figure 4

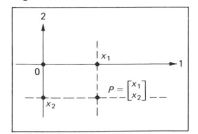

Following ideas of René Descartes (1596–1650), Pierre de Fermat (ca. 1608–1665), and John Wallis (1616–1703), we set up a system of coordinates in a plane by choosing a horizontal line and a vertical line in the plane. We call these lines the 1-**axis** and the 2-**axis**, respectively. We then set up systems of coordinates on each of these lines, as explained above, using the point 0 at which these lines intersect as the origin for both systems. Now suppose P is a point in the plane (Figure 4). Draw the horizontal and vertical lines through P. Suppose the vertical line crosses the 1-axis at a point with coordinate x_1 (read "x one"), and the horizontal line through P crosses the 2-axis at the point with coordinate x_2.

We call these numbers the **coordinates** of P, and we write

$$P = \begin{bmatrix} x_1 \\ x_2 \end{bmatrix} \qquad\qquad 1$$

The coordinates of P form a 2×1 matrix, that is, a column matrix with two entries. (Most areas of mathematics use the notation $P = (x_1, x_2)$; but the matrix notation of equation **1** is best for our purposes.)

In matrix **1** the (1,1)-entry tells us how far P is to the right or left of the 2-axis. (A positive entry indicates "right," a negative entry indicates "left.") The (2,1)-entry tells us how far P is above or below the 1-axis. (A positive entry indicates "above," a negative entry indicates "below.")

When the point x_1 in Figure 4 is considered as a point in the plane, its coordinates are $\begin{bmatrix} x_1 \\ 0 \end{bmatrix}$, since the horizontal line through x_1 (namely, the 1-axis) crosses the 2-axis at the origin. In the same way we see that the point x_2 in Figure 4 has, when considered as a point in the plane, coordinates $\begin{bmatrix} 0 \\ x_2 \end{bmatrix}$. In general, points on the 1-axis have coordinate matrices in which the (2,1)-entry is zero; points on the 2-axis have coordinate matrices in which the (1,1)-entry is zero. Since the origin is on both axes, it has coordinates $\begin{bmatrix} 0 \\ 0 \end{bmatrix}$.

Figure 5

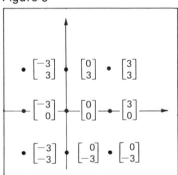

Problem 2   In a plane with a coordinate system indicate the points with coordinates $\begin{bmatrix} x_1 \\ x_2 \end{bmatrix}$, where x_1 and x_2 are each either -3, 0, or 3.

Answer   See Figure 5.

Problem 3   In a plane with a coordinate system, indicate the points with coordinates $\begin{bmatrix} x_1 \\ x_2 \end{bmatrix}$ where x_1 and x_2 are both positive integers, $x_1 < x_2$, and $x_2 < 4$.

For use of the "less than" sign, see Appendix A.

Solution   The second condition implies that x_2 can be either 1, 2, or 3. If $x_2 = 3$, then since x_1 must be less than x_2, x_1 can be 1 or 2. Thus, the points $\begin{bmatrix} 1 \\ 3 \end{bmatrix}$ and $\begin{bmatrix} 2 \\ 3 \end{bmatrix}$ satisfy the conditions. If $x_2 = 2$, the only choice for x_1 is $x_1 = 1$, and we obtain the point $\begin{bmatrix} 1 \\ 2 \end{bmatrix}$.

Finally, if $x_2 = 1$, then x_1 cannot be assigned any value, since there is no positive integer less than 1. These points are indicated in Figure 6.

Linear Equations in Two Variables

Figure 6

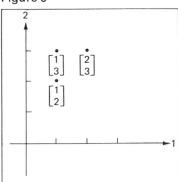

We are now ready to discuss lines in the plane. Consider the equation

$$3x_1 - 2x_2 = 6 \qquad\qquad \mathbf{2}$$

This is an equation involving two **variables**, x_1 and x_2. Recalling the definition of matrix multiplication, we note that equation 2 can be written in matrix form as follows:

$$[3 \quad -2] \begin{bmatrix} x_1 \\ x_2 \end{bmatrix} = 6 \qquad\qquad \mathbf{3}$$

Now there are many matrices $\begin{bmatrix} x_1 \\ x_2 \end{bmatrix}$ that satisfy equation **3**, as we see in the following problem.

Problem 4

Which of the matrices $\begin{bmatrix} 4 \\ 3 \end{bmatrix}$, $\begin{bmatrix} 1 \\ 1 \end{bmatrix}$, $\begin{bmatrix} 2 \\ 0 \end{bmatrix}$, $\begin{bmatrix} 1 \\ -3/2 \end{bmatrix}$, $\begin{bmatrix} 3 \\ 4/3 \end{bmatrix}$ and $\begin{bmatrix} 0 \\ -3 \end{bmatrix}$ satisfy equation **3**?

Solution

$[3 \quad -2] \begin{bmatrix} 4 \\ 3 \end{bmatrix} = 3 \cdot 4 - 2 \cdot 3 = 6$, so $\begin{bmatrix} 4 \\ 3 \end{bmatrix}$ satisfies the equation.

$[3 \quad -2] \begin{bmatrix} 1 \\ 1 \end{bmatrix} = 1 \neq 6$, and hence $\begin{bmatrix} 1 \\ 1 \end{bmatrix}$ does not satisfy the equation. $[3 \quad -2] \begin{bmatrix} 2 \\ 0 \end{bmatrix} = 6$, $[3 \quad -2] \begin{bmatrix} 1 \\ -3/2 \end{bmatrix} = 6$, $[3 \quad -2] \begin{bmatrix} 3 \\ 4/3 \end{bmatrix} = 6^{1}/_{3}$ and $[3 \quad -2] \begin{bmatrix} 0 \\ -3 \end{bmatrix} = 6$. Hence, $\begin{bmatrix} 4 \\ 3 \end{bmatrix}$, $\begin{bmatrix} 2 \\ 0 \end{bmatrix}$, $\begin{bmatrix} 1 \\ -3/2 \end{bmatrix}$, and $\begin{bmatrix} 0 \\ -3 \end{bmatrix}$ satisfy the equation, but $\begin{bmatrix} 3 \\ 4/3 \end{bmatrix}$ and $\begin{bmatrix} 1 \\ 1 \end{bmatrix}$ do not.

Figure 7

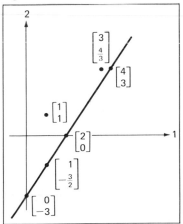

Each of the matrices in Problem 4 can be plotted as a point in a plane with a coordinate system. It can be shown that all of the solutions of equation **3** lie on a straight line and, conversely, that every point on this line has a coordinate matrix that is a solution of equation **3**. We call this line the **graph** of equation **3**, and we say that equation **3**—or, equivalently, equation **2**—is a **linear equation** in the two variables x_1 and x_2 (Figure 7). Note that the points $\begin{bmatrix} 1 \\ 1 \end{bmatrix}$ and $\begin{bmatrix} 3 \\ 4/3 \end{bmatrix}$ which, as we saw in Problem 4, are not solutions of equation **3**, do not lie on the graph.

Problem 5 Consider the linear equation

$$4x_1 + 5x_2 = 10 \qquad \qquad 4$$

(a) For $x_1 = 2$, what value of x_2 satisfies this equation? What is x_2 when $x_1 = -2$?

(b) For $x_2 = 2$, what value of x_1 satisfies this equation? What is x_1 when $x_2 = -2$?

(c) Plot the points $\begin{bmatrix} x_1 \\ x_2 \end{bmatrix}$ obtained in (a) and (b) and draw the graph of equation **4** through them.

Solution

Subtract 8 from each side of the equation. Divide each side of the resulting equation by 5.

(a) For $x_1 = 2$, equation **4** becomes $4 \cdot 2 + 5x_2 = 10$, $8 + 5x_2 = 10$, $5x_2 = 2$, or $x_2 = {}^2/_5$. For $x_1 = -2$ we have $4 \cdot (-2) + 5x_2 = 10$, from which $x_2 = {}^{18}/_5$.

(b) For $x_2 = 2$, we have $4x_1 + 5 \cdot 2 = 10$, $4x_1 = 0$, $x_1 = 0$. For $x_2 = -2$ we have $4x_1 - 10 = 10$, $x_1 = 5$.

(c) We plot the points $\begin{bmatrix} 2 \\ {}^2/_5 \end{bmatrix}$, $\begin{bmatrix} -2 \\ {}^{18}/_5 \end{bmatrix}$, $\begin{bmatrix} 0 \\ 2 \end{bmatrix}$, and $\begin{bmatrix} 5 \\ -2 \end{bmatrix}$, and draw the straight line through them (Figure 8).

Figure 8

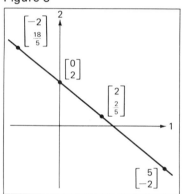

In general it can be shown that any equation of the form

$$ax_1 + bx_2 = c \qquad \qquad 5$$

where a, b, and c are given numbers and a and b are not both zero, is a linear equation in the sense that its graph is a straight line. It is also the case, conversely, that every straight line in a plane with a coordinate system is the graph of some equation of the form of equation **5**, where a and b are not both zero. We may write equation **5** and the requirement that a and b not both be zero in matrix form as follows:

$$[a \quad b] \begin{bmatrix} x_1 \\ x_2 \end{bmatrix} = c, \qquad [a \quad b] \neq [0 \quad 0] \qquad \qquad 6$$

If we wish to use symbols for these matrices, we might let $A = [a \quad b]$, $X = \begin{bmatrix} x_1 \\ x_2 \end{bmatrix}$ and $O = [0 \quad 0]$. The equation and requirement may then be written: $AX = c$, $A \neq O$.

Graphing a Linear Equation in Two Variables

What is the easiest way to draw the graph of equation **3**? We know that two points determine a line; so we need only find two solutions $\begin{bmatrix} x_1 \\ x_2 \end{bmatrix}$ of equation **3**, plot them, and draw the line through them. It turns out that the easiest solutions to find are those in which either x_1 or x_2 is zero. These give, respectively, the points at which the line crosses the 2-axis and the 1-axis. To find these points it is easier to work with equation **2** than with equation **3**. If $x_1 = 0$, equation **2** becomes $3 \cdot 0 - 2x_2 = 6$ or $-2x_2 = 6$; then $x_2 = -3$, so $\begin{bmatrix} 0 \\ -3 \end{bmatrix}$ is a solution. The point $\begin{bmatrix} 0 \\ -3 \end{bmatrix}$ is where the line we are seeking crosses the 2-axis (Figure 7). If $x_2 = 0$, equation **2** becomes $3x_1 - 2 \cdot 0 = 6$ or $3x_1 = 6$; then $x_1 = 2$, so $\begin{bmatrix} 2 \\ 0 \end{bmatrix}$ is a solution. This point is where our line crosses the 1-axis. We connect these two points to obtain the graph.

The above procedure is often the easiest way to obtain the graph of a linear equation. But there are some cases in which the procedure must be modified. We give examples of these cases below.

Problem 6   Graph the equation $\begin{bmatrix} 4 & 5 \end{bmatrix} \begin{bmatrix} x_1 \\ x_2 \end{bmatrix} = 10$.

Solution   This equation is the matrix form of equation **4**. If $x_1 = 0$, we have $5x_2 = 10$, $x_2 = 2$, and we obtain the point $\begin{bmatrix} 0 \\ 2 \end{bmatrix}$ as we did in Problem 5. If $x_2 = 0$, we have $4x_1 = 10$, $x_1 = \frac{5}{2}$, and we obtain $\begin{bmatrix} 5/2 \\ 0 \end{bmatrix}$. The graph of this equation (Figure 9) is the same as the graph in Problem 5 (Figure 8).

Figure 9

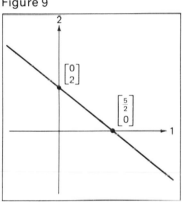

Graph the equation $\begin{bmatrix} 3 & 0 \end{bmatrix} \begin{bmatrix} x_1 \\ x_2 \end{bmatrix} = -5$.

Solution   We have $3x_1 = -5$; so $x_1 = -\frac{5}{3}$. Thus, the graph crosses the 1-axis at $\begin{bmatrix} -5/3 \\ 0 \end{bmatrix}$. Since the (1,2)-entry of the matrix $\begin{bmatrix} 3 & 0 \end{bmatrix}$ is zero, the value we choose for x_2 is irrelevant—whatever it is, it will be multiplied by zero. Thus, while each solution must have $x_1 = -\frac{5}{3}$, its x_2 value may be any number whatsoever. It follows that the graph must be the vertical line through the point $x_1 = -\frac{5}{3}$ (Figure 10).

Figure 10

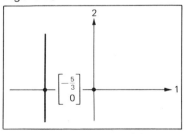

Our previous procedure fails here, since setting $x_1 = 0$ yields the impossibility $3 \cdot 0 = -5$ or $0 = -5$.

Problem 8   Graph the equation $[0 \quad -^2/_3] \begin{bmatrix} x_1 \\ x_2 \end{bmatrix} = -^4/_5.$

Solution   We have $-^2/_3 x_2 = -^4/_5$; so $x_2 = ^6/_5$. Thus, the graph crosses the

Here setting $x_2 = 0$ would give $-^2/_3 \cdot 0 = -^4/_5$, or $0 = -^4/_5$, which is absurd.

2-axis at $\begin{bmatrix} 0 \\ ^6/_5 \end{bmatrix}$. In this case the value of x_1 may be anything we choose, so the graph must be the horizontal line through the point $\begin{bmatrix} 0 \\ ^6/_5 \end{bmatrix}$ (Figure 11).

Figure 11

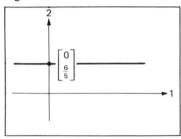

The last two problems above illustrate the following general situation. In the linear equation $ax_1 + bx_2 = c$ we may have $a = 0$ or $b = 0$, but not both $a = 0$ and $b = 0$. If $a = 0$ the equation becomes $bx_2 = c$, or $x_2 = c/b$. If $b = 0$ the equation becomes $ax_1 = c$, or $x_1 = c/a$. In order to interpret correctly an equation such as $x_1 = 2$, it is necessary to know from the context of the problem how many variables are being considered. If two variables are involved then the equation $x_1 = 2$ is to be interpreted as the equation $x_1 + 0x_2 = 2$, whose graph is the vertical line through the point $\begin{bmatrix} 2 \\ 0 \end{bmatrix}$. Note that if this equation had been written in matrix form, $[1 \quad 0] \begin{bmatrix} x_1 \\ x_2 \end{bmatrix} = 2$, then the fact that two variables are involved would be clear.

Problem 9   Graph the equation $[4 \quad 5] \begin{bmatrix} x_1 \\ x_2 \end{bmatrix} = 0.$

Solution   We have $4x_1 + 5x_2 = 0$. The graph crosses both axes at the origin,

Setting $x_1 = 0$ yields $x_2 = 0$ and vice versa.

$\begin{bmatrix} 0 \\ 0 \end{bmatrix}$, and at no other point. We have to find another solution. An easy solution is obtained by switching the 4 and 5 while changing the sign of the 4, making it negative: $[4 \quad 5] \begin{bmatrix} 5 \\ -4 \end{bmatrix} = 0.$ We con-

Figure 12

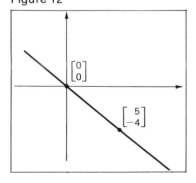

nect these points to obtain the graph (Figure 12).
    The flow chart on p. 67 summarizes this procedure and applies to any linear equation in two variables.

    The opposite problem is also of interest: given a straight-line graph, find an equation for it. The following observations, which are derived from the flow chart on p. 67, are sometimes of use in problems of this sort.

    Let $\begin{bmatrix} r \\ s \end{bmatrix}$ be a given point.

(1) An equation for the horizontal line through $\begin{bmatrix} r \\ s \end{bmatrix}$ is $x_2 = s$,

that is, $\begin{bmatrix} 0 & 1 \end{bmatrix} \begin{bmatrix} x_1 \\ x_2 \end{bmatrix} = s$. (Here it does not matter what r is.)

(2) An equation for the vertical line through $\begin{bmatrix} r \\ s \end{bmatrix}$ is $x_1 = r$, that is,

$\begin{bmatrix} 1 & 0 \end{bmatrix} \begin{bmatrix} x_1 \\ x_2 \end{bmatrix} = r$. (Here it does not matter what s is.)

(3) An equation for the line through $\begin{bmatrix} 0 \\ 0 \end{bmatrix}$ and $\begin{bmatrix} r \\ s \end{bmatrix}$ is

$sx_1 - rx_2 = 0$, that is, $\begin{bmatrix} s & -r \end{bmatrix} \begin{bmatrix} x_1 \\ x_2 \end{bmatrix} = 0$. (Here we must

assume that r and s are not both zero.)

Problem 10   Find an equation for each of the following lines:

(a) the horizontal line through $\begin{bmatrix} 17 \\ -6 \end{bmatrix}$

(b) the vertical line through $\begin{bmatrix} 17 \\ -6 \end{bmatrix}$

(c) the line through $\begin{bmatrix} 17 \\ -6 \end{bmatrix}$ and the origin

To Graph the Equation

$\begin{bmatrix} a & b \end{bmatrix} \begin{bmatrix} x_1 \\ x_2 \end{bmatrix} = c,$

where $\begin{bmatrix} a & b \end{bmatrix} \neq \begin{bmatrix} 0 & 0 \end{bmatrix}$

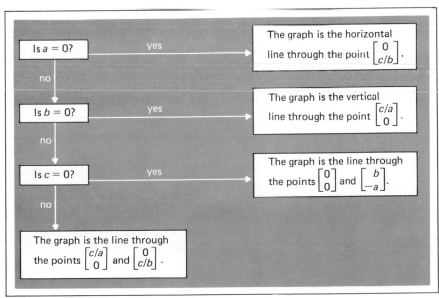

Solution (a) By observation 1, an equation is $x_2 = -6$, that is,

$$[0 \quad 1] \begin{bmatrix} x_1 \\ x_2 \end{bmatrix} = -6.$$

(b) By observation 2, an equation is $x_1 = 17$, that is,

$$[1 \quad 0] \begin{bmatrix} x_1 \\ x_2 \end{bmatrix} = 17.$$

(c) By observation 3, an equation is $-6x_1 - 17x_2 = 0$. If we multiply through this equation by -1, we get the "neater" equation $6x_1 + 17x_2 = 0$, that is, $[6 \quad 17] \begin{bmatrix} x_1 \\ x_2 \end{bmatrix} = 0.$

Of course these observations do not apply to lines that are not horizontal or vertical or do not pass through the origin. For such lines it is best to use the determinant formula developed later on (see formula **7** in Section 6-5).

See also Exercise 13 in the Supplementary Exercises at the end of this Chapter.

Exercises 2.1

1.^A(a) Set up a coordinate system on a horizontal line and indicate the points with coordinates -3, 1.4, 0, $-\frac{4}{3}$, and $\frac{5}{2}$.
(b) Repeat part a for a vertical line.

^A2. Set up a coordinate system in a plane and indicate the points with coordinates $\begin{bmatrix} 0 \\ 0 \end{bmatrix}, \begin{bmatrix} 1 \\ 0 \end{bmatrix}, \begin{bmatrix} 0 \\ 1 \end{bmatrix}, \begin{bmatrix} 2 \\ 3 \end{bmatrix}, \begin{bmatrix} \frac{1}{2} \\ \frac{1}{3} \end{bmatrix}, \begin{bmatrix} -3 \\ 2 \end{bmatrix}, \begin{bmatrix} -2 \\ -3 \end{bmatrix}$, and $\begin{bmatrix} -\frac{1}{2} \\ \frac{8}{3} \end{bmatrix}$.

3. What are the coordinates of the vertices of a square with lower left vertex $\begin{bmatrix} -2 \\ 3 \end{bmatrix}$, sides parallel to the coordinate axes, and sides of length 5?

^A4. A baseball diamond is actually a square 90 feet on a side. If we let 1 unit equal 1 foot and place home base at the origin and second base on the 2-axis, what are the coordinates of third base?

5. Graph each of the following equations:

(a) $[4 \quad 5] \begin{bmatrix} x_1 \\ x_2 \end{bmatrix} = 20$

^A(b) $-3x_1 + x_2 = 6$

(c) $[6 \quad 3] \begin{bmatrix} x_1 \\ x_2 \end{bmatrix} = 18$

^A(d) $-2x_1 + 4x_2 = 5$

(e) $[3 \quad -7] \begin{bmatrix} x_1 \\ x_2 \end{bmatrix} = -4$

^A(f) $[\frac{2}{3} \quad \frac{3}{4}] \begin{bmatrix} x_1 \\ x_2 \end{bmatrix} = \frac{7}{12}$

(g) $\frac{1}{2}x_1 - \frac{2}{3}x_2 = \frac{3}{4}$

(h) $[-\frac{5}{2} \quad \frac{3}{8}] \begin{bmatrix} x_1 \\ x_2 \end{bmatrix} = \frac{1}{16}$

(i) $\frac{2}{3}x_1 - \frac{1}{2}x_2 = \frac{5}{6}$

(j) $-\frac{2}{5}x_1 + \frac{2}{3}x_2 = \frac{1}{3}$

(k) $[2\frac{1}{3} \quad 3\frac{1}{8}] \begin{bmatrix} x_1 \\ x_2 \end{bmatrix} = 5\frac{5}{24}$

^A(l) $[4 \quad 0] \begin{bmatrix} x_1 \\ x_2 \end{bmatrix} = 16$

^A(m) $[0 \quad -5] \begin{bmatrix} x_1 \\ x_2 \end{bmatrix} = 9$

^A(n) $3x_1 + 4x_2 = 0$

(o) $[\,^3/_4 \quad -\,^4/_5\,]\begin{bmatrix} x_1 \\ x_2 \end{bmatrix} = 0$ $\quad$ A(q) $[0 \quad -\,^{64}/_9\,]\begin{bmatrix} x_1 \\ x_2 \end{bmatrix} = 0$

A(p) $[7 \quad 0]\begin{bmatrix} x_1 \\ x_2 \end{bmatrix} = 0$ $\quad$ (r) $[2.3 \quad 3.4]\begin{bmatrix} x_1 \\ x_2 \end{bmatrix} = 7.82$

6. Suppose the line labeled "sleeping" in the graph on p. 60 is given by the equation $x_1 + 40x_2 = 100$. How much insulation is required for each of the following temperatures?

(a) $-10°$
(b) $30°$
C(c) $80°$
C(d) $-80°$

7. Given $3x_1 - 4x_2 = 8$, find the value of the other variable when

A(a) $x_1 = 0$
(b) $x_1 = -2$
A(c) $x_2 = 3$
(d) $x_2 = -4$

8. Find four solutions to the matrix equation $[\,^1/_2 \quad ^2/_3\,]\begin{bmatrix} x_1 \\ x_2 \end{bmatrix} = 20$.

9. Find a linear equation in two variables, in the form of equation **5**, for each of the following.

A(a) the horizontal line through $\begin{bmatrix} 0 \\ 6.2 \end{bmatrix}$

A(b) the vertical line through $\begin{bmatrix} 11 \\ 0 \end{bmatrix}$

(c) the horizontal line through $\begin{bmatrix} 10 \\ -\,^2/_3 \end{bmatrix}$

(d) the vertical line through $\begin{bmatrix} -18 \\ 93 \end{bmatrix}$

A(e) the line through $\begin{bmatrix} 0 \\ 0 \end{bmatrix}$ and $\begin{bmatrix} 2 \\ 3 \end{bmatrix}$

(f) the line through $\begin{bmatrix} 0 \\ 0 \end{bmatrix}$ and $\begin{bmatrix} -2 \\ 4 \end{bmatrix}$

(g) the line through $\begin{bmatrix} 0 \\ 0 \end{bmatrix}$ and $\begin{bmatrix} -\,^2/_3 \\ -\,^3/_5 \end{bmatrix}$

Hint: Make another observation from the flow chart.
Hint: Either guess or see Problem 9, Chapter 6, Section 5.

*(h) the line through $\begin{bmatrix} -4 \\ 0 \end{bmatrix}$ and $\begin{bmatrix} 0 \\ 5 \end{bmatrix}$

*(i) the line through $\begin{bmatrix} 3 \\ 4 \end{bmatrix}$ and $\begin{bmatrix} 5 \\ 6 \end{bmatrix}$

A10. Write the answers to Exercise 4 in the form of equation **6**.
*11. Let P be a point and L a line in a plane. The *reflection* of P in L is the point P' such that the line through P and P' is perpendicular to L and

Figure 13

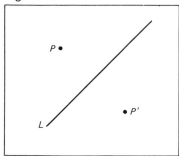

P and P' have the same distance from L (Figure 13). (If P is on L we take $P' = P$.) Suppose the plane has a coordinate system and $P = \begin{bmatrix} x_1 \\ x_2 \end{bmatrix}$. What are the coordinates of each of the following?

(a) the reflection of P in the 1-axis
(b) the reflection of P in the 2-axis
(c) the reflection in the 2-axis of the reflection of P in the 1-axis
(d) the reflection in the 1-axis of the reflection of P in the 2-axis
(e) the reflection of P in the line with equation $x_1 - x_2 = 0$

C12. At a meeting in Ames, Iowa, of the managers of PHE's Midwestern stores, the PHE directors want to show just how daily hamburger sales are related to profits. Each hamburger sold gives a profit of 15 cents. Hamburger profits are expected to contribute $105 toward daily overhead costs. Let x_1 be the daily sale of hamburgers and let x_2 be the hamburger net profit after paying overhead. Deduce the relation between x_1 and x_2 and prepare a graph showing the relation for the sales meeting.

Section 2-2
Systems of Two Linear
Equations in Two
Variables

We learned in Section 2-1 that the equations $x_1 - x_2 = 1$ and $x_1 + x_2 = 5$, for example, each represent a line in the plane. Since two lines will usually intersect in a point, we now ask: what are the coordinates of the point where these two lines intersect? In other words, we seek numbers x_1 and x_2 which will satisfy both equations in the **system**

$$\begin{bmatrix} x_1 - x_2 = 1 \\ x_1 + x_2 = 5 \end{bmatrix} \qquad \textbf{1}$$

This particular system of equations arose at Appalachian Creations when the vice president in charge of freeze-dried foods decided to hire a new chief nutritionist and an assistant nutritionist. After checking existing salaries, she decided to pay the chief nutritionist $10,000 more than his or her assistant. Her budget allowed for a total expenditure of only $50,000. The problem is to determine the salary of each of the new employees. If we let x_1 denote the salary of the chief and x_2 denote the salary of the assistant, then the fact that the chief is paid $10,000 more is expressed by the equation

$$x_1 = x_2 + 10,000$$

or

$$x_1 - x_2 = 10,000$$

The budget allocation translates into the equation
$$x_1 + x_2 = 50,000$$

Finally, if we simplify matters by measuring salaries in units of $10,000 we obtain system **1** above.

If on the same coordinate plane we draw a careful graph of the two lines represented by the equations in our system, they appear to cross at the point $\begin{bmatrix} 3 \\ 2 \end{bmatrix}$ (Figure 14).

Figure 14

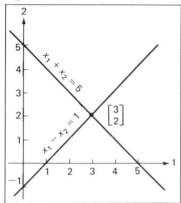

Sure enough, this possible solution, $x_1 = 3$, $x_2 = 2$, checks in both equations of system 1:

$$\begin{bmatrix} 3 - 2 = 1 \\ 3 + 2 = 5 \end{bmatrix} \quad \checkmark \\ \checkmark$$

Thus, the chief nutritionist's salary will be $30,000 and the assistant's, $20,000.

In general, of course, a graph will not be accurate enough to make guessing effective, especially if the actual answer involves fractions. What is needed is an algebraic technique to provide the answer without recourse to graphs.

We introduce this technique by considering the following system.

$$\begin{bmatrix} 3x_1 + 2x_2 - 6 \\ 4x_1 + 3x_2 = 12 \end{bmatrix} \qquad \textbf{2}$$

For problems involving two variables and two equations, there are at least six solution methods, all of which work equally well. For larger problems, no method works as well as the technique we are illustrating here with this simple example. Why this particular technique works will be explained shortly.

We will solve system **2** through the following sequence of steps, illustrating the technique that we will develop in this section. Do not expect to understand immediately the reasons for each of these steps. These reasons will become clearer as you work through more examples. However, you should check that the arithmetic in each step is properly done.

Multiply the first equation by $\tfrac{4}{3}$: $\tfrac{4}{3}(3x_1 + 2x_2 = 6)$ gives $(\tfrac{4}{3})3x_1 + (\tfrac{4}{3})2x_2 = (\tfrac{4}{3})6$, that is, $4x_1 + \tfrac{8}{3}x_2 = 8$. We now have the system

$(\tfrac{4}{3} \text{ eqn. 1}) \begin{bmatrix} 4x_1 + \tfrac{8}{3}x_2 = 8 \\ 4x_1 + 3x_2 = 12 \end{bmatrix} \qquad \textbf{3}$

Now subtract equation 1 of system **3** from equation 2: $(4x_1 + 3x_2) - (4x_1 + \tfrac{8}{3}x_2) = 12 - 8$, that is, $0x_1 + \tfrac{1}{3}x_2 = 4$. We now have

$(\text{eqn. 2} - \text{eqn. 1}) \begin{bmatrix} 4x_1 + \tfrac{8}{3}x_2 = 8 \\ 0x_1 + \tfrac{1}{3}x_2 = 4 \end{bmatrix} \qquad \textbf{4}$

Now multiply equation 2 of system **4** by 8:

$(8 \text{ eqn. 2}) \begin{bmatrix} 4x_1 + \tfrac{8}{3}x_2 = 8 \\ 0x_1 + \tfrac{8}{3}x_2 = 32 \end{bmatrix} \qquad \textbf{5}$

Subtract equation 2 of system **5** from equation 1:

$$(\text{eqn. } 1 - \text{eqn. } 2) \begin{bmatrix} 4x_1 + 0x_2 = -24 \\ 0x_1 + {}^8\!/_3 x_2 = 32 \end{bmatrix} \quad \textbf{6}$$

In system **6** multiply equation 1 by $^1\!/_4$ and equation 2 by $^3\!/_8$:

$$\begin{matrix} (^1\!/_4 \text{ eqn. } 1) \\ (^3\!/_8 \text{ eqn. } 2) \end{matrix} \begin{bmatrix} 1x_1 + 0x_2 = -6 \\ 0x_1 + 1x_2 = 12 \end{bmatrix} \quad \textbf{7}$$

System **7** is the solution to system **2**: $x_1 = -6$ and $x_2 = 12$.

$$\text{Check:} \qquad \begin{bmatrix} 3(-6) + 2(12) = 6 \\ 4(-6) + 3(12) = 12 \end{bmatrix} \checkmark$$

Thus, the two lines which are the graphs of the equations $3x_1 + 2x_2 = 6$ and $4x_1 + 3x_2 = 12$ intersect at the point $\begin{bmatrix} -6 \\ 12 \end{bmatrix}$ (Figure 15).

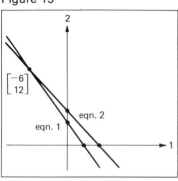

Figure 15

System **2**: $\begin{bmatrix} 3x_1 + 2x_2 = 6 \\ 4x_1 + 3x_2 = 12 \end{bmatrix}$

In solving this system, we first eliminated x_1 from equation **2** and then eliminated x_2 from equation **1**. This resulted in the extremely simple equations of system **6**.

The manipulations carried out above to solve system **2** do not really change the symbols x_1, x_2, the plus signs, or the equal signs; they just change the numbers. So let us leave out the signs and symbols in system **2**. We are left with the **matrix** of system **2**:

$$\begin{bmatrix} 3 & 2 & 6 \\ 4 & 3 & 12 \end{bmatrix} \quad \textbf{2}'$$

In this setting, the references above to equations 1 and 2 of system **2** become references to *rows* 1 and 2 of matrix **2'**; these rows, as we agreed in Chapter 1, can be designated by R_1 and R_2. The steps carried out to solve system **2** can now be described in terms of manipulations with the rows of matrix **2'**, as follows:

$$\begin{bmatrix} 3 & 2 & 6 \\ 4 & 3 & 12 \end{bmatrix} \xrightarrow{\;{}^4\!/_3 R_1\;} \begin{bmatrix} 4 & {}^8\!/_3 & 8 \\ 4 & 3 & 12 \end{bmatrix}$$

$$\xrightarrow{R_2 - R_1} \begin{bmatrix} 4 & {}^8\!/_3 & 8 \\ 0 & {}^1\!/_3 & 4 \end{bmatrix} \xrightarrow{\;8R_2\;} \begin{bmatrix} 4 & {}^8\!/_3 & 8 \\ 0 & {}^8\!/_3 & 32 \end{bmatrix}$$

$$\xrightarrow{R_1 - R_2} \begin{bmatrix} 4 & 0 & -24 \\ 0 & {}^8\!/_3 & 32 \end{bmatrix} \xrightarrow[{}^3\!/_8 R_2]{{}^1\!/_4 R_1} \begin{bmatrix} 1 & 0 & -6 \\ 0 & 1 & 12 \end{bmatrix} \quad \textbf{3}'$$

Actually, there is no need to have so many steps. The first two steps can be compressed into one step: replace row 2 by row 2 minus $\frac{4}{3}$ times row 1, as follows:

$$\begin{bmatrix} 3 & 2 & 6 \\ 4 & 3 & 12 \end{bmatrix} R_2 - \frac{4}{3}R_1 \quad\longrightarrow\quad \begin{bmatrix} 3 & 2 & 6 \\ 0 & \frac{1}{3} & 4 \end{bmatrix}$$

Note that this step changes only R_2. The row R_1 stays the same. We next "eliminate" the (1,2) entry in R_1—that is, we reduce it to zero—by the following step:

$$\begin{bmatrix} 3 & 2 & 6 \\ 0 & \frac{1}{3} & 4 \end{bmatrix} R_1 - 6R_2 \quad\longrightarrow\quad \begin{bmatrix} 3 & 0 & -18 \\ 0 & \frac{1}{3} & 4 \end{bmatrix}$$

Note again that this step changes only one row, R_1, and that the row used to make this change, R_2, is not itself changed at all.

(See Appendix A. Section 10.)

How did we arrive at the $-\frac{4}{3}$ in the expression $R_2 - \frac{4}{3}R_1$? We wanted to multiply R_1 by some number x and add it to R_2 so as to make the 4 at address (2,1) become 0. Thus, we needed the number x to be such that $3x + 4 = 0$. Hence, $x = -\frac{4}{3}$. In the case of the -6 in the expression $R_1 - 6R_2$, we wanted to multiply R_2 by some number x and add it to R_1 so as to make the 2 at address (1,2) become 0. Thus, we needed the number x to be such that $\frac{1}{3}x + 2 = 0$. Hence $x = -6$. We use the simpler but equivalent expression $R_1 - 6R_2$ in place of $R_1 + (-6)R_2$.

Now to complete our work we write

$$\begin{bmatrix} 3 & 0 & -18 \\ 0 & \frac{1}{3} & 4 \end{bmatrix} \begin{matrix} \frac{1}{3}R_1 \\ 3R_2 \end{matrix} \quad\longrightarrow\quad \begin{bmatrix} 1 & 0 & -6 \\ 0 & 1 & 12 \end{bmatrix}$$

which gives the same answer as in system **7** and system **3'**, but in three steps instead of five.

Here is another example.

Problem 1   What is the point of intersection of the lines $2x_1 + 3x_2 = 4$ and $-3x_1 + x_2 = 3$?

Solution   We must solve the system $\begin{bmatrix} 2x_1 + 3x_2 = 4 \\ -3x_1 + x_2 = 3 \end{bmatrix}$, which we do in matrix notation as follows:

$$\begin{bmatrix} 2 & 3 & 4 \\ -3 & 1 & 3 \end{bmatrix} \xrightarrow{R_2 + \frac{3}{2}R_1} \begin{bmatrix} 2 & 3 & 4 \\ 0 & \frac{11}{2} & 9 \end{bmatrix}$$

$$\xrightarrow{R_1 - \frac{6}{11}R_2} \begin{bmatrix} 2 & 0 & -\frac{10}{11} \\ 0 & \frac{11}{2} & 9 \end{bmatrix} \begin{matrix} \frac{1}{2}R_1 \\ \frac{2}{11}R_2 \end{matrix} \quad\longrightarrow\quad \begin{bmatrix} 1 & 0 & -\frac{5}{11} \\ 0 & 1 & \frac{18}{11} \end{bmatrix}$$

Thus, we conclude that $x_1 = -5/11$, $x_2 = 18/11$. The point is $\begin{bmatrix} -5/11 \\ 18/11 \end{bmatrix}$.

Check: $\begin{bmatrix} 2(-5/11) + 3(18/11) = 44/11 = 4 & ✔ \\ -3(-5/11) + \quad 18/11 \quad = 33/11 = 3 & ✔ \end{bmatrix}$

The idea behind each of the above solutions is the same: we write down the matrix of the system to be solved and we work with the rows to "reduce" the matrix to the form

$$\begin{bmatrix} 1 & 0 & a \\ 0 & 1 & b \end{bmatrix} \qquad\qquad \mathbf{8}$$

The solution of the system, then, is just $x_1 = a$, and $x_2 = b$, that is, $\begin{bmatrix} x_1 \\ x_2 \end{bmatrix} = \begin{bmatrix} a \\ b \end{bmatrix}$. In working with the rows we are allowed to perform two types of operations: (1) add a multiple of one row to another row (this includes the possibility of adding a negative multiple of one row to another, that is, subtracting a multiple of one row from another); or (2) multiply a row by a nonzero number (this includes the possibility of multiplying a row by the reciprocal of a nonzero number, that is, dividing a row by a nonzero number).

There are many ways to reduce a matrix to the form of system **8**, but it can be shown that all of these ways lead to the same numbers a and b in system **8**. Here, for example, is another way to reduce the matrix of the previous problem:

$$\begin{bmatrix} 2 & 3 & 4 \\ -3 & 1 & 3 \end{bmatrix} \xrightarrow{\substack{1/2R_1 \\ 1/3R_2}} \begin{bmatrix} 1 & 3/2 & 2 \\ -1 & 1/3 & 1 \end{bmatrix} \xrightarrow{R_2 + R_1} \begin{bmatrix} 1 & 3/2 & 2 \\ 0 & 11/6 & 3 \end{bmatrix}$$

$$\xrightarrow{6/11R_2} \begin{bmatrix} 1 & 3/2 & 2 \\ 0 & 1 & 18/11 \end{bmatrix} \xrightarrow{R_1 - 3/2R_2} \begin{bmatrix} 1 & 0 & -5/11 \\ 0 & 1 & 18/11 \end{bmatrix}$$

The final matrix is the same as before.

One other type of row operation is sometimes of use: switching the rows. This is certainly an allowable operation when working with a system of equations, since it amounts simply to writing down the equations in a different order. This operation is not essential in solving a system, but it can save steps. Here is an example.

See Supplementary Exercise 12, for example.

Problem 2 Find the point of intersection of the lines $[0 \quad 2]\begin{bmatrix} x_1 \\ x_2 \end{bmatrix} = -6$ and

$[3 \quad 1]\begin{bmatrix} x_1 \\ x_2 \end{bmatrix} = 0.$

Solution

$$\begin{bmatrix} 0 & 2 & -6 \\ 3 & 1 & 0 \end{bmatrix} \diagdown \begin{bmatrix} 3 & 1 & 0 \\ 0 & 2 & -6 \end{bmatrix} \xrightarrow{R_1 - \frac{1}{2}R_2} \begin{bmatrix} 3 & 0 & 3 \\ 0 & 2 & -6 \end{bmatrix}$$

$$\xrightarrow[\frac{1}{2}R_2]{\frac{1}{3}R_1} \begin{bmatrix} 1 & 0 & 1 \\ 0 & 1 & -3 \end{bmatrix}$$

Check:

$[0 \quad 2]\begin{bmatrix} 1 \\ -3 \end{bmatrix} = -6$

$[3 \quad 1]\begin{bmatrix} 1 \\ -3 \end{bmatrix} = 0$

Answer: $\begin{bmatrix} 1 \\ -3 \end{bmatrix}$

Not all problems of this kind result in a unique solution. Some problems have no solution at all; others have many solutions. The method developed above also covers these cases, as we see in the following examples.

Problem 3 Find the point of intersection of the lines $[-2 \quad 1]\begin{bmatrix} x_1 \\ x_2 \end{bmatrix} = 2$ and

$[6 \quad -3]\begin{bmatrix} x_1 \\ x_2 \end{bmatrix} = 4.$

Solution

$$\begin{bmatrix} -2 & 1 & 2 \\ 6 & -3 & 4 \end{bmatrix} \xrightarrow{R_2 + 3R_1} \begin{bmatrix} -2 & 1 & 2 \\ 0 & 0 & 10 \end{bmatrix} \xrightarrow{-\frac{1}{2}R_1} \begin{bmatrix} 1 & -\frac{1}{2} & -1 \\ 0 & 0 & 10 \end{bmatrix}$$

One step is really all you need to perform to tell that this system has no solution.

The second row of the final matrix corresponds to the equation $0x_1 + 0x_2 = 10$, which obviously has no solution. Hence, it can be shown that the original problem has no solution, that is, the lines $2x_1 + x_2 = 2$ and $6x_1 - 3x_2 = 4$ have no intersection and are thus *parallel*. (You should graph these lines to see this.)

Problem 4 Find the point of intersection of the lines $[4 \quad -6]\begin{bmatrix} x_1 \\ x_2 \end{bmatrix} = 8$ and

$[-6 \quad 9]\begin{bmatrix} x_1 \\ x_2 \end{bmatrix} = -12.$

Solution

$$\begin{bmatrix} 4 & -6 & 8 \\ -6 & 9 & -12 \end{bmatrix} \xrightarrow{R_2 + \frac{3}{2}R_1} \begin{bmatrix} 4 & -6 & 8 \\ 0 & 0 & 0 \end{bmatrix} \xrightarrow{\frac{1}{4}R_1} \begin{bmatrix} 1 & -\frac{3}{2} & 2 \\ 0 & 0 & 0 \end{bmatrix}$$

The second row of the last matrix corresponds to the equation $0x_1 + 0x_2 = 0$, which holds for all numbers x_1 and x_2. Unlike the situation in Problem 3, this poses no difficulty. Whenever a row of zeros occurs in a problem *of this size* it indicates that the original sytem has many solutions, all of which are given by the equation corresponding to the other row—in this case, $x_1 - \frac{3}{2}x_2 = 2$. To facilitate finding particular solutions, we solve this equation for x_1 in terms of x_2: $x_1 = 2 + \frac{3}{2}x_2$. If, say, $x_2 = 0$, we obtain the point $\begin{bmatrix} 2 \\ 0 \end{bmatrix}$; if $x_2 = 2$, we obtain $\begin{bmatrix} 5 \\ 2 \end{bmatrix}$; if $x_2 = -\frac{4}{3}$, we obtain $\begin{bmatrix} 0 \\ -\frac{4}{3} \end{bmatrix}$. Each of these points and an infinite number of other points are solutions of the system. Since the lines $4x_1 - 6x_2 = 8$ and $-6x_1 + 9x_2 = -12$ have more than one point in common, they are in fact *identical*, as you can easily see by drawing their graphs. We indicate the solution to the problem by the formula

$$\begin{bmatrix} 2 + \frac{3}{2}x_2 \\ x_2 \end{bmatrix}$$

Compare Problem 4 in Section 2-1.

where the presence of the variable x_2 indicates that we are free to assign it any value we wish to obtain a particular solution. This answer does, in fact, check:

$$\begin{bmatrix} 4 & -6 \end{bmatrix} \begin{bmatrix} 2 + \frac{3}{2}x_2 \\ x_2 \end{bmatrix} = 4(2 + \frac{3}{2}x_2) - 6x_2$$
$$= 8 + 6x_2 - 6x_2$$
$$= 8$$

$$\begin{bmatrix} -6 & 9 \end{bmatrix} \begin{bmatrix} 2 + \frac{3}{2}x_2 \\ x_2 \end{bmatrix} = -6(2 + \frac{3}{2}x_2) + 9x_2$$
$$= -12 - 9x_2 + 9x_2$$
$$= -12$$

It is of course possible to rewrite the equation $x_1 - \frac{3}{2}x_2 = 2$ as $-\frac{3}{2}x_2 = 2 - x_1$ or $x_2 = \frac{2}{3}x_1 - \frac{4}{3}$. Doing this results in writing the solution as all points of the form $\begin{bmatrix} x_1 \\ \frac{2}{3}x_1 - \frac{4}{3} \end{bmatrix}$. Since this solution turns out to be equivalent to the other one, we do not actually need to give the answer in this form as well.

We conclude by presenting our general method of solution in flow chart form (p. 77). In this chart we have eliminated the phrase "the entry in position" before each ordered pair. For instance, the box "Is (1,1) zero?" should be "Is the entry in position (1,1) zero?" The word "divide" is used in four places in the chart. In actual practice we multiply by one over the number in question. This has the same effect. Table 1 summarizes the various forms that the reduced matrix may take.

To Solve Any System of Two Linear Equations in Two Variables

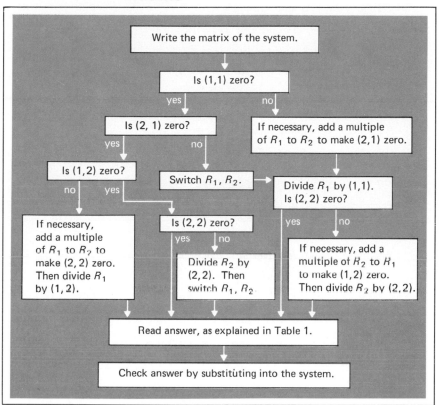

Table 1
Solving Systems: How to Read
the Answer[a]

Reduced Matrix		Answer	
$\begin{bmatrix} 1 & 0 & a \\ 0 & 1 & b \end{bmatrix}$		$\begin{bmatrix} a \\ b \end{bmatrix}$	
$\begin{bmatrix} 1 & c & a \\ 0 & 0 & b \end{bmatrix}$ where $b \neq 0$		no solution	
$\begin{bmatrix} 1 & c & a \\ 0 & 0 & 0 \end{bmatrix}$		$\begin{bmatrix} a - cx_2 \\ x_2 \end{bmatrix}$ where x_2 may be any number	
$\begin{bmatrix} 0 & 1 & a \\ 0 & 0 & b \end{bmatrix}$ where $b \neq 0$		no solution	
$\begin{bmatrix} 0 & 1 & a \\ 0 & 0 & 0 \end{bmatrix}$		$\begin{bmatrix} x_1 \\ a \end{bmatrix}$ where x_1 may be any number	
$\begin{bmatrix} 0 & 0 & a \\ 0 & 0 & b \end{bmatrix}$ where $a \neq 0$ or $b \neq 0$		no solution	
$\begin{bmatrix} 0 & 0 & 0 \\ 0 & 0 & 0 \end{bmatrix}$		$\begin{bmatrix} x_1 \\ x_2 \end{bmatrix}$ where x_1 and x_2 may be any numbers	

[a]The last two cases in this table never occur in a problem about intersecting lines, although they can occur in other contexts.

Exercises 2.2 In each of the following, find the coordinates of the points on the intersection of the two lines with the given equations.

C 1. $x_1 + 2x_2 = 3$ and $3x_1 + 8x_2 = 13$

2. $3x_1 - x_2 = -11$ and $x_1 + 4x_2 = 18$

C 3. $[0 \quad 3] \begin{bmatrix} x_1 \\ x_2 \end{bmatrix} = 16$ and $[1 \quad -6] \begin{bmatrix} x_1 \\ x_2 \end{bmatrix} = 0$

4. $[3 \quad 2] \begin{bmatrix} x_1 \\ x_2 \end{bmatrix} = 4$ and $[1 \quad 2] \begin{bmatrix} x_1 \\ x_2 \end{bmatrix} = 3$

C 5. $4x_1 - x_2 = 2$ and $-20x_1 + 5x_2 = -1$

6. $2x_1 - x_2 = 5$ and $-3x_1 + 4x_2 = 7$

C 7. $[-2 \quad 1] \begin{bmatrix} x_1 \\ x_2 \end{bmatrix} = 3$ and $[4 \quad -2] \begin{bmatrix} x_1 \\ x_2 \end{bmatrix} = -6$

8. $[2 \quad 1] \begin{bmatrix} x_1 \\ x_2 \end{bmatrix} =$ and $[3 \quad -2] \begin{bmatrix} x_1 \\ x_2 \end{bmatrix} = -1$

A 9. $-\frac{1}{2}x_1 + \frac{1}{3}x_2 = 1$ and $x_1 - \frac{1}{4}x_2 = 2$

10. $\frac{2}{5}x_1 + \frac{3}{5}x_2 = 3$ and $\frac{1}{3}x_1 - \frac{1}{2}x_2 = -2$

11. $[-3 \quad 4] \begin{bmatrix} x_1 \\ x_2 \end{bmatrix} = 2$ and $[-2 \quad 5] \begin{bmatrix} x_1 \\ x_2 \end{bmatrix} = 1$

12. $[1 \quad 2] \begin{bmatrix} x_1 \\ x_2 \end{bmatrix} = 4$ and $[-2 \quad -4] \begin{bmatrix} x_1 \\ x_2 \end{bmatrix} = 7$

A 13. $-\frac{2}{3}x_1 + \frac{3}{2}x_2 = -1$ and $4x_1 - 9x_2 = 8$

14. $[1 \quad 2] \begin{bmatrix} x_1 \\ x_2 \end{bmatrix} = 4$ and $[-2 \quad -4] \begin{bmatrix} x_1 \\ x_2 \end{bmatrix} = 8$

15. $[\frac{1}{3} \quad -\frac{1}{4}] \begin{bmatrix} x_1 \\ x_2 \end{bmatrix} = \frac{1}{2}$ and $[\frac{2}{3} \quad -\frac{3}{4}] \begin{bmatrix} x_1 \\ x_2 \end{bmatrix} = -\frac{3}{5}$

16. $\frac{1}{3}x_1 + \frac{1}{2}x_2 = 5$ and $\frac{2}{5}x_1 + \frac{3}{5}x_2 = 6$

A 17. $[1 \quad 0] \begin{bmatrix} x_1 \\ x_2 \end{bmatrix} = 0$ and $[3 \quad 0] \begin{bmatrix} x_1 \\ x_2 \end{bmatrix} = 0$

18. $[\frac{3}{7} \quad -\frac{15}{7}] \begin{bmatrix} x_1 \\ x_2 \end{bmatrix} = 1$ and $[-2 \quad 10] \begin{bmatrix} x_1 \\ x_2 \end{bmatrix} = -14$

19. $[0 \quad 4] \begin{bmatrix} x_1 \\ x_2 \end{bmatrix} = 6$ and $[0 \quad -3] \begin{bmatrix} x_1 \\ x_2 \end{bmatrix} = -\frac{9}{2}$

20. $\frac{9}{5}x_1 - 9x_2 = \frac{24}{5}$ and $-x_1 + 5x_2 = -\frac{7}{3}$

See Exercise 12 in Section 2-1 and,
if necessary, its solution.

Figure 16

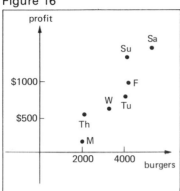

C21. Some of the really big PHE stores at the Ames conference—for instance, the downtown Minneapolis and Chicago stores—have a very different relation between the daily burger sale x_1 and net daily burger profit x_2. At a private meeting with PHE board members, the manager of store 894 in the Chicago Loop, one of the most successful stores in the chain, asked that his store's figures (see Figure 16) be compared with the presentation given at the conference. He checked his daily burger sales and profit for several days. The resulting data were plotted, and while they did not lie on a straight line, a statistical technique, called the method of least squares yielded the line of best fit. This line was $x_1 - 3x_2 = 1500$. Present his figures, and make the comparison.

**Section 2-3
Linear Equations at
Appalachian Creations**

In this section we present some production and supply problems at Appalachian Creations that can be solved by the method discussed in Section 2-2. We then go on to some problems involving a larger number of producers, suppliers, or receivers that seem similar to the previous problems but which we cannot solve at the moment. We will show how to solve these problems and all problems of the same general type in Section 2-4.

Problem   1

The Bangor store orders 29 down jackets and 52 pairs of down booties. In one day of operation, the Springfield factory can make 3 jackets and 6 pairs of booties and the Waterbury factory can make 4 jackets and 5 pairs of booties. How many days should we operate the two factories so as to exactly fill the order from Bangor?

Solution

Let x_1 be the number of days we operate the Springfield factory. In this time this factory will make $3x_1$ jackets and $6x_1$ pairs of booties. Let x_2 be the number of days we operate the Waterbury factory. In this time this factory will make $4x_2$ jackets and $5x_2$ pairs of booties. The total number of jackets made will then be $3x_1 + 4x_2$, and the total number of pairs of booties will be $6x_1 + 5x_2$. The order from Bangor leads us, then, to the following system of two linear equations in two unknowns:

$$\begin{bmatrix} 3x_1 + 4x_2 = 29 \\ 6x_1 + 5x_2 = 52 \end{bmatrix}$$

1

We solve this system by the method of Section 2-2:

$$\begin{bmatrix} 3 & 4 & 29 \\ 6 & 5 & 52 \end{bmatrix} \xrightarrow{R_2 - 2R_1} \begin{bmatrix} 3 & 4 & 29 \\ 0 & -3 & -6 \end{bmatrix} \xrightarrow{R_1 + \frac{4}{3}R_2} \begin{bmatrix} 3 & 0 & 21 \\ 0 & -3 & -6 \end{bmatrix}$$

$$\xrightarrow{-\frac{1}{3}R_2} \begin{bmatrix} 1 & 0 & 7 \\ 0 & 1 & 2 \end{bmatrix} \qquad \text{Answer:} \begin{bmatrix} 7 \\ 2 \end{bmatrix}$$

So we tell the factory managers to operate Springfield for 7 days and Waterbury for 2 days.

Problem 2 The Springfield factory can make 9 cups and 3 canteens per hour while the Danbury factory can make 12 cups and 4 canteens per hour. We regularly receive an order from Falmouth for 108 cups and 36 canteens, which we have been filling by operating the Springfield factory for 8 hours and the Danbury factory for 3. But we are worried about labor problems at both factories. In case of strikes, can we find other ways to fill this order exactly, either by cutting down on the Springfield operation or by cutting down on the Danbury operation?

Solution Let x_1 be the number of hours we operate the Springfield plant, and let x_2 be the number of hours we operate the Danbury plant. As in the previous problem, we are led to a system:

$$\begin{bmatrix} 9x_1 + 12x_2 = 108 \\ 3x_1 + 4x_2 = 36 \end{bmatrix}$$

2

We solve the system:

$$\begin{bmatrix} 9 & 12 & 108 \\ 3 & 4 & 36 \end{bmatrix} \xrightarrow{R_2 - \frac{1}{3}R_1} \begin{bmatrix} 9 & 12 & 108 \\ 0 & 0 & 0 \end{bmatrix} \xrightarrow{\frac{1}{9}R_1} \begin{bmatrix} 1 & \frac{4}{3} & 12 \\ 0 & 0 & 0 \end{bmatrix}$$

Answer: $\begin{bmatrix} 12 - \frac{4}{3}x_2 \\ x_2 \end{bmatrix}$, where x_2 may be any number. Note that if $x_2 = 3$ this gives the solution $\begin{bmatrix} 8 \\ 3 \end{bmatrix}$, which has been used before. Now if we wanted to eliminate the Danbury operation altogether, we would make $x_2 = 0$ to obtain the solution $\begin{bmatrix} 12 \\ 0 \end{bmatrix}$. If we wanted to eliminate the Springfield operation, we would make $12 - \frac{4}{3}x_2 = 0$, that is, $12 = \frac{4}{3}x_2$, $x_2 = (\frac{3}{4})12 = 9$, and obtain the solution $\begin{bmatrix} 0 \\ 9 \end{bmatrix}$. If x_2 exceeded 9, we would be overproducing;

besides, x_1 would become negative, which is meaningless. Similarly, x_1 cannot exceed 12. So x_2 may vary from 0 to 9 and x_1 may vary from 0 to 12, in accordance with the formula

$$x_1 = 12 - \tfrac{4}{3}x_2 \qquad\qquad \textbf{3}$$

An answer to the problem posed by management might take the form of Tables 2 and 3.

Table 2
Operation Schedules: Time at
Danbury Reduced from the
Usual 3 Hours

Danbury	2 hr.	1 hr.	closed
Springfield	9 hr. 20 min.	10 hr. 40 min.	12 hr.

Table 3
Operation Schecules: Time at
Springfield Reduced from the
Usual 8 Hours

Springfield	7 hr.	6 hr.	5 hr.
Danbury	3 hr. 45 min.	4 hr. 30 min.	5 hr. 15 min.
Springfield	4 hr.	3 hr.	2 hr.
Danbury	6 hr.	6 hr. 45 min.	7 hr. 30 min.
Springfield	1 hr.	closed	
Danbury	8 hr. 15 min.	9 hr.	

While Table 3 could have been calculated from formula 3, we first rewrote this formula as follows:

$$\tfrac{4}{3}x_2 = 12 - x_1, \quad x_2 = (\tfrac{3}{4})(12 - x_1) = 9 - \tfrac{3}{4}x_1, \text{ that is,}$$
$$x_2 = 9 - \tfrac{3}{4}x_1 \qquad\qquad \textbf{4}$$

This formula made the preparation of Table 3 easier.

Problem 3　The labor problems in the cup and canteen divisions of Springfield and Danbury have been cleared up, but Falmouth has changed its regular order to 90 cups and 40 canteens. How do we fill the order exactly?

Solution　As in Problem 2, we are led to the system

$$\begin{bmatrix} 9x_1 + 12x_2 = 90 \\ 3x_1 + 4x_2 = 40 \end{bmatrix} \qquad\qquad \textbf{5}$$

We apply the method of Section 2-2:

$$\begin{bmatrix} 9 & 12 & 90 \\ 3 & 4 & 40 \end{bmatrix} \xrightarrow{R_2 - \frac{1}{3}R_1} \begin{bmatrix} 9 & 12 & 90 \\ 0 & 0 & 10 \end{bmatrix} \xrightarrow{\frac{1}{9}R_1} \begin{bmatrix} 1 & \frac{4}{3} & 10 \\ 0 & 0 & 10 \end{bmatrix}$$

See Supplementary Exercise 31 in Chapter 5.

Answer: no solution. We must tell management that the order cannot be filled exactly with existing facilities; they must plan for a surplus of either cups or canteens. We shall have more to say on problems of this type in later chapters.

Problem 4   (See Problem 1.) Bangor has reduced its order to 26 jackets and 28 pairs of booties. Can this be filled exactly?

Solution   We seek a solution for the system

$$\begin{bmatrix} 3x_1 + 4x_2 = 26 \\ 6x_1 + 5x_2 = 28 \end{bmatrix} \qquad \textbf{6}$$

We reduce the matrix:

$$\begin{bmatrix} 3 & 4 & 26 \\ 6 & 5 & 28 \end{bmatrix} \xrightarrow{R_2 - 2R_1} \begin{bmatrix} 3 & 4 & 26 \\ 0 & -3 & -24 \end{bmatrix} \xrightarrow{R_1 + \frac{4}{3}R_2} \begin{bmatrix} 3 & 0 & -6 \\ 0 & -3 & -24 \end{bmatrix}$$

$$\xrightarrow[-\frac{1}{3}R_2]{\frac{1}{3}R_1} \begin{bmatrix} 1 & 0 & -2 \\ 0 & 1 & 8 \end{bmatrix} \qquad \text{Answer: } \begin{bmatrix} -2 \\ 8 \end{bmatrix}$$

This is the only answer for system **6**, but it makes no sense for us, since we cannot operate the Springfield factory for -2 days! Thus, we must announce that the order cannot be filled exactly.

Problem 5   (See Problem 1.) There is some material left over from the jacket and booties operations, which the Springfield and Waterbury factories make into mittens and pillows. In one day of operation, Springfield makes 2 pairs of mittens and 1 pillow and Waterbury makes 3 pairs of mittens and 2 pillows. Bangor orders, in addition to their usual order for jackets and booties, 20 pairs of mittens and 11 pillows. How do we fill this order exactly?

Solution   Using the notation of Problem 1, we have production of $2x_1 + 3x_2$ pairs of mittens and $x_1 + 2x_2$ pillows. Thus, we are asked to solve the system

$$\begin{bmatrix} 3x_1 + 4x_2 = 29 \\ 6x_1 + 5x_2 = 52 \\ 2x_1 + 3x_2 = 20 \\ x_1 + 2x_2 = 11 \end{bmatrix} \qquad \textbf{7}$$

We have not yet developed a systematic method for solving such a large system; we will do so in the next section. However, it happens that we can manage to solve system **7** now. Any solution of this system must, in particular, satisfy the first two equations. But these two equations comprise system **1**, which, as we learned in the solution to Problem 1, has only one solution: $\begin{bmatrix} 7 \\ 2 \end{bmatrix}$. Thus, the only possible solution of system **7** is $\begin{bmatrix} 7 \\ 2 \end{bmatrix}$. This matrix actually is a solution, since this answer checks in the last two equations as well: $2 \cdot 7 + 3 \cdot 2 = 20$ and $7 + 2 \cdot 2 = 11$. Thus, we repeat our instruction of Problem 1: run Springfield for 7 days and Waterbury for 2.

Problem 6   In an attempt to get a solution to Problem 3, management has opened a new factory in Mystic, which makes 6 cups and 4 canteens per hour. Can we now fill the order exactly?

Solution   As before, we let x_1 be the number of hours we operate the Springfield plant and x_2 be the number of hours we operate the Danbury plant. Also, let x_3 be the number of hours we operate the Mystic factory. Then total cup production is $9x + 12x_2 + 6x_3$ and total canteen production is $3x_1 + 4x_2 + 4x_3$. Thus, we are to solve the system

$$\begin{bmatrix} 9x_1 + 12x_2 + 6x_3 = 90 \\ 3x_1 + 4x_2 + 4x_3 = 40 \end{bmatrix} \qquad \textbf{8}$$

Try $\begin{bmatrix} 4 \\ 2 \\ 5 \end{bmatrix}$.   Again, we must wait until Section 2-4 for a method of solving such a system, although we might be able to guess at a solution.

Exercises 2.3   A1. Given the production data of Problem 1, how do we exactly fill each of the following orders?
(a) Kansas City wants 11 jackets and 16 pairs of booties.
(b) Eureka, California, wants 12 jackets and 15 pairs of booties.
(c) Elmont, New York, wants 18 jackets and 2 dozen pairs of booties.
(d) Gerster, Missouri wants 1 jacket, no booties.

A2. The Williams Motor Company with plants at Dearborn and Oshkosh, produces two models of its car: the Civet and the Wildebeest. In one week the Dearborn plant can produce 4200 Civets and 3400 Wildebeests, while Oshkosh can make 2100 Civets and 1700 Wildebeests. We have a production goal of 16,800 Civets and 13,600 Wildebeests. There is a possibility of a strike at either plant. Prepare a table of possible production schedules for each plant, assuming that the struck plant will operate a whole number of weeks, if it operates at all.

A3. AC has been asked to make up packets of freeze-dried food for the American expedition to Sagarmatha. Each packet must contain exactly 86 grams of protein, 88 grams of fats, and 90 grams of carbohydrates. AC has a supply of three types of freeze-dried food, in 12-gram packets, having the following nutritional composition (in grams per package):

Type	Protein	Fats	Carb.	Other
1	3	2	3	4
2	4	1	3	4
3	2	5	3	2

Write the system that must be solved in order to determine how many packets of each type should go into each expedition packet. (This system will be solved in the last example problem of Section 2-4.)

A4. PHE is considering an ad campaign for selected stores in the South, West, and East based on the slogan, "Look what you can buy for . . . !" They want to offer various combinations of items for a fixed total price. The possible combinations must be the same in every region but the total price will vary, because the per-item prices vary from region to region. The items they want to push, with prices shown by region, are:

	Hamburger	Superburger	Fishfeast	Root beer
South	$0.30	$0.40	$1.00	$0.25
West	0.40	0.60	1.40	0.30
East	0.50	0.70	1.70	0.35

The total prices they have in mind, that is, the prices to be used in the slogan, are: South, $3.50; West, $4.80; East, $5.70. How many different combinations can they offer? Just introduce the notation and set up this problem as a system. You will be asked to *solve* this problem in Exercise 8, Section 2-4.

Section 2-4
Gauss-Jordan Reduction

In Section 2-2 we learned how to solve any system of two linear equations in two variables. In Section 2-3 we saw that we are going to need to solve some systems in which there are more than two equations and/or more than two variables. We now present a method to accomplish this: Gauss-Jordan reduction, named after the mathematicians C. F. Gauss (1777–1855) and W. Jordan (1819–1904). Some small systems, with only a few variables and equations, were solved in ancient times, but it was Gauss who first considered systems of arbitrarily large size, and Jordan who first presented the method of solution discussed here.

We call an equation such as $3x_1 - 4x_2 = 5$ a linear equation because its graph is a straight line. The equations to be studied here, some of which we met in Section 2-3, are of the same general type as this. We will also call these new equations linear equations, although we will not show how to graph them. For example, we say that the equation $-2x_1 + \tfrac{2}{3}x_2 - 3x_3 = 7$ is a linear equation in three variables, and that

$$[1 \quad -2 \quad \tfrac{3}{4} \quad 0 \quad 1] \begin{bmatrix} x_1 \\ x_2 \\ x_3 \\ x_4 \\ x_5 \end{bmatrix} = -\tfrac{7}{8}$$

is a linear equation in five variables. We will say a little about the geometric interpretation of these equations in Section 2-5.

Consider now the large systems encountered in Section 2-3. System **7** in that section is a system of four linear equations in two variables; system **8** is a system of two linear equations in three variables. The matrices of these systems are, respectively,

$$\begin{bmatrix} 3 & 4 & 29 \\ 6 & 5 & 52 \\ 2 & 3 & 20 \\ 1 & 2 & 11 \end{bmatrix} \qquad \textbf{1}$$

and

$$\begin{bmatrix} 9 & 12 & 6 & 90 \\ 3 & 4 & 4 & 40 \end{bmatrix} \qquad \textbf{2}$$

These are of size 4×3 and 2×4, respectively. The number of rows equals the number of equations in the system. The number of columns is *one more than* the number of unknowns in the system.

Let us apply the techniques of Section 2-2 to system **7**, Section 2-3. We are going to manipulate the rows of matrix **1** above according to the following rules:

(1) We may add (or subtract) a multiple of one row to (from) another.
(2) We may multiply (or divide) a row by a nonzero number.
(3) We may switch any two rows.
Consider the following manipulations of matrix **1**:

$$\begin{bmatrix} 3 & 4 & 29 \\ 6 & 5 & 52 \\ 2 & 3 & 20 \\ 1 & 2 & 11 \end{bmatrix} \begin{matrix} \\ R_2 - 2R_1 \\ R_3 - \tfrac{2}{3}R_1 \\ R_4 - \tfrac{1}{3}R_1 \end{matrix} \Longrightarrow \begin{bmatrix} 3 & 4 & 29 \\ 0 & -3 & -6 \\ 0 & \tfrac{1}{3} & \tfrac{2}{3} \\ 0 & \tfrac{2}{3} & \tfrac{4}{3} \end{bmatrix} \xrightarrow{\tfrac{1}{3}R_1} \begin{bmatrix} 1 & \tfrac{4}{3} & \tfrac{29}{3} \\ 0 & -3 & -6 \\ 0 & \tfrac{1}{3} & \tfrac{2}{3} \\ 0 & \tfrac{2}{3} & \tfrac{4}{3} \end{bmatrix}$$

$R_1 + \frac{4}{9}R_2$

$R_3 + \frac{1}{9}R_2$
$R_4 + \frac{2}{9}R_2$

$\begin{bmatrix} 1 & 0 & 7 \\ 0 & -3 & -6 \\ 0 & 0 & 0 \\ 0 & 0 & 0 \end{bmatrix}$ $-\frac{1}{3}R_2$ $\begin{bmatrix} 1 & 0 & 7 \\ 0 & 1 & 2 \\ 0 & 0 & 0 \\ 0 & 0 & 0 \end{bmatrix}$

3

The final matrix of sequence **3** represents the system

$$\begin{bmatrix} 1x_1 + 0x_2 = 7 \\ 0x_1 + 7x_2 = 2 \\ 0x_1 + 0x_2 = 0 \\ 0x_1 + 0x_2 = 0 \end{bmatrix}$$

4

that is, $x_1 = 7$ and $x_2 = 2$. The last two equations in system **4** give no additional information. The solution is $\begin{bmatrix} 7 \\ 2 \end{bmatrix}$, as found in Problem 5, Section 2-3.

In the first four steps of sequence **3** we used R_1, the first row, in various ways to change C_1, the first column, from its original form $\begin{bmatrix} 3 \\ 6 \\ 2 \\ 1 \end{bmatrix}$ into $\begin{bmatrix} 1 \\ 0 \\ 0 \\ 0 \end{bmatrix}$. We briefly describe this procedure by saying that we have used R_1 to *clear* C_1. Speaking generally, to use a row R to **clear** a column C means to add or subtract the proper multiples of R to or from all the other rows of the matrix so that each entry in C except the entry in R becomes zero. Next, divide R by whatever number will produce a 1 in column C. In the next four steps of series 3 we used $R_2 = \begin{bmatrix} 0 & -3 & -6 \end{bmatrix}$ to clear $C_2 = \begin{bmatrix} \frac{4}{3} \\ -3 \\ \frac{1}{3} \\ \frac{2}{3} \end{bmatrix}$, that is, to change C_2 to $\begin{bmatrix} 0 \\ 1 \\ 0 \\ 0 \end{bmatrix}$.

As a further illustration, consider the matrix

$$\begin{bmatrix} -2 & 1 & 3 & 2 \\ 4 & 0 & 2 & -2 \\ 0 & -6 & -4 & 0 \\ -5 & -1 & -6 & 3 \end{bmatrix}$$

5

Notice that we cannot, for example, use R_2 of matrix **5** to clear C_2, because the entry at the intersection of R_2 and C_2 (that is, at address (2,2)) is zero, and zero cannot be changed to -1, 6, 1 (or any other number) by multiplication. We can however use R_1 to clear C_2:

$$\begin{bmatrix} -2 & 1 & 3 & 2 \\ 4 & 0 & 2 & -2 \\ 0 & -6 & -4 & 0 \\ -5 & -1 & -6 & 3 \end{bmatrix} \begin{matrix} \\ \\ R_3 + 6R_1 \\ R_4 + \ R_1 \end{matrix} \Longrightarrow \begin{bmatrix} -2 & 1 & 3 & 2 \\ 4 & 0 & 2 & -2 \\ -12 & 0 & 14 & 12 \\ -7 & 0 & -3 & 5 \end{bmatrix}$$

We now describe the method of Gauss-Jordan reduction in flow-chart form. This chart is the generalized version of the chart given in Section 2-2 for systems of two linear equations in two variables. Note that the paragraph below the flow chart entitled "How to Read the Answer" is actually part of the flow

Gauss-Jordan Reduction of a System of Linear Equations

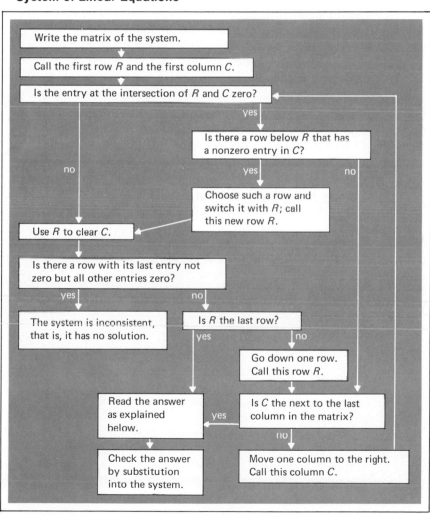

chart. It might not apply to other methods of solution such as those presented in the material on shortcuts in Section 2-5.

How to Read the Answer

If the final matrix obtained from Gauss-Jordan Reduction contains a row with its last entry not zero but all other entries zero, then the system has no solution. For example, if the row 0 0 0 0 0 0 5 occurs in the final matrix (or in any previous matrix for that matter) then the system has no solution.

Suppose we have carried out Gauss-Jordan reduction and the final matrix does not contain a row with all entries zero except the last. Each row that does not consist entirely of zeros corresponds to an equation from which we obtain the value of one variable. This value may be a single number. Suppose, for example, one of the rows is 0 1 0 0 0 0 −5. This corresponds to the equation $0x_1 + 1x_2 + 0x_3 + 0x_4 + 0x_5 + 0x_6 = -5$, that is, $x_2 = -5$. On the other hand, some rows may yield an expression for a variable that involves one or more *free* variables, that is, variables whose values are not determined by the problem. These variables are free in the sense that we may assign them any values whatsoever. Suppose, for example, that one of the rows of the final matrix is

$$0 \quad 0 \quad 1 \quad 0 \quad 2 \quad -4 \quad 3 \qquad \textbf{6}$$

Then the corresponding equation is

$$x_3 + 2x_5 - 4x_6 = 3 \qquad \textbf{7}$$

The first variable occuring in equation **7** is x_3. We solve equation **7** for x_3:

$$x_3 = 3 - 2x_5 + 4x_6 \qquad \textbf{8}$$

This is the final expression for x_3 in this case. The variables x_5 and x_6 are free and they will appear in the answer simply as x_5 and x_6, with the comment that they may be any numbers whatever.

In practice it is easy to move from the row to the answer (that is, from row **6** to equation **8**, for example) without writing any intermediate steps.

For our first illustration of Gauss-Jordan reduction, we solve Problem 6 from Section 2-3. The matrix of system **9** in that section is

$$\begin{array}{c} \quad\quad C \\ R \begin{bmatrix} 9 & 12 & 6 & 90 \\ 3 & 4 & 4 & 40 \end{bmatrix} \end{array} \qquad \textbf{9}$$

We have indicated the first row to be called R and column to be called C. First we use R to clear C:

$$\begin{array}{c} \quad\quad C \\ R \begin{bmatrix} 9 & 12 & 6 & 90 \\ 3 & 4 & 4 & 40 \end{bmatrix} \end{array} \quad R_2 - \tfrac{1}{3}R_1 \quad \begin{bmatrix} 9 & 12 & 6 & 90 \\ 0 & 0 & 2 & 10 \end{bmatrix}$$

$$\tfrac{1}{9}R_1 \quad \begin{bmatrix} 1 & \tfrac{4}{3} & \tfrac{2}{3} & 10 \\ 0 & 0 & 2 & 10 \end{bmatrix}$$

We move down one row to the new R:

$$\begin{array}{c} \quad\quad C \\ \begin{bmatrix} 1 & \tfrac{4}{3} & \tfrac{2}{3} & 10 \\ R\ 0 & 0 & 2 & 10 \end{bmatrix} \end{array}$$

The new column C is C_2. Since the entry at the intersection of R and C is zero, since there is no row below R, and since C is not the next to the last column, the flow chart instructs us to move one column to the right:

$$\begin{array}{c} \quad\quad\quad\quad C \\ \begin{bmatrix} 1 & \tfrac{4}{3} & \tfrac{2}{3} & 10 \\ R\ 0 & 0 & 2 & 10 \end{bmatrix} \end{array}$$

The new C is C_3. Now use R to clear C:

$$\begin{array}{c} \quad\quad\quad\quad C \\ \begin{bmatrix} 1 & \tfrac{4}{3} & \tfrac{2}{3} & 10 \\ R\ 0 & 0 & 2 & 10 \end{bmatrix} \end{array} \quad R_1 - \tfrac{1}{3}R_2 \quad \begin{bmatrix} 1 & \tfrac{4}{3} & 0 & \tfrac{20}{3} \\ 0 & 0 & 2 & 10 \end{bmatrix}$$

$$\tfrac{1}{2}R_2 \quad \begin{bmatrix} 1 & \tfrac{4}{3} & 0 & \tfrac{20}{3} \\ 0 & 0 & 1 & 5 \end{bmatrix} \qquad \textbf{10}$$

This is the final matrix, from which we read the answer: $x_1 = \tfrac{20}{3} - \tfrac{4}{3}x_2$, x_2 is free, $x_3 = 5$. In column form the answer is

$$\begin{bmatrix} {}^{20}/_3 - {}^4/_3 x_2 \\ x_2 \\ 5 \end{bmatrix}$$, where x_2 may be any number. Notice that if we take

$x_2 = 2$ we get the solution $\begin{bmatrix} 4 \\ 2 \\ 5 \end{bmatrix}$, as suggested in Problem 6 of Section 2-3. So, with the new factory at Mystic there are many ways to fill exactly the order from Bangor.

Before continuing with more examples, we offer some explanation of why Gauss-Jordan reduction works. Let us consider the effect of the three types or row operations (p. 85) on a system of equations. If an equation is multiplied by a nonzero number, the resulting equation will have the same solutions as the original equation. For example, any solution of the equation

$$-3x_1 + 4x_2 = -5 \qquad\qquad \textbf{11}$$

will also be a solution of $-\frac{1}{3}$ times the equation

$$x_1 - {}^4/_3 x_2 = {}^5/_3 \qquad\qquad \textbf{12}$$

and, conversely, any solution of equation **12** will also be a solution of equation **11**, since equation **11** can be obtained from equation **12** by multiplying the latter by -3. Also, if we are given two equations and a number k, the system consisting of the first equation and the second equation plus k times the first equation will have the same solutions, if any, as the original system. For example, any solution of the system

$$\begin{array}{l} \text{(eqn. 1)} \\ \text{(eqn. 2)} \end{array} \begin{bmatrix} 4x_1 - 3x_2 + x_3 = 6 \\ -x_1 + 2x_2 - 3x_3 = 4 \end{bmatrix} \qquad\qquad \textbf{13}$$

will also be a solution of the system

Here
$R = R_1$
$k = {}^1/_4$

$$\begin{array}{l} \text{(eqn. 1)} \\ \text{(eqn. 2 + } {}^1/_4 \text{ eqn. 1)} \end{array} \begin{bmatrix} 4x_1 - 3x_2 + x_3 = 6 \\ 0x_1 + {}^5/_4 x_2 - {}^{11}/_4 x_3 = {}^{11}/_2 \end{bmatrix} \qquad\qquad \textbf{14}$$

And conversely, any solution of system **14** will also be a solution of system **13**, since system **13** can be obtained from system **14** by subtracting $\frac{1}{4}$ times eqn. 1 of system **14** from the second equation of system **14**. Finally, given a system of equations, it is obvious that switching the order in which we write the equations will in no way affect the solutions, if any, of the system. Thus, for example, systems **13** and

$$\begin{bmatrix} -x_1 + 2x_2 - 3x_3 = 4 \\ 4x_1 - 3x_2 + x_3 = 6 \end{bmatrix} \qquad\qquad \textbf{15}$$

have exactly the same solutions.

Gauss-Jordan reduction is just a sequence of the three types of operations described above. Since none of these operations affects the solutions of a system, it follows that the solutions, if any, of a given system will be the same as the solutions of the system with the final matrix obtained by Gauss-Jordan reduction. But as we have seen, it is very easy to read the solutions from the final matrix of a Gauss-Jordan reduction. This is why the flow chart above is an effective and practical method for solving systems of equations.

We close this section with more examples and exercises. As you work through these exercises, you may notice some cases in which you can save labor by doing the steps in another order. By all means do so. The Gauss-Jordan method always works but, as we shall see in the next section, it is not always the quickest way.

Problem 1 Solve:

$$\begin{bmatrix} & & 3x_3 & & & = & 9 \\ & & & 4x_4 + 8x_5 & = & 0 \\ -2x_1 - & 6x_2 + 3x_3 & & + 2x_5 & = & 13 \\ 2x_1 & 6x_2 & & + 2x_5 & = & 4 \\ 4x_1 + & 12x_2 - 9x_3 & & - 4x_5 & = & -35 \end{bmatrix} \qquad \textbf{16}$$

Solution

$$\begin{bmatrix} 0 & 0 & 3 & 0 & 0 & 9 \\ 0 & 0 & 0 & 4 & 8 & 0 \\ -2 & -6 & 3 & 0 & 2 & 13 \\ 2 & 6 & 0 & 0 & 2 & 4 \\ 4 & 12 & -9 & 0 & -4 & -35 \end{bmatrix} \xrightarrow{R} \begin{matrix} C \\ \begin{bmatrix} -2 & -6 & 3 & 0 & 2 & 13 \\ 0 & 0 & 0 & 4 & 8 & 0 \\ 0 & 0 & 3 & 0 & 0 & 9 \\ -2 & -6 & 0 & 0 & 2 & 4 \\ 4 & 12 & -9 & 0 & -4 & -35 \end{bmatrix} \end{matrix}$$

$R_4 - R$
$R_5 + 2R$

$$\begin{bmatrix} -2 & -6 & 3 & 0 & 2 & 13 \\ 0 & 0 & 0 & 4 & 8 & 0 \\ 0 & 0 & 3 & 0 & 0 & 9 \\ 0 & 0 & -3 & 0 & 0 & -9 \\ 0 & 0 & -3 & 0 & 0 & -9 \end{bmatrix} \xrightarrow[R]{-\frac{1}{2}R_1} \begin{matrix} C \\ \begin{bmatrix} 1 & 3 & -\frac{3}{2} & 0 & -1 & -\frac{13}{2} \\ 0 & 0 & 3 & 0 & 0 & 9 \\ 0 & 0 & 0 & 4 & 8 & 0 \\ 0 & 0 & -3 & 0 & 0 & -9 \\ 0 & 0 & -3 & 0 & 0 & -9 \end{bmatrix} \end{matrix}$$

$R_1 + \frac{1}{2}R$

$R_4 + R$
$R_5 + R$

$$\begin{bmatrix} 1 & 3 & 0 & 0 & -1 & -2 \\ 0 & 0 & 3 & 0 & 0 & 9 \\ 0 & 0 & 0 & 4 & 8 & 0 \\ 0 & 0 & 0 & 0 & 0 & 0 \\ 0 & 0 & 0 & 0 & 0 & 0 \end{bmatrix} \xrightarrow[R]{\frac{1}{3}R_2} \begin{matrix} C \\ \begin{bmatrix} 1 & 3 & 0 & 0 & -1 & -2 \\ 0 & 0 & 1 & 0 & 0 & 3 \\ 0 & 0 & 0 & 4 & 8 & 0 \\ 0 & 0 & 0 & 0 & 0 & 0 \\ 0 & 0 & 0 & 0 & 0 & 0 \end{bmatrix} \end{matrix}$$

$$\frac{1}{4}R \begin{bmatrix} 1 & 3 & 0 & 0 & -1 & -2 \\ 0 & 0 & 1 & 0 & 0 & 3 \\ 0 & 0 & 0 & 1 & 2 & 0 \\ 0 & 0 & 0 & 0 & 0 & 0 \\ 0 & 0 & 0 & 0 & 0 & 0 \end{bmatrix}$$

Answer: $\begin{bmatrix} -2 - 3x_2 + x_5 \\ x_2 \\ 3 \\ -2x_5 \\ x_5 \end{bmatrix}$, where x_2 and x_5 may be any numbers.

Check:

$$\begin{bmatrix} & & 3(3) & = & 9 \\ & & 4(-2x_5) + 8(x_5) = & 0 \\ -2(-2 - 3x_2 + x_5) - & 6(x_2) + 3(3) & + 2(x_5) = & 13 \\ -2(-2 - 3x_2 + x_5) - & 6(x_2) & + 2(x_5) = & 4 \\ 4(-2 - 3x_2 + x_5) + & 12(x_2) - 9(3) & - 4(x_5) = & -35 \end{bmatrix}$$ ✔ ✔ ✔ ✔ ✔

Problem 2   Solve: $\begin{bmatrix} 3x_1 - & x_2 + 2x_3 = & 7 \\ x_1 - & x_2 + x_3 = & -2 \\ 4x_1 - & 2x_1 + 3x_3 = & 3 \end{bmatrix}$

Solution

$$\begin{bmatrix} 3 & -1 & 2 & 7 \\ 1 & -1 & 1 & -2 \\ 4 & -2 & 3 & 3 \end{bmatrix} \begin{matrix} \\ R_2 - \frac{1}{3}R_1 \\ R_3 - \frac{4}{3}R_1 \end{matrix} \quad \begin{bmatrix} 3 & -1 & 2 & 7 \\ 0 & -\frac{2}{3} & \frac{1}{3} & -\frac{13}{3} \\ 0 & -\frac{2}{3} & \frac{1}{3} & -\frac{19}{3} \end{bmatrix}$$

$$\frac{1}{3}R_1 \begin{bmatrix} 1 & -\frac{1}{3} & \frac{2}{3} & \frac{7}{3} \\ 0 & -\frac{2}{3} & \frac{1}{3} & -\frac{13}{3} \\ 0 & -\frac{2}{3} & \frac{1}{3} & -\frac{19}{3} \end{bmatrix} \begin{matrix} R_1 - \frac{1}{2}R_2 \\ \\ R_3 - R_2 \end{matrix} \quad \begin{bmatrix} 1 & 0 & \frac{1}{2} & \frac{27}{6} \\ 0 & -\frac{2}{3} & \frac{1}{3} & -\frac{13}{3} \\ 0 & 0 & 0 & -2 \end{bmatrix}$$

$$-\frac{3}{2}R_2 \begin{bmatrix} 1 & 0 & \frac{1}{2} & \frac{27}{6} \\ 0 & 1 & -\frac{1}{2} & \frac{13}{2} \\ 0 & 0 & 0 & -2 \end{bmatrix} \qquad \text{Answer: No solution.}$$

Problem 3   Solve Exercise 3 in Section 2-3.

Solution   Let x_1, x_2, x_3, respectively, be the number of packets of types 1, 2, and 3 used to make an expedition packet. We must solve:

$$\begin{bmatrix} 3x_1 + 4x_2 + 2x_3 = 86 \\ 2x_1 + x_2 + 5x_3 = 88 \\ 3x_1 + 3x_2 + 3x_3 = 90 \end{bmatrix}$$

$$\begin{bmatrix} 3 & 4 & 2 & 86 \\ 2 & 1 & 5 & 88 \\ 3 & 3 & 3 & 90 \end{bmatrix} \quad \begin{matrix} R_2 - \frac{2}{3}R_1 \\ R_3 - R_1 \end{matrix} \quad \begin{bmatrix} 3 & 4 & 2 & 86 \\ 0 & -\frac{5}{3} & \frac{11}{3} & \frac{92}{3} \\ 0 & -1 & 1 & 4 \end{bmatrix}$$

$$\frac{1}{3}R_1 \quad \begin{bmatrix} 1 & \frac{4}{3} & \frac{2}{3} & \frac{86}{3} \\ 0 & -\frac{5}{3} & \frac{11}{3} & \frac{92}{3} \\ 0 & -1 & 1 & 4 \end{bmatrix} \quad \begin{matrix} R_1 + \frac{4}{5}R_2 \\ R_3 - \frac{3}{5}R_2 \end{matrix} \quad \begin{bmatrix} 1 & 0 & \frac{18}{5} & \frac{266}{5} \\ 0 & -\frac{5}{3} & \frac{11}{3} & \frac{92}{3} \\ 0 & 0 & -\frac{6}{5} & -\frac{72}{5} \end{bmatrix}$$

$$-\frac{3}{5}R_2 \quad \begin{bmatrix} 1 & 0 & \frac{18}{5} & \frac{266}{5} \\ 0 & 1 & -\frac{11}{5} & -\frac{92}{5} \\ 0 & 0 & -\frac{6}{5} & -\frac{72}{5} \end{bmatrix} \quad \begin{matrix} R_1 + 3R_3 \\ R_2 - \frac{11}{6}R_3 \end{matrix} \quad \begin{bmatrix} 1 & 0 & 0 & 10 \\ 0 & 1 & 0 & 8 \\ 0 & 0 & -\frac{6}{5} & -\frac{72}{5} \end{bmatrix}$$

$$-\frac{5}{6}R_3 \quad \begin{bmatrix} 1 & 0 & 0 & 10 \\ 0 & 1 & 0 & 8 \\ 0 & 0 & 1 & 12 \end{bmatrix}$$

Answer: Use 10 packets of type 1, 8 of type 2, and 12 of type 3 to make one expedition packet.

Check:
$$\begin{bmatrix} 3(10) + 4(8) + 2(12) = 86 \\ 2(10) + 1(8) + 5(12) = 88 \\ 3(10) + 3(8) + 3(12) = 90 \end{bmatrix}$$

Exercises 2-4 1. Solve the following systems of equations:

(a) $\begin{bmatrix} x_1 + 2x_2 + x_3 = 7 \\ 2x_1 + 3x_2 + 3x_3 = 16 \\ 2x_1 + 3x_2 + 4x_3 = 19 \end{bmatrix}$

A(d) $\begin{bmatrix} 9x_1 + 2x_2 + 3x_3 = -40 \\ 2x_1 + x_2 + x_3 = 8 \\ 4x_1 - x_2 = 68 \end{bmatrix}$

A(b) $\begin{bmatrix} x_1 + 3x_2 + x_3 = 3 \\ 3x_1 + 8x_2 + 3x_3 = 7 \\ 2x_1 - 3x_2 + x_3 = -10 \end{bmatrix}$

(e) $\begin{bmatrix} 2x_1 + x_2 + 3x_3 = 9 \\ 3x_1 - 6x_2 - 4x_3 = 7 \\ 5x_1 + 2x_2 + 7x_3 = 22 \end{bmatrix}$

(c) $\begin{bmatrix} 3x_1 + 3x_2 + 5x_3 = 14 \\ x_1 + 6x_2 + 8x_3 = 20 \\ 2x_1 + x_2 + 2x_3 = 6 \end{bmatrix}$

A(f) $\begin{bmatrix} 5x_1 + 9x_2 + 5x_3 = 27 \\ 3x_1 + 3x_2 + 2x_3 = 8 \\ 4x_1 + 5x_2 + 3x_3 = 14 \end{bmatrix}$

(g) $\begin{bmatrix} 2x_1 + 5x_2 + 2x_3 = 4 \\ 2x_1 - 3x_2 + 3x_3 = 6 \\ 3x_1 + 7x_2 + 3x_3 = 6 \end{bmatrix}$
(q) $\begin{bmatrix} 5x_1 + 5x_2 + 9x_3 = 0 \\ 2x_1 - 5x_2 - 2x_3 = 7 \\ 6x_1 + 2x_2 + 4x_3 = 5 \end{bmatrix}$

^{A}(h) $\begin{bmatrix} 5x_1 + 4x_2 + 6x_3 = 0 \\ 2x_1 + 5x_2 + 2x_3 = 0 \\ 2x_1 - 4x_2 + 3x_3 = 0 \end{bmatrix}$
^{A}(r) $\begin{bmatrix} -2x_1 + x_2 + 9x_3 = 8 \\ -6x_1 - 3x_2 + 2x_3 = -6 \\ x_1 - 8x_2 + 4x_3 = 4 \end{bmatrix}$

(i) $\begin{bmatrix} 3x_1 + 3x_2 + 2x_3 = 22 \\ 4x_1 + 5x_2 + 6x_3 = 39 \\ 8x_1 + 9x_2 + 9x_3 = 69 \end{bmatrix}$
(s) $\begin{bmatrix} x_1 + x_2 + x_3 = 5 \\ 2x_1 - x_2 + 2x_3 = 1 \\ -x_1 + 2x_2 - x_3 = 4 \end{bmatrix}$

^{A}(j) $\begin{bmatrix} x_1 + x_2 - x_3 = 0 \\ x_1 + x_2 + x_3 = 6 \\ x_1 - x_2 - x_3 = -4 \end{bmatrix}$
^{A}(t) $\begin{bmatrix} 2x_1 - x_2 - x_3 = 0 \\ x_1 + 2x_2 - 3x_3 = 0 \\ 3x_1 - x_2 - 2x_3 = 0 \end{bmatrix}$

(k) $\begin{bmatrix} 8x_1 + 7x_2 + 2x_3 = 14 \\ 3x_1 + 2x_2 + 5x_3 = 2 \\ x_1 + x_2 - x_3 = -3 \end{bmatrix}$
(u) $\begin{bmatrix} -2x_1 + x_2 + x_3 = 3 \\ -x_2 + x_3 = 1 \\ x_1 + x_2 - 2x_3 = -3 \end{bmatrix}$

^{A}(l) $\begin{bmatrix} 2x_1 + 3x_2 + x_3 = 5 \\ 2x_1 + 7x_2 + x_3 = 5 \\ 3x_1 + 2x_2 + x_3 = 0 \end{bmatrix}$
^{A}(v) $\begin{bmatrix} -\frac{2}{3}x_1 + x_2 - x_3 = \frac{1}{6} \\ \frac{1}{2}x_1 - x_2 - \frac{1}{2}x_3 = -1 \\ \frac{3}{4}x_1 - \frac{1}{2}x_2 + \frac{1}{4}x_3 = \frac{5}{4} \end{bmatrix}$

(m) $\begin{bmatrix} 4x_1 + 7x_2 + 2x_3 = -2 \\ 4x_1 + 5x_3 = -3 \\ 7x_1 + 6x_2 + 7x_3 = 12 \end{bmatrix}$
(w) $\begin{bmatrix} -\frac{3}{4}x_1 - \frac{4}{3}x_2 + x_3 = \frac{2}{3} \\ 9x_1 + 16x_2 - 12x_3 = -8 \\ -3x_1 - \frac{16}{3}x_2 + 4x_3 = \frac{8}{3} \end{bmatrix}$

^{A}(n) $\begin{bmatrix} 7x_1 + 3x_2 + 3x_3 = 10 \\ 2x_1 + x_2 + x_3 = 8 \\ 3x_1 + 7x_2 + 2x_3 = -5 \end{bmatrix}$
^{A}(x) $\begin{bmatrix} 3x_1 + x_2 - 2x_3 + 4x_4 = 7 \\ x_1 - x_2 + 3x_3 - x_4 = 6 \\ 2x_1 - 3x_2 + x_3 - 2x_4 = 5 \end{bmatrix}$

(o) $\begin{bmatrix} x_1 + 2x_2 + 2x_3 = -5 \\ 6x_1 + 5x_2 + 4x_3 = 9 \\ 3x_1 + 3x_2 + 2x_3 = -7 \end{bmatrix}$
(y) $\begin{bmatrix} 5x_1 + 4x_2 + x_3 = 0 \\ 7x_1 + 5x_2 + 6x_3 = 0 \\ 7x_1 + 8x_2 - 3x_3 = 0 \end{bmatrix}$

^{A}(p) $\begin{bmatrix} x_1 - 4x_3 = 2 \\ 2x_1 + 2x_2 + 3x_3 = 5 \\ -2x_1 + 4x_2 + x_3 = 6 \end{bmatrix}$
^{A}(z) $\begin{bmatrix} x_1 + x_2 + x_3 + x_4 + x_5 = 15 \\ x_1 - x_2 + x_3 - x_4 - x_5 = 3 \\ x_1 + x_2 + x_3 - x_4 - x_5 = -3 \\ -x_1 - x_2 + x_3 - x_4 + x_5 = 1 \\ x_1 - x_2 - x_3 - x_4 - x_5 = -5 \end{bmatrix}$

2. Given the information in Exercise 3 of Section 2-3 how does AC make up the *high-altitude* expedition packets, which consist of 28 grams of protein, 37 grams of fats, and 33 grams of carbohydrates?

^{A}3. Find the solutions to Problem 6 of Section 2-3 in which the factory at Danbury is used (a) as little as possible, (b) as much as possible.

^{A}4. A manufacturer makes four types of pocket calculators: A, B, C, and

D. Each calculator must pass through three separate assembly procedures: memory assembly (M), arithmetic unit assembly (U), and final assembly (F). The time in person-hours for each type of calculator is shown in Table 4.

Table 4

Calculator	Procedure		
	M	U	F
A	7	1	9
B	3	2	7
C	4	3	10
D	1	4	9

The manufacturer has available 80 person-hours of memory assembly time, 40 of arithmetic unit assembly, and 100 of final assembly time. Can she use up all of this time?

5. The ACME Company makes three kinds of toys: toy A, toy B and the other toy. It makes $2.00 profit on toy A, $5.00 profit on toy B, and $6.00 profit on the other toy. It takes 4 person-hours to make toy A, 3 to make toy B, and 12 to make the other toy. It takes 1 hour to safety-check each toy A, 1 hour to check each toy B, and 3 hours to check each other toy. The management decides to use 100 person-hours and 50 safety check hours to make $40 profit. How many of each toy should they make?

C 6. I was happy when I thought I had solved the system

$$\begin{bmatrix} -3x_1 & -4x_2 & & -1 \\ & 2x_2 & +3x_3 = 1 \\ -2x_1 & & +4x_3 = 2 \end{bmatrix}$$

by the following sequence of steps:

$$\begin{bmatrix} -3 & -4 & 0 & 1 \\ 0 & 2 & 3 & 1 \\ -2 & 0 & 4 & 2 \end{bmatrix} \xrightarrow[R_3 - {}^2\!/_3 R_1]{-{}^1\!/_3 R_1} \begin{bmatrix} 1 & {}^4\!/_3 & 0 & -{}^1\!/_3 \\ 0 & 2 & 3 & 1 \\ 0 & {}^8\!/_3 & 0 & -{}^4\!/_3 \end{bmatrix}$$

$$\xrightarrow[\substack{R_1 - {}^2\!/_3 R_2 \\ {}^1\!/_2 R_2 \\ R_3 - {}^4\!/_3 R_2}]{} \begin{bmatrix} 1 & 0 & -2 & -1 \\ 0 & 1 & {}^3\!/_2 & {}^1\!/_2 \\ 0 & 0 & -4 & -{}^8\!/_3 \end{bmatrix} \xrightarrow[\substack{R_1 - {}^1\!/_2 R_3 \\ R_2 + {}^3\!/_8 R_3 \\ -{}^1\!/_4 R_3}]{} \begin{bmatrix} 1 & 0 & 0 & {}^1\!/_3 \\ 0 & 1 & 0 & -{}^1\!/_2 \\ 0 & 0 & 1 & {}^2\!/_3 \end{bmatrix}$$

Answer: $x_1 = {}^1\!/_3$, $x_2 = -{}^1\!/_2$, $x_3 = {}^2\!/_3$.

Check: $\begin{bmatrix} -3({}^1\!/_3) & -4(-{}^1\!/_2) & & = -1 + 2 = 1 \\ & 2(-{}^1\!/_2) & +3({}^2\!/_3) = -1 + 2 = 1 \\ -2({}^1\!/_3) & & +4({}^2\!/_3) = -{}^2\!/_3 + {}^8\!/_3 = 2 \end{bmatrix}$ ✔ ✔ ✔

Find my error and give the correct solution.

ᴬ7. A friend of mine has had bit parts in several movies. He also likes to go to the movies with me but only if he's not in the picture. Once he said to me, "Did you notice that last year every movie you went to had me in the theater, either on the screen or in a seat next to you? Furthermore, I went to twice as many movies with you as the number I was in. And the number of times you went without me was equal to three times the number of movies I was in." "Very funny," I replied. How many movies did I go to last year?

8. Solve the PHE problem of the last section (Exercise 4, Section 2-3), that is, give all possible "meaningful combinations. Note: PHE stores will sell half-orders of superburgers (called "semi-burgers") and of fishfeasts.

*9. After the robbery the 3 criminals were able to split the loot according to the following conditions. Since Jimmy drove the car and planned the job, he got twice as much as Lefty and Spike put together. Since Spike did the talking, Lefty got $10,000 less than the difference of the cuts for Jimmy and Spike. How much did they steal and what was each individual's take, assuming that Lefty got $1000 more than Spike?

10. The following matrices arose in an attempt to solve a system of equations. For each matrix, determine if further reduction is required. If so, further reduce the matrix and read the answer. Otherwise, read the answer from the matrix as it stands.

ᴬ(a) $\begin{bmatrix} 1 & 0 & 3 \\ 0 & 1 & 2 \\ 0 & 0 & 3 \end{bmatrix}$

(b) $\begin{bmatrix} 1 & 0 & 1 & 3 \\ 0 & 1 & 0 & 2 \\ 0 & 0 & 1 & 2 \end{bmatrix}$

ᴬ(c) $\begin{bmatrix} 1 & 0 & 0 & 0 & 2 \\ 0 & 0 & 1 & 0 & 3 \\ 0 & 0 & 0 & 1 & 2 \\ 0 & 0 & 0 & 0 & 0 \end{bmatrix}$

(d) $\begin{bmatrix} 1 & 1 & 0 & 0 & 0 & 2 \\ 0 & 0 & 1 & 1 & 0 & 3 \\ 0 & 0 & 0 & 0 & 1 & 4 \\ 0 & 0 & 0 & 0 & 0 & 0 \end{bmatrix}$

ᴬ(e) $\begin{bmatrix} 0 & 0 & 0 & 1 & 0 & 4 \\ 0 & 1 & 0 & 0 & 2 & 3 \\ 1 & 0 & 3 & 0 & 0 & 2 \end{bmatrix}$

(f) $\begin{bmatrix} 2 & 3 & 0 & 2 & 4 \\ 1 & 0 & 1 & 0 & 2 \\ 0 & 0 & 0 & 1 & 3 \\ 0 & 0 & 0 & 0 & 1 \end{bmatrix}$

ᴬ(g) $\begin{bmatrix} 2 & 3 & 0 & 2 & 4 \\ 1 & 0 & 1 & 0 & 2 \\ 0 & 0 & 0 & 1 & 3 \\ 0 & 0 & 0 & 0 & 0 \end{bmatrix}$

(h) $\begin{bmatrix} 1 & 1 & 1 & 1 \\ 1 & 1 & 1 & 2 \\ 1 & 1 & 1 & 3 \\ 1 & 1 & 1 & 4 \end{bmatrix}$

ᴬ(i) $\begin{bmatrix} 1 & 0 & 2 & 0 & 1 & 3 \\ 0 & 1 & 1 & 0 & 0 & 2 \\ 0 & 0 & 0 & 1 & 3 & 4 \\ 0 & 0 & 0 & 0 & 1 & 2 \end{bmatrix}$

(j) $\begin{bmatrix} 1 & 0 & 1 & 0 & 3 \\ 2 & 1 & 0 & 0 & 4 \\ 3 & 0 & 0 & 1 & 5 \end{bmatrix}$

Section 2-5
Shortcuts in Gauss-Jordan Reduction

Gaussian reduction (see below) is faster for computers by a ratio of 3 to 2.

While other techniques for solving systems of equations are known, none are nearly as fast or efficient for hand calculation as Gauss-Jordan reduction. As efficient as Gauss-Jordan reduction is, however, it often can be improved significantly by several slight modifications, which we are about to describe below. Since these shortcuts depend upon Gauss-Jordan reduction, you cannot use them properly without first having a good command of the technique described in the flow chart on p. 87. In fact, the un-considered use of these techniques is totally ineffective. The effective use of shortcuts depends on taking advantage of the particular arrangement of numbers in a given problem. This is why the following methods could not be included in the flow chart.

Once you begin to understand the way in which Gauss-Jordan reduction successively eliminates variables, you will be ready to begin to use the following tricks. It should be stated, however, that none of these tricks are needed to solve any system; they merely help to shorten the work for some types of students. If you are confused by this section, retreat to the flow chart on p. 87.

Some of these shortcuts are of greater psychological value than actual computational value; it is a matter of personal taste whether or not you will want to use them. Others actually do shorten the work.

Shortcut 1   *Do not routinely require there to be a 1 in a cleared column.*

Problem 1   Solve $\begin{bmatrix} 3x_1 + x_2 = 7 \\ 2x_1 + 3x_2 = 9 \end{bmatrix}$

Solution
$$\begin{bmatrix} 3 & 1 & 7 \\ 2 & 3 & 9 \end{bmatrix} \xrightarrow{R_2 - \frac{2}{3}R_1} \begin{bmatrix} 3 & 1 & 7 \\ 0 & \frac{7}{3} & \frac{13}{3} \end{bmatrix} \xrightarrow{R_1 - \frac{3}{7}R_2} \begin{bmatrix} 3 & 0 & \frac{36}{7} \\ 0 & \frac{7}{3} & \frac{13}{2} \end{bmatrix}$$ **1**

The first row corresponds to

$$3x_1 = \frac{36}{7}$$

or

$$x_1 = \frac{12}{7}$$

The second row corresponds to

$$\frac{7}{3}x_2 = \frac{13}{3}$$

Check:
$3(\frac{12}{7}) + \frac{13}{7} = 7$ ✔
$2(\frac{12}{7}) + 3(\frac{13}{7}) = 9$ ✔

or

$$x_2 = \frac{13}{7}$$

The point here is that after a little practice the answer can simply be read from the final matrix. Do the division in your head. As a further example, suppose the final matrix is:

$$\begin{bmatrix} 4 & 3 & 0 & 7 & 2 \\ 0 & 0 & 3 & -8 & 5 \end{bmatrix}$$

Again, we divide mentally and simply write the answer:

$$x_1 = \tfrac{1}{4}(2 - 3x_2 - 7x_4)$$
$$x_3 = \tfrac{1}{3}(5 + 8x_4)$$

Shortcut 2   *Multiply to avoid fractions.*

Problem 2   Solve $\begin{bmatrix} 2x_1 - x_2 + x_3 = -2 \\ x_1 + 2x_2 - x_3 = 3 \\ -x_1 + x_2 + 2x_3 = 2 \end{bmatrix}$

Solution

$$\begin{bmatrix} 2 & -1 & 1 & -2 \\ 1 & 2 & -1 & 3 \\ -1 & 1 & 2 & 2 \end{bmatrix} \xrightarrow[2R_3]{2R_2} \begin{bmatrix} 2 & -1 & 1 & -2 \\ 2 & 4 & -2 & 6 \\ -2 & 2 & 4 & 4 \end{bmatrix}$$

$$\xrightarrow[R_3 + R_1]{R_2 - R_1} \begin{bmatrix} 2 & -1 & 1 & -2 \\ 0 & 5 & -3 & 8 \\ 0 & 1 & 5 & 2 \end{bmatrix}$$

$$\xrightarrow[5R_3]{5R_1} \begin{bmatrix} 10 & -5 & 5 & -10 \\ 0 & 5 & -3 & 8 \\ 0 & 5 & 25 & 10 \end{bmatrix} \xrightarrow[R_3 - R_2]{R_1 + R_2} \begin{bmatrix} 10 & 0 & 2 & -2 \\ 0 & 5 & -3 & 8 \\ 0 & 0 & 28 & 2 \end{bmatrix}$$

$$\xrightarrow[28R_2]{14R_1} \begin{bmatrix} 140 & 0 & 28 & -28 \\ 0 & 140 & -84 & 224 \\ 0 & 0 & 28 & 2 \end{bmatrix} \xrightarrow[R_2 + 3R_3]{R_1 - R_3} \begin{bmatrix} 140 & 0 & 0 & -30 \\ 0 & 140 & 0 & 230 \\ 0 & 0 & 28 & 2 \end{bmatrix}$$

Thus $x_1 = -\tfrac{3}{14}$, $x_2 = \tfrac{23}{14}$, $x_3 = \tfrac{1}{14}$.

Check: $\begin{bmatrix} 2(-\tfrac{3}{14}) - 1(\tfrac{23}{14}) + 1(\tfrac{1}{14}) = -2 \\ 1(-\tfrac{3}{14}) + 2(\tfrac{23}{14}) - 1(\tfrac{1}{14}) = 3 \\ -1(-\tfrac{3}{14}) + 1(\tfrac{23}{14}) + 2(\tfrac{1}{14}) = 2 \end{bmatrix}$ ✔ ✔ ✔

Shortcut 3   *Substitute to find the other variables.*

This technique is called Gaussian reduction. While this method is quicker for computers, for hand computation it is generally more time consuming than Gauss-Jordan reduction. To use it, clear only the part of column C that is below R.

Problem 3   Solve $\begin{bmatrix} x_1 + 2x_2 - x_3 = 3 \\ 2x_1 + x_2 + x_3 = 0 \\ 3x_1 + 6x_2 + 2x_3 = 14 \end{bmatrix}$

Solution

$$\begin{bmatrix} 1 & 2 & -1 & 3 \\ 2 & 1 & 1 & 0 \\ 3 & 6 & 2 & 14 \end{bmatrix} \begin{array}{l} \\ R_2 - 2R_1 \\ R_3 - 3R_1 \end{array} \longrightarrow \begin{bmatrix} 1 & 2 & -1 & 3 \\ 0 & -3 & 3 & -6 \\ 0 & 0 & 5 & 5 \end{bmatrix}$$

From the last row we see that $x_3 = 1$. Substituting this into the equation corresponding to row 2 yields

$$-3x_2 + 3 \cdot 1 = -6$$
$$-3x_2 = -9$$
$$x_2 = 3$$

Finally, substituting the two known values into the equation corresponding to row 1 yields

$$x_1 + 2 \cdot 3 - 1 \cdot 1 = 3$$
$$x_1 + 5 - \quad\quad 3$$
$$x_1 = -2$$

Check: $\begin{bmatrix} 1(-2) + 2(3) - 1(1) = 3 \\ 2(-2) + 1(3) + 1(1) = 0 \\ 3(-2) + 6(3) + 2(1) = 14 \end{bmatrix}$ ✔ ✔ ✔

Shortcut 4   *Combine row operations.*

Problem 4   Solve $\begin{bmatrix} 3x_1 + 2x_2 \quad x_3 = 1 \\ 2x_1 + 4x_2 - 2x_3 = -1 \\ 4x_1 - 5x_2 + x_3 = 3 \end{bmatrix}$

Solution

$$\begin{bmatrix} 3 & 2 & -1 & 1 \\ 2 & 4 & -2 & -1 \\ 4 & -5 & 1 & 3 \end{bmatrix} \begin{array}{l} \\ 3R_2 - 2R_1 \\ 3R_3 - 4R_1 \end{array} \longrightarrow \begin{bmatrix} 3 & 2 & -1 & 1 \\ 0 & 8 & -4 & -5 \\ 0 & -23 & 7 & 5 \end{bmatrix}$$

$$\begin{array}{l} 8R_1 - 2R_2 \\ \\ 8R_3 + 23R_2 \end{array} \longrightarrow \begin{bmatrix} 24 & 0 & 0 & 18 \\ 0 & 8 & -4 & -5 \\ 0 & 0 & -36 & -75 \end{bmatrix}$$

$$\begin{array}{l} 36R_2 - 4R_3 \end{array} \longrightarrow \begin{bmatrix} 24 & 0 & 0 & 18 \\ 0 & 288 & 0 & 120 \\ 0 & 0 & -36 & -75 \end{bmatrix}$$

Thus $x_1 = \frac{9}{12}$, $x_2 = \frac{5}{12}$, $x_3 = \frac{25}{12}$. Rather than reducing $\frac{9}{12}$ to $\frac{3}{4}$ it is easier to keep a common denominator for checking:

$$\begin{bmatrix} 3(\frac{9}{12}) + 2(\frac{5}{12}) - \frac{25}{12} & = & 1 \\ 2(\frac{9}{12}) + 4(\frac{5}{12}) - 2(\frac{25}{12}) & = & -1 \\ 4(\frac{9}{12}) - 5(\frac{5}{12}) + \frac{25}{12} & = & 3 \end{bmatrix} \; \checkmark \checkmark \checkmark$$

Shortcut 5   *Use 1's effectively.*

Shortcut 6   *Never switch rows.*

These techniques work together and are by far the most useful shortcuts available. They are easy to use when the system has a unique solution.

Problem 5   Solve $\begin{bmatrix} 2x_1 + 3x_2 + 7x_3 = 15 \\ 5x_1 + 4x_2 - 4x_3 = -9 \\ -2x_1 + x_2 + 2x_3 = 7 \end{bmatrix}$

Solution   The matrix of the system is $\begin{bmatrix} 2 & 3 & 7 & 15 \\ 5 & 4 & -4 & -9 \\ -2 & 1 & 2 & 7 \end{bmatrix}$

Instead of using the $(1,1)$ entry to clear the first column, we use the $(3,2)$ entry to clear the second column. This is an allowable operation and since it uses a 1 to clear the column, it will not involve fractions:

$\begin{matrix} R_1 - 3R_3 \\ R_2 - 4R_3 \end{matrix}$ $\begin{bmatrix} 8 & 0 & 1 & -6 \\ 13 & 0 & -12 & -37 \\ -2 & 1 & 2 & 7 \end{bmatrix}$

We now use the $(1,3)$ entry to clear the third column:

$\begin{matrix} R_2 + 12R_1 \\ R_3 - 2R_1 \end{matrix}$ $\begin{bmatrix} 8 & 0 & 1 & -6 \\ 109 & 0 & 0 & -109 \\ -18 & 1 & 0 & 19 \end{bmatrix}$

We now perform $\frac{1}{109}R_2$ and then use the $(2,1)$ entry to clear the first column:

$\frac{1}{109}R_2$ $\begin{bmatrix} 8 & 0 & 1 & -6 \\ 1 & 0 & 0 & -1 \\ -18 & 1 & 0 & 19 \end{bmatrix}$ $\begin{matrix} R_1 - 8R_2 \\ \\ R_2 + 18R_2 \end{matrix}$ $\begin{bmatrix} 0 & 0 & 1 & 2 \\ 1 & 0 & 0 & -1 \\ 0 & 1 & 0 & 1 \end{bmatrix}$

Without switching rows, we obtain the answer:

$$x_1 = -1, \; x_2 = 1, \; x_3 = 2$$

$$\text{Check:} \begin{bmatrix} 2(-1) + 3(1) + 7(2) = & 15 \\ 5(-1) + 4(1) - 4(2) = & -9 \\ -2(-1) + 1(1) + 2(2) = & 7 \end{bmatrix} \begin{matrix} ✔ \\ ✔ \\ ✔ \end{matrix}$$

It is important to remember that once you have used a row to clear a column, you can never use that row again.

This same idea can be used even though the initial matrix does not contain a 1.

Problem 6   Solve $\begin{bmatrix} 3x_1 + 4x_2 + 3x_3 = & 7 \\ 4x_1 + 6x_2 - 3x_3 = -11 \\ 2x_1 - 8x_2 + 4x_3 = & 0 \end{bmatrix}$

Solution   The matrix of the system is $\begin{bmatrix} 3 & 4 & 3 & 7 \\ 4 & 6 & -3 & -11 \\ 2 & -8 & 4 & 0 \end{bmatrix}$

There are many ways to proceed. We decide to use a preliminary operation, $R_2 - R_1$, to obtain a 1 at address $(2,1)$:

$R_2 - R_1$ $\begin{bmatrix} 3 & 4 & 3 & 7 \\ 1 & 2 & -6 & -18 \\ 2 & -8 & 4 & 0 \end{bmatrix}$

Using this 1 as before, we obtain

$R_1 - 3R_2$ $\begin{bmatrix} 0 & -2 & 21 & 61 \\ 1 & 2 & -6 & -18 \\ 0 & -12 & 16 & 36 \end{bmatrix}$
$R_3 - 2R_2$

We now use the $(1,2)$ entry to clear the second column:

$R_2 + R_1$ $\begin{bmatrix} 0 & -2 & 21 & 61 \\ 1 & 0 & 15 & 43 \\ 0 & 0 & -110 & -330 \end{bmatrix}$
$R_3 - 6R_1$

$\begin{bmatrix} 0 & -2 & 21 & 61 \\ 1 & 0 & 15 & 43 \\ 0 & 0 & 1 & 3 \end{bmatrix}$ $\begin{matrix} R_1 - 21R_3 \\ R_2 - 15R_3 \end{matrix}$ $\begin{bmatrix} 0 & -2 & 0 & -2 \\ 1 & 0 & 0 & -2 \\ 0 & 0 & 1 & 3 \end{bmatrix}$
$-\frac{1}{110}R_3$

Thus $x_1 = -2$, $x_2 = 1$, $x_3 = 3$.

$$\text{Check:} \begin{bmatrix} 3(-2) + 4\cdot1 + 3\cdot3 = & 7 \\ 4(-2) + 6\cdot1 - 3\cdot3 = -11 \\ 2(-2) - 8\cdot1 + 4\cdot3 = & 0 \end{bmatrix} \begin{matrix} ✔ \\ ✔ \\ ✔ \end{matrix}$$

Preliminary row operations are useful not only for obtaining 1's but also for reducing the size of the numbers in the matrix. A row that is used in such a preliminary operation may, of course, be used later to clear a column.

Shortcuts 5 and 6 can also be used when the system has multiple answers. However, in this case the answer is not read according to the instructions following the flow chart on p. 87. While it is possible to give a very formal description of how to proceed, we will limit ourselves to some general advice:

(a) It is never necessary to clear more columns than there are rows.

(b) Keeping this in mind, clear columns until further reduction no longer simplifies the matrix.

(c) A column is cleared if it is not the last column and if it contains exactly one nonzero entry. Choose as many cleared columns as possible subject to the condition that no two chosen columns may have their nonzero entry in the same row.

(d) The variables corresponding to the columns chosen in (c) are determined by the remaining variables (which are free).

Problem 7 Here is the final matrix that arose in trying to solve a system of equations. Read the answer.

$$\begin{bmatrix} 3 & 0 & 0 & 2 & 0 & -11 \\ 2 & -3 & 4 & 0 & 0 & 7 \\ 1 & 0 & 0 & 0 & 1 & 18 \end{bmatrix}$$

Solution Columns C_2, C_3, C_4, and C_5 are cleared. According to (c) above, we may choose either C_2, C_4, and C_5 or C_3, C_4, and C_5. The first choice determines the variables x_2, x_4, x_5 in terms of the other variables, x_1 and x_3.

x_2 is given by R_2:

$$x_2 = -\tfrac{1}{3}(7 - 2x_1 - 4x_3)$$

x_4 is given by R_1:

$$x_4 = \tfrac{1}{2}(-11 - 3x_1)$$

x_5 is given by R_3:

$$x_5 = 18 - x_1$$

This gives the solution $\begin{bmatrix} x_1 \\ -\tfrac{1}{3}(7 - 2x_1 - 4x_3) \\ x_3 \\ \tfrac{1}{2}(-11 - 3x_1) \\ 18 - x_1 \end{bmatrix}$, where x_1 and x_3 are free.

The second choice determines the variables x_3, x_4, and x_5 in terms of the other variables, x_1 and x_2.

x_3 is given by R_2:

$$x_3 = \tfrac{1}{4}(7 - 2x_1 + 3x_2)$$

x_4 is given by R_1:

$$x_4 = \tfrac{1}{2}(-11 - 3x_1)$$

x_5 is given by R_3:

$$x_5 = 18 - x_1$$

This gives the solution $\begin{bmatrix} x_1 \\ x_2 \\ \tfrac{1}{4}(7 - 2x_1 + 3x_2) \\ \tfrac{1}{2}(-11 - 3x_1) \\ 18 - x_1 \end{bmatrix}$, where x_1 and x_2 are free.

In general, problems with multiple answers will have many equivalent answers which appear in quite different forms. However, any two equivalent answers will have the same number of free variables.

Shortcut 7   *Use "Error Chasing."*

This technique can save a lot of time. Let us consider the following sequence of steps to "solve" Problem 2:

Problem 2′   Solve $\begin{bmatrix} 2x_1 - x_2 + x_3 = -2 \\ x_1 + 2x_2 - x_3 = 3 \\ -x_1 + x_2 + 2x_3 = 2 \end{bmatrix}$

Solution

$$\begin{bmatrix} 2 & -1 & 1 & -2 \\ 1 & 2 & -1 & 3 \\ -1 & 1 & 2 & 2 \end{bmatrix} \begin{array}{l} \\ 2R_2 \\ 2R_3 \end{array} \rightarrow \begin{bmatrix} 2 & -1 & 1 & -2 \\ 2 & 4 & -2 & 6 \\ -2 & 2 & 4 & 4 \end{bmatrix}$$

$$\begin{array}{l} \\ R_2 - R_1 \\ R_3 + R_1 \end{array} \begin{bmatrix} 2 & -1 & 1 & -2 \\ 0 & 5 & -3 & 8 \\ 0 & 1 & 5 & 2 \end{bmatrix} \begin{array}{l} 5R_1 \\ \\ 5R_3 \end{array} \rightarrow \begin{bmatrix} 10 & -5 & 5 & -10 \\ 0 & 5 & -3 & 8 \\ 0 & 5 & 25 & 10 \end{bmatrix}$$

$$\begin{array}{l} R_1 + R_2 \\ \\ R_3 - R_2 \end{array} \begin{bmatrix} 10 & 0 & 2 & -2 \\ 0 & 5 & -3 & 8 \\ 0 & 0 & 28 & 2 \end{bmatrix} \begin{array}{l} 14R_1 \\ 28R_2 \\ \\ \end{array} \rightarrow \begin{bmatrix} 140 & 0 & 28 & -28 \\ 0 & 140 & -84 & 234 \\ 0 & 0 & 28 & 2 \end{bmatrix}$$

$$\begin{array}{l} R_1 - R_3 \\ R_2 + 3R_3 \end{array} \begin{bmatrix} 140 & 0 & 0 & -30 \\ 0 & 140 & 0 & 240 \\ 0 & 0 & 28 & 2 \end{bmatrix}$$

Thus, $x_1 = -3/14$, $x_2 = 24/14$, $x_3 = 1/14$.

Check:
$$\begin{bmatrix} 2(-3/14) - 1(24/14) + 1(1/14) = -29/14 \end{bmatrix} \text{ Oops!}$$

Most people would grit their teeth and start over again. This approach is generally a mistake. Most of this solution is probably all right. To do these seven steps over again might just introduce *new* errors. Even rechecking the arithmetic is not as good as substituting the answer into each matrix starting with the last. The answer obviously checks in the last matrix. The next to last matrix corresponds to the system

$$\begin{bmatrix} 140x_1 & + 28x_3 = -28 \\ 140x_2 - 84x_3 = 234 \\ 28x_3 = 2 \end{bmatrix}$$

Since this system must have the same answer as each system in the chain, our answer must check here as well:

$$\begin{bmatrix} 140(-3/14) & + 28(1/14) = -30 + 2 = -28 & \checkmark \\ 140(24/14) - 84(1/14) = 240 - 6 = 234 & \checkmark \\ 28(1/14) = 2 & \checkmark \end{bmatrix}$$

Thus no error was made in reducing this system. We check our answer in the next matrix:

$$\begin{bmatrix} 140(-3/14) & + 2(1/14) = (-30+2)/14 = -2 & \checkmark \\ 5(24/14) - 3(1/14) = 117/14 & = 85/14 & \text{Wrong} \end{bmatrix}$$

An error has been caught. A mistake was made in going from this matrix to the next. The correct step is

$$\begin{bmatrix} 10 & 0 & 2 & -2 \\ 0 & 5 & -3 & 8 \\ 0 & 0 & 28 & 2 \end{bmatrix} \begin{matrix} 14R_1 \\ 28R_2 \\ \end{matrix} \Rightarrow \begin{bmatrix} 140 & 0 & 0 & -28 \\ 0 & 140 & -84 & 224 \\ 0 & 0 & 28 & 2 \end{bmatrix}$$

We complete the solution:

$$\begin{matrix} R_1 - R_3 \\ R_2 + 3R_3 \end{matrix} \Rightarrow \begin{bmatrix} 140 & 0 & 0 & -30 \\ 0 & 140 & 0 & 230 \\ 0 & 0 & 28 & 2 \end{bmatrix},$$

find that $x_1 = -3/14$, $x_2 = 23/14$, and $x_3 = 1/14$, and check again.

Check:
$$\begin{bmatrix} 2(-3/14) - 1(23/14) + 1(1/14) = -2 & \checkmark \\ 1(-3/14) + 2(23/14) - 1(1/14) = 3 & \checkmark \\ -1(-3/14) + 1(23/14) + 2(1/14) = 2 & \checkmark \end{bmatrix}$$

This time we have the correct answer. If it failed to check again, we would proceed as before until we found another error.

Obviously, if you have made ten arithmetic mistakes, this procedure is slower than starting over. However, it is often the case that a wrong answer occurs because of a single mistake near the end of the process.

Exercises 2-5

1. Select five phone numbers at random from the directory. Discarding the prefixes, write four of the remaining four-digit numbers as rows in a coefficient matrix. Use the fifth number for the requirement column. (For example, the numbers 999-3838, 555-1212, 203-7629,

$$312\text{-}8207, 205\text{-}3291 \text{ result in the matrix} \begin{bmatrix} 3 & 8 & 3 & 8 & 3 \\ 1 & 2 & 1 & 2 & 2 \\ 7 & 6 & 2 & 9 & 9 \\ 8 & 2 & 0 & 7 & 1 \end{bmatrix}.) \text{ Solve the}$$

resulting system of equations.

2. Repeat Exercise 1 a few times. An easier problem results if you use only four phone numbers, decide upon an answer, say $x_1 = -1$, $x_2 = 3$, $x_3 = -2$, $x_4 = 1$, and substitute to find the last column. For ex-

$$\text{ample, in the previous problem take} \begin{bmatrix} 3 & 8 & 3 & 8 \\ 1 & 2 & 1 & 2 \\ 7 & 6 & 2 & 9 \\ 8 & 2 & 0 & 7 \end{bmatrix} \text{ and substitute}$$

an answer picked at random:

$$3(-1) + 8(3) + 3(-2) + 8(1) = 23$$
$$1(-1) + 2(3) + 1(-2) + 2(1) = 5$$
$$7(-1) + 6(3) + 2(-2) + 9(1) = 16$$
$$8(-1) + 2(3) + 0(-2) + 7(1) = 5$$

Now solve the system

$$\begin{bmatrix} 3x_1 + 8x_2 + 3x_3 + 8x_4 = 23 \\ x_1 + 2x_2 + x_3 + 2x_4 = 5 \\ 7x_1 + 6x_2 + 2x_3 + 9x_4 = 16 \\ 8x_1 + 2x_2 + 7x_4 = 5 \end{bmatrix}$$

Making up and solving exercises is actually the best way to learn mathematics.

In this way you will have a problem in which you not only know the answer but also are sure that the answer does not involve fractions (although the reduction may).

3. As a variation of Problem 2, introduce some minus signs in the coefficient matrix.

ᴬ4. PHE is preparing to ship some display materials to an exhibitor who requires 181 flags, 150 banners, 142 shields, 155 windmills, and 106 pennants. It has on hand five types of packets of display materials, stocked as shown in Table 5. How many of each type do they ship?

Table 5

Packet type	Flags	Banners	Shields	Windmills	Pennants
I	20	5	11	7	16
II	16	12	4	10	5
III	4	11	15	12	7
IV	3	10	12	13	4
V	8	11	19	11	2

*5. Solve:

$$\begin{bmatrix} & & & & x_5 & & + x_7 - x_8 = & 2 \\ & x_2 & & - x_4 & & + x_6 & = & 1 \\ & & x_3 & & + x_5 & & - x_7 & = & 0 \\ x_1 + x_2 & & & & & & = & -1 \\ & & x_3 & & - x_5 + x_6 & & = & -2 \\ & & & x_4 & & - x_6 & + x_8 = & 0 \\ & x_2 - x_3 + x_4 & & & & = & -1 \\ & & & & - x_5 & & + x_7 - x_8 = & 2 \\ x_1 - x_2 + x_3 - x_4 + x_5 - x_6 & & & & = & -1 \\ & & - x_3 & & + x_5 + x_6 - x_7 & & = & 1 \end{bmatrix}$$

Section 2-6
The Geometry of Linear Equations in Three Variables

This section is designed to increase your understanding of the various possible outcomes of Gauss-Jordan reduction. It will also be helpful in your study of the simplex method (Chapter 5). There are no new techniques of practical importance in this section; you are not necessarily expected to become proficient at drawing graphs in three-dimensional space. Linear equations in four, five, or more variables also have geometrical interpretations as graphs (called "hyperplanes") in four-dimensional, five-dimensional, or higher-dimensional "space," but since no one can draw higher-dimensional pictures, we will not explore that topic!

Coordinates in Space

Figure 17

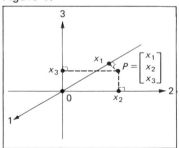

Just as we extended our discussion of coordinates for a line (the one-variable case) to a discussion of coordinates for the plane (the two-variable case), we will now go further, setting up a coordinate system in space (the three-variable case) by taking three lines through a point 0, each line perpendicular to the other two (Figure 17). Establish a coordinate system on each of these lines, each with origin 0, and call these lines, in some order, the 1-axis, 2-axis, and 3-axis. Now suppose P is a point in space. Draw the lines from P that are perpendicular to each of the three axes. Say these lines meet the 1-, 2-, and 3-axes at points with coordinates x_1, x_2, and x_3, respectively. These numbers are called the *coordinates* of P, and we write

$$P = \begin{bmatrix} x_1 \\ x_2 \\ x_3 \end{bmatrix} \qquad\qquad \mathbf{1}$$

In most areas of mathematics it is customary or more convenient to write $P = (x_1, x_2, x_3)$, but for our work the notation in equation **1** is best.

Points on the 1-, 2-, and 3-axis have coordinates of the form $\begin{bmatrix} x_1 \\ 0 \\ 0 \end{bmatrix}$, $\begin{bmatrix} 0 \\ x_2 \\ 0 \end{bmatrix}$, and $\begin{bmatrix} 0 \\ 0 \\ x_3 \end{bmatrix}$, respectively. In particular, the origin has coordinates

$$\begin{bmatrix} 0 \\ 0 \\ 0 \end{bmatrix} \qquad\qquad \mathbf{2}$$

Figure 18

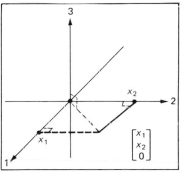

The plane containing the 1-axis and the 2-axis is called the **1-2 plane**. The points in the 1-2 plane all have coordinates of the form $\begin{bmatrix} x_1 \\ x_2 \\ 0 \end{bmatrix}$ (Figure 18). Similarly, the points in the 1-3 plane and the 2-3 plane all have coordinates of the form $\begin{bmatrix} x_1 \\ 0 \\ x_3 \end{bmatrix}$ and $\begin{bmatrix} 0 \\ x_2 \\ x_3 \end{bmatrix}$, respectively.

Consider the linear equation

$$3x_1 - 2x_2 + \tfrac{1}{2}x_3 = 2 \qquad\qquad \mathbf{3}$$

The **graph** of this equation consists of all points $\begin{bmatrix} x_1 \\ x_2 \\ x_3 \end{bmatrix}$ whose coordinates satisfy equation **3**; in other words, it consists of all points satisfying

$$\begin{bmatrix} 3 & -2 & \tfrac{1}{2} \end{bmatrix} \begin{bmatrix} x_1 \\ x_2 \\ x_3 \end{bmatrix} = 2 \qquad\qquad \mathbf{3'}$$

Are there any such points on the 1-axis? To find out, set $x_2 = 0$ and $x_3 = 0$ in equation **3**. Then equation **3** becomes $3x_1 = 2$, that is, $x_1 = \tfrac{2}{3}$. So there is one such point on the 1-axis: $\begin{bmatrix} \tfrac{2}{3} \\ 0 \\ 0 \end{bmatrix}$. On the 2-axis, we set $x_1 = 0$ and $x_3 = 0$ in equation **3** to obtain $-2x_2 = 2$,

Figure 19

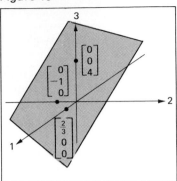

$x_2 = -1$; the point $\begin{bmatrix} 0 \\ -1 \\ 0 \end{bmatrix}$ is on the graph. On the 3-axis we set

$x_1 = 0$ and $x_2 = 0$ in equation **3** to obtain $\frac{1}{2}x_3 = 2$, $x_3 = 4$; the

point $\begin{bmatrix} 0 \\ 0 \\ 4 \end{bmatrix}$ is on the graph. These three points determine a plane

(Figure 19). It can be shown that *this plane is the graph of equation 3*. In general, the graph of any linear equation in three variables,

$$ax_1 + bx_2 + cx_3 = d \qquad \qquad \textbf{4}$$

where a, b, c, and d are given numbers and a, b, and c are not all zero, is always a plane. (Compare this with the statement made about equation **5** in Section 2-1). In matrix form this becomes

$$[a \quad b \quad c] \begin{bmatrix} x_1 \\ x_2 \\ x_3 \end{bmatrix} = d, \text{ where } [a \quad b \quad c] \neq [0 \quad 0 \quad 0]$$

If in equation **4** it happens that $a = 0$, $b = 0$, or $c = 0$, then the plane is parallel to the 1-axis, 2-axis, or 3-axis, respectively. For example, the plane with equation

$$2x_1 - 3x_3 = 6 \qquad \qquad \textbf{5}$$

Figure 20

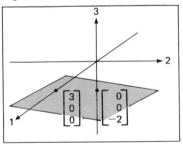

crosses the 1-axis at $\begin{bmatrix} 3 \\ 0 \\ 0 \end{bmatrix}$, is parallel to the 2-axis, and crosses the

3-axis at $\begin{bmatrix} 0 \\ 0 \\ -2 \end{bmatrix}$ (Figure 20). Another example: the plane with equa-

tion

$$[0 \quad 1 \quad 0] \begin{bmatrix} x_1 \\ x_2 \\ x_3 \end{bmatrix} = 4 \qquad \qquad \textbf{6}$$

Figure 21

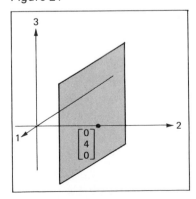

is parallel to the 1-axis and the 3-axis (and hence is parallel to the

1-3 plane), and it crosses the 2-axis at $\begin{bmatrix} 0 \\ 4 \\ 0 \end{bmatrix}$ (Figure 21). Note that

the cautions given on p. 66 regarding missing variables also apply here.

Now a system of two linear equations in three variables represents two planes in space. There are three possibilities:

(i) the planes are parallel,

(ii) the planes intersect in a line, or

(iii) the planes are identical, that is, the two equations represent the same plane.

These possibilities correspond to the possible solutions of the system:

(i') no solution,

(ii') a solution involving one free variable, or

(iii') a solution involving two free variables.

Problem 1 Solve and interpret geometrically the following systems.

(a) $\begin{bmatrix} x_1 - 2x_2 + x_3 = -2 \\ 2x_1 - 4x_2 + 2x_3 = 3 \end{bmatrix}$

(b) $\begin{bmatrix} x_1 - 2x_2 + x_3 = -2 \\ 2x_1 + 4x_2 + x_3 = 0 \end{bmatrix}$

(c) $\begin{bmatrix} x_1 - 2x_2 + x_3 = -2 \\ 2x_1 \quad 4x_2 + 2x_3 = -4 \end{bmatrix}$

Solution (a) $\begin{bmatrix} 1 & -2 & 1 & -2 \\ 2 & -4 & 2 & 3 \end{bmatrix}$ ⟹ $\begin{bmatrix} 1 & -2 & 1 & -2 \\ 0 & 0 & 0 & 7 \end{bmatrix}$

No solution. The planes are parallel (Figure 22).

Figure 22

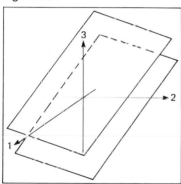

(b) $\begin{bmatrix} 1 & -2 & 1 & -2 \\ 2 & 4 & 1 & 0 \end{bmatrix}$ $\begin{bmatrix} 1 & -2 & 1 & -2 \\ 0 & 8 & -1 & 4 \end{bmatrix}$

$\begin{bmatrix} 1 & 0 & 3/4 & -1 \\ 0 & 1 & -1/8 & 1/2 \end{bmatrix}$

Answer: $\begin{bmatrix} -1 - 3/4 x_3 \\ 1/2 + 1/8 x_3 \\ x_3 \end{bmatrix}$ where x_3 is free. The planes intersect in a line (Figure 23).

Figure 23

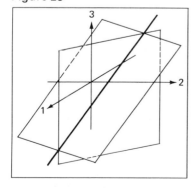

(c) $\begin{bmatrix} 1 & -2 & 1 & -2 \\ 2 & -4 & 2 & -4 \end{bmatrix}$ $\begin{bmatrix} 1 & -2 & 1 & -2 \\ 0 & 0 & 0 & 0 \end{bmatrix}$

Answer: $\begin{bmatrix} -2 + 2x_2 - x_3 \\ x_2 \\ x_3 \end{bmatrix}$, where x_2 and x_3 are free. The planes are identical (both are the lower plane in Figure 22).

Figure 24

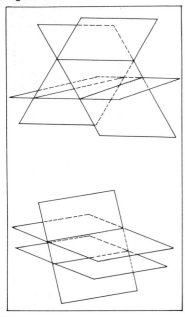

If we consider three or more planes in space, then there are two further cases:

(iv) the planes intersect in one point

This corresponds to the following possibility for a system of three or more equations in three variables:

(iv′) the system has exactly one solution, that is, there are no free variables.

The final case is

(v) the planes are not all parallel, but they have no common point

Case v can occur in many different ways (see, for example, Figure 24). Fortunately we do not have to worry about these possibilities in connection with the corresponding system of three or more linear equations in three variables, because they all correspond to possibility i′ above: the system has no solution.

Problem 2 Solve and interpret geometrically:

$$\begin{bmatrix} x_1 - 2x_2 + x_3 = -2 \\ x_1 - 2x_2 - x_3 = 0 \\ x_1 + 2x_2 + x_3 = 1 \end{bmatrix}$$

Solution

$$\begin{bmatrix} 1 & -2 & 1 & -2 \\ 1 & -2 & -1 & 0 \\ 1 & 2 & 1 & 1 \end{bmatrix} \Rightarrow \begin{bmatrix} 1 & -2 & 1 & -2 \\ 0 & 0 & -2 & 2 \\ 0 & 4 & 0 & 3 \end{bmatrix}$$

$$\Rightarrow \begin{bmatrix} 1 & -2 & 1 & -2 \\ 0 & 4 & 0 & 3 \\ 0 & 0 & -2 & 2 \end{bmatrix} \Rightarrow \begin{bmatrix} 1 & 0 & 1 & -1/2 \\ 0 & 4 & 0 & 3 \\ 0 & 0 & -2 & 2 \end{bmatrix}$$

$$\Rightarrow \begin{bmatrix} 1 & 0 & 1 & -1/2 \\ 0 & 1 & 0 & 3/4 \\ 0 & 0 & -2 & 2 \end{bmatrix} \Rightarrow \begin{bmatrix} 1 & 0 & 0 & 1/2 \\ 0 & 1 & 0 & 3/4 \\ 0 & 0 & -2 & 2 \end{bmatrix}$$

$$\Rightarrow \begin{bmatrix} 1 & 0 & 0 & 1/2 \\ 0 & 1 & 0 & 3/4 \\ 0 & 0 & 1 & -1 \end{bmatrix}$$

Figure 25

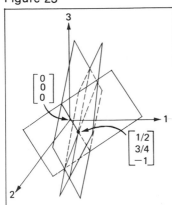

Answer: $\begin{bmatrix} 1/2 \\ 3/4 \\ -1 \end{bmatrix}$.

The planes meet in one point (Figure 25).

Exercises 2-6   Solve and interpret geometrically each of the following systems. You need not actually draw any planes.

1. $\begin{bmatrix} x_1 - x_2 + x_3 = 4 \\ -2x_1 + 12x_2 - 2x_3 = -6 \end{bmatrix}$

A2. $\begin{bmatrix} 4x_1 - 3x_2 + 6x_3 = 12 \\ 1/3x_1 - 1/4x_2 + 1/2x_3 = 1 \end{bmatrix}$

3. $\begin{bmatrix} -2/3x_1 + x_3 = 1 \\ 3x_1 - 2x_2 = 0 \end{bmatrix}$

A4. $\begin{bmatrix} x_1 + 2x_2 + 3x_3 = 1 \\ 4x_1 + 5x_2 + 6x_3 = 2 \\ 7x_1 + 8x_2 + 9x_3 = 3 \end{bmatrix}$

5. $\begin{bmatrix} x_1 + 2x_2 + 3x_3 = 3 \\ 4x_1 + 5x_2 + 6x_3 = 2 \\ 7x_1 + 8x_2 + 9x_3 = 2 \end{bmatrix}$

A6. $\begin{bmatrix} 2x_1 + x_2 + x_3 = 0 \\ x_1 - x_2 + x_3 = 1 \\ x_1 + x_2 - 2x_3 = -1 \\ 3x_1 + 2x_2 + x_3 = 0 \end{bmatrix}$

7. $\begin{bmatrix} 2x_1 + x_2 + x_3 = 0 \\ x_1 - x_2 + x_3 = 1 \\ x_1 + x_2 - 2x_3 = -1 \\ 4x_1 + x_2 = 0 \end{bmatrix}$

A8. $\begin{bmatrix} 2x_1 + x_2 + x_3 = 0 \\ x_1 - x_2 + x_3 = 1 \\ 3x_1 + 2x_3 = 1 \\ x_1 + 2x_2 = -1 \end{bmatrix}$

*9. Proceeding by analogy with the discussion in this section, give a geometrical interpretation of the system of equations and its solution resulting from the PHE advertising campaign (Exercise 4 in Section 2-3 and Exercise 8 in Section 2-4).

Review Exercises   Reread Chapter 2, then do the following exercises without looking back. Answers to all of these exercises may be found in Appendix B.

A1. One day last month Bill suggested that we could make some extra money doing odd jobs around our housing development. Everybody knew that Bill was lazy but since he agreed to put up the money for the necessary equipment, I agreed to talk to Jane and Fred, the twins, about joining our company. During our short business meeting we decided to offer three services:

(i) We would mow lawns. In the development each lawn is exactly the same size. I can do 1 lawn in an afternoon; Jane can do 2; Bill is allergic to grass; and Fred, being Jane's twin, can also do 2.

(ii) We would trim hedges. In the development each yard has the same number of hedges. I can trim 2 sets of hedges the same day I mow the lawns. Jane can do 3 in addition to her mowing. Bill reluctantly agrees to do 1. Fred does the same as Jane.

(iii) We would wash windows. Each house in our development is the same. I decided that even with the mowing and trimming I could handle 3 window-washing jobs after supper. Bill said he would do 2 window-washing jobs a day. Jane will not wash windows so Fred will not either.

(a) The first week's orders are in: mow 13 lawns; trim 29 sets of hedges; wash 33 sets of windows. While Jane and Fred do not care how many days they work, neither will work for a fractional part of a day. What are the possible work schedules?

(b) In the second week we are asked to mow 18 lawns, trim 36 sets of hedges, and wash 30 sets of windows. Fred catches a cold but agrees to work exactly half as much as Jane. What are the possible work schedules?

(c) In the third week, the orders are for 19 mowings, 34 trimmings, and 21 washings. Our employers, not understanding matrix algebra, began to think that Jane was lazier than Bill. For this reason, Jane demands to work at least 1 day more than Bill. Fred will not work for a fractional part of a day. What are the possible work schedules?

(d) In the fourth week a black sedan with plain clothes detectives begins following Bill around. Our orders fall off and we decide to work fractional parts of days if we must. That is, all of us except Jane. The orders are for 8 mowings, 17 trimmings, and 17 washings. What are the work schedules now?

The fifth week finds Bill is in jail on grand larceny charges and all our equipment impounded. We quit.

A2. Plot the following 5 points:

(a) $\begin{bmatrix} 0 \\ 3\frac{1}{2} \end{bmatrix}$ (b) $\begin{bmatrix} 4 \\ 0 \end{bmatrix}$ (c) $\begin{bmatrix} 2\frac{1}{2} \\ -3 \end{bmatrix}$ (d) $\begin{bmatrix} -1 \\ -3 \end{bmatrix}$ (e) $\begin{bmatrix} 2\frac{1}{2} \\ 3 \end{bmatrix}$

A3. Graph the following 5 lines:

(a) $3x_1 + 2x_2 = 6$

(b) $[2\frac{1}{2} \quad 0] \begin{bmatrix} x_1 \\ x_2 \end{bmatrix} = 3$

(c) $5x_2 = -7$

(d) $[-2 \quad -4] \begin{bmatrix} x_1 \\ x_2 \end{bmatrix} = 9$

(e) $3x_1 - 4x_2 = 0$

A 4. Solve the following 10 systems of equations:

(a) $\begin{bmatrix} 2x_1 + x_2 = 7 \\ x_1 + x_2 = 5 \end{bmatrix}$

(b) $\begin{bmatrix} 2x_1 + x_2 = 7 \\ 4x_1 + 2x_2 = 8 \end{bmatrix}$

(c) $\begin{bmatrix} 2x_1 + x_2 = 7 \\ 4x_1 + 2x_2 = 14 \end{bmatrix}$

(d) $\begin{bmatrix} \frac{1}{2}x_1 - \frac{1}{3}x_2 = 3 \\ \frac{2}{5}x_1 + \frac{3}{8}x_2 = 4 \end{bmatrix}$

(e) $\begin{bmatrix} 2x_1 + x_2 + 2x_3 = 5 \\ 2x_1 + 2x_2 + 2x_3 = 10 \\ x_1 + 2x_2 + x_3 = 4 \end{bmatrix}$

(f) $\begin{bmatrix} 2x_1 + 2x_2 + x_3 = 5 \\ 2x_1 + 2x_2 + 2x_3 = 6 \\ x_1 + x_2 + 2x_3 = 4 \end{bmatrix}$

(g) $\begin{bmatrix} 3x_1 + 3x_2 + 2x_3 = 5 \\ x_1 + 2x_2 + x_3 = 1 \\ 4x_1 + 3x_2 + 4x_3 = 7 \end{bmatrix}$

(h) $\begin{vmatrix} 2x_1 + 2x_2 + x_3 = 1 \\ 2x_1 + 3x_2 + 2x_3 = 5 \\ x_1 + x_2 + x_3 = 2 \end{vmatrix}$

(i) $\begin{bmatrix} 10x_1 + 8x_2 + 4x_3 + 3x_4 = 27 \\ 7x_1 + 4x_2 + 3x_3 + 3x_4 = 19 \\ 4x_1 + 4x_2 + 2x_3 = 12 \end{bmatrix}$

(j) $\begin{bmatrix} 7x_2 + x_3 - 6x_4 = 4 \\ x_1 - x_3 + x_4 = 2 \\ -x_1 - 5x_2 + 4x_4 = 5 \\ - 3x_2 + 2x_4 = 7 \end{bmatrix}$

Supplementary Exercises

1. Graph the points $\begin{bmatrix} 2 \\ -\frac{3}{2} \end{bmatrix}$, $\begin{bmatrix} -\frac{3}{2} \\ 2 \end{bmatrix}$, $\begin{bmatrix} -2 \\ \frac{3}{2} \end{bmatrix}$, and $\begin{bmatrix} -\frac{3}{2} \\ -2 \end{bmatrix}$.

*2. Indicate all points with coordinates $\begin{bmatrix} x_1 \\ x_2 \end{bmatrix}$ such that x_1 and x_2 are integers, $-3 < x_1 < 4$, $-2 < x_2 < 10$, and $(x_1)^2 < x_2$.

3. Graph the lines with the following equations:
   (a) $x_1 + 4x_2 = 8$
   A (b) $-4x_1 + x_2 = 12$
   (c) $x_1 + 3x_2 = 0$

A(d) $\frac{1}{2}x_1 - \frac{2}{3}x_2 = \frac{3}{4}$

(e) $-\frac{7}{12}x_1 + \frac{12}{7}x_2 = 1$

A (f) $-\frac{7}{12}x_1 + \frac{12}{7}x_2 = 0$

(g) $x_1 + x_2 + 2 = 0$

A(h) $3x_1 + 4 = 5x_2 + 6$

4. Find an equation for each of the following lines:

(a) the vertical line through $\begin{bmatrix} -2 \\ -3 \end{bmatrix}$

A (b) the horizontal line through $\begin{bmatrix} -2 \\ -3 \end{bmatrix}$

(c) the line through $\begin{bmatrix} 0 \\ 0 \end{bmatrix}$ and $\begin{bmatrix} -2 \\ -3 \end{bmatrix}$

A(d) the line through $\begin{bmatrix} -2 \\ 0 \end{bmatrix}$ and $\begin{bmatrix} 0 \\ -3 \end{bmatrix}$

5. Find the point of intersection of the following pairs of lines:

(a) $2x_1 - 3x_2 = -5$, $x_1 + x_2 = 5$

A(b) $6x_1 + 2x_2 = 3$, $3x_1 + x_2 = 8$

(c) $6x_1 + 2x_2 = 3$, $4x_1 - 2x_2 = 5$

A(d) $5x_1 = 4$, $6x_2 = 5$

(e) $-9x_1 + 12x_2 = 27$, $3x_1 - 4x_2 = -9$

6. Solve:

(a) $\begin{bmatrix} 2x_1 + x_2 = 7 \\ x_1 + x_2 = 5 \end{bmatrix}$

A(b) $\begin{bmatrix} x_1 + x_2 = 10 \\ x_1 - x_2 = 4 \end{bmatrix}$

(c) $\begin{bmatrix} 5x_1 + 7x_2 = -2 \\ 3x_1 + 4x_2 = -1 \end{bmatrix}$

A(d) $\begin{bmatrix} 2x_1 + 2x_2 = 10 \\ x_1 - 3x_2 = -15 \end{bmatrix}$

(e) $\begin{bmatrix} 3x_1 - x_2 = 2 \\ 2x_1 + x_2 = 3 \end{bmatrix}$

A(f) $\begin{bmatrix} 2x_1 + x_2 + x_3 = 3 \\ x_1 - x_2 + 3x_3 = 8 \\ 2x_1 + x_2 - x_3 = -1 \end{bmatrix}$

(g) $\begin{bmatrix} -x_1 + 2x_2 - 2x_3 = 10 \\ x_1 - x_2 + x_3 = -5 \\ 2x_1 + 2x_2 + x_3 = 4 \end{bmatrix}$

A(h) $\begin{bmatrix} 3x_1 + x_2 + x_3 = 6 \\ x_1 + 2x_2 + x_3 = 6 \\ x_1 + x_2 + x_3 = 4 \end{bmatrix}$

(i) $\begin{bmatrix} 2x_1 + x_4 = 3 \\ x_1 + 2x_3 = 0 \\ 2x_1 + x_2 + x_3 = 8 \end{bmatrix}$

A(j) $\begin{bmatrix} x_1 + x_2 + 2x_3 = 3 \\ -x_1 - x_2 + 3x_3 = 2 \\ x_1 + 2x_2 + 3x_3 = 7 \end{bmatrix}$

(k) $\begin{bmatrix} 2x_1 + 3x_2 + 2x_3 = 5 \\ 3x_1 + 2x_2 + 2x_3 = 5 \\ 5x_1 + 4x_2 + 4x_3 = 9 \end{bmatrix}$

A(l) $\begin{bmatrix} x_1 - 3x_2 + x_3 = -1 \\ x_1 - 4x_2 + 2x_3 = 0 \\ 2x_1 + x_2 - 5x_3 = -9 \end{bmatrix}$

$$\text{(m)} \begin{bmatrix} 2x_1 + x_2 + 2x_3 = 5 \\ 2x_1 + 2x_2 + 2x_3 = 10 \\ x_1 + 2x_2 + x_3 = 4 \end{bmatrix}$$

$$\text{(s)} \begin{bmatrix} 5x_1 + 2x_2 + x_3 = 1 \\ 5x_1 + 4x_2 + 5x_3 = 5 \\ 3x_1 + 3x_2 + 4x_3 = 4 \end{bmatrix}$$

$$^{A}\text{(n)} \begin{bmatrix} 2x_1 + 2x_2 + x_3 = 5 \\ 2x_1 + 2x_2 + 2x_3 = 6 \\ x_1 + x_2 + 2x_3 = 4 \end{bmatrix}$$

$$^{A}\text{(t)} \begin{bmatrix} 3x_1 + 2x_2 + 4x_3 = 3 \\ 3x_1 + 2x_2 + x_3 = 2 \\ 2x_1 + 2x_2 + 2x_3 = 7 \end{bmatrix}$$

$$\text{(o)} \begin{bmatrix} 3x_1 + 3x_2 + 2x_3 = 5 \\ x_1 + 2x_2 + x_3 = 1 \\ 4x_1 + 3x_2 + 4x_3 = 7 \end{bmatrix}$$

$$\text{(u)} \begin{bmatrix} 3x_1 + 2x_2 + 3x_3 = 1 \\ 2x_1 + 3x_2 + 2x_3 = 2 \\ 3x_1 + 3x_2 + 2x_3 = 1 \end{bmatrix}$$

$$^{A}\text{(p)} \begin{bmatrix} 2x_1 + 2x_2 + x_3 = 1 \\ 2x_1 + 3x_2 + 2x_3 = 5 \\ x_1 + x_2 + x_3 = 2 \end{bmatrix}$$

$$^{A}\text{(v)} \begin{bmatrix} 2x_1 + 3x_2 = 13 \\ x_1 + 2x_2 = 8 \\ 3x_1 + x_2 = 9 \end{bmatrix}$$

$$\text{(q)} \begin{bmatrix} 3x_1 + x_2 - 2x_3 = -1 \\ 3x_1 + 2x_2 + 3x_3 = 13 \\ 2x_1 + x_2 + 2x_3 = 9 \end{bmatrix}$$

$$\text{(w)} \begin{bmatrix} 2x_1 + x_2 + x_3 = 8 \\ x_1 + 2x_2 + x_3 = 7 \\ 3x_1 + x_2 + 2x_3 = 13 \\ 2x_1 + 2x_2 + x_3 = 11 \end{bmatrix}$$

$$^{A}\text{(r)} \begin{bmatrix} 2x_1 + 2x_2 - x_3 = 4 \\ 5x_1 + 4x_2 + 5x_3 = 24 \\ 3x_1 + 2x_2 + 3x_3 = 14 \end{bmatrix}$$

7. Solve:

$$\text{(a)} \begin{bmatrix} x_1 + 2x_2 - x_3 & = 2 \\ x_2 + 2x_3 - x_4 & = 4 \\ x_3 + 2x_4 - x_5 = 6 \\ x_1 + x_2 & + x_5 = 8 \\ 3x_2 & + x_4 & = 10 \end{bmatrix}$$

$$^{A}\text{(b)} \begin{bmatrix} x_1 & + x_3 & = -1 \\ x_2 + x_3 + x_4 = 2 \\ x_1 + x_2 + x_3 & = 1 \\ x_1 + x_2 + x_3 + x_4 = 2 \end{bmatrix}$$

$$\text{(c)} \begin{bmatrix} 6x_1 + x_2 + 3x_3 + 2x_4 = 13 \\ 4x_1 + x_2 + 2x_3 + 2x_4 = 10 \end{bmatrix}$$

$$^{A}\text{(d)} \begin{bmatrix} x_1 - 4x_2 - 2x_3 + 6x_4 - 4 \\ 3x_1 + 2x_2 + 3x_3 = 15 \\ x_1 + 2x_2 + 2x_3 - 2x_4 = 5 \end{bmatrix}$$

$$\text{(e)} \begin{bmatrix} 3x_1 + 2x_2 + 4x_3 = 4 \\ x_1 + 3x_2 + 2x_3 = 9 \end{bmatrix}$$

$$^{A}\text{(f)} \begin{bmatrix} 4x_1 + 6x_2 + 2x_3 + 12x_4 = 35 \\ 2x_1 + 5x_2 + 2x_3 + 10x_4 = 28 \\ 4x_1 + 7x_2 + 2x_3 + 4x_4 = 39 \end{bmatrix}$$

$$\text{(g)} \begin{bmatrix} 2x_1 + 3x_2 + x_3 = 7 \\ 4x_1 + 6x_2 + 3x_3 = 17 \end{bmatrix}$$

$$^A\text{(h)} \begin{bmatrix} 2x_1 + 3x_2 = 5 \\ 2x_1 + 2x_2 = 4 \\ 4x_1 + x_2 = 6 \end{bmatrix}$$

$$\text{(i)} \begin{bmatrix} 2x_1 + 4x_2 + 3x_4 = 7 \\ 6x_1 + 4x_2 + 3x_3 + 5x_4 = 15 \\ 4x_1 + 4x_2 + 4x_4 = 10 \end{bmatrix}$$

$$^A\text{(j)} \begin{bmatrix} 3x_1 - x_2 + 2x_3 + 9x_4 = 2 \\ 4x_1 + x_2 - 2x_3 - 3x_4 = -11 \\ x_1 + x_2 - 2x_3 - 6x_4 = -7 \\ x_1 - x_2 + 3x_4 = -2 \end{bmatrix}$$

$$\text{(k)} \begin{bmatrix} 3x_1 + 2x_2 + x_3 = 1 \\ 2x_1 + x_2 + x_3 = 2 \\ x_1 + x_2 + 2x_3 = 3 \\ 2x_1 + x_2 + 2x_3 = 4 \end{bmatrix}$$

$$^A\text{(l)} \begin{bmatrix} 10x_1 + 8x_2 + 4x_3 + 3x_4 = 27 \\ 7x_1 + 4x_2 + 3x_3 + 3x_4 = 19 \\ 4x_1 + 4x_2 + 2x_3 = 12 \end{bmatrix}$$

$$\text{(m)} \begin{bmatrix} 2x_1 + 2x_2 + x_3 + 3x_4 = 4 \\ 2x_1 + 2x_2 + 2x_3 + x_4 = 2 \\ 5x_1 + 4x_2 + 4x_3 + 5x_4 = 3 \\ 3x_1 + 3x_2 + 3x_3 + 2x_4 = 1 \end{bmatrix}$$

C8. The staff numerologist at AC has decided to adjust the relative numbers of employees. For some mystical reason, he arbitrarily decides they should have 5 times as many junior executives as senior executives and 50 times as many workers as junior executives. The budget department allows $107,000 for weekly salaries, and the contract stipulates salaries at $200 a week for worker, $500 a week for junior executives, and $1000 a week for senior executives. Determine the size of the new staff.

9. Radio station WJBJ has four kinds of programming: advertising, public service announcements, music, and news. They are on the air 24 hours a day. They must net at least $73,400 a day over expenses, which are $150 an hour to produce the music and $1600 an hour for the news. The only way to meet these expenses is through selling advertising at a rate of $10,000 an hour. The station manager decides to spend $5/8$ of the day on news and music and $1/3$ of the day on advertising. How much time will be spent on each of the four areas?

A10. I just joined the faculty and I want my grade distribution to fit in with everyone else's. In my class of 32 students, I decide that only 5 should get A's or B's (just as on Professor Smith's roster); I decide to give as many C's as A's and F's put together (for balance). I decide to give 6 more D's than B's (to prove I'm tough). Finally, I decide to give 6 more F's than A's. What is my grade distribution?

11. I follow a daily exercise program in which the idea is to exercise six

mornings a week. Either I swim or I jog or I bicycle. But I hate jogging, so I make certain that I swim more than I jog and that I bicycle exactly one day more than I jog. What are my possible exercise schedules?

* 12. We remarked in Section 2-2 that the operation of switching two rows was not as fundamental to Gauss-Jordan reduction as the operations of adding a multiple of a row to another row and multiplying a row by a nonzero number. Show, for example, how to change the matrix

$$\begin{bmatrix} 0 & 1 & 2 \\ 3 & 1 & 0 \\ 1 & 2 & 3 \end{bmatrix} \text{ into the matrix } \begin{bmatrix} 1 & 2 & 3 \\ 3 & 1 & 0 \\ 0 & 1 & 2 \end{bmatrix} \text{ by using only these last two}$$

types of row operations.

13. Here is another way to find an equation for the line through two points. We illustrate by finding an equation for the line through

$$\begin{bmatrix} 2 \\ -3 \end{bmatrix} \text{ and } \begin{bmatrix} -3 \\ 4 \end{bmatrix}:$$

(1) Write the system of equations $x_1 a + x_2 b - c = 0$ for the given

points $\begin{bmatrix} x_1 \\ x_2 \end{bmatrix}$:

$$\begin{bmatrix} 2a - 3b - c = 0 \\ -3a + 4b - c = 0 \end{bmatrix}$$

(2) Solve (here the variables are a, b, and c!):

$$\begin{bmatrix} 2 & -3 & -1 & 0 \\ -3 & 4 & -1 & 0 \end{bmatrix} \implies \begin{bmatrix} 1 & -3/2 & -1/2 & 0 \\ 0 & -1/2 & -5/2 & 0 \end{bmatrix} \implies \begin{bmatrix} 1 & 0 & 7 & 0 \\ 0 & 1 & 5 & 0 \end{bmatrix}.$$

Answer: $\begin{bmatrix} -7c \\ -5c \\ c \end{bmatrix}$, where c is free.

(3) Choose a specific solution where a, b, and c are not all zero: $a = 7$, $b = 5$, $c = -1$.

(4) An equation for the line is $ax_1 + bx_2 = c$: $7x_1 + 5x_2 = -1$.

(5) Check: $7 \cdot 2 + 5 \cdot (-3) = -1$; $7 \cdot (-3) + 5 \cdot 4 = -1$.

Use this procedure to find equations for the lines determined by the following pairs of points:

^{A}(a) $\begin{bmatrix} 1 \\ 2 \end{bmatrix}, \begin{bmatrix} 3 \\ 4 \end{bmatrix}$ (d) $\begin{bmatrix} 0 \\ 0 \end{bmatrix}, \begin{bmatrix} 9 \\ 12 \end{bmatrix}$

(b) $\begin{bmatrix} -7 \\ 6 \end{bmatrix}, \begin{bmatrix} 5 \\ 10 \end{bmatrix}$ (e) $\begin{bmatrix} 4 \\ 0 \end{bmatrix}, \begin{bmatrix} 19 \\ 0 \end{bmatrix}$

^{A}(c) $\begin{bmatrix} 7/3 \\ -7/4 \end{bmatrix}, \begin{bmatrix} -5 \\ 15/4 \end{bmatrix}$ ^{A}(f) $\begin{bmatrix} 0 \\ 6 \end{bmatrix}, \begin{bmatrix} -4 \\ 0 \end{bmatrix}$

Summary

Terms   Coordinates of a point in the plane   pp. 61–62

Linear equation   pp. 63–64

System of linear equations   pp. 70, 85

Matrix of a system of linear equations   pp. 72, 85

Coordinates of a point in space   pp. 106–107

Computational Technique   To find all solutions, if any, of a system of equations: the three row operations (p. 85) are used to clear columns (p. 86) as described in the flow chart (p. 87). Shortcuts (Section 2-5) may help.

Chapter 3
Inverse
Matrices

As we mentioned in Section 1-6, there is another matrix operation to be introduced: matrix division. This operation is of great importance in theoretical mathematics and it finds practical application in connection with certain kinds of business situations—for example, one in which many different orders are to be filled from the same inventories (see Section 3-4). Matrix division is introduced through the concept of the inverse of a matrix. It is this concept, rather than the operation of matrix division, that is of central importance. The actual method of calculation of an inverse matrix is merely a modification of Gauss-Jordan reduction. This chapter continues our study of systems of linear equations and the use of matrices in their solution.

Section 3-1
Matrix Notation for a
System of Linear
Equations

Consider the system

$$\begin{bmatrix} 6x_1 - x_2 - 5x_3 = 4 \\ -7x_1 + x_2 + 5x_3 = -3 \\ -10x_1 + 2x_2 + 11x_3 = 8 \end{bmatrix} \qquad \mathbf{1}$$

In Chapter 2 we saw how each of these equations can be written in matrix form:

$$\begin{bmatrix} 6 & -1 & -5 \end{bmatrix} \begin{bmatrix} x_1 \\ x_2 \\ x_3 \end{bmatrix} = 4, \quad \begin{bmatrix} -7 & 1 & 5 \end{bmatrix} \begin{bmatrix} x_1 \\ x_2 \\ x_3 \end{bmatrix} = -3,$$

$$\begin{bmatrix} -10 & 2 & 11 \end{bmatrix} \begin{bmatrix} x_1 \\ x_2 \\ x_3 \end{bmatrix} = 8 \qquad \mathbf{2}$$

Notice now that these three matrix equations can be written as *one* matrix equation:

$$\begin{bmatrix} 6 & -1 & -5 \\ -7 & 1 & 5 \\ -10 & 2 & 11 \end{bmatrix} \begin{bmatrix} x_1 \\ x_2 \\ x_3 \end{bmatrix} = \begin{bmatrix} 4 \\ -3 \\ 8 \end{bmatrix} \qquad \mathbf{3}$$

It follows from the definition of matrix multiplication (Section 1-5) that equation **3** is equivalent to system **1**. Equation **3** is called the **matrix form** of system **1**. The matrix

$$\begin{bmatrix} 6 & -1 & -5 \\ -7 & 1 & 5 \\ -10 & 2 & 11 \end{bmatrix} \qquad \mathbf{4}$$

of equation **3** is called the **coefficient matrix** of system **1**. The number of rows of the coefficient matrix equals the number of equations in the system; the number of columns of the coefficient matrix equals the number of different variables in the system. Thus, since system **1** is a system of **3** equations in **3** variables, the coefficient matrix **4** is a 3×3 matrix. The system

$$\begin{bmatrix} 3x_2 = -2 \\ -4x_1 - x_2 = 1 \\ 3x_1 + x_2 = -1 \end{bmatrix} \qquad \textbf{5}$$

Notice that in system **5** we do not bother to write the variable x_1 in the first equation, but in system **6** we must include a zero to fill its space in the first row.

of three equations in two variables has a 3×2 coefficient matrix:

$$\begin{bmatrix} 0 & 3 \\ -4 & -1 \\ 3 & 1 \end{bmatrix} \qquad \textbf{6}$$

The system

$$\begin{bmatrix} -x_1 + x_2 \quad - x_4 = 1 \\ x_2 + x_3 \quad = 3 \end{bmatrix} \qquad \textbf{7}$$

of two equations in four variables has the 2×4 coefficient matrix

$$\begin{bmatrix} -1 & 1 & 0 & -1 \\ 0 & 1 & 1 & 0 \end{bmatrix} \qquad \textbf{8}$$

Conversely, any matrix may be regarded as the coefficient matrix of a system of linear equations. For example, the matrix

$$\begin{bmatrix} 2 & -1 \\ -3 & 4 \end{bmatrix} \qquad \textbf{9}$$

is the coefficient matrix of a system of the form

$$\begin{bmatrix} 2x_1 - x_2 = ? \\ -3x_1 + 4x_2 = ? \end{bmatrix} \qquad \textbf{10}$$

We will not know what numbers to put for the question marks until we are given the **requirement matrix**. The requirement matrix for system **1** is

$$\begin{bmatrix} 4 \\ -3 \\ 8 \end{bmatrix} \qquad \textbf{11}$$

For system **5**, it is

$$\begin{bmatrix} -2 \\ 1 \\ -1 \end{bmatrix} \qquad \textbf{12}$$

For system **7**, it is

$$\begin{bmatrix} 1 \\ 3 \end{bmatrix}$$

13

The requirement matrix must be a column matrix, and the number of entries (rows) in the column must equal the number of equations in the system. Thus, the number of entries in the requirement matrix must equal the number of rows in the coefficient matrix. For example, the requirement matrix for a system with coefficient matrix **9** must be a 2×1 matrix. If matrix **13** is the requirement matrix, then the system with coefficient matrix **9** is

$$\begin{bmatrix} 2x_1 - x_2 = 1 \\ -3x_1 + 4x_2 = 3 \end{bmatrix}$$

14

If we denote coefficient matrix **9** by A, the "unknown matrix" $\begin{bmatrix} x_1 \\ x_2 \end{bmatrix}$ by X, and the requirement matrix **13** by B, then equation **15** becomes simply $AX = B$.

or, in matrix form,

$$\begin{bmatrix} 2 & -1 \\ -3 & 4 \end{bmatrix} \begin{bmatrix} x_1 \\ x_2 \end{bmatrix} = \begin{bmatrix} 1 \\ 3 \end{bmatrix}$$

15

Now suppose we apply Gauss-Jordan reduction to these systems. For system **1**, we have

$$\begin{bmatrix} 6 & -1 & -5 & 4 \\ -7 & 1 & 5 & -3 \\ -10 & 2 & 11 & 8 \end{bmatrix} \Rightarrow \begin{bmatrix} 1 & -1/6 & -5/6 & 2/3 \\ 0 & -1/6 & -5/6 & 5/3 \\ 0 & 1/3 & 8/3 & 44/3 \end{bmatrix}$$

$$\Rightarrow \begin{bmatrix} 1 & 0 & 0 & -1 \\ 0 & 1 & 5 & -10 \\ 0 & 0 & 1 & 18 \end{bmatrix} \Rightarrow \begin{bmatrix} 1 & 0 & 0 & -1 \\ 0 & 1 & 0 & -100 \\ 0 & 0 & 1 & 18 \end{bmatrix}$$

In our new notation, this reduced system is written

$$\begin{bmatrix} 1 & 0 & 0 \\ 0 & 1 & 0 \\ 0 & 0 & 1 \end{bmatrix} \begin{bmatrix} x_1 \\ x_2 \\ x_3 \end{bmatrix} = \begin{bmatrix} -1 \\ -100 \\ 18 \end{bmatrix}$$

16

Since

$$\begin{bmatrix} 1 & 0 & 0 \\ 0 & 1 & 0 \\ 0 & 0 & 1 \end{bmatrix} \begin{bmatrix} x_1 \\ x_2 \\ x_3 \end{bmatrix} = \begin{bmatrix} x_1 \\ x_2 \\ x_3 \end{bmatrix}$$

equation **16** becomes

$$\begin{bmatrix} x_1 \\ x_2 \\ x_3 \end{bmatrix} = \begin{bmatrix} -1 \\ -100 \\ 18 \end{bmatrix}$$

that is, $x_1 = -1$, $x_2 = -100$, $x_3 = 18$. For system **5** we have

$$\begin{bmatrix} 0 & 3 & -2 \\ -4 & -1 & 1 \\ 3 & 1 & -1 \end{bmatrix} \Longrightarrow \begin{bmatrix} -4 & -1 & 1 \\ 0 & 3 & -2 \\ 3 & 1 & -1 \end{bmatrix}$$

$$\Longrightarrow \begin{bmatrix} 1 & \frac{1}{4} & -\frac{1}{4} \\ 0 & 3 & -2 \\ 0 & \frac{1}{4} & -\frac{1}{4} \end{bmatrix} \Longrightarrow \begin{bmatrix} 1 & 0 & -\frac{1}{12} \\ 0 & 1 & -\frac{2}{3} \\ 0 & 0 & -\frac{1}{12} \end{bmatrix}$$

that is,

$$\begin{bmatrix} 1 & 0 \\ 0 & 1 \\ 0 & 0 \end{bmatrix} \begin{bmatrix} x_1 \\ x_2 \end{bmatrix} = \begin{bmatrix} -\frac{1}{12} \\ -\frac{2}{3} \\ -\frac{1}{12} \end{bmatrix} \qquad\qquad \textbf{17}$$

Since equation **17** requires $0x_1 + 0x_2 = -\frac{1}{12}$, it has no solution. Hence, system **5** has no solution.

For system **7**, we have

$$\begin{bmatrix} -1 & 1 & 0 & -1 & 1 \\ 0 & 1 & 1 & 0 & 3 \end{bmatrix} \Longrightarrow \begin{bmatrix} 1 & -1 & 0 & 1 & -1 \\ 0 & 1 & 1 & 0 & 3 \end{bmatrix}$$

$$\Longrightarrow \begin{bmatrix} 1 & 0 & 1 & 1 & 2 \\ 0 & 1 & 1 & 0 & 3 \end{bmatrix}$$

that is,

$$\begin{bmatrix} 1 & 0 & 1 & 1 \\ 0 & 1 & 1 & 0 \end{bmatrix} \begin{bmatrix} x_1 \\ x_2 \\ x_3 \\ x_4 \end{bmatrix} = \begin{bmatrix} 2 \\ 3 \end{bmatrix} \qquad\qquad \textbf{18}$$

which gives the "family" of solutions $x_1 = 2 - x_3 - x_4$, $x_2 = 3 - x_3$, where x_3 and x_4 may be any numbers.

Look at the coefficient matrices for the "reduced" systems **16**, **17**, and **18**:

$$\begin{bmatrix} 1 & 0 & 0 \\ 0 & 1 & 0 \\ 0 & 0 & 1 \end{bmatrix} \qquad\qquad \textbf{19}$$

$$\begin{bmatrix} 1 & 0 \\ 0 & 1 \\ 0 & 0 \end{bmatrix} \qquad\qquad \textbf{20}$$

$$\begin{bmatrix} 1 & 0 & 1 & 1 \\ 0 & 1 & 1 & 0 \end{bmatrix} \qquad\qquad \textbf{21}$$

The system with coefficient matrix **19** had a *unique* solution, that is, a solution with no free variables. The systems with coefficient matrices **20** and **21** did not have unique solutions; they either had no solution (system **17**) or many solutions (system **18**). Not only did system **16**, with coefficient matrix **19**, have a unique solution, but *any* system with matrix **19** as its coefficient matrix has a unique solution. For example, the system

$$\begin{bmatrix} 1 & 0 & 0 \\ 0 & 1 & 0 \\ 0 & 0 & 1 \end{bmatrix} \begin{bmatrix} x_1 \\ x_2 \\ x_3 \end{bmatrix} = \begin{bmatrix} 2 \\ 0 \\ {}^1\!/_5 \end{bmatrix} \qquad \textbf{22}$$

has the unique solution $x_1 = 2$, $x_2 = 0$, $x_3 = {}^1\!/_5$.

Matrix **19** is called the 3×3 **identity matrix**, for reasons to be explained in Section 3-2. The 2×2 identity matrix is the matrix

$$\begin{bmatrix} 1 & 0 \\ 0 & 1 \end{bmatrix} \qquad \textbf{23}$$

Just as with the 3×3 identity matrix, it is the case that *any* system with coefficient matrix **23** has a unique solution. For example, the system

$$\begin{bmatrix} 1 & 0 \\ 0 & 1 \end{bmatrix} \begin{bmatrix} x_1 \\ x_2 \end{bmatrix} = \begin{bmatrix} -800 \\ {}^4\!/_{159} \end{bmatrix} \qquad \textbf{24}$$

has the unique solution $x_1 = -800$ and $x_2 = {}^4\!/_{159}$. Similarly, the 4×4 and 5×5 identity matrices are defined to be the matrices

$$\begin{bmatrix} 1 & 0 & 0 & 0 \\ 0 & 1 & 0 & 0 \\ 0 & 0 & 1 & 0 \\ 0 & 0 & 0 & 1 \end{bmatrix} \text{ and } \begin{bmatrix} 1 & 0 & 0 & 0 & 0 \\ 0 & 1 & 0 & 0 & 0 \\ 0 & 0 & 1 & 0 & 0 \\ 0 & 0 & 0 & 1 & 0 \\ 0 & 0 & 0 & 0 & 1 \end{bmatrix} \qquad \textbf{25}$$

respectively. In general, an identity matrix is a square matrix with 1's in the **main diagonal** (that is, the northwest to southeast diagonal) and 0's everywhere else. If we perform Gauss-Jordan reduction on a system, the coefficient matrix of the reduced system is called (naturally enough) the **reduced coefficient matrix** of the system. We summarize the previous remarks as follows.

> If the reduced coefficient matrix of a system of linear equations is an identity matrix, then the system has a unique solution. If the reduced coefficient matrix is square but is not an identity matrix, then the system does not

We give two examples to illustrate the truth of the second sentence of this summary. If the reduced matrix of a system is

$$\begin{bmatrix} 1 & 0 & 0 & -2 \\ 0 & 0 & 0 & a \\ 0 & 0 & 1 & 2 \end{bmatrix}$$

where a is yet to be given, then the coefficient matrix

$$\begin{bmatrix} 1 & 0 & 0 \\ 0 & 0 & 0 \\ 0 & 0 & 1 \end{bmatrix}$$

is square but is not the identity matrix. If a is not zero, the system has no solution. If a is zero, the system has an infinite number of solutions.

have a unique solution; that is, it has either no solution or an infinite number of solutions.

The situation with nonsquare coefficient matrices is more complicated. System **17** had no solution, but the system

$$\begin{bmatrix} 1 & 0 \\ 0 & 1 \\ 0 & 0 \end{bmatrix} \begin{bmatrix} x_1 \\ x_2 \end{bmatrix} = \begin{bmatrix} -\frac{1}{2} \\ -\frac{2}{3} \\ 0 \end{bmatrix}$$

which has the same coefficient matrix, has the unique solution $x_1 = -\frac{1}{2}$, $x_2 = -\frac{2}{3}$. So we cannot tell merely from reduced matrix **20** whether the system has a unique solution or not. On the other hand, any system with reduced coefficient matrix **21** will have an infinite number of solutions. We may state the following general result:

If a system of linear equations has more variables than equations, then the system does not have a unique solution; that is, it has either no solution or an infinite number of solutions.

Sometimes we are given a system of equations and asked only to determine whether or not it has a unique solution. Even if a system does have a unique solution, we may not need to know what the solution actually is. In such a case we may apply these results in the following flow-chart form.

To Determine Whether or Not a Given System of Linear Equations Has a Unique Solution

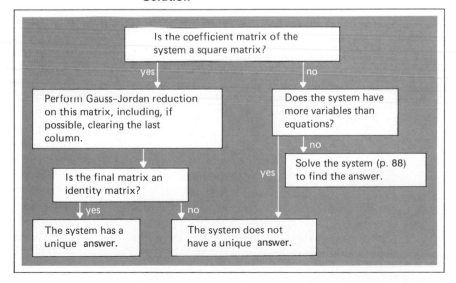

Problem 1   Does the system $\begin{bmatrix} 3x_1 - x_2 + 4x_3 & - x_5 = & 6 \\ x_2 - x_3 + 3x_4 & = & -2 \\ x_1 \quad - 4x_3 & + 2x_5 = & 0 \\ 6x_1 & - x_5 = & 2 \end{bmatrix}$ have a unique solution?

Solution   No, because it has five variables but only four equations.

Problem 2   Does the system $\begin{bmatrix} -3x_1 + 2x_2 & = & 3 \\ 4x_3 = & 6 \\ 6x_1 - 4x_2 + 4x_3 = & 0 \\ 3x_1 - 2x_2 + 12x_3 = & 15 \end{bmatrix}$ have a unique solution?

Solution   There are more equations than variables, so we must solve the system:

$$\begin{bmatrix} -3 & 2 & 0 & 3 \\ 0 & 0 & 4 & 6 \\ 6 & -4 & 4 & 0 \\ 3 & -2 & 12 & 15 \end{bmatrix} \Rightarrow \begin{bmatrix} -3 & 2 & 0 & 3 \\ 0 & 0 & 4 & 6 \\ 0 & 0 & 4 & 6 \\ 0 & 0 & 12 & 18 \end{bmatrix}$$

$$\Rightarrow \begin{bmatrix} -3 & 2 & 0 & 3 \\ 0 & 0 & 4 & 6 \\ 0 & 0 & 0 & 0 \\ 0 & 0 & 0 & 0 \end{bmatrix}$$

No, the system has an infinite number of solutions.

Problem 3   Does system 1 have a unique solution?

Solution $\begin{bmatrix} 6 & -1 & -5 \\ -7 & 1 & 5 \\ -10 & 2 & 11 \end{bmatrix} \Rightarrow \begin{bmatrix} 1 & -1/6 & -5/6 \\ 0 & -1/6 & -5/6 \\ 0 & 1/3 & 8/3 \end{bmatrix} \Rightarrow \begin{bmatrix} 1 & 0 & 0 \\ 0 & 1 & 5 \\ 0 & 0 & 1 \end{bmatrix}$

$$\Rightarrow \begin{bmatrix} 1 & 0 & 0 \\ 0 & 1 & 0 \\ 0 & 0 & 1 \end{bmatrix}$$

This last matrix is an identity matrix. Thus, this system has a unique solution.

Problem 4   Does the system $\begin{bmatrix} 2x_1 - x_2 + 2x_3 = & 478.6 \\ x_1 - 2x_2 + 2x_3 = & 214.9 \\ -x_1 - x_2 - x_3 = & -917.6 \end{bmatrix}$ have a unique

solution?

Solution

$$\begin{bmatrix} 2 & -1 & 2 \\ 1 & -2 & 2 \\ -1 & -1 & -1 \end{bmatrix} \Rightarrow \begin{bmatrix} 1 & -1/2 & 1 \\ 0 & -3/2 & 1 \\ 0 & -3/2 & 0 \end{bmatrix} \Rightarrow \begin{bmatrix} 1 & 0 & 2/3 \\ 0 & 1 & -2/3 \\ 0 & 0 & -1 \end{bmatrix}$$

$$\Rightarrow \begin{bmatrix} 1 & 0 & 0 \\ 0 & 1 & 0 \\ 0 & 0 & 1 \end{bmatrix}$$

This is an identity matrix; the system has a unique solution. (Fortunately we are not asked to find what the solution is.)

Problem 5   Does the system $\begin{bmatrix} 2 & 0 & 2 & 4 \\ 4 & 0 & -2 & 1 \\ 0 & 0 & 4 & 3 \\ 2 & 0 & 3 & 1 \end{bmatrix} \begin{bmatrix} x_1 \\ x_2 \\ x_3 \\ x_4 \end{bmatrix} = \begin{bmatrix} 190 \\ 0 \\ -204 \\ 1/73 \end{bmatrix}$ have a

unique solution?

Solution

$$\begin{bmatrix} 2 & 0 & 2 & 4 \\ 4 & 0 & -2 & 1 \\ 0 & 0 & 4 & 3 \\ -2 & 0 & 3 & 1 \end{bmatrix} \Rightarrow \begin{bmatrix} 1 & 0 & 1 & 2 \\ 0 & 0 & -6 & -7 \\ 0 & 0 & 4 & 3 \\ 0 & 0 & 5 & 5 \end{bmatrix} \Rightarrow \begin{bmatrix} 1 & 0 & 0 & 5/6 \\ 0 & 0 & 1 & 7/6 \\ 0 & 0 & 0 & -5/3 \\ 0 & 0 & 0 & -5/6 \end{bmatrix}$$

$$\Rightarrow \begin{bmatrix} 1 & 0 & 0 & 0 \\ 0 & 0 & 1 & 0 \\ 0 & 0 & 0 & 1 \\ 0 & 0 & 0 & 0 \end{bmatrix}$$

The reduced matrix is not an identity matrix, so the system does not have a unique solution.

Exercises 3.1

A 1. Given the system $\begin{bmatrix} x_2 - x_3 + x_4 = 3 \\ 2x_1 + x_2 \qquad - 3x_4 = 2 \\ -x_1 - x_2 + 3x_3 - x_4 = 0 \end{bmatrix}$, write:

(a) the matrix of the system.
(b) the coefficient matrix of the system.
(c) the requirement matrix of the system.

(d) the system in the form of a matrix equation.

2. Given the system
$$\begin{bmatrix} 2x_1 - \frac{1}{3}x_2 && = \frac{3}{2} \\ x_1 & + \frac{1}{3}x_3 = \frac{2}{3} \\ -3x_1 - & x_2 + x_3 = 4 \\ & \frac{2}{3}x_2 - \frac{4}{5}x_3 = \frac{5}{6} \end{bmatrix}, \text{ write:}$$

   (a) the coefficient matrix of the system.
   (b) the requirement matrix of the system.
   (c) the system in the form of a matrix equation.
   (d) the matrix of the system.

A 3. Consider the system of Exercise 1.
   (a) Write the reduced coefficient matrix of the system.
   (b) What does the flow chart on p. 125 tell us about this system.
   (c) Solve the system.

4. Consider the system of Exercise 2.
   (a) Write the reduced coefficient matrix of the system.
   (b) What does the flow chart tell us about this system?
   (c) Solve the system.

5. Write out the system $AX = B$, where:

(a) $A = \begin{bmatrix} 0 & 3 \\ -1 & 2 \\ 3 & 0 \end{bmatrix}$, $B = \begin{bmatrix} 6 \\ 2 \\ 0 \end{bmatrix}$

(b) $A = \begin{bmatrix} 0 & 3 & -1 \\ 2 & 3 & 0 \end{bmatrix}$, $B = \begin{bmatrix} 6 \\ 2 \end{bmatrix}$

(c) $A = \begin{bmatrix} 3 & 4 & \frac{1}{2} & -\frac{3}{4} \end{bmatrix}$, $B = \begin{bmatrix} \frac{7}{2} \end{bmatrix}$

6. For each of the following matrices, determine whether or not a system having the matrix as its coefficient matrix will have a unique solution.

A(a) $\begin{bmatrix} 1 & 2 \\ 3 & 4 \end{bmatrix}$

(b) $\begin{bmatrix} -1 & 2 \\ 3 & 1 \\ -3 & 4 \end{bmatrix}$

A(c) $\begin{bmatrix} 1 & 2 & 3 \\ 4 & 5 & 6 \\ 7 & 8 & 9 \end{bmatrix}$

(d) $\begin{bmatrix} 1 & 2 & 3 & 4 \\ 5 & 6 & 7 & 8 \\ 9 & 10 & 11 & 12 \\ 13 & 14 & 15 & 16 \end{bmatrix}$

(e) $\begin{bmatrix} 3 & -\frac{1}{2} & 2 & -\frac{2}{5} \\ 1 & 3 & -2 & 4 \\ 0 & 0 & 1 & 2 \end{bmatrix}$

A (f) $\begin{bmatrix} 1 & -1 & 1 & -1 & 1 \\ 1 & -1 & -1 & 1 & -1 \\ 1 & -1 & -1 & -1 & 1 \\ 1 & -1 & -1 & -1 & -1 \\ 1 & 1 & 1 & 1 & 1 \end{bmatrix}$

7. Determine which of the following systems of equations have unique solutions.

(a) $\begin{bmatrix} 3x_1 - 2x_2 = 4 \\ 6x_1 - 4x_2 = 8 \end{bmatrix}$

$$^A(b) \quad \begin{bmatrix} 5x_1 - x_2 = 0 \\ 3x_1 + x_2 = 0 \end{bmatrix}$$

$$^A(d) \quad \begin{bmatrix} \frac{2}{3}x_1 - \frac{7}{8}x_2 + \frac{7}{9}x_3 = \frac{11}{14} \\ \frac{7}{11}x_1 + \frac{7}{3}x_2 - \frac{13}{21}x_3 = \frac{19}{21} \end{bmatrix}$$

$$(c) \quad \begin{bmatrix} x_1 - x_2 + x_3 = 1 \\ x_1 + x_2 + x_3 = 3 \\ x_1 + 2x_2 - x_3 = 2 \\ x_1 - x_2 - x_3 = 4 \end{bmatrix}$$

$$(e) \quad \begin{bmatrix} x_1 - x_2 = 0 \\ x_1 + x_2 = 2 \\ 2x_1 - x_2 = 1 \\ x_1 + 2x_2 = 3 \end{bmatrix}$$

8. PHE's philosophy has always been to stick to what it knows best: the fast-food business. Given its size and volume (sales last year: $2.3 billion), it might be   expected to have purchased beef ranches and slaughterhouses. But the managers still prefer to deal with a vast number of suppliers, constantly adjusting their orders to get the best quality-price ratios. Of course,   this method of operation involves making a great many decisions about order quantities for many stores from many suppliers. Decisions of this sort often lead to systems of linear equations. One of the systems that arose recently was the following, having to do with orders from northern Texas for pickles, catsup, and other condiments.

$$\begin{bmatrix} 2x_1 + 3x_2 + x_3 \quad\quad + x_5 = 947 \\ x_1 + 2x_2 + 4x_3 + x_4 + 3x_5 = 1860 \\ 3x_2 \quad\quad + 4x_4 + 2x_5 = 870 \\ x_1 \quad\quad + 3x_3 + 3x_4 \quad\quad = 698 \\ x_2 + 2x_3 \quad\quad + 3x_5 - 1240 \end{bmatrix}$$

25

(a) Write system **25** in matrix form.
(b) Does the system have a unique solution?

Section 3-2
Identity Matrices and the
Inverse of a Matrix

Let us multiply the 3×3 identity matrix (matrix **19** in Section 3-1) by a few matrices:

$$\begin{bmatrix} 1 & 0 & 0 \\ 0 & 1 & 0 \\ 0 & 0 & 1 \end{bmatrix} \begin{bmatrix} 2 \\ 3 \\ 4 \end{bmatrix} = \begin{bmatrix} 2 \\ 3 \\ 4 \end{bmatrix}, \quad \begin{bmatrix} 1 & 0 & 0 \\ 0 & 1 & 0 \\ 0 & 0 & 1 \end{bmatrix} \begin{bmatrix} 2 & 5 \\ 3 & 6 \\ 4 & 7 \end{bmatrix} = \begin{bmatrix} 2 & 5 \\ 3 & 6 \\ 4 & 7 \end{bmatrix},$$

$$\begin{bmatrix} 1 & 0 & 0 \\ 0 & 1 & 0 \\ 0 & 0 & 1 \end{bmatrix} \begin{bmatrix} 2 & 5 & 1 & 3 \\ 3 & 6 & 0 & 0 \\ 4 & 7 & 2 & 0 \end{bmatrix} = \begin{bmatrix} 2 & 5 & 1 & 3 \\ 3 & 6 & 0 & 0 \\ 4 & 7 & 2 & 0 \end{bmatrix}$$

Let us also multiply some matrices by the 3×3 identity matrix in the other order:

$$\begin{bmatrix} 2 & 3 & 4 \end{bmatrix} \begin{bmatrix} 1 & 0 & 0 \\ 0 & 1 & 0 \\ 0 & 0 & 1 \end{bmatrix} = \begin{bmatrix} 2 & 3 & 4 \end{bmatrix}$$

$$\begin{bmatrix} 2 & 3 & 4 \\ 3 & 2 & 4 \\ 4 & 3 & 2 \end{bmatrix} \begin{bmatrix} 1 & 0 & 0 \\ 0 & 1 & 0 \\ 0 & 0 & 1 \end{bmatrix} = \begin{bmatrix} 2 & 3 & 4 \\ 3 & 2 & 4 \\ 4 & 3 & 2 \end{bmatrix}$$

$$\begin{bmatrix} 1 & -3 & 6 \\ -1 & 4 & -6 \\ 2 & -4 & 7 \\ -2 & 5 & -7 \\ 3 & -5 & 0 \end{bmatrix} \begin{bmatrix} 1 & 0 & 0 \\ 0 & 1 & 0 \\ 0 & 0 & 1 \end{bmatrix} = \begin{bmatrix} 1 & -3 & 6 \\ -1 & 4 & -6 \\ 2 & -4 & 7 \\ -2 & 5 & -7 \\ 3 & -5 & 0 \end{bmatrix}$$

In general, whenever we multiply a given matrix and the 3×3 identity matrix in either order, we get the given matrix as the answer (assuming, of course, that the matrices can be multiplied). The same results hold true for the other identity matrices. For example,

$$\begin{bmatrix} 2 & 0 \\ 0 & 3 \\ 1 & 2 \\ 3 & -5 \\ 7 & 6 \end{bmatrix} \begin{bmatrix} 1 & 0 \\ 0 & 1 \end{bmatrix} = \begin{bmatrix} 2 & 0 \\ 0 & 3 \\ 1 & 2 \\ 3 & -5 \\ 7 & 6 \end{bmatrix}$$

and

$$\begin{bmatrix} 1 & 0 & 0 & 0 & 0 \\ 0 & 1 & 0 & 0 & 0 \\ 0 & 0 & 1 & 0 & 0 \\ 0 & 0 & 0 & 1 & 0 \\ 0 & 0 & 0 & 0 & 1 \end{bmatrix} \begin{bmatrix} 2 & 0 \\ 0 & 3 \\ 1 & 2 \\ 3 & -5 \\ 7 & 6 \end{bmatrix} = \begin{bmatrix} 2 & 0 \\ 0 & 3 \\ 1 & 2 \\ 3 & -5 \\ 7 & 6 \end{bmatrix}$$

This is why we call these matrices "identity" matrices.

Identity matrices are usually denoted by the letter I. If in a given discussion it is desirable to emphasize the size, then a subscript may be appended to I giving the number of rows (or columns) in the particular identity matrix being considered. Thus, for example,

> The identity matrix plays the same role in matrix multiplication as does the number 1 in the multiplication of numbers. Multiplication by them leaves everything the same.

$$I_2 = \begin{bmatrix} 1 & 0 \\ 0 & 1 \end{bmatrix}, \ I_3 = \begin{bmatrix} 1 & 0 & 0 \\ 0 & 1 & 0 \\ 0 & 0 & 1 \end{bmatrix}, \ I_4 = \begin{bmatrix} 1 & 0 & 0 & 0 \\ 0 & 1 & 0 & 0 \\ 0 & 0 & 1 & 0 \\ 0 & 0 & 0 & 1 \end{bmatrix}$$

> If C is a square matrix such that CI is defined, then IC will also be defined. We may say then: If C is any square matrix and I is the identity matrix of the same size, then
> $$CI = IC = C$$

and so forth. We may express our discovery of the "identity" properties of these matrices as follows: Let I be an identity matrix and let A and B be any matrices such that AI and IB are defined. Then

$$AI = A \text{ and } IB = B$$

Now what about dividing one matrix by another? It might be useful if we could do this. We learned in the last section that any system of linear equations could be written in the form $AX = B$, where A is the coefficient matrix, X is the unknown matrix and B is the requirement matrix. Symbolically, the *matrix* equation $AX = B$ looks a lot like the *number* equation $ax = b$, which we solve by division. For example: we solve the equation $3x = 6$ by dividing 6 by 3 to obtain $x = 2$. Why not solve say, matrix equation **3** in Section 3-1—

$$\begin{bmatrix} 6 & -1 & -5 \\ -7 & 1 & 5 \\ -10 & 2 & 11 \end{bmatrix} \begin{bmatrix} x_1 \\ x_2 \\ x_3 \end{bmatrix} = \begin{bmatrix} 4 \\ -3 \\ 8 \end{bmatrix} \qquad \textbf{1}$$

—by dividing $\begin{bmatrix} 4 \\ -3 \\ 8 \end{bmatrix}$ by $\begin{bmatrix} 6 & -1 & -5 \\ -7 & 1 & 5 \\ -10 & 2 & 11 \end{bmatrix}$? We proceed to do this as follows.

It happens that the matrix

$$\begin{bmatrix} -1 & -1 & 0 \\ -27 & -16 & -5 \\ 4 & 2 & 1 \end{bmatrix} \qquad \textbf{2}$$

when multiplied by the coefficient matrix

$$\begin{bmatrix} 6 & -1 & -5 \\ -7 & 1 & 5 \\ -10 & 2 & 11 \end{bmatrix} \qquad \textbf{3}$$

results in the 3×3 identity matrix:

$$\begin{bmatrix} -1 & -1 & 0 \\ -27 & -16 & -5 \\ 4 & 2 & 1 \end{bmatrix} \begin{bmatrix} 6 & -1 & 5 \\ -7 & 1 & 5 \\ -10 & 2 & 11 \end{bmatrix} = \begin{bmatrix} 1 & 0 & 0 \\ 0 & 1 & 0 \\ 0 & 0 & 1 \end{bmatrix} \qquad \textbf{4}$$

We use this discovery to do our "division." We multiply matrix **2** by each side of equation **1**:

$$\begin{bmatrix} -1 & -1 & 0 \\ -27 & -16 & -5 \\ 4 & 2 & 1 \end{bmatrix} \begin{bmatrix} 6 & -1 & -5 \\ -7 & 1 & 5 \\ -10 & 2 & 11 \end{bmatrix} \begin{bmatrix} x_1 \\ x_2 \\ x_3 \end{bmatrix}$$

$$= \begin{bmatrix} -1 & -1 & 0 \\ -27 & -16 & -5 \\ 4 & 2 & 1 \end{bmatrix} \begin{bmatrix} 4 \\ -3 \\ 8 \end{bmatrix} \qquad \textbf{5}$$

Using equation **4** and carrying out the multiplication on the right side, we convert equation **5** to

$$\begin{bmatrix} 1 & 0 & 0 \\ 0 & 1 & 0 \\ 0 & 0 & 1 \end{bmatrix} \begin{bmatrix} x_1 \\ x_2 \\ x_3 \end{bmatrix} = \begin{bmatrix} -1 \\ -100 \\ 18 \end{bmatrix} \qquad \textbf{6}$$

Similarly, it happens that the number $\frac{1}{3}$ when multiplied by the coefficient 3 yields the identity 1. If we multiply both sides of the equation $3x = 6$ by $\frac{1}{3}$, we obtain $\frac{1}{3}(3x) = \frac{1}{3}(6)$ or $1 \cdot x = 2$ or $x = 2$.

that is,

$$\begin{bmatrix} x_1 \\ x_2 \\ x_3 \end{bmatrix} = \begin{bmatrix} -1 \\ -100 \\ 18 \end{bmatrix} \qquad \textbf{7}$$

This agrees with the solution found in Section 3-1 by Gauss-Jordan reduction.

How did we discover matrix **2**? you may ask. We will discuss this, and give you a practical procedure for finding such matrices, in Section 3-3. For now, let us discuss this discovery. Matrix **2** is called the **inverse** of matrix **3**. The inverse of a given *square* matrix is defined as the square matrix, necessarily of the same size, which when multiplied by the given matrix yields the identity matrix. For example,

$$\begin{bmatrix} -3 & 2 \\ 1 & -1 \end{bmatrix} \qquad \textbf{8}$$

is the inverse of

$$\begin{bmatrix} -1 & -2 \\ -1 & -3 \end{bmatrix} \qquad \textbf{9}$$

because

$$\begin{bmatrix} -1 & -2 \\ -1 & -3 \end{bmatrix} \begin{bmatrix} -3 & 2 \\ 1 & -1 \end{bmatrix} = \begin{bmatrix} 1 & 0 \\ 0 & 1 \end{bmatrix} \qquad \textbf{10}$$

Recall from Section 1-6 that the order in which two matrices are multiplied usually makes a difference. It comes as a pleasant surprise then to note that besides equations **4** and **10** we have

$$\begin{bmatrix} 6 & -1 & -5 \\ -7 & 1 & 5 \\ -10 & 2 & 11 \end{bmatrix} \begin{bmatrix} -1 & -1 & 0 \\ -27 & -16 & -5 \\ 4 & 2 & 1 \end{bmatrix} = \begin{bmatrix} 1 & 0 & 0 \\ 0 & 1 & 0 \\ 0 & 0 & 1 \end{bmatrix} \qquad \textbf{11}$$

and

$$\begin{bmatrix} -3 & 2 \\ 1 & -1 \end{bmatrix} \begin{bmatrix} -1 & -2 \\ -1 & -3 \end{bmatrix} = \begin{bmatrix} 1 & 0 \\ 0 & 1 \end{bmatrix} \qquad \textbf{12}$$

In general, the product *in either order* of a square matrix and its inverse is the same identity matrix. In view of this, we can say that matrix **9** is the inverse of matrix **8** as well. In general, if we find two square matrices whose product is an identity matrix, then each matrix is the inverse of the other. For example, check that

$$\begin{bmatrix} 0 & 5 & -7 & 1 \\ -2 & 0 & 0 & 1 \\ 1 & -2 & 3 & -1 \\ 1 & -1 & 1 & 0 \end{bmatrix} \begin{bmatrix} 1 & 1 & 2 & 1 \\ 4 & 5 & 9 & 1 \\ 3 & 4 & 7 & 1 \\ 2 & 3 & 4 & 2 \end{bmatrix} = \begin{bmatrix} 1 & 0 & 0 & 0 \\ 0 & 1 & 0 & 0 \\ 0 & 0 & 1 & 0 \\ 0 & 0 & 0 & 1 \end{bmatrix} \quad 13$$

This being true, we may say that the first matrix is the inverse of the second. We may also say that the second matrix is the inverse of the first and confidently write (although you may wish to check)

$$\begin{bmatrix} 1 & 1 & 2 & 1 \\ 4 & 5 & 9 & 1 \\ 3 & 4 & 7 & 1 \\ 2 & 3 & 4 & 2 \end{bmatrix} \begin{bmatrix} 0 & 5 & -7 & 1 \\ -2 & 0 & 0 & 1 \\ 1 & -2 & 3 & -1 \\ 1 & -1 & 1 & 0 \end{bmatrix} = \begin{bmatrix} 1 & 0 & 0 & 0 \\ 0 & 1 & 0 & 0 \\ 0 & 0 & 1 & 0 \\ 0 & 0 & 0 & 1 \end{bmatrix} \quad 14$$

Notice that we have only been discussing the inverses of *square* matrices. Nonsquare matrices are not very satisfactory as regards inverses. For example,

$$\begin{bmatrix} 1 & 2 & -1 \\ 2 & 2 & 3 \end{bmatrix} \begin{bmatrix} -9 & -3 \\ 6 & 2 \\ 2 & 1 \end{bmatrix} = \begin{bmatrix} 1 & 0 \\ 0 & 1 \end{bmatrix} \quad 15$$

but

$$\begin{bmatrix} -9 & -3 \\ 6 & 2 \\ 2 & 1 \end{bmatrix} \begin{bmatrix} 1 & 2 & -1 \\ 2 & 2 & 3 \end{bmatrix} = \begin{bmatrix} -15 & -24 & 0 \\ 10 & 16 & 0 \\ 4 & 6 & 1 \end{bmatrix} \quad 16$$

Thus, the two matrices on the left side of equation **15** are, we might say, "semi-inverses" of each other. Also, a given nonsquare matrix may have many such "semi-inverses." For example, compare equation **15** with

$$\begin{bmatrix} 1 & 2 & -1 \\ 2 & 2 & 3 \end{bmatrix} \begin{bmatrix} 3 & 1 \\ -3/_2 & -1/_2 \\ -1 & 0 \end{bmatrix} = \begin{bmatrix} 1 & 0 \\ 0 & 1 \end{bmatrix} \quad 17$$

It can be shown that situations of this type always occur with any nonsquare matrix. We shall avoid these difficulties by declaring, simply, that *nonsquare matrices do not have inverses*.

Do not think that every square matrix has an inverse. Consider for example the matrix

$$\begin{bmatrix} 1 & 2 \\ 0 & 0 \end{bmatrix} \quad 18$$

Its inverse would have to be a 2×2 matrix $\begin{bmatrix} x & y \\ z & w \end{bmatrix}$ such that

$$\begin{bmatrix} 1 & 2 \\ 0 & 0 \end{bmatrix} \begin{bmatrix} x & y \\ z & w \end{bmatrix} = \begin{bmatrix} 1 & 0 \\ 0 & 1 \end{bmatrix} \qquad \textbf{19}$$

But the $(2,2)$ entry in the product on the left side of system **19** is $0y + 0w$, and that cannot be made equal to 1 no matter what values we assign to y and w. Thus, there is no matrix that can make "equation" **19** hold. That is, matrix **18,** although square has no inverse. "Oh well," you may say, "matrix **18** has a row of zeros. Surely a square matrix with no row or column of zeros will always have an inverse." Not so. It can be shown, for example, that the matrix

$$\begin{bmatrix} 1 & 2 \\ 3 & 6 \end{bmatrix} \qquad \textbf{20}$$

See Exercise 11 in this section, or Section 3-6.

has no inverse. To summarize: nonsquare matrices do not have inverses. Some square matrices have inverses and some do not, and there appears to be no easy way to tell whether or not a given square matrix has an inverse.

Exercises 3.2

1. Write down the 5×5 identity matrix.
^{A}2. (a) How many 1's are in the 100×100 identity matrix?
    (b) How many 0's are in the 100×100 identity matrix?
3. What do you get if you multiply an identity matrix by itself?
4. Carry out the following multiplications:

^{A}(a) $\begin{bmatrix} 1 & 0 & 0 \\ 0 & 1 & 0 \\ 0 & 0 & 1 \end{bmatrix} \begin{bmatrix} 2 & 3 & 4 \\ 5 & 6 & 7 \\ 8 & 9 & 10 \end{bmatrix}$ ^{A}(e) $\begin{bmatrix} 1 & 0 & 0 \\ 0 & 0 & 0 \\ 0 & 0 & 0 \end{bmatrix} \begin{bmatrix} 2 & 3 & 4 \\ 5 & 6 & 7 \\ 8 & 9 & 10 \end{bmatrix}$

(b) $\begin{bmatrix} 1 & 0 & 0 \\ 0 & 1 & 0 \\ 0 & 0 & 0 \end{bmatrix} \begin{bmatrix} 2 & 3 & 4 \\ 5 & 6 & 7 \\ 8 & 9 & 10 \end{bmatrix}$ (f) $\begin{bmatrix} 2 & 3 & 4 \\ 5 & 6 & 7 \\ 8 & 9 & 10 \end{bmatrix} \begin{bmatrix} 1 & 0 & 0 \\ 0 & 0 & 0 \\ 0 & 0 & 0 \end{bmatrix}$

^{A}(c) $\begin{bmatrix} 2 & 3 & 4 \\ 5 & 6 & 7 \\ 8 & 9 & 10 \end{bmatrix} \begin{bmatrix} 1 & 0 & 0 \\ 0 & 1 & 0 \\ 0 & 0 & 0 \end{bmatrix}$ ^{A}(g) $\begin{bmatrix} 0 & 0 & 1 \\ 0 & 1 & 0 \\ 1 & 0 & 0 \end{bmatrix} \begin{bmatrix} 2 & 3 & 4 \\ 5 & 6 & 7 \\ 8 & 9 & 10 \end{bmatrix}$

(d) $\begin{bmatrix} 1 & 0 & 0 \\ 0 & 0 & 0 \\ 0 & 0 & 1 \end{bmatrix} \begin{bmatrix} 2 & 3 & 4 \\ 5 & 6 & 7 \\ 8 & 9 & 10 \end{bmatrix}$ (h) $\begin{bmatrix} 2 & 3 & 4 \\ 5 & 6 & 7 \\ 8 & 9 & 10 \end{bmatrix} \begin{bmatrix} 0 & 0 & 1 \\ 0 & 1 & 0 \\ 1 & 0 & 0 \end{bmatrix}$

5. Use the facts given by equations **10, 12, 13,** and **14** to solve, without Gauss-Jordan reduction, the following systems:

^{A}(a) $\begin{bmatrix} -3x_1 + 2x_2 = 4 \\ x_1 - x_2 = 7 \end{bmatrix}$ (b) $\begin{bmatrix} -x_1 - 2x_2 = \tfrac{1}{3} \\ -x_1 - 3x_2 = -\tfrac{1}{4} \end{bmatrix}$

(c) $\begin{bmatrix} x_1 + x_2 + 2x_3 + x_4 = 0 \\ 4x_1 + 5x_2 + 9x_3 + x_4 = 1 \\ 3x_1 + 4x_2 + 7x_3 + x_4 = -1 \\ 2x_1 + 3x_2 + 4x_3 + 2x_4 = 2 \end{bmatrix}$

A(d) $\begin{bmatrix} 5x_2 - 7x_3 + x_4 = -6 \\ -2x_1 + x_4 = \frac{2}{3} \\ x_1 - 2x_2 + 3x_3 - x_4 = 18 \\ x_1 - x_2 + x_3 = -41 \end{bmatrix}$

6. Use the facts that $\begin{bmatrix} 7 & 5 \\ 4 & 3 \end{bmatrix}$ is the inverse of $\begin{bmatrix} 3 & -5 \\ -4 & 7 \end{bmatrix}$,

$\frac{1}{18}\begin{bmatrix} 5 & -2 \\ -1 & 4 \end{bmatrix}$ is the inverse of $\begin{bmatrix} 4 & 2 \\ 1 & 5 \end{bmatrix}$, and $\begin{bmatrix} 1 & -1 & 1 \\ -1 & 1 & 0 \\ 0 & -1 & 1 \end{bmatrix}$

is the inverse of $\begin{bmatrix} 1 & 0 & -1 \\ 1 & 1 & -1 \\ 1 & 1 & 0 \end{bmatrix}$ to solve:

(a) $\begin{bmatrix} 7x_1 + 5x_2 = 1 \\ 4x_1 + 3x_2 = 2 \end{bmatrix}$

(d) $\begin{bmatrix} x_1 - x_2 + x_3 = 2 \\ -x_1 + x_2 = 4 \\ -x_2 + x_3 = 6 \end{bmatrix}$

(b) $\begin{bmatrix} 3x_1 - 5x_2 = -7 \\ -4x_1 + 7x_2 = 20 \end{bmatrix}$

(e) $\begin{bmatrix} x_1 - x_2 + x_3 = -1 \\ -x_1 + x_2 = 3 \\ -x_2 + x_3 = -5 \end{bmatrix}$

(c) $\begin{bmatrix} 4x_1 + 2x_2 = 10 \\ x_1 + 5x_2 = 25 \end{bmatrix}$

7. Solve:

(a) $\begin{bmatrix} 2x_1 + 4x_2 + 3x_3 + 2x_4 = 0 \\ 3x_1 + 6x_2 + 5x_3 + 2x_4 = 0 \\ 2x_1 + 5x_2 + 2x_3 - 3x_4 = -1 \\ 4x_1 + 5x_2 + 14x_3 + 14x_4 = 2 \end{bmatrix}$

A(b) $\begin{bmatrix} -115x_1 + 145x_2 - 64x_3 - 18x_4 = 1 \\ 50x_1 - 60x_2 + 26x_3 + 7x_4 = 0 \\ 5x_1 - 10x_2 + 6x_3 + 2x_4 = -1 \\ 10x_1 - 10x_2 + 3x_3 + x_4 = 2 \end{bmatrix}$

Note: the inverse of $\begin{vmatrix} 2 & 4 & 3 & 2 \\ 3 & 6 & 5 & 2 \\ 2 & 5 & 2 & -3 \\ 4 & 5 & 14 & 14 \end{vmatrix}$ is

$\frac{1}{5}\begin{bmatrix} -115 & 145 & -64 & -18 \\ 50 & -60 & 26 & 7 \\ 5 & -10 & 6 & 2 \\ 10 & -10 & 3 & 1 \end{bmatrix}$.

A 8. Is $\begin{bmatrix} 3 & 0 & -5 \\ -14 & 1 & 24 \\ -1 & 0 & 2 \end{bmatrix}$ the inverse of $\begin{bmatrix} 2 & 0 & 5 \\ 5 & 1 & 1 \\ 1 & 0 & 3 \end{bmatrix}$? Why or why not?

9. In I_{5000}:
   (a) What is the (2476,2476) entry?
   (b) What is the entry at address (4974,4749)?
   (c) List all addresses of the entry 2.

10. The second row of the inverse of the coefficient matrix of the PHE system **25** in Exercise 8 in Section 3-1 is $[-5/19 \quad 23/19 \quad 4/19 \quad -13/19 \quad 24/19]$. In the solution of system **25**, what is the value of x_2, which represents the total order of catsup (14-oz. bottles)?

*11. Suppose matrix **20** has an inverse. Then there is a 2×2 matrix $\begin{bmatrix} x & y \\ z & w \end{bmatrix}$ satisfying

$$\begin{bmatrix} 1 & 2 \\ 3 & 6 \end{bmatrix} \begin{bmatrix} x & y \\ z & w \end{bmatrix} \stackrel{?}{=} \begin{bmatrix} 1 & 0 \\ 0 & 1 \end{bmatrix} \qquad \textbf{21}$$

   (a) What are the (1,1) and (2,1) entries of the product on the left side of equation **21**?
   (b) If equation **21** is to hold, write the system of equations that would result from (a).
   (c) Apply Gauss-Jordan reduction to the system of (b). What happens?
   (d) In view of all this, what do you conclude about matrix **20**?

*12. Use the fact that the inverse of the matrix $\begin{bmatrix} 7 & -8 & 3 \\ 3 & -4 & 2 \\ 4 & -3 & 2 \end{bmatrix}$ is

$\frac{1}{9}\begin{bmatrix} 2 & -7 & 4 \\ 2 & 2 & -5 \\ -7 & 11 & 4 \end{bmatrix}$ to obtain the solution of the system

$$\begin{bmatrix} 4x_1 - 14x_2 + 8x_3 = 0 \\ 2x_1 \quad\; 2x_2 - 5x_3 = 8 \\ -7x_1 + 11x_2 + 4x_3 = 4 \end{bmatrix}.$$

Section 3-3
Calculating the Inverse of a Matrix

Our problem is: given a square matrix, does it have an inverse and, if it does, how is the inverse found? In some cases, mostly of theoretical interest only, we may want to know only whether a matrix has an inverse, not what the inverse is. There is a method to determine just this, which we discuss in Section 6-5. In the method to be presented in this section, we begin to calculate the inverse of the matrix without knowing whether or not it has one. If our calculation succeeds, we have the inverse; and it can be shown that if it fails, the inverse does not exist.

   We develop this method by calculating the inverse of the matrix

$$\begin{bmatrix} 6 & -1 & -5 \\ -7 & 1 & 5 \\ -10 & 2 & 11 \end{bmatrix} \qquad \textbf{1}$$

which we studied in the previous section. We seek a 3×3 matrix [?] such that

$$\begin{bmatrix} 6 & -1 & -5 \\ -7 & 1 & 5 \\ -10 & 2 & 11 \end{bmatrix} \begin{bmatrix} & & \\ & ? & \\ & & \end{bmatrix} = \begin{bmatrix} 1 & 0 & 0 \\ 0 & 1 & 0 \\ 0 & 0 & 1 \end{bmatrix} \qquad \textbf{2}$$

We calculate this unknown matrix one column at a time. Say the first column is $\begin{bmatrix} x_1 \\ x_2 \\ x_3 \end{bmatrix}$. Then matrix **1** times this column must give the first column of the 3×3 identity matrix in equation **2**:

$$\begin{bmatrix} 6 & -1 & -5 \\ 7 & 1 & 5 \\ -10 & 2 & 11 \end{bmatrix} \begin{bmatrix} x_1 \\ x_2 \\ x_3 \end{bmatrix} = \begin{bmatrix} 1 \\ 0 \\ 0 \end{bmatrix} \qquad \textbf{3}$$

We recognize equation **3** as the matrix form of a system of three linear equations in three variables (Section 3-1). We solve system **3** by Gauss-Jordan reduction:

$$\begin{bmatrix} 6 & -1 & -5 & 1 \\ -7 & 1 & 5 & 0 \\ 10 & 2 & 11 & 0 \end{bmatrix} \begin{matrix} \\ R_2 + \tfrac{7}{6}R_1 \\ R_3 + \tfrac{5}{3}R_1 \end{matrix} \Rightarrow \begin{bmatrix} 6 & -1 & -5 & 1 \\ 0 & -\tfrac{1}{6} & -\tfrac{5}{6} & \tfrac{7}{6} \\ 0 & \tfrac{1}{3} & \tfrac{8}{3} & \tfrac{5}{3} \end{bmatrix}$$

$$\tfrac{1}{6}R_1 \Rightarrow \begin{bmatrix} 1 & -\tfrac{1}{6} & -\tfrac{5}{6} & \tfrac{1}{6} \\ 0 & -\tfrac{1}{6} & -\tfrac{5}{6} & \tfrac{7}{6} \\ 0 & \tfrac{1}{3} & \tfrac{8}{3} & \tfrac{5}{3} \end{bmatrix} \begin{matrix} R_1 - R_2 \\ \\ R_3 + 2R_2 \end{matrix} \Rightarrow \begin{bmatrix} 1 & 0 & 0 & -1 \\ 0 & -\tfrac{1}{6} & -\tfrac{5}{6} & \tfrac{7}{6} \\ 0 & 0 & 1 & 4 \end{bmatrix}$$

$$-6R_2 \Rightarrow \begin{bmatrix} 1 & 0 & 0 & -1 \\ 0 & 1 & 5 & -7 \\ 0 & 0 & 1 & 4 \end{bmatrix} \begin{matrix} \\ R_2 - 5R_3 \\ \end{matrix} \Rightarrow \begin{bmatrix} 1 & 0 & 0 & -1 \\ 0 & 1 & 0 & -27 \\ 0 & 0 & 1 & 4 \end{bmatrix}$$

Hence, $x_1 = -1$, $x_2 = -27$, $x_3 = 4$, and so the first column of the inverse matrix is

$$C_1 = \begin{bmatrix} -1 \\ -27 \\ 4 \end{bmatrix} \qquad \textbf{4}$$

Having found the first column of the inverse, we now proceed in a similar fashion to find the second column. Suppose the second column of the inverse matrix is $\begin{bmatrix} x_1 \\ x_2 \\ x_3 \end{bmatrix}$. Then matrix **1** times this column must give the second column of the identity matrix of equation **2**:

$$\begin{bmatrix} 6 & -1 & -5 \\ -7 & 1 & 5 \\ -10 & 2 & 11 \end{bmatrix} \begin{bmatrix} x_1 \\ x_2 \\ x_3 \end{bmatrix} = \begin{bmatrix} 0 \\ 1 \\ 0 \end{bmatrix} \qquad \textbf{5}$$

We solve system **5** by Gauss-Jordan reduction:

$$\begin{bmatrix} 6 & -1 & -5 & 0 \\ -7 & 1 & 5 & 1 \\ -10 & 2 & 11 & 0 \end{bmatrix} \begin{matrix} \\ R_2 + \frac{7}{6}R_1 \\ R_3 + \frac{5}{3}R_1 \end{matrix} \longrightarrow \begin{bmatrix} 6 & -1 & -5 & 0 \\ 0 & -\frac{1}{6} & -\frac{5}{6} & 1 \\ 0 & \frac{1}{3} & \frac{8}{3} & 0 \end{bmatrix}$$

$$\frac{1}{6}R_1 \longrightarrow \begin{bmatrix} 1 & -\frac{1}{6} & -\frac{5}{6} & 0 \\ 0 & -\frac{1}{6} & -\frac{5}{6} & 1 \\ 0 & \frac{1}{3} & \frac{8}{3} & 0 \end{bmatrix} \begin{matrix} R_1 - R_2 \\ \\ R_3 + 2R_2 \end{matrix} \longrightarrow \begin{bmatrix} 1 & 0 & 0 & -1 \\ 0 & -\frac{1}{6} & -\frac{5}{6} & 1 \\ 0 & 0 & 1 & 2 \end{bmatrix}$$

$$-6R_2 \longrightarrow \begin{bmatrix} 1 & 0 & 0 & -1 \\ 0 & 1 & 5 & -6 \\ 0 & 0 & 1 & 2 \end{bmatrix} \begin{matrix} \\ R_2 - 5R_3 \\ \end{matrix} \longrightarrow \begin{bmatrix} 1 & 0 & 0 & -1 \\ 0 & 1 & 0 & -16 \\ 0 & 0 & 1 & 2 \end{bmatrix}$$

The second column of the inverse matrix is thus

$$C_2 = \begin{bmatrix} -1 \\ -16 \\ 2 \end{bmatrix} \qquad \textbf{6}$$

Finally, we consider the third column. Suppose the third column of the inverse matrix is $\begin{bmatrix} x_1 \\ x_2 \\ x_3 \end{bmatrix}$. Then, proceeding as before, we must solve

$$\begin{bmatrix} 6 & -1 & -5 \\ -7 & 1 & 5 \\ -10 & 2 & 11 \end{bmatrix} \begin{bmatrix} x_1 \\ x_2 \\ x_3 \end{bmatrix} = \begin{bmatrix} 0 \\ 0 \\ 1 \end{bmatrix} \qquad \textbf{7}$$

$$\begin{bmatrix} 6 & -1 & -5 & 0 \\ -7 & 1 & 5 & 0 \\ -10 & 2 & 11 & 1 \end{bmatrix} \begin{matrix} \\ R_2 + \frac{7}{6}R_1 \\ R_3 + \frac{5}{3}R_1 \end{matrix} \longrightarrow \begin{bmatrix} 6 & -1 & -5 & 0 \\ 0 & -\frac{1}{6} & -\frac{5}{6} & 0 \\ 0 & \frac{1}{3} & \frac{8}{3} & 1 \end{bmatrix}$$

$$\frac{1}{6}R_1 \longrightarrow \begin{bmatrix} 1 & -\frac{1}{6} & -\frac{5}{6} & 0 \\ 0 & -\frac{1}{6} & -\frac{5}{6} & 0 \\ 0 & \frac{1}{3} & \frac{8}{3} & 1 \end{bmatrix} \begin{matrix} R_1 - R_2 \\ \\ R_3 + 2R_2 \end{matrix} \longrightarrow \begin{bmatrix} 1 & 0 & 0 & 0 \\ 0 & -\frac{1}{6} & -\frac{5}{6} & 0 \\ 0 & 0 & 1 & 1 \end{bmatrix}$$

$$-6R_2 \quad \begin{bmatrix} 1 & 0 & 0 & 0 \\ 0 & 1 & 5 & 0 \\ 0 & 0 & 1 & 1 \end{bmatrix} \quad R_2 - 5R_3 \quad \begin{bmatrix} 1 & 0 & 0 & 0 \\ 0 & 1 & 0 & -5 \\ 0 & 0 & 1 & 1 \end{bmatrix}$$

The third column of the inverse matrix is thus

$$C_3 = \begin{bmatrix} 0 \\ -5 \\ 1 \end{bmatrix} \qquad \qquad \textbf{8}$$

We have computed each column; from equations **4**, **6**, and **8**, the inverse matrix is

$$\begin{bmatrix} -1 & -1 & 0 \\ -27 & -16 & -5 \\ 4 & 2 & 1 \end{bmatrix} \qquad \qquad \textbf{9}$$

as was checked in equation **4** of Section 3-2.

Now notice this: since the coefficient matrix for each of systems **3**, **5**, and **7** is the same matrix, namely matrix **1**, the steps performed in each of the three Gauss-Jordan reductions above were exactly the same. We could have carried out all three Gauss-Jordan reductions at the same time, simply carrying along, in order, the computations with all three columns of the identity matrix of equation **2**, like this:

$$\begin{bmatrix} 6 & -1 & -5 & 1 & 0 & 0 \\ -7 & 1 & 5 & 0 & 1 & 0 \\ -10 & 2 & 11 & 0 & 0 & 1 \end{bmatrix} \quad \begin{matrix} R_2 + \frac{7}{6}R_1 \\ R_3 + \frac{5}{3}R_1 \end{matrix} \quad \begin{bmatrix} 6 & -1 & -5 & 1 & 0 & 0 \\ 0 & -\frac{1}{6} & -\frac{5}{6} & \frac{7}{6} & 1 & 0 \\ 0 & \frac{1}{3} & \frac{8}{3} & \frac{5}{3} & 0 & 1 \end{bmatrix}$$

$$\frac{1}{6}R_1 \quad \begin{bmatrix} 1 & -\frac{1}{6} & -\frac{5}{6} & \frac{1}{6} & 0 & 0 \\ 0 & -\frac{1}{6} & -\frac{5}{6} & \frac{7}{6} & 1 & 0 \\ 0 & \frac{1}{3} & \frac{8}{3} & \frac{5}{3} & 0 & 1 \end{bmatrix}$$

$$\begin{matrix} R_1 - R_2 \\ \\ R_3 + 2R_2 \end{matrix} \quad \begin{bmatrix} 1 & 0 & 0 & -1 & -1 & 0 \\ 0 & -\frac{1}{6} & -\frac{5}{6} & \frac{7}{6} & 1 & 0 \\ 0 & 0 & 1 & 4 & 2 & 1 \end{bmatrix}$$

$$-6R_2 \quad \begin{bmatrix} 1 & 0 & 0 & -1 & -1 & 0 \\ 0 & 1 & 5 & -7 & -6 & 0 \\ 0 & 0 & 1 & 4 & 2 & 1 \end{bmatrix}$$

$$R_2 - 5R_3 \quad \begin{bmatrix} 1 & 0 & 0 & -1 & -1 & 0 \\ 0 & 1 & 0 & -27 & -16 & -5 \\ 0 & 0 & 1 & 4 & 2 & 1 \end{bmatrix}$$

Note: if you do not perform Gauss-Jordan reduction as presented in Section 2-4 but use instead some modification such as the shortcuts in Section 2-5, add to the third box of the flow chart the statement, "Do not clear any column in the right half of the matrix."

The answer, matrix **9**, appears "all put together" at the end (outlined in dashes).

It can be shown that this method for finding the inverse of a matrix will work for any square matrix that has an inverse. If the method fails, it is either because the given matrix is not square or because it is square but has no inverse. We present the general method in the following flow chart.

**To Find the Inverse of a
Matrix, If It Exists**

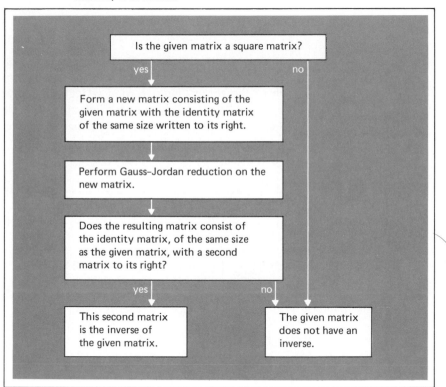

Problem 1   Find the inverse, if it exists, of $\begin{bmatrix} 0 & 5 & -7 & 1 \\ -2 & 0 & 0 & 1 \\ 1 & -2 & 3 & -1 \\ 1 & -1 & 1 & 0 \end{bmatrix}$

Solution

$$\left[\begin{array}{cccc|cccc} 0 & 5 & -7 & 1 & 1 & 0 & 0 & 0 \\ -2 & 0 & 0 & 1 & 0 & 1 & 0 & 0 \\ 1 & -2 & 3 & -1 & 0 & 0 & 1 & 0 \\ 1 & -1 & 1 & 0 & 0 & 0 & 0 & 1 \end{array}\right]$$

$$\left[\begin{array}{cccc|cccc} -2 & 0 & 0 & 1 & 0 & 1 & 0 & 0 \\ 0 & 5 & -7 & 1 & 1 & 0 & 0 & 0 \\ 1 & -2 & 3 & -1 & 0 & 0 & 1 & 0 \\ 1 & -1 & 1 & 0 & 0 & 0 & 0 & 1 \end{array}\right]$$

$$\left[\begin{array}{cccc|cccc} 1 & 0 & 0 & -\frac{1}{2} & 0 & -\frac{1}{2} & 0 & 0 \\ 0 & 5 & -7 & 1 & 1 & 0 & 0 & 0 \\ 0 & -2 & 3 & -\frac{1}{2} & 0 & \frac{1}{2} & 1 & 0 \\ 0 & -1 & 1 & \frac{1}{2} & 0 & \frac{1}{2} & 0 & 1 \end{array}\right]$$

$$\left[\begin{array}{cccc|cccc} 1 & 0 & 0 & -\frac{1}{2} & 0 & -\frac{1}{2} & 0 & 0 \\ 0 & 1 & -\frac{7}{5} & \frac{1}{5} & \frac{1}{5} & 0 & 0 & 0 \\ 0 & 0 & \frac{1}{5} & -\frac{1}{10} & \frac{2}{5} & \frac{1}{2} & 1 & 0 \\ 0 & 0 & -\frac{2}{5} & \frac{7}{10} & \frac{1}{5} & \frac{1}{2} & 0 & 1 \end{array}\right]$$

$$\left[\begin{array}{cccc|cccc} 1 & 0 & 0 & -\frac{1}{2} & 0 & -\frac{1}{2} & 0 & 0 \\ 0 & 1 & 0 & -\frac{1}{2} & 3 & \frac{7}{2} & 7 & 0 \\ 0 & 0 & 1 & -\frac{1}{2} & 2 & \frac{5}{2} & 5 & 0 \\ 0 & 0 & 0 & \frac{1}{2} & 1 & \frac{3}{2} & 2 & 1 \end{array}\right]$$

$$\left[\begin{array}{cccc|cccc} 1 & 0 & 0 & 0 & 1 & 1 & 2 & 1 \\ 0 & 1 & 0 & 0 & 4 & 5 & 9 & 1 \\ 0 & 0 & 1 & 0 & 3 & 4 & 7 & 1 \\ 0 & 0 & 0 & 1 & 2 & 3 & 4 & 2 \end{array}\right]$$

The inverse is $\begin{bmatrix} 1 & 1 & 2 & 1 \\ 4 & 5 & 9 & 1 \\ 3 & 4 & 7 & 1 \\ 2 & 3 & 4 & 2 \end{bmatrix}$ as was checked in equations **13** and **14** in Section 3-2.

Problem 2 Find the inverse, if it exists, of $\begin{bmatrix} 0 & 1 & 2 & 3 \\ 4 & 7 & 2 & 1 \end{bmatrix}$

Solution The matrix is not square, so it has no inverse.

Problem 3   Find the inverse, if it exists, of

$$\begin{bmatrix} -1 & 2 & 4 \\ 3 & -4 & -3 \\ 2 & -2 & 1 \end{bmatrix} \qquad \textbf{10}$$

Solution

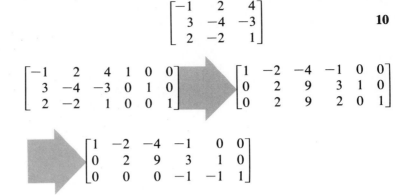

$$\begin{bmatrix} -1 & 2 & 4 & 1 & 0 & 0 \\ 3 & -4 & -3 & 0 & 1 & 0 \\ 2 & -2 & 1 & 0 & 0 & 1 \end{bmatrix} \implies \begin{bmatrix} 1 & -2 & -4 & -1 & 0 & 0 \\ 0 & 2 & 9 & 3 & 1 & 0 \\ 0 & 2 & 9 & 2 & 0 & 1 \end{bmatrix}$$

$$\implies \begin{bmatrix} 1 & -2 & -4 & -1 & 0 & 0 \\ 0 & 2 & 9 & 3 & 1 & 0 \\ 0 & 0 & 0 & -1 & -1 & 1 \end{bmatrix}$$

The point is: if at any point in the reduction you see that you cannot achieve the identity matrix in the left half, QUIT. This illustrates one advantage the mind has over the computer, or at least we like to think it does.

Halt! It is now clear that we are not going to be able to achieve the 3 × 3 identity matrix in the left half of this matrix; so we may now say that matrix **10** has no inverse. Perhaps you saw that this was going to happen one step before we did. If so, that's fine; no need to go further. Perhaps, on the other hand, you still do not see what is going to happen. Then you must proceed:

$$\implies \begin{bmatrix} 1 & 0 & 5 & 2 & 1 & 0 \\ 0 & 1 & 9/2 & 3/2 & 1/2 & 0 \\ 0 & 0 & 0 & -1 & -1 & 1 \end{bmatrix} \implies \begin{bmatrix} 1 & 0 & 5 & 0 & -1 & 2 \\ 0 & 1 & 9/2 & 0 & -1 & 3/2 \\ 0 & 0 & 0 & 1 & 1 & -1 \end{bmatrix}$$

This is the final result of Gauss-Jordan reduction, from which we conclude that the matrix has no inverse.

All of the shortcuts for Gauss-Jordan reduction that were discussed in Section 2-5 may be employed in calculating inverse matrices, with three exceptions: (1) do not clear any column in the right half of the matrix being reduced; (2) rows may have to be switched in order to achieve the identity matrix in the left side; and (3) the cleared columns in the left half must contain 1's. As in Section 2-5, we recommend shortcuts only to students who have a thorough understanding of Gauss-Jordan reduction.

Exercises 3.3   Find the inverse, if it exists, of each of the following matrices:

A1. $\begin{bmatrix} 2 & 0 \\ 0 & 3 \end{bmatrix}$

A3. $\begin{bmatrix} 2 & 4 \\ 0 & 3 \end{bmatrix}$

2. $\begin{bmatrix} 0 & 2 \\ 3 & 0 \end{bmatrix}$

4. $\begin{bmatrix} 2 & 4 \\ 5 & 3 \end{bmatrix}$

A5. $\begin{bmatrix} -2 & 4 \\ -5 & 3 \end{bmatrix}$

6. $\begin{bmatrix} 1/2 & 1/4 \\ 5 & 3 \end{bmatrix}$

A7. $\begin{bmatrix} -2/3 & -3/4 \\ 4/5 & -3/7 \end{bmatrix}$

8. $\begin{bmatrix} 2 & 0 & 0 \\ 0 & -3 & 0 \\ 0 & 0 & 4/5 \end{bmatrix}$

A9. $\begin{bmatrix} 1 & 2 & 3 \\ 4 & 5 & 6 \\ 7 & 8 & 9 \end{bmatrix}$

10. $\begin{bmatrix} 1 & 2 & 3 \\ 4 & 5 & 6 \\ 9 & 8 & 7 \end{bmatrix}$

A11. $\begin{bmatrix} 1 & 2 & 3 & 4 \\ 5 & 6 & 7 & 8 \end{bmatrix}$

12. $\begin{bmatrix} 5 & 6 & 6 \\ 3 & 3 & 2 \\ 4 & 3 & 3 \end{bmatrix}$

A13. $\begin{bmatrix} 3 & -2 & 4 \\ 2 & 1 & 2 \\ 5 & 3 & 5 \end{bmatrix}$

14. $\begin{bmatrix} 3 & 3 & 5 \\ 1 & 6 & 8 \\ 2 & 1 & 2 \end{bmatrix}$

15. $\begin{bmatrix} 4 & 6 & 1 \\ 9 & 8 & 5 \\ 4 & 4 & 2 \end{bmatrix}$

16. $\begin{bmatrix} 2 & 3 & 1 \\ 2 & 7 & 1 \\ 3 & 2 & 1 \end{bmatrix}$

17. $\begin{bmatrix} 1 & 2 & 3 \\ 3 & 1 & 2 \\ 2 & 3 & 1 \end{bmatrix}$

*18. $\begin{bmatrix} -1/3 & 0 & 2/3 \\ 3 & -1/2 & 2 \\ 2/3 & -3/4 & 0 \end{bmatrix}$

*19. $\begin{bmatrix} 0 & -3 & 2 & 1 \\ 0 & 2 & 1 & -1 \\ 4 & -2 & 0 & 3 \\ 2 & 0 & 0 & -7 \end{bmatrix}$

*20. $\begin{bmatrix} 2 & 4 & 3 & 2 \\ 3 & 6 & 5 & 2 \\ 2 & 5 & 2 & -3 \\ 4 & 5 & 14 & 14 \end{bmatrix}$

A21. Find a 2×2 matrix X that satisfies the following equation:

$$\begin{bmatrix} 2 & 3 \\ 1 & 2 \end{bmatrix} X - \begin{bmatrix} 3 & -1 \\ -1 & 2 \end{bmatrix} = \begin{bmatrix} 7 & -1 \\ 5 & 4 \end{bmatrix}.$$

*22. Calculate the inverse of the coefficient matrix of the PHE system **25** of Exercise 8, Section 3-1 (the second row of this inverse is given in Exercise 10, Section 3-2). This problem is starred not because it is hard but because it is computationally tedious. The current indoor record for hand calculation of this inverse is 30 minutes. Perhaps you can beat this time.

Section 3-4
Applications of Inverse
Matrices

For solving a single system of linear equations, using Gauss-Jordan reduction is shorter than first finding the inverse of the coefficient matrix. For example, to solve the system

$$\begin{bmatrix} 2x_1 + 5x_2 = 9 \\ x_1 + 3x_2 = 5 \end{bmatrix}$$

1

by inversion we must first find the inverse of the coefficient matrix $\begin{bmatrix} 2 & 5 \\ 1 & 3 \end{bmatrix}$:

$$\begin{bmatrix} 2 & 5 & 1 & 0 \\ 1 & 3 & 0 & 1 \end{bmatrix} \Rightarrow \begin{bmatrix} 1 & 5/2 & 1/2 & 0 \\ 0 & 1/2 & -1/2 & 1 \end{bmatrix} \Rightarrow \begin{bmatrix} 1 & 0 & 3 & -5 \\ 0 & 1 & -1 & 2 \end{bmatrix}$$

Answer: $\begin{bmatrix} 3 & -5 \\ -1 & 2 \end{bmatrix}$. Check: $\begin{bmatrix} 2 & 5 \\ 1 & 3 \end{bmatrix}\begin{bmatrix} 3 & -5 \\ -1 & 2 \end{bmatrix} = \begin{bmatrix} 1 & 0 \\ 0 & 1 \end{bmatrix}$.

Then we must rewrite system **1** in matrix form:

$$\begin{bmatrix} 2 & 5 \\ 1 & 3 \end{bmatrix}\begin{bmatrix} x_1 \\ x_2 \end{bmatrix} = \begin{bmatrix} 9 \\ 5 \end{bmatrix}$$

and multiply through by the inverse found above:

$$\begin{bmatrix} 3 & -5 \\ -1 & 2 \end{bmatrix}\begin{bmatrix} 2 & 5 \\ 1 & 3 \end{bmatrix}\begin{bmatrix} x_1 \\ x_2 \end{bmatrix} = \begin{bmatrix} 3 & -5 \\ -1 & 2 \end{bmatrix}\begin{bmatrix} 9 \\ 5 \end{bmatrix}$$

that is,

$$\begin{bmatrix} 1 & 0 \\ 0 & 1 \end{bmatrix}\begin{bmatrix} x_1 \\ x_2 \end{bmatrix} = \begin{bmatrix} 2 \\ 1 \end{bmatrix}$$

that is,

$$\begin{bmatrix} x_1 \\ x_2 \end{bmatrix} = \begin{bmatrix} 2 \\ 1 \end{bmatrix} \qquad\qquad \mathbf{2}$$

On the other hand, the solution of system **1** by Gauss-Jordan reduction runs as follows:

$$\begin{bmatrix} 2 & 5 & 9 \\ 1 & 3 & 5 \end{bmatrix} \Rightarrow \begin{bmatrix} 1 & 5/2 & 9/2 \\ 0 & 1/2 & 1/2 \end{bmatrix} \Rightarrow \begin{bmatrix} 1 & 0 & 2 \\ 0 & 1 & 1 \end{bmatrix}$$

from which we obtain system **2**. This is certainly a shorter solution!

The practical importance of matrix inversion comes in dealing with many systems of linear equations, each having the same coefficient matrix (assuming that this matrix has an inverse). One example will suffice to illustrate this.

Appalachian Creations often fills orders for long and short tent stakes. In their Boston warehouse they have two types of boxes of tent stakes: type 1 contains 2 dozen short and 1 dozen long stakes; type 2 contains 5 dozen short and 3 dozen long stakes. In filling orders it is easier to ship these boxes than to break them up. For some orders this is possible and for others it seems impossible.

Problem 1   How can the order department of AC determine whether or not it is possible to fill a given order with whole boxes, and if it is, how many of each type are needed to fill the order? Table 1 shows a few of the vast number of incoming orders.

Table 1

	Order number						
	51	52	53	54	55	56	57
Short stakes (doz.)	9	5	4	7	98	74	40
Long stakes (doz.)	5	3	2	4	52	30	40

Solution   Let x_1 be the number of type 1 boxes to be shipped and x_2 be the number of type 2 boxes to be shipped in filling a given order. A shipment of x_1 type 1 boxes and x_2 type 2 boxes contains $2x_1 + 5x_2$ dozen short stakes and $x_1 + 3x_2$ dozen long stakes. In order to fill order number 51, we need to solve the system

$$\left[\begin{array}{l} 2x_1 + 5x_2 = 9 \\ x_1 + 3x_2 = 5 \end{array}\right]$$

that is, system **1**. We solved this system most easily by Gauss-Jordan reduction, but now we have to solve the system for order number 52,

$$\left|\begin{array}{l} 2x_1 + 5x_2 = 5 \\ x_1 + 3x_2 = 3 \end{array}\right]$$

then the system for order number 53, and so on. Is there an easier way to proceed, that is, a way involving fewer computations than are needed with repeated Gauss-Jordan reductions?

We have to solve many systems of the form

$$\left[\begin{array}{l} 2x_1 + 5x_2 = b \\ x_1 + 3x_2 = c \end{array}\right] \qquad \textbf{3}$$

for various orders of b dozen short stakes and c dozen long stakes. Rewrite system **3** in matrix form:

$$\begin{bmatrix} 2 & 5 \\ 1 & 3 \end{bmatrix} \begin{bmatrix} x_1 \\ x_2 \end{bmatrix} = \begin{bmatrix} b \\ c \end{bmatrix} \qquad \textbf{4}$$

Multiply the inverse of the coefficient matrix, found above, by both sides of equation **4**:

$$\begin{bmatrix} 3 & -5 \\ -1 & 2 \end{bmatrix} \begin{bmatrix} 2 & 5 \\ 1 & 3 \end{bmatrix} \begin{bmatrix} x_1 \\ x_2 \end{bmatrix} = \begin{bmatrix} 3 & -5 \\ -1 & 2 \end{bmatrix} \begin{bmatrix} b \\ c \end{bmatrix}$$

$$\begin{bmatrix} x_1 \\ x_2 \end{bmatrix} = \begin{bmatrix} 3 & -5 \\ -1 & 2 \end{bmatrix} \begin{bmatrix} b \\ c \end{bmatrix} \qquad \textbf{5}$$

We may use equation **5** again and again to determine quickly solutions to system **3** for various orders. Then we can see if the orders can be realized, that is, if the numbers x_1 and x_2 obtained are non-negative. (This is obviously a requirement of this problem, since Boston cannot ship a negative number of packages of either type.) We may arrange our work as shown in Table 2.

Table 2

Order #	Order	Result of using equation 5	Instructions or comment
51	$\begin{bmatrix} 9 \\ 5 \end{bmatrix}$	$\begin{bmatrix} 2 \\ 1 \end{bmatrix}$	ship 2 type 1 and 1 type 2
52	$\begin{bmatrix} 5 \\ 3 \end{bmatrix}$	$\begin{bmatrix} 0 \\ 1 \end{bmatrix}$	ship 1 type 2
53	$\begin{bmatrix} 4 \\ 2 \end{bmatrix}$	$\begin{bmatrix} 2 \\ 0 \end{bmatrix}$	ship 2 type 1
54	$\begin{bmatrix} 7 \\ 4 \end{bmatrix}$	$\begin{bmatrix} 1 \\ 1 \end{bmatrix}$	ship 1 type 1 and 1 type 2
55	$\begin{bmatrix} 98 \\ 52 \end{bmatrix}$	$\begin{bmatrix} 34 \\ 6 \end{bmatrix}$	ship 34 type 1 and 6 type 2
56	$\begin{bmatrix} 74 \\ 30 \end{bmatrix}$	$\begin{bmatrix} 72 \\ -14 \end{bmatrix}$	cannot be done with whole boxes
57	$\begin{bmatrix} 40 \\ 40 \end{bmatrix}$	$\begin{bmatrix} -80 \\ 40 \end{bmatrix}$	cannot be done with whole boxes

Instructions for this procedure could easily be put on a small computer using the flow chart below.

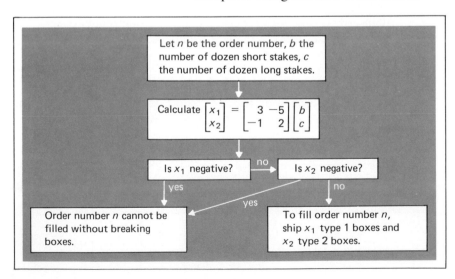

Exercises 3-4 ^1. In the earlier days at AC the two types of boxes of tent stakes (these were wooden stakes) were packed as follows: Type 1 boxes contained 5 pairs of long stakes and 7 pairs of short stakes; type 2 boxes contained 2 pairs of long stakes and 3 of short. Please fill each of the orders in Table 3, except for those that cannot be filled without opening boxes:

Table 3

	Order#								
	1	2	3	4	5	6	7	8	9
Pairs of long	6	7	114	76	137	80	9815	5	2
Pairs of short	9	10	162	106	204	124	14179	7	3

2. Solve the systems $\begin{bmatrix} x_1 + x_2 + 2x_3 + x_4 = a \\ 4x_1 + 5x_2 + 9x_3 + x_4 = b \\ 3x_1 + 4x_2 + 7x_3 + x_4 = c \\ 2x_1 + 3x_2 + 4x_3 + 2x_4 = d \end{bmatrix}$ for:

(a) $a = 1, b = c = d = 0$
(b) $a = b = 1, c = d = 0$
(c) $a = b = c = 1, d = 0$
(d) $a = b = c = d = 1$
(e) $a = 0, b = c = d = 1$
(f) $a = b = 0, c = d = 1$
(g) $a = b = c = 0, d = 1$
(h) $a = -19, b = 8, c = 14, d = -12$

3. Solve each of the following systems:

(a) $\begin{bmatrix} 6 & 11 \\ 1 & 2 \end{bmatrix} \begin{bmatrix} x_1 \\ x_2 \end{bmatrix} = \begin{bmatrix} 7 \\ 8 \end{bmatrix}$

(d) $\begin{bmatrix} 6x_1 + 11x_2 \\ x_1 + 2x_2 \end{bmatrix} = \begin{bmatrix} -47 \\ 0 \end{bmatrix}$

(b) $\begin{bmatrix} 6x_1 + 11x_2 \\ x_1 + 2x_2 \end{bmatrix} = \begin{bmatrix} 8 \\ 7 \end{bmatrix}$

c(e) $\begin{bmatrix} 11x_1 + 6x_2 \\ 2x_1 + x_2 \end{bmatrix} = \begin{bmatrix} 2 \\ 3 \end{bmatrix}$

c(c) $\begin{bmatrix} x_1 + 2x_2 \\ 6x_1 + 11x_2 \end{bmatrix} = \begin{bmatrix} 9 \\ 10 \end{bmatrix}$

4. Solve $\begin{bmatrix} 2 & 5 \\ 1 & 3 \end{bmatrix} \begin{bmatrix} x_1 \\ x_2 \end{bmatrix} = \begin{bmatrix} -3 & 2 \\ 1 & 3 \end{bmatrix} \begin{bmatrix} 2 \\ 4 \end{bmatrix}$

c5. The Acme Candy Company has a large supply of three types of boxes of candy left over from last Christmas. Type I contains 4 rolls of mints, 4 chocolate bars, and 7 jawbreakers; type II contains 7 rolls of mints and 6 jawbreakers; type III contains 2 rolls of mints, 5 chocolate bars, and 6 jawbreakers. The company is receiving orders for these candies which it would like to fill using these boxes, unopened. For which of the orders in Table 4 can this be done, and how?

Table 4

	71	72	73	74	75	76	77
Mints	45	50	50	40	62	30	450
Chocolates	80	85	90	70	40	39	800
Jawbreakers	110	115	120	100	80	75	1080

6. My cousin claims to be an expert bug exterminator. He tells me that the best insecticide is made from a combination of three chemicals: PX17, 3R27, and 2Z16. He wants to test 8 different insecticides with the quantities (in ounces) of each of these chemicals listed in Table 5. His problem is that PX17, 3R27, and 2Z16 are not available in pure form. They can only be obtained mixed together in an oil base according to Table 6, where the numbers give ounces per quart. Which combinations can be produced by mixing chemicals A, B, and C, and how many quarts of each chemical should be used in each case?

Table 5

	a	b	c	d	e	f	g	h
PX17	20	30	10	45	65	45	50	100
3R27	10	15	20	20	15	10	20	15
2Z16	10	10	30	35	10	15	30	20

Table 6

	PX17	3R27	2Z16
Mixture A	7	8	3
Mixture B	3	4	2
Mixture C	4	3	0

A7. Some of the systems of equations obtained at PHE in connection with order placement have solutions involving fractions, which do not correspond to reality. One cannot, for example, order 244 $^{15}/_{43}$ bottles of catsup. The standard PHE policy in cases like this is always to order the next higher whole number; so in this case the order is placed for 245 bottles. On the other hand, a negative value of one of the variables indicates that the region already has an oversupply of that item, and then the policy is, of course, to order no more items. Thus, with these policies, a solution to a given problem such as, say,

$$\begin{bmatrix} x_1 \\ x_2 \\ x_3 \\ x_4 \end{bmatrix} = \begin{bmatrix} 49^1/_5 \\ 180 \\ -16^2/_3 \\ 8^7/_8 \end{bmatrix},$$ would be adjusted to the "practical" solution

$$\begin{bmatrix} x_1 \\ x_2 \\ x_3 \\ x_4 \end{bmatrix} = \begin{bmatrix} 50 \\ 180 \\ 0 \\ 9 \end{bmatrix}.$$ With these policies in mind, give the "practical" solu-

tions to the systems AX = B, which arose in connection with supply of various types of hamburger and hot-dog buns to several eastern

regions, where $A = \begin{bmatrix} 3 & 2 & 4 \\ 7 & 0 & 3 \\ 6 & 4 & 9 \end{bmatrix}$ and where:

(a) $B = \begin{bmatrix} 8 \\ 9 \\ 10 \end{bmatrix}$ (c) $B = \begin{bmatrix} 60 \\ 50 \\ 100 \end{bmatrix}$

(b) $B = \begin{bmatrix} 10 \\ 11 \\ 70 \end{bmatrix}$ (d) $B = \begin{bmatrix} 200 \\ 500 \\ 400 \end{bmatrix}$

Section 3-5
Notation and Properties of
Matrix Inversion

The inverse of the matrix

$$\begin{bmatrix} 2 & 5 \\ 1 & 3 \end{bmatrix} \tag{1}$$

if it exists, is denoted

$$\begin{bmatrix} 2 & 5 \\ 1 & 3 \end{bmatrix}^{-1} \tag{2}$$

which is read " $\begin{bmatrix} 2 & 5 \\ 1 & 3 \end{bmatrix}$ inverse." This notation is analogous to

the notation for the reciprocal (multiplicative inverse) of a non-

See Appendix A, Section 4-4.   zero number. For example, $3^{-1} = \frac{1}{3}$ and $(-\frac{4}{5})^{-1} = -\frac{5}{4}$. We discovered in Section 3-4 that the inverse of matrix 1 does exist and
is

$$\begin{bmatrix} 3 & -5 \\ -1 & 2 \end{bmatrix} \tag{3}$$

We may use this new notation to express this fact by writing

$$\begin{bmatrix} 2 & 5 \\ 1 & 3 \end{bmatrix}^{-1} = \begin{bmatrix} 3 & -5 \\ -1 & 2 \end{bmatrix} \tag{4}$$

Recall that since matrix **3** is the inverse of matrix **1**, we may also say that matrix **1** is the inverse of matrix **3**. This fact is expressed by writing

$$\begin{bmatrix} 3 & -5 \\ -1 & 2 \end{bmatrix}^{-1} = \begin{bmatrix} 2 & 5 \\ 1 & 3 \end{bmatrix} \qquad \textbf{5}$$

Another example: equations **13** and **14** in Section 3-2 tell us that

$$\begin{bmatrix} 0 & 5 & -7 & 1 \\ -2 & 0 & 0 & 1 \\ 1 & -2 & 3 & -1 \\ 1 & -1 & 1 & 0 \end{bmatrix}^{-1} = \begin{bmatrix} 1 & 1 & 2 & 1 \\ 4 & 5 & 9 & 1 \\ 3 & 4 & 7 & 1 \\ 2 & 3 & 4 & 2 \end{bmatrix} \qquad \textbf{6}$$

and

$$\begin{bmatrix} 1 & 1 & 2 & 1 \\ 4 & 5 & 9 & 1 \\ 3 & 4 & 7 & 1 \\ 2 & 3 & 4 & 2 \end{bmatrix}^{-1} = \begin{bmatrix} 0 & 5 & -7 & 1 \\ -2 & 0 & 0 & 1 \\ 1 & -2 & 3 & -1 \\ 1 & -1 & 1 & 0 \end{bmatrix} \qquad \textbf{7}$$

Symbolically, the inverse of a square matrix A, if it exists, is denoted A^{-1}. If A^{-1} exists, then
$$AA^{-1} = A^{-1}A = I$$
where I is the identity matrix of the same size as A.

How is inversion related to multiplication of a matrix by a number? Given equation **7**, what is the inverse of, say,

$$(-\tfrac{4}{5}) \begin{bmatrix} 1 & 1 & 2 & 1 \\ 4 & 5 & 9 & 1 \\ 3 & 4 & 7 & 1 \\ 2 & 3 & 4 & 2 \end{bmatrix} \qquad \textbf{8}$$

Equation **7** tells us the inverse of the matrix; the inverse of the number $-\tfrac{4}{5}$ is $(-\tfrac{4}{5})^{-1} = -\tfrac{5}{4}$. Multiply the inverse of the number by the inverse of the matrix to obtain

We could actually perform the indicated multiplications; however, it would needlessly complicate the matrix.

$$(-\tfrac{5}{4}) \begin{bmatrix} 0 & 5 & -7 & 1 \\ -2 & 0 & 0 & 1 \\ 1 & -2 & 3 & -1 \\ 1 & -1 & 1 & 0 \end{bmatrix} \qquad \textbf{9}$$

This is the inverse of matrix **8**, as we may easily check by use of matrix arithmetic.

See Section 1-6, result **10**.

$$(-\tfrac{4}{5}) \begin{bmatrix} 1 & 1 & 2 & 1 \\ 4 & 5 & 9 & 1 \\ 3 & 4 & 7 & 1 \\ 2 & 3 & 4 & 2 \end{bmatrix} (-\tfrac{5}{4}) \begin{bmatrix} 0 & 5 & -7 & 1 \\ -2 & 0 & 0 & 1 \\ 1 & -2 & 3 & -1 \\ 1 & -1 & 1 & 0 \end{bmatrix}$$

$$= (-\tfrac{4}{5})(-\tfrac{5}{4}) \begin{bmatrix} 1 & 1 & 2 & 1 \\ 4 & 5 & 9 & 1 \\ 3 & 4 & 7 & 1 \\ 2 & 3 & 4 & 2 \end{bmatrix} \begin{bmatrix} 0 & 5 & -7 & 1 \\ -2 & 0 & 0 & 1 \\ 1 & -2 & 4 & -1 \\ 1 & -1 & 1 & 0 \end{bmatrix}$$

$$= (1) \begin{bmatrix} 1 & 0 & 0 & 0 \\ 0 & 1 & 0 & 0 \\ 0 & 0 & 1 & 0 \\ 0 & 0 & 0 & 1 \end{bmatrix} = \begin{bmatrix} 1 & 0 & 0 & 0 \\ 0 & 1 & 0 & 0 \\ 0 & 0 & 1 & 0 \\ 0 & 0 & 0 & 1 \end{bmatrix}$$

Thus, matrix **9** is the inverse of matrix **8**. This may be very neatly expressed in our new notation as follows:

$$\left((-\tfrac{4}{5}) \begin{bmatrix} 1 & 1 & 2 & 1 \\ 4 & 5 & 9 & 1 \\ 3 & 4 & 7 & 1 \\ 2 & 3 & 4 & 2 \end{bmatrix} \right)^{-1} = (-\tfrac{4}{5})^{-1} \begin{bmatrix} 1 & 1 & 2 & 1 \\ 4 & 5 & 9 & 1 \\ 3 & 4 & 7 & 1 \\ 2 & 3 & 4 & 2 \end{bmatrix}^{-1} \qquad \textbf{10}$$

We may express this fact in our capital-letter symbolism by stating that if A is a matrix that has an inverse and r is any number other than zero, then $(rA)^{-1} = r^{-1}A^{-1}$.

It can be shown that this result holds for any nonzero number and any square matrix that has an inverse. We state the result in words: given a nonzero number and a matrix that has an inverse, *the inverse of the product of the number and the matrix equals the inverse of the number times the inverse of the matrix.* Suppose, for example, that the number is 3 and the matrix is matrix **1**. Then

$$\left(3 \begin{bmatrix} 2 & 5 \\ 1 & 3 \end{bmatrix} \right)^{-1} = 3^{-1} \begin{bmatrix} 2 & 5 \\ 1 & 3 \end{bmatrix}^{-1} = \tfrac{1}{3} \begin{bmatrix} 3 & -5 \\ -1 & 2 \end{bmatrix}$$

by equation **4**. In other words,

$$\begin{bmatrix} 6 & 15 \\ 3 & 9 \end{bmatrix}^{-1} = \begin{bmatrix} 1 & -\tfrac{5}{3} \\ -\tfrac{1}{3} & \tfrac{2}{3} \end{bmatrix}$$

as may easily be checked:

$$\begin{bmatrix} 6 & 15 \\ 3 & 9 \end{bmatrix} \begin{bmatrix} 1 & -\tfrac{5}{3} \\ -\tfrac{1}{3} & \tfrac{2}{3} \end{bmatrix} = \begin{bmatrix} 1 & 0 \\ 0 & 1 \end{bmatrix} \checkmark$$

Problem 1   Using equation **6** and the general result above, find

$$\begin{bmatrix} 0 & \tfrac{10}{3} & -\tfrac{14}{3} & \tfrac{2}{3} \\ -\tfrac{4}{3} & 0 & 0 & \tfrac{2}{3} \\ \tfrac{2}{3} & -\tfrac{4}{3} & 2 & -\tfrac{2}{3} \\ \tfrac{2}{3} & -\tfrac{2}{3} & \tfrac{2}{3} & 0 \end{bmatrix}^{-1}$$

Solution

$$\begin{bmatrix} 0 & \tfrac{10}{3} & -\tfrac{14}{3} & \tfrac{2}{3} \\ -\tfrac{4}{3} & 0 & 0 & \tfrac{2}{3} \\ \tfrac{2}{3} & -\tfrac{4}{3} & 2 & -\tfrac{2}{3} \\ \tfrac{2}{3} & -\tfrac{2}{3} & \tfrac{2}{3} & 0 \end{bmatrix}^{-1} = \left(\tfrac{2}{3} \begin{bmatrix} 0 & 5 & -7 & 1 \\ -2 & 0 & 0 & 1 \\ 1 & -2 & 3 & -1 \\ 1 & -1 & 1 & 0 \end{bmatrix} \right)^{-1}$$

$$= \tfrac{3}{2} \begin{bmatrix} 1 & 1 & 2 & 1 \\ 4 & 5 & 9 & 1 \\ 3 & 4 & 7 & 1 \\ 2 & 3 & 4 & 2 \end{bmatrix} = \begin{bmatrix} \tfrac{3}{2} & \tfrac{3}{2} & 3 & \tfrac{3}{2} \\ 6 & \tfrac{15}{2} & \tfrac{27}{2} & \tfrac{3}{2} \\ \tfrac{9}{2} & 6 & \tfrac{21}{2} & \tfrac{3}{2} \\ 3 & \tfrac{9}{2} & 6 & 3 \end{bmatrix}$$

How is inversion related to multiplication of one matrix by another matrix? We see from equation **10** in Section 3-2 that

$$\begin{bmatrix} -3 & 2 \\ 1 & -1 \end{bmatrix}^{-1} = \begin{bmatrix} -1 & -2 \\ -1 & -3 \end{bmatrix} \qquad \textbf{11}$$

From equations **4** and **11** we now have

$$\begin{bmatrix} 2 & 5 \\ 1 & 3 \end{bmatrix}^{-1} = \begin{bmatrix} 3 & -5 \\ -1 & 2 \end{bmatrix} \qquad \textbf{12}$$

$$\begin{bmatrix} -3 & 2 \\ 1 & -1 \end{bmatrix}^{-1} = \begin{bmatrix} -1 & -2 \\ -1 & -3 \end{bmatrix} \qquad \textbf{13}$$

The product of $\begin{bmatrix} 2 & 5 \\ 1 & 3 \end{bmatrix}$ and $\begin{bmatrix} -3 & 2 \\ 1 & -1 \end{bmatrix}$ is $\begin{bmatrix} -1 & -1 \\ 0 & -1 \end{bmatrix}$

We find the inverse of this matrix:

$$\begin{bmatrix} -1 & -1 & 1 & 0 \\ 0 & -1 & 0 & 1 \end{bmatrix} \Longrightarrow \begin{bmatrix} -1 & 0 & 1 & -1 \\ 0 & -1 & 0 & 1 \end{bmatrix}$$

$$\Longrightarrow \begin{bmatrix} 1 & 0 & -1 & 1 \\ 0 & 1 & 0 & -1 \end{bmatrix}$$

so

$$\begin{bmatrix} -1 & -1 \\ 0 & -1 \end{bmatrix}^{-1} = \begin{bmatrix} -1 & 1 \\ 0 & -1 \end{bmatrix} \qquad \textbf{14}$$

How is this matrix related to the inverses of the matrices $\begin{bmatrix} 2 & 5 \\ 1 & 3 \end{bmatrix}$ and $\begin{bmatrix} -3 & 2 \\ 1 & -1 \end{bmatrix}$? It is *not* the inverse of the matrix $\begin{bmatrix} 2 & 5 \\ 1 & 3 \end{bmatrix}$ times the inverse of the matrix $\begin{bmatrix} -3 & 2 \\ 1 & -1 \end{bmatrix}$:

$$\begin{bmatrix} 3 & -5 \\ -1 & 2 \end{bmatrix}\begin{bmatrix} -1 & -2 \\ -1 & -3 \end{bmatrix} = \begin{bmatrix} 2 & 9 \\ -1 & -4 \end{bmatrix} \neq \begin{bmatrix} -1 & 1 \\ 0 & -1 \end{bmatrix}$$

but it is the product of the inverses *the other way around:*

$$\begin{bmatrix} -1 & -2 \\ -1 & -3 \end{bmatrix}\begin{bmatrix} 3 & -5 \\ -1 & 2 \end{bmatrix} = \begin{bmatrix} -1 & 1 \\ 0 & -1 \end{bmatrix} \qquad \textbf{15}$$

We may also express equation **15** thus:

$$\left(\begin{bmatrix} 2 & 5 \\ 1 & 3 \end{bmatrix}\begin{bmatrix} -3 & 2 \\ 1 & -1 \end{bmatrix}\right)^{-1} = \begin{bmatrix} -3 & 2 \\ 1 & -1 \end{bmatrix}^{-1}\begin{bmatrix} 2 & 5 \\ 1 & 3 \end{bmatrix}^{-1} \qquad \textbf{16}$$

This translates into our capital-letter notation as follows: Let A and B be square matrices of the same size. Suppose that A has an inverse and B has an inverse. Then AB has an inverse and
$$(AB)^{-1} = B^{-1}A^{-1}$$
Warning: It is hardly ever the case that $(AB)^{-1}$ is equal to $A^{-1}B^{-1}$.

It can be shown that this result holds for any two square matrices of the same size, provided each matrix has an inverse. We state the result in words: given any two matrices of the same size, each having an inverse, *the inverse of their product is the product of their inverses* IN THE OPPOSITE ORDER.

Problem 2

Given that

$$\begin{bmatrix} -1 & 1 & 1 \\ 1 & 4 & -5 \\ 1 & -2 & 0 \end{bmatrix}^{-1} = \begin{bmatrix} 10 & 2 & 9 \\ 5 & 1 & 4 \\ 6 & 1 & 5 \end{bmatrix}$$

and

$$\begin{bmatrix} 7 & 3 & 4 \\ 2 & 2 & 1 \\ 9 & 4 & 5 \end{bmatrix}^{-1} = \begin{bmatrix} -6 & -1 & 5 \\ 1 & 1 & -1 \\ 10 & 1 & -8 \end{bmatrix}$$

find

$$\left(\begin{bmatrix} -1 & 1 & 1 \\ 1 & 4 & -5 \\ 1 & -2 & 0 \end{bmatrix} \begin{bmatrix} 7 & 3 & 4 \\ 2 & 2 & 1 \\ 9 & 4 & 5 \end{bmatrix} \right)^{-1}$$

Solution

$$\left(\begin{bmatrix} -1 & 1 & 1 \\ 1 & 4 & -5 \\ 1 & -2 & 0 \end{bmatrix} \begin{bmatrix} 7 & 3 & 4 \\ 2 & 2 & 1 \\ 9 & 4 & 5 \end{bmatrix} \right)^{-1}$$

$$= \begin{bmatrix} 7 & 3 & 4 \\ 2 & 2 & 1 \\ 9 & 4 & 5 \end{bmatrix}^{-1} \begin{bmatrix} -1 & 1 & 1 \\ 1 & 4 & -5 \\ 1 & -2 & 0 \end{bmatrix}^{-1}$$

$$= \begin{bmatrix} -6 & -1 & 5 \\ 1 & 1 & -1 \\ 10 & 1 & -8 \end{bmatrix} \begin{bmatrix} 10 & 2 & 9 \\ 5 & 1 & 4 \\ 6 & 1 & 5 \end{bmatrix} = \begin{bmatrix} -35 & -8 & -33 \\ 9 & 2 & 8 \\ 57 & 13 & 54 \end{bmatrix}$$

To check our answer we calculate

$$\begin{bmatrix} -1 & 1 & 1 \\ 1 & 4 & -5 \\ 1 & -2 & 0 \end{bmatrix} \begin{bmatrix} 7 & 3 & 4 \\ 2 & 2 & 1 \\ 9 & 4 & 5 \end{bmatrix} = \begin{bmatrix} 4 & 3 & 2 \\ -30 & -9 & -17 \\ 3 & -1 & 2 \end{bmatrix}$$

and multiply by the supposed inverse

$$\begin{bmatrix} 4 & 3 & 2 \\ -30 & -9 & -17 \\ 3 & -1 & 2 \end{bmatrix} \begin{bmatrix} -35 & -8 & -33 \\ 9 & 2 & 8 \\ 57 & 13 & 54 \end{bmatrix} = \begin{bmatrix} 1 & 0 & 0 \\ 0 & 1 & 0 \\ 0 & 0 & 1 \end{bmatrix} \quad ✔$$

Problem 3   Given equations **11** and **14**, find $\left(\begin{bmatrix} 4 & 4 \\ 0 & 4 \end{bmatrix} \begin{bmatrix} -1 & ^2\!/_3 \\ ^1\!/_3 & -^1\!/_3 \end{bmatrix} \right)^{-1}$

Solution   We use both results stated above.

$$\left(\begin{bmatrix} 4 & 4 \\ 0 & 4 \end{bmatrix} \begin{bmatrix} -1 & ^2\!/_3 \\ ^1\!/_3 & -^1\!/_3 \end{bmatrix} \right)^{-1} = \begin{bmatrix} -1 & ^2\!/_3 \\ ^1\!/_3 & -^1\!/_3 \end{bmatrix}^{-1} \begin{bmatrix} 4 & 4 \\ 0 & 4 \end{bmatrix}^{-1}$$

$$= \left((^1\!/_3) \begin{bmatrix} -3 & 2 \\ 1 & -1 \end{bmatrix} \right)^{-1} \left((-4) \begin{bmatrix} -1 & -1 \\ 0 & -1 \end{bmatrix} \right)^{-1}$$

$$= (3) \begin{bmatrix} -1 & -2 \\ -1 & -3 \end{bmatrix} (-^1\!/_4) \begin{bmatrix} -1 & 1 \\ 0 & -1 \end{bmatrix} = (-^3\!/_4) \begin{bmatrix} 1 & 1 \\ 1 & 2 \end{bmatrix}$$

$$= \begin{bmatrix} -^3\!/_4 & -^3\!/_4 \\ -^3\!/_4 & -^3\!/_2 \end{bmatrix}$$

Check:
$$\begin{bmatrix} 4 & 4 \\ 0 & 4 \end{bmatrix} \begin{bmatrix} -1 & ^2\!/_3 \\ ^1\!/_3 & -^1\!/_3 \end{bmatrix} = \begin{bmatrix} -^8\!/_3 & ^4\!/_3 \\ ^4\!/_3 & -^4\!/_3 \end{bmatrix},$$

$$\begin{bmatrix} -^8\!/_3 & ^4\!/_3 \\ ^4\!/_3 & -^4\!/_3 \end{bmatrix} \begin{bmatrix} -^3\!/_4 & -^3\!/_4 \\ -^3\!/_4 & -^3\!/_2 \end{bmatrix} = \begin{bmatrix} 1 & 0 \\ 0 & 1 \end{bmatrix}$$

so the answer really is $\begin{bmatrix} -^3\!/_4 & -^3\!/_4 \\ -^3\!/_4 & -^3\!/_2 \end{bmatrix}$.

One final property: if a given matrix has an inverse, then *the inverse of the inverse of the matrix is just the given matrix.* For example:

$$\begin{bmatrix} 1 & 2 \\ 3 & 4 \end{bmatrix}^{-1} = \begin{bmatrix} -2 & 1 \\ ^3\!/_2 & -^1\!/_2 \end{bmatrix}$$

and so

$$\left(\begin{bmatrix} 1 & 2 \\ 3 & 4 \end{bmatrix}^{-1} \right)^{-1} = \begin{bmatrix} -2 & 1 \\ ^3\!/_2 & -^1\!/_2 \end{bmatrix}^{-1} = \begin{bmatrix} 1 & 2 \\ 3 & 4 \end{bmatrix}$$

that is, the inverse of the inverse of $\begin{bmatrix} 1 & 2 \\ 3 & 4 \end{bmatrix}$ is just $\begin{bmatrix} 1 & 2 \\ 3 & 4 \end{bmatrix}$ itself.

On the other hand, $\left(\begin{bmatrix} -1 & 2 & 4 \\ 3 & -4 & -3 \\ 2 & -2 & 1 \end{bmatrix}^{-1} \right)^{-1}$ is *not* equal to

Symbolically we may say that if a matrix A has an inverse, then so does A⁻¹, and
$$(A^{-1})^{-1} = A$$

$\begin{bmatrix} -1 & 2 & 4 \\ 3 & -4 & -3 \\ 2 & -2 & 1 \end{bmatrix}$ because this matrix, first seen as matrix **10** in Section 3-3, does not have an inverse.

What about properties connecting inversion with matrix addition and subtraction? There are no particularly useful properties; for example, two matrices may have inverses while neither their sum nor their difference have inverses. While $\begin{bmatrix} 2 & 0 \\ 0 & 3 \end{bmatrix}$ and

$\begin{bmatrix} 2 & 0 \\ 0 & -3 \end{bmatrix}$ have inverses, neither $\begin{bmatrix} 2 & 0 \\ 0 & 3 \end{bmatrix} + \begin{bmatrix} 2 & 0 \\ 0 & -3 \end{bmatrix} = \begin{bmatrix} 4 & 0 \\ 0 & 0 \end{bmatrix}$

nor $\begin{bmatrix} 2 & 0 \\ 0 & 3 \end{bmatrix} - \begin{bmatrix} 2 & 0 \\ 0 & -3 \end{bmatrix} = \begin{bmatrix} 0 & 0 \\ 0 & 6 \end{bmatrix}$ have inverses.

Exercises 3.5

Given that $\begin{bmatrix} 7 & 8 & 3 \\ 3 & 4 & 2 \\ 4 & 3 & 0 \end{bmatrix}^{-1} = \begin{bmatrix} -6 & 9 & 4 \\ 8 & -12 & -5 \\ -7 & 11 & 4 \end{bmatrix}$ and $\begin{bmatrix} 1 & 0 & 1/2 \\ 0 & 1/3 & 2 \\ -2 & -1 & 1/2 \end{bmatrix}^{-1}$

$= 1/15 \begin{bmatrix} 13 & -3 & -1 \\ -24 & 9 & -12 \\ 4 & 6 & 2 \end{bmatrix}$, find each of the following:

A1. $\begin{bmatrix} 13 & -3 & -1 \\ -24 & 9 & -12 \\ 4 & 6 & 2 \end{bmatrix}^{-1}$

A2. $\begin{bmatrix} -7/2 & -4 & -3/2 \\ -3/2 & -2 & -1 \\ -2 & -3/2 & 0 \end{bmatrix}^{-1}$

3. $\left(\begin{bmatrix} -6 & 9 & 4 \\ 8 & -12 & -5 \\ -7 & 11 & 4 \end{bmatrix}^{-1} \right)^{-1}$

4. $\left(\begin{bmatrix} 7 & 8 & 3 \\ 3 & 4 & 2 \\ 4 & 3 & 0 \end{bmatrix} \begin{bmatrix} 1 & 0 & 1/2 \\ 0 & 1/3 & 2 \\ -2 & -1 & 1/2 \end{bmatrix} \right)^{-1}$

A5. $\left(\begin{bmatrix} 6 & 0 & 3 \\ 0 & 2 & 12 \\ -12 & -6 & 3 \end{bmatrix} \begin{bmatrix} -6 & 9 & 4 \\ 8 & -12 & -5 \\ 7 & 11 & 4 \end{bmatrix} \right)^{-1}$

A6. $\left(\begin{bmatrix} 13 & -3 & -1 \\ -24 & 9 & -12 \\ 4 & 6 & 2 \end{bmatrix} \begin{bmatrix} 15 & 0 & 15/2 \\ 0 & 5 & 30 \\ -30 & -15 & 15/2 \end{bmatrix} \right)^{-1}$

7. True or false:

C(a) $\left(\begin{bmatrix} 1 & 2 \\ 3 & 6 \end{bmatrix}^{-1} \right)^{-1} = \begin{bmatrix} 1 & 2 \\ 3 & 6 \end{bmatrix}$

A(b) $\left(\begin{bmatrix} 1 & 2 \\ 3 & 4 \end{bmatrix}^{-1} \right)^{-1} = \begin{bmatrix} 1 & 2 \\ 3 & 4 \end{bmatrix}$

A(c) $\begin{bmatrix} 0 & 0 \\ 0 & 0 \end{bmatrix}^{-1} = \begin{bmatrix} 0 & 0 \\ 0 & 0 \end{bmatrix}$

$$^A(d) \quad \begin{bmatrix} 1 & 1 \\ 1 & 1 \end{bmatrix}^{-1} = \begin{bmatrix} 1 & 1 \\ 1 & 1 \end{bmatrix}$$

^{A}(e) If I is an identity matrix, then $I^{-1} = I$.

^{A}8. Suppose A, B, and C are square matrices of the same size, each of which has an inverse. Express $(ABC)^{-1}$ in terms of A^{-1}, B^{-1}, C^{-1}.

*9. If A, B, and C are square matrices of the same size, each having an inverse, express $((3A^{-1})(^2/_3B))^{-1}(2CA)^{-1}$ in terms of A, B, C and their inverses, as simply as possible.

*10. In solving the systems $AX = B$ of Exercise 7 in Section 3-4 in connection with Eastern supplies of buns, PHE calculated that $A^{-1} = {}^1/_{14} \begin{bmatrix} 12 & 2 & -6 \\ 45 & -3 & -19 \\ -28 & 0 & 14 \end{bmatrix}$. At the same time, similar supply problems in the West led to systems $CX = D$, for various requirement matrices D, where $C = \begin{bmatrix} 7 & -8 & 3 \\ 3 & -4 & 2 \\ 4 & -3 & 2 \end{bmatrix}$. Later it turned out that similar supply problems in the Midwest led to systems $CAX = E$, for various requirement matrices $E = \begin{bmatrix} a \\ b \\ c \end{bmatrix}$. Write a flow chart giving the "practical" solutions to these systems, in the sense of Exercise 7 in Section 3-4.

Section 3-6
A Formula for the 2 × 2 Inverse

A 2 × 2 matrix is so small that we can carry out the calculations for the inverse of a *general* 2 × 2 matrix, that is, the matrix

$$\begin{bmatrix} a & b \\ c & d \end{bmatrix} \qquad \qquad 1$$

where a, b, c, and d may be any numbers. (It is of course quite possible to carry out similar calculations for general matrices of size 3 × 3, 4 × 4, and so on, but the resulting formulas are too complicated to be of practical use.)

We will not actually carry out these calculations here (they appear in the answers to Exercises 3, 4, and 5), but will present the final results. In these calculations we encounter the following matrix, which is known as the *adjoint* of matrix 1:

"Adjoint" is not an important term for us beyond this section.

$$\begin{bmatrix} d & -b \\ -c & a \end{bmatrix} \qquad \qquad 2$$

In words, the adjoint of a 2 × 2 matrix is obtained by switching the entries in the main diagonal and changing the sign of the other entries. For example, the adjoint of

$$\begin{bmatrix} 1 & 2 \\ 3 & 4 \end{bmatrix} \qquad \mathbf{3}$$

is

$$\begin{bmatrix} 4 & -2 \\ -3 & 1 \end{bmatrix} \qquad \mathbf{4}$$

and the adjoint of

$$\begin{bmatrix} 1 & -2 \\ 2 & -4 \end{bmatrix} \qquad \mathbf{5}$$

is

$$\begin{bmatrix} -4 & 2 \\ -2 & 1 \end{bmatrix} \qquad \mathbf{6}$$

The importance of the adjoint arises from the following formula:

$$\begin{bmatrix} a & b \\ c & d \end{bmatrix} \begin{bmatrix} d & -b \\ -c & a \end{bmatrix} = (ad - bc) \begin{bmatrix} 1 & 0 \\ 0 & 1 \end{bmatrix} \qquad \mathbf{7}$$

which we may easily verify:

$$\begin{bmatrix} a & b \\ c & d \end{bmatrix} \begin{bmatrix} d & -b \\ -c & a \end{bmatrix} = \begin{bmatrix} ad - bc & -ab + ba \\ cd - dc & -cb + da \end{bmatrix}$$
$$= \begin{bmatrix} ad - bc & 0 \\ 0 & ad - bc \end{bmatrix} = (ad - bc) \begin{bmatrix} 1 & 0 \\ 0 & 1 \end{bmatrix}$$

The number $ad - bc$ in formula **7** is called the **determinant** of matrix **1**. Determinants will be discussed further in Section 6-5. Formula **7** may be stated in words: the product of a 2 × 2 matrix and its adjoint equals the determinant of the matrix times the 2 × 2 identity matrix. Applying this formula to matrices **3** and **5**, for example, we have

$$\begin{bmatrix} 1 & 2 \\ 3 & 4 \end{bmatrix} \begin{bmatrix} 4 & -2 \\ -3 & 1 \end{bmatrix} = (1 \cdot 4 - 2 \cdot 3) \begin{bmatrix} 1 & 0 \\ 0 & 1 \end{bmatrix} = \begin{bmatrix} -2 & 0 \\ 0 & -2 \end{bmatrix} \qquad \mathbf{8}$$

$$\begin{bmatrix} 1 & -2 \\ 2 & -4 \end{bmatrix} \begin{bmatrix} -4 & 2 \\ -2 & 1 \end{bmatrix} = (1(-4) - (-2)2) \begin{bmatrix} 1 & 0 \\ 0 & 1 \end{bmatrix} = \begin{bmatrix} 0 & 0 \\ 0 & 0 \end{bmatrix} \qquad \mathbf{9}$$

A matrix with determinant different from 0, such as matrix **3**, has an inverse. Consider the matrix $(^1/_{-2}) \begin{bmatrix} 4 & -2 \\ -3 & 1 \end{bmatrix}$. By equation **8**, $\begin{bmatrix} 1 & 2 \\ 3 & 4 \end{bmatrix} (^1/_{-2}) \begin{bmatrix} 4 & -2 \\ -3 & 1 \end{bmatrix} = (^1/_{-2}) \begin{bmatrix} 1 & 2 \\ 3 & 4 \end{bmatrix} \begin{bmatrix} 4 & -2 \\ -3 & 1 \end{bmatrix}$

$$= (^1/_{-2}) \begin{bmatrix} -2 & 0 \\ 0 & -2 \end{bmatrix} = \begin{bmatrix} 1 & 0 \\ 0 & 1 \end{bmatrix}.$$ In other words,

$$\begin{bmatrix} 1 & 2 \\ 3 & 4 \end{bmatrix}^{-1} = -^1/_2 \begin{bmatrix} 4 & -2 \\ -3 & 1 \end{bmatrix} = \begin{bmatrix} -2 & 1 \\ ^3/_2 & -^1/_2 \end{bmatrix}.$$

A matrix with determinant 0, such as matrix **5**, cannot have an inverse. Suppose, for example, that matrix **5** did have an inverse—say, $\begin{bmatrix} p & q \\ r & s \end{bmatrix} \begin{bmatrix} 1 & -2 \\ 2 & -4 \end{bmatrix} = \begin{bmatrix} 1 & 0 \\ 0 & 1 \end{bmatrix}$. Then, multiplying this inverse by each side of equation **9**, we would have $\begin{bmatrix} p & q \\ r & s \end{bmatrix} \begin{bmatrix} 1 & -2 \\ 2 & -4 \end{bmatrix} \begin{bmatrix} -4 & 2 \\ -2 & 1 \end{bmatrix} = \begin{bmatrix} p & q \\ r & s \end{bmatrix} \begin{bmatrix} 0 & 0 \\ 0 & 0 \end{bmatrix}$, that is to say, $\begin{bmatrix} 1 & 0 \\ 0 & 1 \end{bmatrix} \begin{bmatrix} -4 & 2 \\ -2 & 1 \end{bmatrix} = \begin{bmatrix} 0 & 0 \\ 0 & 0 \end{bmatrix}$, which is obviously wrong. So it must be that matrix **5** does not have an inverse.

We present the above results for the inverse of the general 2×2 matrix in the following flow chart.

To Find the Inverse of $\begin{bmatrix} a & b \\ c & d \end{bmatrix}$

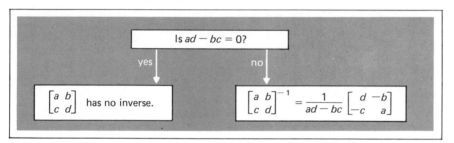

Is $ad - bc = 0$?

yes → $\begin{bmatrix} a & b \\ c & d \end{bmatrix}$ has no inverse.

no → $\begin{bmatrix} a & b \\ c & d \end{bmatrix}^{-1} = \dfrac{1}{ad - bc} \begin{bmatrix} d & -b \\ -c & a \end{bmatrix}$

Problem 1 Find the inverses, if they exist, of each of the following matrices:

(a) $\begin{bmatrix} 7 & 8 \\ 9 & 10 \end{bmatrix}$ (b) $\begin{bmatrix} 6 & 2 \\ 1 & ^1/_3 \end{bmatrix}$ (c) $\begin{bmatrix} ^1/_2 & ^2/_3 \\ ^3/_4 & ^4/_5 \end{bmatrix}$ (d) $\begin{bmatrix} -3 & ^1/_2 \\ -^1/_3 & ^2/_3 \end{bmatrix}$

Solution (a) $7 \cdot 10 - 8 \cdot 9 = -2$; so $\begin{bmatrix} 7 & 8 \\ 9 & 10 \end{bmatrix}^{-1} = ^1/_{-2} \begin{bmatrix} 10 & -8 \\ -9 & 7 \end{bmatrix} =$

$\begin{bmatrix} -5 & 4 \\ ^9/_2 & -^7/_2 \end{bmatrix}$. Check: $\begin{bmatrix} 7 & 8 \\ 9 & 10 \end{bmatrix} \begin{bmatrix} -5 & 4 \\ ^9/_2 & -^7/_2 \end{bmatrix} = \begin{bmatrix} 1 & 0 \\ 0 & 1 \end{bmatrix}$ ✔

(b) $6 \cdot ^1/_3 - 2 \cdot 1 = 0$; so $\begin{bmatrix} 6 & 2 \\ 1 & ^1/_3 \end{bmatrix}$ has no inverse.

(c) $(^1/_2)(^4/_5) - (^2/_3)(^3/_4) = ^2/_5 - ^1/_2 = -^1/_{10}$, and so $\begin{bmatrix} ^1/_2 & ^2/_3 \\ ^3/_4 & ^4/_5 \end{bmatrix}^{-1}$

$= \dfrac{1}{-^1/_{10}} \begin{bmatrix} ^4/_5 & -^2/_3 \\ -^3/_4 & ^1/_2 \end{bmatrix} = -10 \begin{bmatrix} ^4/_5 & -^2/_3 \\ -^3/_4 & ^1/_2 \end{bmatrix} = \begin{bmatrix} -8 & ^{20}/_3 \\ ^{15}/_2 & -5 \end{bmatrix}.$

Check: $\begin{bmatrix} 1/2 & 2/3 \\ 3/4 & 4/5 \end{bmatrix}\begin{bmatrix} -8 & 20/3 \\ 15/2 & -5 \end{bmatrix} = \begin{bmatrix} 1 & 0 \\ 0 & 1 \end{bmatrix}$ ✔

(d) $(-3)(2/3) - (1/2)(-1/3) = -2 + 1/6 = -11/6;$ so $\begin{bmatrix} -3 & 1/2 \\ -1/3 & 2/3 \end{bmatrix}^{-1}$

$= \dfrac{1}{-11/6}\begin{bmatrix} 2/3 & -1/2 \\ 1/3 & -3 \end{bmatrix} = -6/11\begin{bmatrix} 2/3 & -1/2 \\ 1/3 & -3 \end{bmatrix} = \begin{bmatrix} -4/11 & 3/11 \\ -2/11 & 18/11 \end{bmatrix}.$

Check: $\begin{bmatrix} -3 & 1/2 \\ -1/3 & 2/3 \end{bmatrix}\begin{bmatrix} -4/11 & 3/11 \\ -2/11 & 18/11 \end{bmatrix} = \begin{bmatrix} 1 & 0 \\ 0 & 1 \end{bmatrix}$ ✔

Exercises 3.6

1. Without using Gauss-Jordan reduction, find the inverse, it it exists, of each of the following matrices:

A(a) $\begin{bmatrix} 1 & 2 \\ 3 & 4 \end{bmatrix}$

A(d) $\begin{bmatrix} -1 & 2 \\ 3/2 & -3 \end{bmatrix}$

A(b) $\begin{bmatrix} 1 & 2 \\ 3 & 6 \end{bmatrix}$

*(e) $\begin{bmatrix} 7/9 & -11/13 \\ 17/24 & 3/16 \end{bmatrix}$

A(c) $\begin{bmatrix} -1 & 2 \\ 3 & -4 \end{bmatrix}$

*(f) $\begin{bmatrix} 6.81 & 3.21 \\ -4.31 & 2.22 \end{bmatrix}$

2. Develop flow charts, similar to the one given in this section, for finding the inverses of the following matrices:

A(a) $\begin{bmatrix} a & 0 \\ 0 & d \end{bmatrix}$ A(c) $\begin{bmatrix} a & b \\ 0 & d \end{bmatrix}$

(b) $\begin{bmatrix} 0 & b \\ c & 0 \end{bmatrix}$ (d) $\begin{bmatrix} 0 & b \\ c & d \end{bmatrix}$

3. Using Gauss-Jordan reduction, calculate $\begin{bmatrix} 0 & b \\ c & d \end{bmatrix}^{-1}$, assuming $b \neq 0$ and $c \neq 0$.

A 4. Using Gauss-Jordan reduction, calculate $\begin{bmatrix} a & b \\ 0 & d \end{bmatrix}^{-1}$, assuming $a \neq 0$ and $d \neq 0$.

*5. Using Gauss-Jordan reduction, calculate $\begin{bmatrix} a & b \\ c & d \end{bmatrix}^{-1}$, assuming $a \neq 0$, $b \neq 0$, $c \neq 0$, $d \neq 0$, and $ad - bc \neq 0$.

*6. Suppose $ad - bc \neq 0$ and $k \neq 0$. Find a formula for $\begin{bmatrix} k & 0 & 0 \\ 0 & a & b \\ 0 & c & d \end{bmatrix}^{-1}$.

7. PHEM, the computer-model company invented by PHE, is assumed to have available two types of cases of buns. Type A contains 8 boxes of hamburger buns and 7 boxes of superburger buns; type B contains 7 boxes of hamburger buns and 6 boxes of superburger buns. The company receives various orders for boxes of hamburger buns and boxes of superburger buns. Design a flow chart to determine which orders can be filled without opening cases.

Review Exercises   Reread Chapter 3, then do the following exercises without looking back. Answers to all of these exercises are in Appendix B.

^A1. The Big Profit Oil Company has four small special refineries that produce three types of petroleum products on demand: ZK28 Engine Oil, 3P25 Lubricant, and 24ST Transmission Fluid. While their main product is gasoline, they do spend some time running these refineries to fill weekly orders. In order to do this they must be able to spare hours from their main plant. Thus, the total number of hours they can run these refineries is limited. Tables 7 and 8 contain the relevant data. Determine how many hours they should operate each factory each week to fill the orders and use all available hours.

Table 7
Refinery Production (in
hundreds of gallons per hour)

Oil	Refinery			
	I	II	III	IV
ZK28	3	4	5	2
3P25	1	5	6	0
24ST	5	5	6	5

Table 8
Weekly Orders (in hundreds of
gallons) and Available Hours

Orders	Week of					
	May 4	May 11	May 18	May 25	June 1	June 8
ZK28	345	183	335	200	210	266
3P25	245	148	275	158	180	175
24ST	565	288	545	296	315	490
Hours	110	55	105	57	60	98

^A2. Find the inverses of the following matrices:

(a) $\begin{bmatrix} 29 & 12 \\ 12 & 5 \end{bmatrix}$

(b) $\begin{bmatrix} 9 & 27 \\ 4 & 12 \end{bmatrix}$

$$(c) \begin{bmatrix} 0 & 0 & 1 \\ 1 & 0 & 0 \\ 0 & 1 & 0 \end{bmatrix} \qquad (e) \begin{bmatrix} 1 & 2 & 1 \\ 2 & 1 & 1 \\ 1 & 1 & 2 \end{bmatrix}$$

$$(d) \begin{bmatrix} 1 & 2 & 3 \\ 4 & 5 & 6 \\ 7 & 8 & 9 \end{bmatrix}$$

A3. Which of the following systems of equations have unique solutions?

(a) $$\begin{bmatrix} 3x_1 + 2x_2 = 37 \\ 2x_1 + 5x_2 = 29 \end{bmatrix}$$

(b) $$\begin{bmatrix} x_1 + x_2 + x_3 = {}^{21}/_3 \\ x_1 + 2x_2 + x_3 = {}^{32}/_5 \\ x_1 - 3x_2 + 2x_3 = {}^{17}/_6 \end{bmatrix}$$

(c) $$\begin{bmatrix} x_1 + x_2 + x_4 = 3.7 \\ x_2 + x_3 = 102.4 \\ x_1 + x_3 + x_4 = 28.6 \\ x_1 + x_2 + x_3 + x_4 = 37.2 \end{bmatrix}$$

(d) $$\begin{bmatrix} x_1 + x_2 + x_4 = 19 \\ x_2 + x_3 = 1204 \\ x_1 + x_3 = 386 \\ x_1 + x_2 + x_3 + x_4 = 192 \end{bmatrix}$$

(e) $$\begin{bmatrix} x_1 + x_2 + x_4 = 3 \\ x_2 + x_3 = 2 \\ x_1 + x_3 = -1 \end{bmatrix}$$

Supplementary Exercises   1. Find the inverse, if it exists, of each of the following:

A (a) $$\begin{bmatrix} 99 & 100 \\ 101 & 102 \end{bmatrix}$$

(e) $$\begin{bmatrix} 1 & 3 & 1 \\ 2 & 8 & 3 \\ 2 & -3 & 1 \end{bmatrix}$$

(b) $$\begin{bmatrix} {}^1/_3 & -{}^1/_4 \\ {}^2/_5 & -{}^3/_7 \end{bmatrix}$$

A(f) $$\begin{bmatrix} 3 & 4 & -1 \\ -2 & -1 & -3 \\ 2 & 2 & 1 \end{bmatrix}$$

A(c) $$\begin{bmatrix} {}^8/_9 & {}^9/_8 \\ -{}^9/_8 & -{}^8/_9 \end{bmatrix}$$

(g) $$\begin{bmatrix} 1 & 2 & 1 \\ 2 & 3 & 3 \\ 2 & 3 & 4 \end{bmatrix}$$

(d) $$\begin{bmatrix} 4 & 5 & 4 \\ 0 & 6 & -1 \\ -1 & 5 & -2 \end{bmatrix}$$

(h) $$\begin{bmatrix} 4 & 6 & 1 \\ 9 & 8 & 5 \\ 4 & 4 & 2 \end{bmatrix}$$

A(i) $\begin{bmatrix} -2 & -3 & 1 \\ 5 & 5 & 4 \\ 4 & 3 & 7 \end{bmatrix}$ *(n) $\begin{bmatrix} -1 & -5 & 0 \\ -4 & -6 & 5 \\ 4 & 8 & 3 \end{bmatrix}$

(j) $\begin{vmatrix} 7 & 3 & 2 \\ 2 & 1 & 1 \\ 3 & 7 & 2 \end{vmatrix}$ *(o) $\begin{bmatrix} 5 & 5 & 9 \\ 2 & -5 & -2 \\ 6 & 2 & 4 \end{bmatrix}$

(k) $\begin{bmatrix} 5 & 4 & 3 \\ 7 & 5 & 6 \\ 7 & 8 & -3 \end{bmatrix}$ *(p) $\begin{bmatrix} 4 & 5 & 7 \\ 6 & 9 & 3 \\ 6 & 6 & -7 \end{bmatrix}$

A(l) $\begin{bmatrix} 2 & 3 & 1 \\ 2 & 7 & 1 \\ 3 & 2 & 1 \end{bmatrix}$ *(q) $\begin{bmatrix} -2 & 1 & 9 \\ -6 & -3 & 2 \\ 1 & -8 & 4 \end{bmatrix}$

(m) $\begin{bmatrix} 1 & 2 & 3 \\ 3 & 1 & 2 \\ 2 & 3 & 1 \end{bmatrix}$ (r) $\begin{bmatrix} 4 & 3 & 2 & 4 \\ 1 & 1 & 2 & 3 \\ 2 & 2 & 9 & 6 \\ 2 & 2 & 6 & 7 \end{bmatrix}$

2. Solve each of the following equations for the matrix X:

C(a) $\begin{bmatrix} -1 & 1 \\ 1 & -1 \end{bmatrix} X + \begin{bmatrix} 3 & 2 & 0 \\ -1 & 2 & 4 \end{bmatrix} = \begin{bmatrix} 3 & 2 & 1 \\ 4 & 2 & 5 \end{bmatrix}$

(b) $\begin{bmatrix} 2 & 1 & 3 \\ 3 & -6 & -4 \\ 5 & 2 & 7 \end{bmatrix} X + \begin{bmatrix} -1 & 7 \\ 2 & 4 \\ 3 & 6 \end{bmatrix} = \begin{bmatrix} 2 & -1 \\ 4 & 2 \\ 7 & 0 \end{bmatrix}$

C(c) $X \begin{bmatrix} -1 & 1 \\ 1 & -1 \end{bmatrix} + \begin{bmatrix} 3 & 2 & 0 \\ -1 & 2 & 4 \end{bmatrix} = \begin{bmatrix} 3 & 2 & 1 \\ 4 & 2 & 5 \end{bmatrix}$

(d) $\begin{bmatrix} 1 & 0 \\ 2 & 1 \end{bmatrix} X + \begin{bmatrix} 3 \\ 4 \end{bmatrix} = \begin{bmatrix} 2 & -1 \\ 7 & 4 \end{bmatrix}$

C(e) $X \begin{bmatrix} -2 & -3 & 1 \\ 5 & 5 & 4 \\ 4 & 3 & 7 \end{bmatrix} - \begin{bmatrix} 4 & 5 & 0 \\ 2 & 0 & 1 \\ -3 & -2 & 3 \end{bmatrix} = \begin{bmatrix} 8 & -1 & 0 \\ 0 & 2 & 1 \\ 0 & 3 & 0 \end{bmatrix}$

(f) $\begin{bmatrix} 1 & 0 \\ 0 & 1 \end{bmatrix} X + X \begin{bmatrix} 1 & 0 \\ 0 & 1 \end{bmatrix} = \begin{bmatrix} 1 & 0 \\ 0 & 1 \end{bmatrix}$

3. Use the fact that $\begin{bmatrix} 1 & 1 & 2 & 1 \\ 4 & 5 & 9 & 1 \\ 3 & 4 & 7 & 1 \\ 2 & 3 & 4 & 2 \end{bmatrix}^{-1} = \begin{bmatrix} 0 & 5 & -7 & 1 \\ -2 & 0 & 0 & 1 \\ 1 & -2 & 3 & -1 \\ 1 & -1 & 1 & 0 \end{bmatrix}$ to solve the following systems:

(a) $\begin{bmatrix} & 5x_2 - 7x_3 + x_4 = & 0 \\ -2x_1 & + x_4 = -1 \\ x_1 - 2x_2 + 3x_3 - x_4 = -2 \\ x_1 - x_2 + x_3 & = -3 \end{bmatrix}$

$$(b) \quad \begin{bmatrix} x_1 + x_2 + 2x_3 + x_4 = \frac{1}{2} \\ 4x_1 + 5x_2 + 9x_3 + x_4 = -\frac{2}{3} \\ 3x_1 + 4x_2 + 7x_3 + x_4 = \frac{3}{4} \\ 2x_1 + 3x_2 + 4x_3 + 2x_4 = -\frac{7}{8} \end{bmatrix}$$

4. Solve the following systems:

$$(a) \quad \begin{bmatrix} x_1 + 3x_2 + 3x_3 + 2x_4 + x_5 = -2 \\ x_1 + 4x_2 + 3x_3 + 3x_4 - x_5 = 3 \\ x_1 + 3x_2 + 4x_3 + x_4 + x_5 = -1 \\ x_1 + x_2 + x_3 + x_4 - x_5 = 2 \\ x_1 - 2x_2 - x_3 + 2x_4 + 2x_5 = 4 \end{bmatrix}$$

$$^A(b) \quad \begin{bmatrix} 30x_1 - 20x_2 - 15x_3 + 25x_4 - 5x_5 = -2 \\ 30x_1 - 11x_2 - 18x_3 + 7x_4 - 8x_5 = 3 \\ -30x_1 + 12x_2 + 21x_3 - 9x_4 + 6x_5 = -1 \\ -15x_1 + 12x_2 + 6x_3 - 9x_4 + 6x_5 = 2 \\ 15x_1 - 7x_2 - 6x_3 - x_4 - x_5 = 4 \end{bmatrix}$$

Note:
$$\begin{bmatrix} 1 & 3 & 3 & 2 & 1 \\ 1 & 4 & 3 & 3 & -1 \\ 1 & 3 & 4 & 1 & 1 \\ 1 & 1 & 1 & 1 & -1 \\ 1 & -2 & -1 & 2 & 2 \end{bmatrix}^{-1} = \frac{1}{15}\begin{bmatrix} 30 & -20 & -15 & 25 & -5 \\ 30 & -11 & -18 & 7 & -8 \\ -30 & 12 & 21 & -9 & 6 \\ -15 & 12 & 6 & -9 & 6 \\ 15 & -7 & -6 & -1 & -1 \end{bmatrix}$$

5. Which of the following matrices are inverses of which?

$$A = \begin{bmatrix} \frac{2}{3} & 5 \\ \frac{3}{25} & \frac{6}{5} \end{bmatrix}, \quad B = \begin{bmatrix} \frac{10}{3} & -25 \\ -\frac{3}{5} & 6 \end{bmatrix},$$

$$C = \begin{bmatrix} 0 & -\frac{2}{3} \\ -\frac{3}{2} & 0 \end{bmatrix}, \quad D = \begin{bmatrix} 6 & -25 \\ -\frac{3}{5} & \frac{10}{3} \end{bmatrix}$$

6. Write the matrix form of the system of linear equations whose
matrix is
$$\begin{bmatrix} 0 & 0 & 6 & -2 & 3 & 4 & \frac{1}{2} \\ 0 & -\frac{3}{2} & -1 & 0 & 0 & -\frac{7}{9} & 0 \\ 14 & 8 & 0 & 0 & 0 & 62 & -90 \end{bmatrix}.$$

$^A7.$ Write the third equation in the system $AX = B$, where

$$A = \begin{bmatrix} 9 & 10 & 11 & 12 \\ 13 & 14 & 15 & 16 \\ 17 & 18 & 19 & 20 \\ 21 & 22 & 23 & 24 \end{bmatrix} \quad \text{and} \quad B = \begin{bmatrix} 21 \\ 22 \\ 23 \\ 24 \end{bmatrix}.$$

8. If A is a 4×4 matrix and B is the inverse of A, write out the matrix
$A(\frac{2}{3}B)$.

9. If $\begin{bmatrix} -21 & -8 & 19 \\ 12 & 16 & -20 \\ 15 & -8 & 7 \end{bmatrix}$ is the inverse of $\begin{bmatrix} 48 & 96 & 144 \\ 384 & 432 & 192 \\ 336 & 288 & 240 \end{bmatrix}$, then what is

the inverse of $\begin{bmatrix} 1 & 2 & 3 \\ 8 & 9 & 4 \\ 7 & 6 & 5 \end{bmatrix}$?

10. Calculate the inverse of $\begin{bmatrix} -4 & -3 & -2 \\ -1 & 0 & 1 \\ 2 & 3 & 4 \end{bmatrix}$.

11. Invert $\begin{bmatrix} 2 & 2 & -2 & -2 \\ 0 & 0 & -2 & -2 \\ 2 & 0 & 0 & -2 \\ 0 & 2 & 0 & -2 \end{bmatrix}$.

A12. Find a 3×5 matrix $[?]$ such that $\begin{bmatrix} -2 & 2 & 1 \\ 1 & -\frac{1}{2} & -\frac{1}{2} \\ 1 & -1 & -1 \end{bmatrix} \begin{bmatrix} & & & ? & \\ & & & & \end{bmatrix}$

$$= \begin{bmatrix} 1 & 2 & -1 & 0 & 2 \\ 2 & 0 & 1 & -2 & 0 \\ 0 & -1 & -1 & 2 & 1 \end{bmatrix}.$$

13. Find $\begin{bmatrix} 117482 & (117482 \cdot 9643 - 769)/87415 \\ 87415 & 9643 \end{bmatrix}^{-1}$.

*14. Let k be any number at all. Find $\begin{bmatrix} k+1 & k+2 \\ k+3 & k+4 \end{bmatrix}^{-1}$

*15. Find a matrix $[?]$ such that

$$\begin{bmatrix} 0 & 2 & -1 \\ 1 & 2 & 0 \\ -1 & 0 & -2 \end{bmatrix} \begin{bmatrix} & & \\ & ? & \\ & & \end{bmatrix} \begin{bmatrix} 0 & 2 & -1 \\ 1 & 2 & 0 \\ -1 & 0 & -2 \end{bmatrix} = \begin{bmatrix} 1 & 1 & 1 \\ 1 & 1 & 2 \\ 1 & 2 & 2 \end{bmatrix}$$

*16. If a square matrix A has an inverse, we speak of *the* inverse of A. This is a proper usage because no square matrix can have more than one inverse. Why?

*17. Find three different 2×2 matrices A, B, C such that $AA = BB = CC = I_2$, the 2×2 identity matrix.

*18. Let $A = [2 \quad -3 \quad 4]$, $B = \begin{bmatrix} \frac{3}{2} \\ \frac{2}{3} \\ -\frac{8}{5} \end{bmatrix}$. Write out the system $BAX = B$.

C19. True or false: if A and B are matrices and $(AB)^{-1}$ exists, then $(AB)^{-1} = B^{-1}A^{-1}$.

*20. Solve the system $\begin{bmatrix} \begin{bmatrix} 2 & -1 \\ 0 & 1 \end{bmatrix} X_1 + \begin{bmatrix} -1 & 2 \\ 3 & 1 \end{bmatrix} X_2 = \begin{bmatrix} 3 & 2 \\ 7 & 0 \end{bmatrix} \\ \begin{bmatrix} 0 & 1 \\ 3 & 2 \end{bmatrix} X_1 + \begin{bmatrix} 2 & -3 \\ 1 & 1 \end{bmatrix} X_2 = \begin{bmatrix} -2 & 1 \\ 10 & 11 \end{bmatrix} \end{bmatrix}$, where

the variables X_1 and X_2 are both 2×2 matrices.

Summary

Terms   Coefficient matrix of a system of linear equations   p. 120
Requirement matrix of a system of linear equations   p. 121
Identity matrix   pp. 124, 130
Inverse matrix of a square matrix   p. 132

Computational Techniques

1. To compute the inverse of a matrix, use the flow chart on p. 140.

2. To solve efficiently many systems of linear equations all having the same coefficient matrix, see Problem 1, Section 3-4. (Note that the coefficient matrix must have an inverse.)

3. To compute the inverse of a 2×2 matrix, use the flow chart on p. 158.

Results

1. The inverse of a number times a matrix equals the inverse of the number times the inverse of the matrix (p. 151).

2. The inverse of the product of two matrices equals the product of the inverses IN THE OPPOSITE ORDER (p. 153).

3. If a matrix has an inverse, then that inverse has an inverse, namely, the original matrix (p. 154).

Chapter 4
Transportation
Problems

For a brief general description of linear programming, see p. 224.

Section 4-1
Introduction

In this chapter we begin the study of linear programming by considering an important application, the transportation problem. Early in the history of this subject, Frank L. Hitchcock (1941) and T. C. Koopmans (1947) independently formulated and solved the following type of problem: A company produces a product in different quantities at several factories. The company has many customers, each with a different demand. The shipping costs from each factory to each customer are different and there are many ways in which the total factory output can be distributed to the customers. How many of each item should each factory ship to each customer so that all the company's orders will be filled and the shipping costs will be as small as possible?

Like most companies, Appalachian Creations also has to cope with this situation. For example, consider its camp stove production. While this is not its most popular item, there is some demand for it at three locations: the Bangor and Falmouth stores each order 40 stoves a year, and the Hartford store orders 20 stoves a year. It is convenient for AC to restrict production of stoves to its Springfield and Waterbury factories. Since the Springfield factory is three times as large, the production is split; 75 at Springfield and 25 at Waterbury. The shipping costs from each factory to each store are given in Table 1.

Table 1
Shipping Cost per Stove
(in cents)

From factory	To store		
	Bangor	Falmouth	Hartford
Springfield	160	40	40
Waterbury	120	80	120

The entries in this table could reflect more than just the transportation costs. They could also involve any other costs that the company would incur by supplying a particular store from a particular factory. In any event, the question is: How can AC meet its demands at minimum cost? In a problem of this size it may be possible to proceed by trial and error, checking all possibilities. But since this approach is certainly impractical for larger problems, a systematic method is called for.

While the details of the technique for solving the transportation problem are special to it, the general philosophy is the same here as it is for essentially all linear programming situations. If we agree to call a *solution* to the problem any shipping schedule that supplies the demand, then we are to find a cheapest solution.

Beginning with some solution, we will derive a sequence of successively cheaper solutions until a cheapest solution is reached. There are three distinct steps in this process. The initial step involves finding some solution to the problem. We will obtain this solution by using the *northwest corner method,* a technique introduced by Abraham Charnes and William Cooper in 1954. At each subsequent step the solution at hand must be checked to see if it can be improved. For this, we use a method developed by George Dantzig in 1951. Finally, if a particular solution can be improved, there must be a means by which such improvement actually takes place. Here a method called *patterns of change* is used. It was also introduced by Charnes and Cooper in 1954. The northwest corner method will be discussed in Section 4-2. Patterns of change and Dantzig's technique are treated in Sections 4-3 and 4-4. In Section 4-5 we show how to deal with situations in which supply and demand are not the same, and in Section 4-6 we formulate and solve a type of linear programming problem called the assignment problem, which is closely related to the transportation problem.

You may have noticed that we referred above to *a* cheapest solution. Some transportation problems, as we shall see, may have several different cheapest solutions, each costing the same amount.

Exercises 4.1   A1. Petaluma Hamburger Emporium, Inc., does not operate any factories as such, but it does have to solve many linear programming problems, and in particular many transportation problems. Consider, for example, seafood supply to stores in four New Mexico cities: Gallup, Silver City, Carlsbad, and Raton. Frozen shrimp, clams, and fillets are sent from storage centers in Los Angeles and Houston. Trucking rates are the same from either center, $1/4$¢ per pound per mile, but purchase costs per pound are different at the two centers: shrimp costs 12¢ per pound more at Los Angeles than at Houston. Fillets cost 9¢ more per pound at Houston. Clams are the same at both centers. These differences should be included in the "shipping costs". Using the following distances (given in miles), prepare the shipping cost tables, as in Table 1, for shrimp, clams, and fillets.

	Gallup	Silver City	Carlsbad	Raton
Los Angeles	570	600	830	810
Houston	900	800	580	740

2. Try to find a cheapest solution to the stove shipment problem above by trial and error, combined with reasoning.

<table>
<tr><td></td><td>Section 4-2
The Northwest Corner
Method</td><td>In this section we ignore cost considerations and concentrate on the first step in solving the transportation problem, namely, finding *some* means by which the demand can be met. In a sense, this problem is opposite to the others we have studied. We are given a blank table and asked to fill in the entries (see Table 2). Of course, we do have some information. Since Bangor has ordered 40 stoves, the entries in the first column should add up to 40. Similarly, Falmouth's requirement of 40 stoves means that the second column should also total 40. To meet the demand at Hartford, the third column should add up to 20. As for the rows, since Springfield makes 75 stoves, the entries in the first row should add up to 75 and since Waterbury makes 25 stoves, the second row should total 25. To set up problems of this type for solution we enlarge the previous table to include this information about the row and column sums (see Table 3).</td></tr>
</table>

Table 2
Number of Stoves Shipped From Each Factory to Each Store

From factory	To store		
	Bangor	Falmouth	Hartford
Springfield			
Waterbury			

Table 3
Number of Stoves Shipped from Each Factory to Each Store

From factory	To store			Production
	Bangor	Falmouth	Hartford	
Springfield				75
Waterbury				25
Demand	40	40	20	

The following blank matrix results:

$$\begin{bmatrix} & & \end{bmatrix}\begin{matrix}75\\25\end{matrix}$$
$$\quad 40 \quad 40 \quad 20$$

Note that the total demand equals the total supply: $40 + 40 + 20 = 75 + 25$.

Now the problem is to supply entries for this matrix in such a way that the column entries add up to the numbers in the bottom row and the row entries add up to the numbers in the column at the right. In a problem of this size, trial and error could be used. Here are three solutions arrived at by trial and error:

$$\begin{bmatrix} 25 & 35 & 15 \\ 15 & 5 & 5 \end{bmatrix}\begin{matrix} 75 \\ 25 \end{matrix} \qquad \begin{bmatrix} 30 & 30 & 15 \\ 10 & 10 & 5 \end{bmatrix}\begin{matrix} 75 \\ 25 \end{matrix} \qquad \begin{bmatrix} 40 & 15 & 20 \\ & 25 & \end{bmatrix}\begin{matrix} 75 \\ 25 \end{matrix}$$
$$\begin{matrix} 40 & 40 & 20 \end{matrix} \qquad\qquad \begin{matrix} 40 & 40 & 20 \end{matrix} \qquad\qquad \begin{matrix} 40 & 40 & 20 \end{matrix}$$

In the last matrix we have left blanks instead of inserting zeros. When a matrix solution is rewritten as a shipping table, these blanks are to be interpreted as zeros. Thus, according to the third matrix, for example, all of the production at Waterbury is to be shipped to Falmouth, and Waterbury ships nothing to Bangor or Hartford. In the next section, however, when the patterns of change technique is used to improve a given solution, it will be necessary to distinguish between a blank position and a position containing the entry zero.

For larger problems, trial and error is not effective. In fact it rarely is effective, since not every solution can be successively changed into a cheapest solution by the standard technique developed in the next sections. While none of the trial-and-error solutions given above is a cheapest solution, the third can be altered to yield a cheapest solution, while the others cannot be so altered using our technique (see Rule 1, p. 174). The northwest corner method not only is a systematic technique for obtaining an initial solution, but also provides a solution that can be changed (via patterns of change) into a cheapest solution.

Page 172 contains a flow chart for the northwest corner method. We will illustrate the method with two examples, which we will discuss in conjunction with the chart.

Problem 1   Use the northwest corner method to find a solution to:

$$\begin{bmatrix} & & \\ & & \end{bmatrix}\begin{matrix} 75 \\ 25 \end{matrix}$$
$$\begin{matrix} 40 & 40 & 20 \end{matrix}$$

Solution   (1) The flow chart instructs us to begin at address (1,1), the "northwest corner." We insert 40, the smaller of 40 and 75:

$$\begin{bmatrix} 40 & & \\ & & \end{bmatrix}\begin{matrix} 75 \\ 25 \end{matrix}$$
$$\begin{matrix} 40 & 40 & 20 \end{matrix}$$

(2) The entries in the first row add up to 40, and the entries in the

first column add up to 40. Since the row sum is less than 75, we still have supply to allocate in the first row. According to the flow chart, we move to the right:

$$\begin{bmatrix} 40 & * & \\ & & \end{bmatrix}\begin{matrix} 75 \\ 25 \end{matrix}$$
$$\quad 40 \quad 40 \quad 20$$

We have marked the position with a *. While the second store demands 40, we cannot insert 40 since to do so would make the first

The Northwest Corner Method for Finding a First Solution to a Transportation Problem

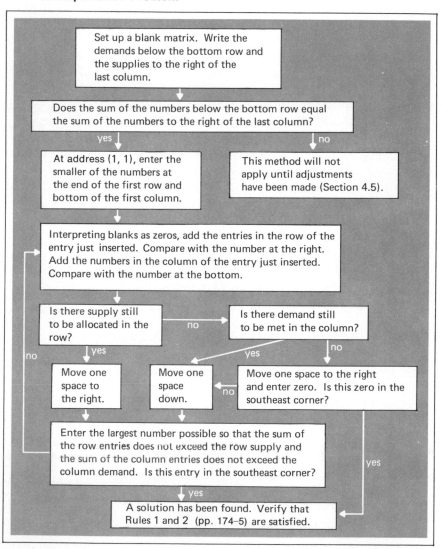

row total 80, which would exceed the available supply. The largest allowable entry is 35, which we insert:

$$\begin{bmatrix} 40 & 35 & \\ & & \end{bmatrix}\begin{matrix}75 \\ 25\end{matrix}$$
$$40 \quad 40 \quad 20$$

(3) The row of the 35 now adds up to 75. Its column adds up to 35. Since 35 is less than 40, we move one square down to the space marked with a *:

$$\begin{bmatrix} 40 & 35 & \\ & * & \end{bmatrix}\begin{matrix}75 \\ 25\end{matrix}$$
$$40 \quad 40 \quad 20$$

The most we can enter in the space is 5:

$$\begin{bmatrix} 40 & 35 & \\ & 5 & \end{bmatrix}\begin{matrix}75 \\ 25\end{matrix}$$
$$40 \quad 40 \quad 20$$

(4) The column of the 5 adds up to 40. The second demand is satisfied. The row of the 5 totals only 5, so the available supply has not been completely utilized. We move to the right:

$$\begin{bmatrix} 40 & 35 & \\ & 5 & * \end{bmatrix}\begin{matrix}75 \\ 25\end{matrix}$$
$$40 \quad 40 \quad 20$$

Entering 20 in the southwest corner completes the solution:

$$\begin{bmatrix} 40 & 35 & \\ & 5 & 20 \end{bmatrix}\begin{matrix}75 \\ 25\end{matrix}$$
$$40 \quad 40 \quad 20$$

Problem 2   A company has five factories that produce an item in the quantities shown in Table 4. Eight stores demand this item, as shown in Table 5. Use the northwest corner method to find a possible shipping schedule.

Table 4

Factory	Number of Items
1	60
2	20
3	70
4	20
5	40

Table 5

Store	Demand
A	20
B	30
C	40
D	50
E	10
F	30
G	20
H	10

Solution We first note that total supply and total demand both equal 210. Here is the matrix solution, with arrows indicating the order in which the numbers were obtained:

$$
\begin{bmatrix}
20 & 30 & 10 & & & & & & \\
& & 20 & & & & & & \\
& & 10 & 50 & 10 & 0 & & & \\
& & & & & 20 & & & \\
& & & & & 10 & 20 & 10 &
\end{bmatrix}
\begin{matrix}
60 \\
20 \\
70 \\
20 \\
40
\end{matrix}
$$

$$20 \quad 30 \quad 40 \quad 50 \quad 10 \quad 30 \quad 20 \quad 10$$

At position (3,5), the third row summed to 70 and the fifth column summed to 10. Thus, the first five demands were met and the first three factories' outputs were totally allocated. When this happens, the flow chart instructions require that a zero be inserted as indicated and the problem be continued diagonally toward the southeast. The (3,6) entry is the number zero and is not blank. This matrix would be interpreted as in Table 6, where the zero has been deleted to simplify the presentation to management.

Table 6
Shipping Allocation Table

From factory	To store							
	A	B	C	D	E	F	G	H
1	20	30	10					
2			20					
3			10	50	10			
4						20		
5						10	20	10

In 1959 Simonnard and Hadley determined how many entries would appear in the final cheapest solution matrix (after the patterns of change technique is applied—see Section 4-3). This number is the same as the number of entries in the initial solution, since the patterns of change technique does not change the number of entries. The following two rules are useful checks, which each solution must satisfy:

Rule 1 **The number of entries in any solution must equal the number of rows plus the number of columns minus 1. (Do not include the demand row or supply column.)**

The first problem involved 2 rows and 3 columns (not including the demand row or supply column). Thus, there were $2 + 3 - 1 = 4$ entries in the solution. The second problem involved 5 rows and 8 columns. Thus, there were $5 + 8 - 1 = 12$ entries in the solution. Notice that in the second problem the zero entry must be counted in order for this rule to be satisfied.

Rule 2   **In any solution, the entries in any row must add up to the supply number at the right and the entries in any column must add up to the demand number at the bottom.**

The following sections describe a technique for successively replacing one solution with a cheaper solution. After each application of this technique, the new solution should be checked to be sure it satisfies Rule 1 and Rule 2.

Exercises 4.2   Use the northwest corner method to find a solution to the following transportation problems.

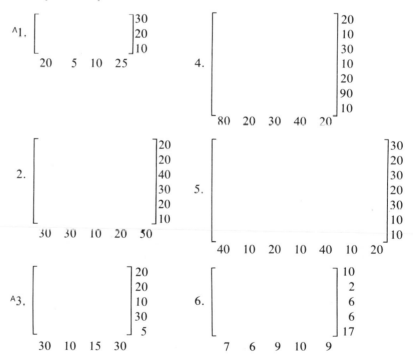

^7. A company has eight factories which produce an item according to Table 7. There is demand for this item at eight stores according to Table 8. Write the shipping table obtained by the northwest corner method.

Table 7

Factory	Production
1	30
2	10
3	40
4	20
5	20
6	40
7	20
8	20

Table 8

Store	Demand
A	20
B	30
C	40
D	30
E	40
F	20
G	10
H	10

*8. Another company has four distribution centers that have supplies available according to Table 9. The centers supply four stores which demand according to Table 10. Write the shipping table obtained by the northwest corner method.

Table 9

Distribution Center	Available Supplies
1	902
2	296
3	1223
4	788

Table 10

Store	Demand
A	528
B	1497
C	708
D	476

9. PHE is using its model company PHEM in a feasibility study involving charitable contributions of Petalunches (consisting of a cheeseburger, fries, and a shake in a box) to various organizations, namely: A—orphanage; B—home for aged; C—addict treatment center; D—scout troop. The supply and demand data in Tables 11 and 12 are to be assumed. Write the chart obtained by the northwest corner method.

Table 11

Store	Supply
1	70
2	80
3	50

Table 12

Organization	Demand
A	55
B	40
C	87
D	18

Section 4-3
Patterns of Change

A pattern of change is the means by which a given solution is changed into a less costly one. In this section we begin by considering short patterns of change. Longer patterns are similar, but discussion of them will be postponed until the next section. Rules 1 and 2, along with the rules given in this section, have been formulated to apply to the next sections as well.

In the last section we left AC with a shipping schedule that was obtained without any consideration of cost. In order to deal with costs, it is necessary to have all the available data at hand in the same chart. A transportation problem always involves the following three kinds of data:

(1) Available supply: in this case we have Table 13.

Table 13
Available Supply

Factory	Supply
Springfield	75
Waterbury	25

(2) Current demand: in this case we have Table 14.

Table 14
Current Demand

Store	Demand
Bangor	40
Falmouth	40
Hartford	20

(3) Shipping and distribution costs: in this case we have Table 15.

Table 15
Shipping Costs per Stove (in cents)

From factory	To store		
	Bangor	Falmouth	Hartford
Springfield	160	40	40
Waterbury	120	80	120

We have seen how to put the information of Tables 13 and 14 into matrix form. In order to record the shipping costs as well, it is necessary to abandon matrices and allow two numbers to occupy the same position. The standard format for setting up a transpor-

Chart 1

160	40	40	
			75
120	80	120	
			25
40	40	20	

Chart 2

160	40	40	
(40)	(35)		75
120	80	120	
	(5)	(20)	25
40	40	20	

Chart 3

160	40	40	
(40)	(35)	(20)	75
120	80	120	
	(5)		25
40	40	20	

Chart 4

160	40	40	
(40)	(15)	(20)	75
120	80	120	
	(5)		25
40	40	20	

Chart 5

160	40	40	
(40)	(15)	(20)	75
120	80	120	
	(25)		25
40	40	20	

tation problem is illustrated in Chart 1. As before, the right column represents supply and the bottom row represents demand. The other numbers are the transportation costs. Applying the northwest corner method results in Chart 2.

From now on, the numbers in a solution will be circled. This will distinguish them from other numbers that will be inserted in the blanks later to help arrive at a minimal cost solution.

With the aid of Chart 2, it is particularly easy to determine the total shipping costs that this solution will entail. Shipping 40 items at 160 cents per item costs $40 \times 1.60 = \$64$. Shipping 35 items at 40 cents per item costs $35 \times .40 = \$14$. Similarly, to ship 5 items at 80 cents costs $4 and 20 items at 120 cents costs $24. We tabulate this arithmetic as follows:

$$
\begin{aligned}
40 \times 1.60 &= \$\ 64 \\
35 \times .40 &= \quad 14 \\
5 \times .80 &= \quad\ 4 \\
20 \times 1.20 &= \quad 24 \\
\text{Total} &\quad\ \$106
\end{aligned}
$$

This shipping schedule costs the company $106.

Notice that it would be much cheaper to supply Hartford from Springfield than from Waterbury. The former costs only 40 cents per item, while the latter costs 120. If we do this, we obtain Chart 3.

Of course, this is not a solution since we are asking Springfield to supply 95 stoves when it has only 75. We adjust by subtracting 20 from the 35 in the (1,2) postion (Chart 4). Now the first row is correct, but the second column is not adequate to supply the 40 items required at Falmouth. We make a final adjustment by adding 20 to the 5 in the (2,2) position (Chart 5).

We check that Rules 1 and 2 are satisifed, and then tabulate the cost of this shipping scheme:

$$
\begin{aligned}
40 \times 1.60 &= \$\ 64 \\
15 \times .40 &= \quad\ 6 \\
20 \times .40 &= \quad\ 8 \\
25 \times .80 &= \quad 20 \\
\text{Total} &\quad \$\ 98
\end{aligned}
$$

This schedule is $8 cheaper than the previous one.

We have just performed a pattern of change. The procedure involved here is not as arbitrary as it may seem. In fact, the following rules governing patterns of change completely determine the course of action and leave no significant choices undecided:

Rule 3   **A pattern of change must involve adding and subtracting the same quantity alternately to and from squares (that is, positions in the chart) that are arranged around a loop. Each time a pattern of change is performed, one blank square becomes occupied and one occupied square becomes blank.**

In the pattern of change performed above, the 20 was

(a) removed from the (2,3) position,
(b) placed in the (1,3) position,
(c) subtracted from the 35 in the (1,2) position,
(d) added to the 5 in the (2,2) position.

The loop in question can be indicated with arrows on the original chart (see Chart 6). The dotted arrow is inserted to complete the loop; no actual transfer occurs along it.

All the patterns of change encountered in this section will require four steps. The next rule is phrased to apply here as well as to the longer patterns of change in the next section.

Chart 6

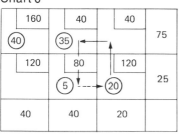

Rule 4   **The moves in a pattern of change must be alternately vertical and horizontal (including the first and last moves) through occupied squares and exactly one blank square. (There is, therefore, always an even number of moves.)**

Moving alternately horizontally and vertically and alternately adding and subtracting guarantee that the row and column sums will be the same before and after a pattern of change is completed. We present two other examples (see Charts 7 and 8) of loops for patterns of change involving charts different from the above shipping problem. Notice that in both instances the indicated pattern of change passes through but does not involve certain occupied squares. This is allowed. In the first case, the 10 is removed from the (3,3) position and inserted in the (3,1) position. Then 10 is subtracted from the 20 in the (1,1) position and added to the 30 in the (1,3) position. The 10 in the (1,2) position is unaffected. These patterns of change result in Charts 9 and 10.

Chart 7

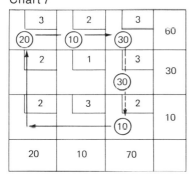

Performing a pattern of change in a given problem may increase, decrease, or leave unchanged the total transportation costs. It can be shown that given any blank square, there will always be one and only one pattern of change that will result in its becoming an occupied square. Thus, a standard procedure is to consider each blank, consider the unique pattern of change that fills it, and perform a simple calculation (to be described below) to determine whether or not the change will save money. Having performed this calculation for each blank, we then perform a pattern of change that results in a decrease in cost.

Chart 8

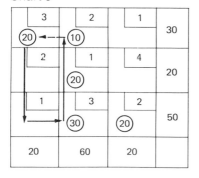

Chart 9 (From Chart 7)

3	2	3	60
(10)	(10)	(40)	
2	1	3	30
		(30)	
2	3	2	10
(10)			
20	10	70	

Chart 10 (From Chart 8)

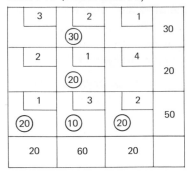

3	2	1	30
	(30)		
2	1	4	20
	(20)		
1	3	2	50
(20)	(10)	(20)	
20	60	20	

Chart 11

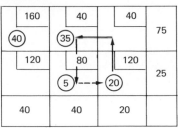

160	40	40	75
(40)	(35)		
120	80	120	25
	(5)	(20)	
40	40	20	

In the AC example, Chart 2, we performed the pattern of change indicated in Chart 11. By shifting the 20 out we saved $120 \cdot 20$. This move changed the total cost by $-120 \cdot 20$. However, adding 20 to the blank increased cost by $40 \cdot 20$. This insertion changed the total cost by $+40 \cdot 20$. This move was offset by deleting 20 from the 35. Here the cost changed by $-40 \cdot 20$. Finally, to make the row and column sums correct, 20 was added to the 5. The cost was changed by $+80 \cdot 20$. The net effect of this change was

$$-120 \cdot 20 + 40 \cdot 20 - 40 \cdot 20 + 80 \cdot 20$$

which is the same as

$$(-120 + 40 - 40 + 80) \cdot 20 = (-40) \cdot 20.$$

The number in parentheses is called the **cost factor**. Since this number is negative, performing this pattern of change will result in a cheaper solution. To compute the cost factor we do not need to deal with the number being shifted, we need only alternately add and subtract the transportation costs. Also, in making this computation it does not matter whether one goes clockwise or counterclockwise around the loop. Thus, to compute the cost factor one need not determine which number might be shifted.

There is one other blank in the AC example. The pattern of change shown in Chart 12 fills it. The cost factor is

$$-80 + 120 - 160 + 40 = -80$$

Notice that doing the calculation around the loop in the other direction gives the same result: $-160 + 120 - 80 + 40 = -80$.

Having computed the cost factors, without actually performing any pattern of change, we are now ready to obtain a new solution by actually performing that pattern of change which involves

Since the costs in this problem were given in cents, each of these products is a number of cents.

The cost factor is the amount saved per item shifted. The total amount saved is the number shifted times the cost factor.

Chart 12

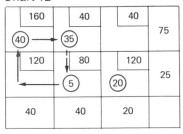

160	40	40	75
(40)	(35)		
120	80	120	25
	(5)	(20)	
40	40	20	

A longer pattern of change is slightly more complicated; the more general rule for deciding which number to move is given below (Rule 5).

the most negative factor. However, there is one final question to be answered: which number gets shifted? That is, in the example just considered, should the 40 or the 5 be moved into the blank square at address (2,1)? Answer: In a four-move pattern of change, the smaller number must be shifted. In this case we must insert 5 in the blank. If we insert 40 in the blank, we will not be allowed to subtract 40 from the 5, since a negative number of stoves cannot be shipped.

We are now in a position to solve the AC example. After tabulating the data, performing the northwest corner method, and inserting the cost factors, we get Chart 13. The cost factors are not circled. As explained above, we perform the pattern of change to fill the blank, here marked *, with the most negative cost factor. This results in Chart 14. (For purposes of reference in this example, we have lettered the squares. This is not necessary in practice.)

The pattern of change that fills blank c involves the loop

$$f\ c\ a\ d$$

The cost factor for this change is

$$-40 + 40 - 160 + 120 = -40$$

The pattern that fills blank e involves the loop

$$d\ e\ b\ a$$

The cost factor here is

$$-120 + 80 - 40 + 160 = 80$$

Recording this information yields Chart 15. The most (and only) negative cost factor here is -40. Performing the pattern of change to fill this blank and calculating the cost factors after the change yields Chart 16. Since performing any more patterns of change will increase cost by factors of 80 or 40, no further im-

Chart 13

160	40	40	75
(40)	(35)	−40	
120	80	120	25
−80*	(5)	(20)	
40	40	20	

Chart 14

160	40	40	75
(35) a	(40) b	c	
120	80	40	25
(5) d	e	(20) f	
40	40	20	

Chart 15

160	40	40	75
(35) a	(40) b	−40* c	
120	80	120	25
(5) d	80 e	(20) f	
40	40	20	

Chart 16

160	40	40	75
(15)	(40)	(20)	
120	80	40	25
(25)	80	40	
40	40	20	

provement of this sort is possible. It can be shown that in fact no further improvement of *any* sort is possible. Thus, this is the optimal (cheapest) solution. The final chart should always be interpreted as a table (see Table 16). Total cost: $15 \cdot 1.60 + 40 \cdot 40 + 20 \cdot 40 + 25 \cdot 1.20 = \$78.00$.

Table 16
Optimal Shipping Schedule

From factory	To store		
	Bangor	Falmouth	Hartford
Springfield	15	40	20
Waterbury	25	0	0

The flow chart for solving the transportation problem requires one more concept. Independently of the direction traveled around a loop in a pattern of change, additions will be made to certain squares and subtractions will be made from other squares. The same number will be added or subtracted throughout the entire change. Squares to which additions will be made are called **A-squares.** Squares from which subtractions will be made are called **S-squares.** The blank square is always an A-square, and the A's and S's alternate around the loop. In a four-move pattern of change, the blank square and the square across the diagonal from it are A-squares. The other two are S-squares.

Rule 5   **The quantity to be shifted in a pattern of change must be the smallest number in an S-square.**

Note that if there is a choice for the smallest circled number in an S-square, then the pattern of change will introduce circled zeros.

We can now present the flow chart for the transportation problem (see p. 183). The examples following the flow chart are intended to illustrate the use of this flow chart for problems involving four-move patterns of change only. This same flow chart applies to problems in which the solution involves patterns of change requiring more than four moves. Such patterns will be discussed in the next section.

Flow Chart for Solving the Transportation Problem

It is only necessary to choose a negative cost factor, not a *most* negative one. However, our approach can reduce the number of steps needed.

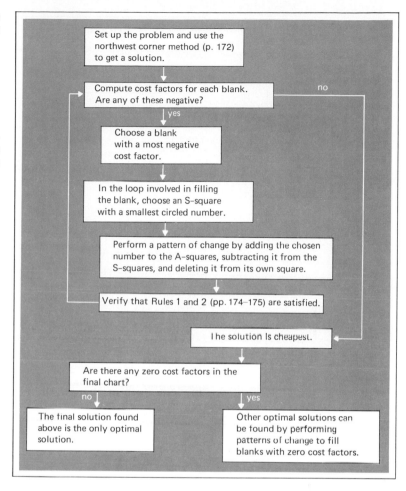

Set up the problem and use the northwest corner method (p. 172) to get a solution.

Compute cost factors for each blank. Are any of these negative?   no

yes

Choose a blank with a most negative cost factor.

In the loop involved in filling the blank, choose an S-square with a smallest circled number.

Perform a pattern of change by adding the chosen number to the A-squares, subtracting it from the S-squares, and deleting it from its own square.

Verify that Rules 1 and 2 (pp. 174-175) are satisfied.

The solution is cheapest.

Are there any zero cost factors in the final chart?

no   yes

The final solution found above is the only optimal solution.

Other optimal solutions can be found by performing patterns of change to fill blanks with zero cost factors.

Problem 1   AC receives goose down from Norway at Boston and New York City. From these ports of entry it is shipped by truck to Danbury, Springfield, and Rutland. Half of the down arrives at Boston and half at New York. The factories utilize down according to Table 17. The shipping costs from port of entry to factory are shown in Table 18. (The shipping costs from Boston to Springfield include tolls on the Massachusetts Turnpike.) How should AC supply its factories with down?

Table 17

Factory	Percentage down utilization
Danbury	35
Springfield	40
Rutland	25

Table 18
Shipping Costs (in dollars)

From	To		
	Danbury	Springfield	Rutland
Boston	3	2	3
New York	1	3	4

Solution

Chart 17

3	2	3	50
(35) a	(15) b	0 c	
1	3	4	50
−3* d	(25) e	(25) f	
35	40	25	

Patterns of change generally change most of the cost factors.

Following the flow chart, we set up the problem, find a solution via the Northwest Corner Method, and insert the cost factors (Chart 17). Square c would be filled via the pattern of change b c f e. The cost factor for such a change is $-2 + 3 - 4 + 3 = 0$. Similarly, to fill square d we would use the loop e d a b with cost factor $-3 + 1 - 3 + 2 = -3$. Since -3 is the only negative cost factor, we perform the second change. The S-squares for this change are a and e, which contain 35 and 25. We must shift the 25 and compute the new cost factors. This yields Chart 18.

Now square c is filled by the change a c f d with cost factor $-3 + 3 - 4 + 1 = -3$, while square e is filled by the change d e b a with cost factor $-1 + 3 - 2 + 3 = 3$. Filling square c and computing the new cost factors yields Chart 19. While this is a cheapest solution, the zero cost factor at address (2,2) indicates that performing a pattern of change to fill this square will not alter the cost. Hence, there is another cheapest solution, shown in Chart 20. The shipping manager is provided with the alternatives shown in Tables 19 and 20.

Chart 18

3	2	3	50
(10) a	(40) b	−3* c	
1	3	4	50
(25) d	3 e	(25) f	
35	40	25	

Chart 19

3	2	3	50
3 a	(40) b	(10) c	
1	3	4	50
(35) d	0 e	(15) f	
35	40	25	

Chart 20

3	2	3	50
3	(25)	(25)	
1	3	4	50
(35)	(15)	0	
35	40	25	

Table 19
Down Shipping Schedule #1
(% of total)

From	To		
	Danbury	Springfield	Rutland
Boston	0	40	10
New York	35	0	15

Table 20
Down Shipping Schedule #2
(% of total)

From	To		
	Danbury	Springfield	Rutland
Boston	0	25	25
New York	35	15	0

The numbers in these tables represent the percentage of the total amount of down available. One could also write these shipping schedules with numbers that reflected the percentage of the total down available in each port. Since $^{40}/_{50} = 80\%$, $^{10}/_{50} = 20\%$, etc., the tables would read as shown in Tables 21 and 22.

Table 21
Down Shipping Schedule #1
(% in port)

From	To		
	Danbury	Springfield	Rutland
Boston	0	80	20
New York	70	0	30

Table 22
Down Shipping Schedule #2
(% in port)

(The shipping manager decided on the second schedule since his nephew was on the New York to Springfield run.)

From	To		
	Danbury	Springfield	Rutland
Boston	0	50	50
New York	70	30	0

Problem 2

A company has three factories that produce the same commodity, in the quantities shown in Table 23. There is demand for this product at three locations, in the quantities shown in Table 24. The shipping costs are given by Table 25. How can the company most economically meet its demands?

Table 23

Factory	Quantity
A	65
B	20
C	30

Table 24

Location	Quantity
I	40
II	20
III	55

Table 25
Shipping Costs

From	To		
	I	II	III
A	1	2	1
B	3	2	4
C	1	4	5

Solution

Chart 21 is the first chart, together with the northwest corner method solution and cost factors. Table 26 shows how the various cost factors were computed.

Chart 21

1 (40) a	2 (20) b	1 (5) c	65
3 −1 d	2 −3 e	4 (20) f	20
1 −4* g	4 −2 h	5 (30) i	30
40	20	55	

Table 26

Blank	Pattern of change	Sum	Cost factor
d	f d a c	−4 + 3 − 1 + 1	−1
e	f e b c	−4 + 2 − 2 + 1	−3
g	i g a c	−5 + 1 − 1 + 1	−4*
h	b h i c	−2 + 4 − 5 + 1	−2

Chart 22

1 (10) a	2 (20) b	1 (35) c	65
3 −1 d	2 −3* e	4 (20) f	20
1 (30) g	4 2 h	5 4 i	30
40	20	55	

Since −4 is the most negative cost factor, we fill its blank first, as Chart 22 shows. The new cost-factor calculations are shown in Table 27.

Table 27

Blank	Pattern of change	Sum	Cost factor
d	a d f c	−1 +3 −4 +1	−1
e	same as before	same	−3*
h	b h g a	−2 +4 −1 +1	2
i	g i c a	−1 +5 −1 +1	4

Chart 23

1 (10) a	2 (0) b	1 (55) c	65
3 2 d	2 (20) e	4 3 f	20
1 (30) g	4 2 h	5 4 i	30
40	20	55	

Since −3 is the most negative, we wish to fill its square in the next pattern of change. Squares b and f are S-squares and they both contain 20. We arbitrarily decide to shift the 20 from f as shown in Chart 23. Notice that b now contains a zero and not a blank. Only one new blank is created during any pattern of change. Table 28 is the cost factor table for this chart. Since all the cost factors are positive this is the optimal solution. Table 29 is the corresponding shipping schedule. Total cost: $135.

Table 28

Blank	Pattern of change	Sum	Cost factor
d	a d e b	−1 +3 −2 +2	2
f	e f c b	−2 +4 −1 +2	3
h	b h g a	−2 +4 −1 +1	2
i	same as before	same	4

Table 29
Shipping Schedule

From	To		
	I	II	III
A	10	0	55
B	0	20	0
C	30	0	0

Exercises 4.3

1. A company produces an item at two factories (Table 30) and sells it at three stores (Table 31). The shipping charges are given in Table 32. How can the company meet its demands most economically, and what will be the total shipping costs?

Table 30

Factory	Quantity
Boston	30
Chicago	25

Table 31

Store	Quantity
Columbus	10
Harrisburg	30
Boston	15

Table 32
Shipping Charges
(in dollars)

From factory	To store		
	Columbus	Harrisburg	Boston
Boston	2	3	0
Chicago	1	3	2

C2. Find the minimal cost solution associated with the transportation problem in Tables 33, 34, and 35.

Table 33

Factory	Supply
I	17
II	25

Table 34

Store	Demand
A	7
B	17
C	8
D	10

Table 35
Shipping Costs (in cents)

From factory	To store			
	A	B	C	D
I	23	25	18	14
II	16	20	21	20

C3. Solve the transportation problem associated with the data in Tables 36, 37, and 38.

Table 36

Factory	Supply
I	20
II	50

Table 37

Store	Demand
A	20
B	30
C	20

Table 38
Shipping Costs (in dollars)

From factory	To store		
	A	B	C
I	1	2	3
II	2	1	1

A4. Solve the transportation problem associated with the data in Tables 39, 40, and 41.

Table 39

Factory	Supply
I	47
II	27
III	43

Table 40

Store	Demand
A	15
B	77
C	25

Table 41
Shipping Costs (in dollars)

From factory	To store		
	A	B	C
I	3	1	2
II	1	2	4
III	2	2	4

C5. The You-Lug-It truck rental company has 20 trucks in Columbus, Ohio, 15 trucks in Columbus, Georgia, and 75 trucks in Columbus, Texas. There are people willing to rent these trucks in three cities: 45 in Baton Rouge, Louisiana, 30 in Little Rock, Arkansas, and 135 in Columbus, Ohio. To get the trucks to the people who want them, the company must hire drivers. Besides paying these drivers by the hour, You-Lug-It must also pay for gas and oil. Table 42 lists these charges. How should they distribute their trucks? What is the total cost of the cheapest solution?

Table 42
Total Charges (in dollars)

From	To		
	Baton Rouge	Little Rock	Columbus, OH
Columbus, OH	40	30	0
Columbus, GA	30	18	40
Columbus, TX	20	10	30

6. Two farmers, Abner and Bert, supply potatoes to the three PHE stores in Wichita. Demand and supply, in pounds per week, are:

Store			
	I	II	III
Demand	52	41	67

Farmer		
	A	B
Supply	90	70

Table 43 shows the trucking charges. Minimize the total trucking charge.

Table 43
Trucking Charges
(cents per lb.)

Farmer	Store		
	I	II	III
A	4	3	5
B	2	6	3

Section 4-4
Patterns of Change
(continued)

We have discussed only one way to change a given solution into a less costly one—by performing a pattern of change to fill a blank with a negative cost factor. However, adhering to the rules previously given may mean that a pattern of change involves more than four moves. For example, consider the pattern of change required to fill the blank labeled g in Chart 24: Since a pattern of change involves only one blank, all the squares involved in it (except square g) must be occupied. Rule 4 limits us to horizontal and vertical moves; thus, we must move from i to g or from a to g. Suppose we choose to move horizontally from i. Our next move must be vertical, to the only occupied square in the first column, square a. Our horizontal move must then be to b. The vertical move must be to e, and finally we are forced to move horizontally back to f (Chart 25).

Chart 24

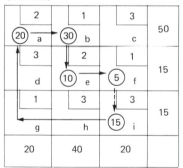

The cost factor for square g is computed in the same way as before, only now there are six terms involved in the sum:

$$-3 + 1 - 2 + 1 - 2 + 1 = -4$$

Had we moved from a to g, the path would have been a g i f e b with cost factor

$$-2 + 1 - 3 + 1 - 2 + 1 = -4$$

As before, it does not matter in which direction the loop is traversed when computing the cost factor. All the other blanks in this chart are filled using four-move patterns of change. Their cost factors are as indicated in Chart 26.

We should perform the change that fills square g. To do this, we first locate the A- and S-squares throughout the pattern. This is easily done: label the blank A and then alternate S and A around the loop (Chart 27). The S-squares contain the numbers 20, 10, and 15. Since 10 is the smallest, its square becomes blank and 10 is added and subtracted (as indicated by A's and S's) around the loop (Chart 28). In computing the new cost factors, four-move patterns of change are required to fill squares c, d, and h. Square e is filled with the six-move pattern of change f e b a g i,

Chart 25

2	1	3	
(20) a	(30) b	c	50
3	2	1	
d	(10) e	(5) f	15
1	3	3	
g	h	(15) i	15
20	40	20	

Chart 26

2	1	3	
(20) a	(30) b	3 c	50
3	2	1	
0 d	(10) e	(5) f	15
1	3	3	
−4* g	−1 h	(15) i	15
20	40	20	

Chart 27

S 2	A 1	5	50
⟨20⟩ a	⟨30⟩ b	5 c	
3	S 2	A 1	15
0 d	⟨10⟩ e	⟨5⟩ f	
A 1	3	S 3	15
−4 g	−1 h	⟨15⟩ i	
20	40	20	

Chart 28

2	1	5	20
⟨10⟩ a	⟨40⟩ b	1 c	
3	2	1	15
4 d	4 e	⟨15⟩ f	
1	3	3	15
⟨10⟩ g	3 h	⟨5⟩ i	
20	40	20	

Chart 29

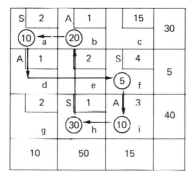

with cost factor $-1 + 2 - 1 + 2 - 1 + 3 = 4$. Since all cost factors are positive, this is the final chart.

The only surprise left is that the loop involved in a pattern of change may cross over itself, as in Chart 29. The path for this change is a d f i h b, with cost factor $-2 + 1 - 4 + 3 - 1 + 1 = -2$. The smallest number in an S-square is the 5 in square f. Performing this change yields Chart 30.

To indicate how Rules 3, 4, and 5 lead one to the unique pattern of change for any given blank, we present the following problem, solved in complete detail.

Chart 30   Problem 1

2	1	5	30
⟨5⟩	⟨25⟩		
1	2	4	5
⟨5⟩			
2	1	3	40
	⟨25⟩	⟨15⟩	
10	50	15	

A company produces desk calculators at four factories according to table 44. There is a demand for its product at five locations (Table 45). The shipping and installation costs are not part of the purchase price but are paid for by the company. These charges (Table 46) vary according to distance and the nature of the locally negotiated union contract. How can the company meet its demands most economically? What will the total shipping and installation cost be?

Table 44

Factory	Production
1	50
2	50
3	70
4	10

Table 45

Location	Demand
1	30
2	50
3	50
4	10
5	40

Table 46
Shipping and Installation Costs
(in dollars)

From factory	To location				
	1	2	3	4	5
1	3	5	5	4	5
2	5	2	7	4	2
3	1	4	1	3	5
4	5	4	4	3	2

Table 46 / Shipping and Installation Costs (in dollars)

Solution

Using the northwest corner method, we arrive at Chart 31. Before computing the cost factors, we will indicate how the rules for a pattern of change help determine the path involved in filling blank k. Since we must move horizontally or vertically from an occupied square, the only choices are to move from a, m, n, or o. We consider each of these possibilities:

(1) Move from a to k. Since this is a vertical move, our next move is horizontal. We do not move to n because there would be no (vertical) next move from n to an occupied square. If we move to o, our next move would have to be down to t, from which there is no horizontal exit to an occupied square. Thus, we move to m. Once at m, the only choice for a vertical move is to h. Next we must move horizontally to g and finally back to b. It can be shown that only one pattern of change will fill any particular blank, and thus we don't need to worry about other possibilities. However, we want to indicate what would happen had we not been so lucky and had chosen m, n, or o instead of a.

(2) Move from m to k. This move results in the same pattern of change as before (the clockwise version): m k a b g h.

(3) Move from n to k. Our next move is to a, then horizontally to b, vertically to g, horizontally over to h, vertically down to m. This pattern, indicated in Chart 32, violates Rule 4 in that the first and last moves are both horizontal. Also, there are seven moves in the loop, but a pattern of change always has an even number of moves.

(4) Move from o to k. Here the path is o k a b g h m, and again the first and last moves are both horizontal.

Chart 31

3 (30) a	5 (20) b	5 c	4 d	5 e	50
5 f	2 (30) g	7 (20) h	4 i	2 j	50
1 k	4 l	1 (30) m	3 (10) n	5 (30) o	70
5 p	4 q	4 r	3 s	2 (10) t	10
30	50	50	10	40	

Chart 32

3 (30) a →	5 (20) b	5 c	4 d	5 e	50
5 f	2 (30) g	7 (20) h	4 i	2 j	50
1 k	4 l	1 (30) m	3 (10) n	5 (30) o	70
5 p	4 q	4 r	3 s	2 (10) t	10
30	50	50	10	40	

Chart 33

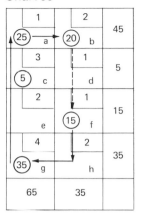

	65	35	

In the last two cases, backtracking was involved in the proposed pattern of change. To illustrate another *non*-pattern of change, in which backtracking is not involved, we present Chart 33, which shows a different problem. The path f h g a b shown in Chart 33 is not a pattern of change because it involves the consecutive vertical moves b f and f h. Also, the path consists of five moves, which is not an even number of moves. The correct pattern of change to fill blank square h is b h g a.

We now proceed to solve the problem at hand by computing the cost factors, resulting in Chart 34.

Table for Chart 34

Blank	Pattern of change	Sum	Cost factor
c	h c b g	$-7 + 5 - 5 + 2$	-5
d	n d b g h m	$-3 + 4 - 5 + 2 - 7 + 1$	-8
e	b e o m h g	$-5 + 5 - 5 + 1 - 7 + 2$	-9
f	a f g b	$-3 + 5 - 2 + 5$	5
i	n i h m	$-3 + 4 - 7 + 1$	-5
j	h j o m	$-7 + 2 - 5 + 1$	$-9*$
k	a k m h g b	$-3 + 1 - 1 + 7 - 2 + 5$	7
l	g l m h	$-2 + 4 - 1 + 7$	8
p	t p a b g h m o	$-2 + 5 - 3 + 5 - 2 + 7 - 1 + 5$	14
q	t q g h m o	$-2 + 4 - 2 + 7 - 1 + 5$	11
r	t r m o	$-2 + 4 - 1 + 5$	6
s	n s t o	$-3 + 3 - 2 + 5$	3

Chart 34

3	5	5	4	5	
(30) a	(20) b	-5 c	-8 d	-9 e	50
5	2	7	4	2	
5 f	(30) g	(20) h	5 i	$-9*$ j	50
1	4	1	3	5	
7 k	8 l	(30) m	(10) n	(30) o	70
5,	4	4	3	2	
14 p	11 q	6 r	3 s	(10) t	10
30	50	50	10	40	

The lowest cost factor is -9, which occurs in squares e and j. We arbitrarily decide to fill square j. The result of this pattern of change is Chart 35.

Table for Chart 35

Blank	Pattern of change	Sum	Cost factor
c	b c m o j g	$-5 + 5 - 1 + 5 - 2 + 2$	4
d	n d b g j o	$-3 + 4 - 5 + 2 - 2 + 5$	1
e	b e j g	$-5 + 5 - 2 + 2$	0
f	same as before		5
h	j h m o	$-2 + 7 - 1 + 5$	9
i	n i j o	$-3 + 4 - 2 + 5$	4
k	o k a b g j	$-5 + 1 - 3 + 5 - 2 + 2$	$-2*$
l	o l g j	$-5 + 4 - 2 + 2$	-1
p	t p a b g j	$-2 + 5 - 3 + 5 - 2 + 2$	5
q	t q g j	$-2 + 4 - 2 + 2$	2
r	same as before		6
s	same as before		3

Chart 35

S 3	A 5	5	4	5	
(30) a	(20) b	4 c	1 d	0 e	50
5	S 2	7	4	A 2	
5 f	(30) g	9 h	4 i	(20) j	50
A 1	4	1	3	S 5	
$-2*$ k	-2 l	(50) m	(10) n	(10) o	70
5	4	4	3	2	
5 p	2 q	6 r	3 s	(10) t	10
30	50	50	10	40	

The flow chart requires us to fill square k. The A- and S-squares for the appropriate pattern of change are indicated on Chart 35. Since 10 is the smallest number in an S-square, the pattern of change is performed clockwise. Chart 36 results.

Chart 36

3	5	5	4	5	
(20) a	(30) b	2 c	−1* d	0 e	50
5	2	7	4	2	
5 f	(20) g	7 h	2 i	(30) j	50
1	4	1	3	5	
(10) k	1 l	(50) m	(10) n	2 o	70
5	4	4	3	2	
5 p	2 q	4 r	1 s	(10) t	10
30	50	50	10	40	

Table for Chart 36

Blank	Pattern of change	Sum	Cost factor
c	a c m k	$-3 + 5 - 1 + 1$	2
d	n d a k	$-3 + 4 - 3 + 1$	−1*
e		same as before	0
f		same as before	5
h	g h m k a b	$-2 + 7 - 1 + 1 - 3 + 5$	7
i	n i g b a k	$-3 + 4 - 2 + 5 - 3 + 1$	2
l	k l b a	$-1 + 4 - 5 + 3$	1
o	k o j g b a	$-1 + 5 - 2 + 2 - 5 + 3$	2
p		same as before	5
q		same as before	2
r	t r m k a b g j	$-2 + 4 - 1 + 1 - 3 + 5 - 2 + 2$	4
s	t s n k a b g j	$-2 + 3 - 3 + 1 - 3 + 5 - 2 + 2$	1

Chart 37

3	5	5	4	5	
(10) a	(30) b	2 c	(10) d	0 e	50
5	2	7	4	2	
5 f	(20) g	7 h	3 i	(30) j	50
1	4	1	3	5	
(20) k	1 l	(50) m	1 n	2 o	70
5	4	4	3	2	
5 p	2 q	4 r	2 s	(10) t	10
30	50	50	10	40	

One more pattern of change is required; Chart 37 results. Since all the cost factors are positive or zero, we know that we have reached an optimal solution. The zero in square e indicates that there is another optimal solution. In this case, it happens that if we perform the pattern of change to fill square e we will reduce the number of shipments, although we will not save more money. Chart 38 is the final chart. Table 47 gives the answer in tabular form.

Chart 38

3	5	5	4	5	
(10) a	0 b	2 c	(10) d	(30) e	50
5	2	7	4	2	
5 f	(50) g	7 h	3 i	(0) j	50
1	4	1	3	5	
(20) k	1 l	(50) m	1 n	2 o	70
5	4	4	3	2	
5 p	2 q	4 r	2 s	(10) t	10
30	50	50	10	40	

Table 47
Shipping Schedule

From factory	To location				
	1	2	3	4	5
1	10	0	0	10	30
2	0	50	0	0	0
3	20	0	50	0	0
4	0	0	0	0	10

The total shipping costs will be:

$$3 \times 10 = \$\ 30 \quad \text{(square a)}$$
$$4 \times 10 = \ \ 40 \quad \text{(square d)}$$
$$5 \times 30 = \ 150 \quad \text{(square e)}$$
$$2 \times 50 = \ 100 \quad \text{(square g)}$$
$$1 \times 20 = \ \ 20 \quad \text{(square k)}$$
$$1 \times 50 = \ \ 50 \quad \text{(square m)}$$
$$2 \times 10 = \ \ \underline{20} \quad \text{(square t)}$$
$$\text{Total} \qquad \$410$$

There is a shortcut that will usually reduce the number of patterns of change needed to solve a particular transportation problem. Before applying the northwest corner method, arrange to have one of the smallest shipping costs in the northwest corner. This can always be done by rearranging the input data. In the previous problem, the lowest shipping costs are the 1's at addresses (3,1) and (3,3). Thus if we interchange rows 1 and 3 we will bring a 1 into the (1,1) position. (We could also interchange columns 1 and 3 and then rows 1 and 3 to move the 1 in the (3,3) position into the (1,1) position. While a double switch may be necessary for some problems, here a single switch suffices.) Table 48 shows the new shipping and installation costs.

Table 48
Shipping and Installation Costs
(in dollars)

From factory	To location				
	1	2	3	4	5
3	1	4	1	3	5
2	5	2	7	4	2
1	3	5	5	4	5
4	5	4	4	3	2

Chart 39 is the new chart.

Chart 39

1	4	1	3	5	70
5	2	7	4	2	50
3	5	5	4	5	50
5	4	4	3	2	10
30	50	50	10	40	

When interpreting the final answer, care must be taken to recall that the first row represents the third factory and the third row represents the first factory.

Exercises 4.4

Chart 40

(20)	(110)	(30)	(25)	(30)	(20)	
1 ⟨20⟩a	2 ⟨30⟩b	2 c	3	1	1 d	50
3 e	2 ⟨40⟩f	2 g	3 h	2	2	40
5 i	1 ⟨30⟩j	3 ⟨5⟩k	2	4	1 ⟨10⟩l	45
4 m	3 ⟨40⟩n	5	4 ⟨20⟩o	1 ⟨30⟩p	2	90
2 q	2 r	3	1	5 s	1 ⟨10⟩t	10

Exercises 1 and 2 refer to Chart 40.

A1. Describe the patterns of change required to fill the indicated blanks:

(a) e	(d) m	(g) d	(j) q
(b) h	(e) g	(h) s	
(c) c	(f) r	(i) i	

A2. Find the cost factors for each of the blanks indicated in the previous exercise.

A3. What actual change in total cost will result from applying the patterns of change required to fill the blanks listed in Exercise 1? (Interpret the numbers in the upper right corners as dollar amounts.)

In Exercises 4 through 10, solve the transportation problem for the data supplied.

4.

Factory	Supply
1	30
2	55
3	15

Location	Demand
1	20
2	35
3	40
4	5

Shipping Costs

From factory	To location			
	1	2	3	4
1	3	1	2	3
2	2	3	1	2
3	1	3	4	1

A5.

Factory	Supply
1	35
2	70
3	20
4	5

Location	Demand
1	15
2	35
3	30
4	50

Shipping Costs

From factory	To location			
	1	2	3	4
1	5	4	1	2
2	1	3	4	5
3	2	1	3	4
4	1	2	1	3

6.

Factory	Supply
1	30
2	35
3	35
4	35
5	40

Location	Demand
1	10
2	25
3	75
4	65

Shipping Costs

From factory	To location			
	1	2	3	4
1	1	3	2	1
2	1	4	2	3
3	3	4	1	5
4	4	2	1	2
5	3	1	5	2

A7.

Factory	Supply
1	35
2	15
3	35
4	25
5	15
6	30

Location	Demand
1	15
2	25
3	25
4	90

Shipping Costs

From factory	To location			
	1	2	3	4
1	1	2	4	3
2	4	4	3	4
3	2	3	1	4
4	4	3	2	1
5	4	1	2	3
6	2	3	4	1

C8.

Factory	Supply
1	33
2	21
3	21
4	8

Location	Demand
1	15
2	24
3	24
4	20

Shipping Costs

From factory	To location			
	1	2	3	4
1	1	1	2	3
2	3	4	1	2
3	4	3	3	1
4	2	5	4	4

9.

Factory	Supply
1	14
2	12
3	6
4	6

Location	Demand
1	6
2	12
3	12
4	8

Shipping Costs

From factory	To location			
	1	2	3	4
1	1	1	3	1
2	2	4	1	2
3	1	4	3	3
4	2	5	4	4

A10.

Factory	Supply
1	15
2	14
3	7
4	6

Location	Demand
1	7
2	13
3	3
4	19

Shipping Costs

From factory	To location			
	1	2	3	4
1	1	1	2	3
2	3	4	2	1
3	2	3	4	1
4	1	2	5	4

11. Return to the PHEM feasibility study for the free Petalunch program, Exercise 9 in Section 4-2. Suppose preparation costs are the same at all three stores and distribution costs are as shown in Table 49. Find all possible cheapest ways to carry out the distribution. What common feature is shared by all solutions?

Table 49

Store	Organization			
	A	B	C	D
1	1	3	4	7
2	2	4	6	8
3	2	3	4	5

**Section 4-5
Unequal Supply and
Demand**

It is relatively easy to treat cases of the transportation problem in which supply and demand are unequal. Here is how AC handled a case in which supply exceeded demand. The fuel shortage of 1974 hurt the sales of AC as much as it hurt those of any other company. Their four factories turned out snowmobiles as usual (Table 50). But the demand at the four retail stores fell sharply (Table 51). The supply was 140 while the demand was only 65. Thus, 75 snowmobiles had to be stored. At Concord and Hartford there was no problem about storage. Both factories had empty warehouses on the premises. Houlton had no extra space, but the excess snowmobiles could be shipped to Concord at a cost of $10 per vehicle. Boston had storage facilities, but the security system amounted to $3 per item. While this cost was not a transportation cost, it was a cost to be minimized. Table 52 shows these "transportation" costs.

Table 50
1974 Snowmobile Production

Concord	30
Hartford	40
Boston	50
Houlton	20
Total supply	**140**

Table 51
Snowmobile Demand

Boston	10
Orono	30
Albany	20
Falmouth	5
Total demand	**65**

Table 52
1974 Transportation Costs
(in dollars per item)

From	To				
	Boston	Orono	Albany	Falmouth	Warehouses
Concord	10	15	5	15	0
Hartford	15	25	5	10	0
Boston	1	15	10	5	3
Houlton	15	5	20	15	10

Using the trick described at the end of Section 4-4, we switch the first and fifth columns to get one of the lowest costs in the (1,1) position. Beginning with the northwest corner method, we get the sequence of patterns of change shown in Charts 41–45. Since all the cost factors are now positive or zero, this is an optimal solution. The shipping schedule is given in Table 53.

Chart 41

cost	0	15	5	15	10	supply
alloc	(30)	3	−2	13	8	30
cost	0	25	5	10	15	
alloc	(40)	13	−2	8	13	40
cost	3	15	10	5	1	
alloc	(5)	(30)	(15)	0	−4	50
cost	10	5	20	15	15	
alloc	−3	−20*	(5)	(5)	(10)	20
demand	75	30	20	5	10	

Chart 42

cost	0	15	5	15	10	supply
alloc	(30)	3	−2	−7	−12	30
cost	0	25	5	10	15	
alloc	(40)	13	−2	−12	−7	40
cost	3	15	10	5	1	
alloc	(5)	(25)	(20)	−20	−24*	50
cost	10	5	20	15	15	
alloc	17	(5)	20	(5)	(10)	20
demand	75	30	20	5	10	

Chart 43

cost	0	15	5	15	10	supply
alloc	(30)	3	−2	−7	12	30
cost	0	25	5	10	15	
alloc	(40)	13	−2	−12	17	40
cost	3	15	10	5	1	
alloc	(5)	(15)	(20)	−20*	(10)	50
cost	10	5	20	15	15	
alloc	17	(15)	20	(5)	24	20
demand	75	30	20	5	10	

Chart 44

cost	0	15	5	15	10	supply
alloc	(30)	3	−2	13	12	30
cost	0	25	5	10	15	
alloc	(40)	13	−2*	8	17	40
cost	3	15	10	5	1	
alloc	(5)	(10)	(20)	(5)	(10)	50
cost	10	5	20	15	15	
alloc	17	(20)	20	20	24	20
demand	75	30	20	5	10	

Table 53
Shipping Schedule

From	To				
	Boston	Orono	Albany	Falmouth	Warehouses
Concord	0	0	0	0	30
Hartford	0	0	20	0	20
Boston	10	10	0	5	25
Houlton	0	20	0	0	0

Chart 45

0	15	5	15	10	
(30)	3	0	13	12	30
0	25	5	10	15	
(20)	13	(20)	8	17	40
3	15	10	5	1	
(25)	(10)	2	(5)	(10)	50
10	5	20	15	15	
17	(20)	22	20	24	20
75	30	20	5	10	

Whenever supply exceeds demand, some action is required with respect to the extra items produced. They may be stored, discarded, sold at discount, disassembled, etc. In these cases one introduces one or more new demands called "warehouse," "discount sale," etc., and includes in the per-item "transportation" costs all extra costs and losses. In this example, the fact that storage was so cheap at Concord meant that all the output at Concord would be stored.

After some careful consideration, AC decided that it would not be sensible to store so many snowmobiles, particularly since a new 1975 model was on the drawing board. They contracted with a discount chain in New York to sell 40 snowmobiles at a reduced price without the AC label. The loss per vehicle, including the additional delivery and handling costs, is shown in Table 54.

Table 54
Loss due to discount sales

Concord	$50
Hartford	$55
Boston	$65
Houlton	$75

The president of AC also decided to donate 10 machines to the Muskrat Club of Gulliver, Michigan (see Table 55). This left only 25 for the warehouse. These modifications require adding two more columns to the chart. You will be asked to solve this problem in Exercise 6.

Table 55
Dollar loss due to donation

Concord	300
Hartford	295
Boston	350
Houlton	400

We now turn to the other type of problem: when demand exceeds supply.

In the early seventies, the snowmobile situation was quite different. Transportation costs were lower (Table 56), as was total production (Table 57).

Table 56
1971 Transportation Costs
(in dollars per item)

From	To			
	Boston	Orono	Albany	Falmouth
Concord	8	12	4	12
Hartford	12	20	4	9
Boston	1	12	8	4
Houlton	14	4	18	12

Table 57
1971 Snowmobile Production

Concord	20
Hartford	30
Boston	15
Houlton	40
Total supply	105

However, an increase in demand caught the company by surprise (Table 58). Due to both mismanagement and computer failures, this situation did not come to light until it was too late to do anything to actually meet the demand. Production schedules had been set and raw materials ordered, so that overtime was out of the question. It was clear that some stores were not going to be supplied.

Table 58
1971 Snowmobile Demand

Boston	40
Orono	50
Albany	40
Falmouth	20
Total demand	150

The AC management's first response was one of indifference. They decided simply to supply whichever stores they could. The technique of the previous section was modified by introducing an imaginary factory that would account for the 45 extra snowmobiles. The "output" of this factory represented nonsupply. Obviously, it costs nothing to ship nothing, so the associated transportation costs were zero. Chart 46 is the final chart, and Table 59 shows the resulting shipping schedule.

Table 59
Shipping Schedule

From	To			
	Boston	Orono	Albany	Falmouth
Concord	10		10	
Hartford			30	
Boston	15			
Houlton		40		

Chart 46

	0		0		0		0		
	(15)		(10)		4		(20)		45
	8		12		4		12		
	(10)		4		(10)		4		20
	12		20		4		9		
	4		12		(30)		1		30
	1		12		8		4		
	(15)		11		11		3		15
	14		4		18		12		
	10		(40)		18		8		40
	40		50		40		20		

The AC managers were ready to adopt the shipping schedule. They sent it to the president for final approval, and she responded as follows: "Under ordinary circumstances, this would be an acceptable shipping schedule. However, it just so happens that this year in Falmouth our competition is introducing a new model snowmobile. To fail to supply Falmouth will mean losing that market forever. Since we cannot afford that, we must supply Falmouth."

It is possible to modify the transportation problem to handle this particular variation. In fact, it is relatively easy to guarantee that a particular demand will be supplied. However, this model is rather sensitive, and more sophisticated modifications might result in situations that cannot sensibly be handled by these techniques. Here is an outline indicating the three types of modification we will consider.

When the Demand Cannot Be Met

(a) To treat each demand equally, set the transportation costs from the imaginary factory equal to zero.

(b) To assure that one particular demand will be met, set the transportation costs from the imaginary factory to all other demands equal zero, but assign a very large shipping cost to the demand that is to be met. (For example, set this cost equal to +$1,000,000, or any other number significantly larger than every other transportation cost in the table.) Of course, it might happen that one particular demand exceeds the total supply so that it cannot be met under any circumstances. This technique will, however, guarantee that it receives as much as possible.

Chart 47

0	0	0	1000	45
(35)	(10)	2	997	
8	12	4	12	20
2	6	(20)	3	
12	20	4	9	30
6	14	(20)	(10)	
1	12	8	4	15
(5)	14	9	(10)	
14	4	18	12	40
10	(40)	16	5	
40	50	40	20	

(c) To assure that one particular demand is *not* met, proceed as in (b) but assign a very large negative (say −$1,000,000) shipping cost from the imaginary factory to the demand not to be met.

Be careful: we are asserting only that this method works when (a), (b), *or* (c) is used *once,* and in the case of (b) or (c), when there is exactly one special demand.

In our practical example, we arbitrarily assign $1000 as the shipping cost from nonsupply to Falmouth, and obtain the revised Chart 47 and the associated shipping schedule shown in Table 60. This schedule was better for everybody (except Boston).

Table 60

From	To			
	Boston	Orono	Albany	Falmouth
Concord			20	
Hartford			20	10
Boston	5			10
Houlton		40		

After the 1971 fiasco, the AC managers improved their prediction skills and also arranged for overtime capability. They increased regular production as much as they could (Table 61). When the 1972 demand figures were made available, there was still time to arrange for overtime. Again, the demand increased sharply (Table 62). Per-item overtime costs varied from factory to factory (Table 63). These costs were costs to be minimized, and were thus included in the shipping costs—which were the same as in 1971. Tables 64 and 65 show the total transportation costs for 1972 overtime production.

Table 61
1972 Snowmobile Production

Concord	30
Hartford	40
Boston	20
Houlton	50
Total supply	140

Table 62
1972 Snowmobile Demand

Boston	50
Orono	70
Albany	50
Falmouth	30
Total demand	200

Table 63
Per Item Overtime Costs
(in dollars)

Concord	20
Hartford	40
Boston	50
Houlton	15

Table 64
1972 Transportation Costs (in
dollars per item produced in
overtime)

From	To			
	Boston	Orono	Albany	Falmouth
Concord	8 + 20	12 + 20	4 + 20	12 + 20
Hartford	12 + 40	20 + 40	4 + 40	9 + 40
Boston	1 + 50	12 + 50	8 + 50	4 + 50
Houlton	14 + 15	4 + 15	18 + 15	12 + 15

Table 65
1972 Transportation Costs (in
dollars per item produced in
overtime)

From	To			
	Boston	Orono	Albany	Falmouth
Concord	28	32	24	32
Hartford	52	60	44	49
Boston	51	62	58	54
Houlton	29	19	33	27

Table 64 gives the breakdown of regular shipping costs (taken from Table 56) and overtime production costs (from Table 63). In Table 65, these costs have been added.

From Table 65, we note that if we are going to supply Boston from overtime production it is cheapest to use Concord. Similarly, Orono should be supplied from Houlton, Albany from Concord, and Falmouth from Houlton. It does not make sense even to consider any other possibility. We now handle the problem by introducing a fifth supplier called "overtime." The transportation costs from overtime to Boston will be $28, since if Boston is supplied with any overtime at all it will be overtime from Concord shipped at $28 per item. The cost to ship from overtime to Orono is $19, since the overtime items will actually originate at Houlton. Similarly, overtime to Albany is $24 and overtime to Falmouth is

Chart 48

1 (20)	12	8	4	20
12 (30)	20 (10)	4	9	40
8	12 (30)	4	12	30
14	4 (30)	18 (20)	12	50
C 28	H 19	C 24 (30)	H 27 (30)	60
50	70	50	30	

$27. The problem is set up as shown in Chart 48 (we have moved the 1 to the top left and written overtime supply in the last row).

The C's and H's indicate that any overtime shipments that end up in these boxes will originate at Concord and Houlton, respectively. Chart 49 is the final chart, and the shipping schedule is shown in Table 66.

Table 66
Shipping Schedule

From	To			
	Boston	Orono	Albany	Falmouth
Concord	30		10[a]	
Hartford			40	
Boston	20			
Houlton		70[b]		30[a]

[a] Overtime production.
[b] 20 of these represent overtime.

Chart 49

1 (20) / 21	12 / 12	8 / 5	4	20
12 / 3	20 / 21	4 (40)	9 / 2	40
8 (30) / 14	12 / 1	4 / 6	12	30
14 (0)	4 (50) / 9	18 / 0	12	50
C 28 / 1	H 19 (20)	C 24 (10)	H 27 (30)	60
50	70	50	30	

Houlton is required to produce $20 + 30 = 50$ snowmobiles in overtime, and Concord must produce 10. Fortunately for AC, such overtime schedules could be arranged. If for some reason overtime at Houlton had been limited, we would have had to deal with a much more difficult problem—one we will not even discuss.

We summarize the techniques just discussed in the following outline:

Some Modifications for Unequal Supply and Demand

I. *Supply exceeds demand:* In this case, introduce new demands such as warehouses and discount sales to take up the extra supply.

II. *Demand exceeds supply*

   A. *Overtime impossible:* Create a fake factory to represent nonsupply.

     (1) To treat each customer equally, make all the new shipping costs zero.

(2) To guarantee that one particular customer is supplied, make the cost to supply that customer from the fake factory extraordinarily large (and positive).

(3) To guarantee that one particular customer is not supplied make the cost to supply that customer from the fake factory extraordinarily large (and negative).

B. *Overtime possible:* Compute all the new costs associated with overtime production. Determine which producer should be used to supply each demand with overtime. In the new chart, introduce a line called "overtime" and use as shipping costs the overtime plus shipping costs that are cheapest for each demand. Solve the transportation problem and institute overtime at the factories indicated on the final chart.

There are obviously situations in which none of these models is appropriate. It is hoped that this brief discussion will be of use in most cases and will serve to indicate some of the possibilities.

Exercises 4.5

Exercises 1-4 refer to the Ridem Bicycle Corporation, which produces its standard model at three different locations (Table 67). Demand varies from month to month. When faced with excess supply, rather than store the extra bikes, Ridem changes the brand name and sells the excess at a substantial loss to a mail-order house in Chicago (see Table 68). The overtime costs per bicycle vary from factory to factory (Table 69).

Table 67

Factory	Quantity produced (per month)
Charleston WV	24
Pittsburgh, PA	32
Montpeilier, VT	18

Table 68
Cost to company per bicycle
sold to Chicago (in dollars)

Factory	Cost
Charleston	42
Pittsburgh	38
Montpelier	58

<table>
<tr><td>Table 69
Extra cost per item made on
overtime (in dollars)</td></tr>
</table>

Factory	Cost
Charleston	6
Pittsburgh	5
Montpelier	6

A 1. In May, Ridem receives orders from three stores in the quantities shown in Table 70. The company determines the shipping charges as shown in Table 71. How does it supply the demand most economically?

Table 70

Store	Quantity
Columbus, OH	24
Boston, MA	25
Rochester, NY	25

Table 71
Shipping Charges (in dollars)

From	To		
	Columbus	Boston	Rochester
Charleston	2	5	4
Pittsburgh	2	4	3
Montpelier	8	2	3

A 2. In June the word gets around to the stores that the bicycles produced at the Montpelier factory are of superior quality. The Columbus store is willing to pay the company $7 per bicycle for bicycles obtained from Montpelier. The Boston store agrees to pay $2 and the Rochester store decides to pay $4. Assuming that each store places the same order as in May, what will be the new shipping schedule?

3. Due to unusually cold weather, the demand for bicycles slacks off in July. The stores not only refuse to pay extra for the Montpelier bicycles, but do not even place very large orders (see Table 72). What will be the shipping schedule for July?

Table 72

Store	Quantity
Columbus	18
Boston	15
Rochester	12

C 4. In August business picks up (see Table 73). The orders are large but the stores are unwilling to pay extra for the Montpelier bicycles (see Exercise 2). Given these orders, how will the company fill them most economically?

Table 73

Store	Quantity
Columbus	50
Boston	25
Rochester	36

5. Three factories have made do-it-yourself beer-brewing kits (Table 74). Three hobby stores have ordered the kits (Table 75). Shipping costs (in dollars) are shown in Table 76. A discount store will pick up a total of 30 kits from any of the factories, but it will pay *less* than what the other stores pay (Table 77—amounts are in dollars). The rest of the kits must be shipped to a warehouse, at the dollar cost per kit shown in Table 78. Minimize shipping and discounting costs.

Table 74

Factory	Quantity
1	100
2	100
3	50

Table 75

Store	Order
A	50
B	60
C	70

Table 76

Factory	Store		
	A	B	C
1	2	3	4
2	4	3	5
3	3	5	6

Table 77

Factory	Discount
1	8
2	9
3	10

Table 78

Factory	Cost
1	3
2	4
3	3

A 6. Solve the modified snowmobile problem on p. 202.

7. The data for the PHE feasibility study of Exercise 9, Section 4-2 and Exercise 11, Section 4-4, were obtained from real stores and organizations in Nashville. Based on the study, the PHE managers decided to carry out the project. After the Petalunches were prepared, however, they discovered that demand had fallen at the organizations by the following amounts: A, 10; B, 5; C, 12; D, 3. The extra lunches had to be thrown out, resulting in a loss (not tax deductible as a contribution) per lunch, as follows: Store 1, 70¢; Store 2, 65¢; Store 3, 60¢. These losses represent preparation costs. Minimize the total loss and transportation costs.

Section 4-6
The Assignment Problem

When the advertising agency for Appalachian Creations read that another sporting-goods company had wilderness tested its gear as an advertising gimmick, the agency convinced the president of AC that sales would improve considerably if AC could say that its equipment had been used on expeditions by famous mountaineers.

While AC had provided food for the Sagarmatha expedition, to test all of its gear in Nepal would have been too costly. Thus, three peaks in the Appalachian Mountains were selected for the test: Mount Mansfield (4393 feet), Mount Washington (6288 feet), and Mount Mitchell (6684 feet). AC was able to obtain the services of three famous mountaineer-authors: Thomas Black of Pittsburg, Mary Green of New York City, and John White of Philadelphia.

Each expedition was to last one week in January and would consist of the leader and one of the sales clerks at AC. Since each team was to be identically equipped, the only variable was the cost of getting to base camp. These costs were computed and tabulated as shown in Table 79.

Table 79
Transportation Costs
(in dollars)

Leader	Mountain		
	Mt. Mansfield	Mt. Washington	Mt. Mitchell
Black	$510	$595	$610
Green	315	425	625
White	450	495	585

The obvious question is: Which leader should be assigned to which mountain to minimize the cost? This problem can be viewed as a special type of transportation problem in which each

Chart 50

315 ①	425 ⓪	625 185	1
510 25	595 ①	610 ⓪	1
450 −10	495 −75*	585 ①	1
1	1	1	

Chart 51

315 ①	425 ⓪	625 185	1
510 25	595 ⓪	610 ①	1
450 65	495 ①	585 75	1
1	1	1	

Note that in any assignment problem of a given size the northwest corner method always yields the same result: 1's on the main diagonal and 0's just above.

supply location supplies one item (each leader supplies himself or herself) and each demand location requires one item (each expedition requires one leader). Such a problem is called an *assignment problem*. While special techniques have been developed specifically for this problem (for example, the Hungarian method as given by H. W. Kuhn in the *Naval Research Logistics Quarterly*, vol. 2, 1955, pp. 83–97), the techniques which we have already developed will solve this problem in two steps (Charts 50 and 51): the Black expedition scales Mount Mitchell; the Green expedition climbs Mount Mansfield; the White expedition ascends Mount Washington. The total cost to the company is $1420.

In the following summer, AC management found it necessary to solve another assignment problem. To improve community relations, they decided to hire high-school students as administrative assistants for the five senior vice presidents. The personnel department screened applicants and came up with five suitable candidates: the president's son, Malcolm P. Pennypincher, III, Joe College, Sally Sophmore, Willy Phlunkett, and Helen Helpus. These were each interviewed and ranked by the VPs as shown in Table 80. Now the question is: How should the students be assigned to create maximum satisfaction? The problem is set up in Chart 52. Another aspect of using patterns of change on an assignment problem is that many changes will involve a shift of zero. For example, assuming that you made the same choice of first move as we did, the first two charts are Charts 53 and 54.

Chart 52

1 ①	1 ⓪	1	2	1	1
2	4 ①	3 ⓪	1	2	1
5	2	2 ①	4 ⓪	3	1
3	5	4	3 ①	4 ⓪	1
4	3	5	5	5 ①	1
1	1	1	1	1	

Chart 53

1 ①	1 ⓪	1	2	1	1
2	4 ①	3	1 ⓪	2	1
5	2	2 ①	4 ⓪	3	1
3	5	4	3 ①	4 ⓪	1
4	3	5	5	5 ①	1
1	1	1	1	1	

Chart 54

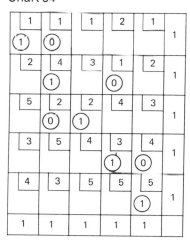

1 ①	1 ⓪	1	2	1	1
2	4 ①	3	1 ⓪	2	1
5	2 ⓪	2 ①	4	3	1
3	5	4	3 ①	4 ⓪	1
4	3	5	5	5 ①	1
1	1	1	1	1	

Table 80
Ranking of Candidates

Student	Vice President of				
	Production	Sales	Shipping	Personnel	Public rel.
Malcolm	1	1	1	2	1
Joe	2	4	3	1	2
Sally	5	2	2	4	3
Willy	3	5	4	3	4
Helen	4	3	5	5	5

After several more patterns of change we arrive at the solution (Table 81).

Table 81
Assignments

Student	Vice President of				
	Production	Sales	Shipping	Personnel	Public rel.
Malcolm					X
Joe				X	
Sally			X		
Willy	X				
Helen		X			

We used the assignment problem model here since it was the only technique available to us. However, in this problem as in Exercise 5, following this section, the model may not be appropriate. Suppose, for example, that the public relations VP felt that all of the candidates were essentially equal and assigned rankings arbitrarily. Suppose on the other hand that the sales VP felt that Malcolm and Sally were exceptionally well qualified while the last three were equally bad. Now if we assign Malcolm to sales and Helen to public relations, we will not disappoint public relations, but we will significantly increase the satisfaction at Sales. One way to improve the model would be to ask the VPs to assign a number grade, say from 0 to 100, to each of the candidates. These problems give some indication of the difficulties involved in quantifying happiness. They also point out deficiencies in the model. Most mathematical models have deficiencies. Before using any mathematical model, be sure you understand both the model and the real-world problem. You cannot expect to conclude much from a model that does not represent your problem adequately.

Authors' Message.

As a final example, consider the problem the AC managers had when they were forced by the union to relocate four factory workers whose jobs had been phased out. Fortunately, there were four other openings involving typing, sewing, cutting, and packing. AC commissioned a psychological testing firm to devise aptitude tests for these four positions. Each of the individuals took these tests; the results are shown in Table 82.

Table 82
Aptitude Test Results (100 is a perfect score)

Person	Typing	Sewing	Cutting	Packing
A	52	95	38	94
B	75	23	55	75
C	80	75	76	64
D	32	90	88	55

Now the problem is to assign the jobs so as to maximize the total "score." Since we have developed minimization techniques only, the data at hand must be altered to fit the previous pattern. The following is a standard technique.

Pick a highest score. Replace each score with the difference between it and this highest score.

In our case, the highest score is 95. Table 83 is the new table.

Table 83
Adjusted Aptitude Scores

Person	Typing	Sewing	Cutting	Packing
A	43	0	57	1
B	20	72	40	20
C	15	20	19	31
D	63	5	7	40

Obviously, a low score on this table indicates that the original score was close to the highest score on the original table. Our job is now one of minimization, which we handle in the standard way (Chart 55). Patterns of change yield the assignments (see Table 84).

Chart 55

0	43	57	1	
(1)	(0)	−6	−74	1
72	20	40	20	
95	(1)	(0)	−32	1
20	15	19	31	
64	−16	(1)	(0)	1
5	03	7	40	
40	55	−21	(1)	1
1	1	1	1	

Table 84
Job Assignments

Person	Job			
	Typing	Sewing	Cutting	Packing
A		X		
B				X
C	X			
D			X	

The average grade for a person in this assignment is 84.5.

Exercises 4.6   [A]1. Three people have been selected for three jobs. Their scores on aptitude tests are shown in Table 85. A higher score indicates a better performance. How should these jobs be assigned?

Table 85
Test Scores

Person	Job		
	I	II	III
A	95	80	40
B	90	85	35
C	93	67	40

2. A history teacher has four tutorials. He wants to assign a different final term paper to each student with the idea of helping the student bring up his or her lowest mark. The students have been previously tested in four areas, and the term paper is to cover one of these. Table 86 shows the prior grade. How should the topics be assigned?

Table 86
Grades (out of 10)

Student	Topic			
	I	II	III	IV
A	8	7	5	6
B	7	6	4	5
C	7	7	6	5
D	8	8	7	7

A 3. I have been in charge of part of a charity drive for several years. We have five donors who generally will give large sums if properly approached. In the past the task of approaching these donors has been randomly assigned to the same five individuals. Table 87 shows the result. How should the assignments be made to maximize our success?

Table 87
Percentage of Success

Collector	Donor				
	I	II	III	IV	V
A	95	95	0	0	30
B	90	0	0	50	75
C	90	50	30	20	0
D	100	90	100	95	100
E	85	20	10	70	30

4. Carry out the patterns of change for:
   (a) the student assistant problem discussed above.
   (b) the worker relocation problem discussed above.
* 5. The senior executive vice president of PHE wants to keep his five key vice presidents as happy with their jobs as possible. These vice presidents are in charge of (1) store siting, (2) meat acquisition, (3) product development, (4) personnel, and (5) advertising. The SEVP asks each VP to list, in descending order of preference, the responsibilities he or she would like to have. The results are shown in Table 88.

Table 88

(For example: the store siting VP would most like job 3 (product development) and least like job 2 (meat acquistion).)

	Preferences
VP store siting	3, 4, 1, 5, 2
VP meat acquisition	5, 1, 4, 3, 2
VP product development	2, 3, 5, 1, 4
VP personnel	4, 2, 1, 3, 5
VP advertising	5, 4, 3, 1, 2

What changes, if any, should be made so as to maximize the VPs' collective happiness?

Review Exercises   Restudy the chapter, then do these exercises without looking back. Answers to all of these exercises are in Appendix B.

A1. Chart 56 is the first transportation problem ever formulated, solved, and published (F. L. Hitchcock, *Journal of Mathematical Physics,* vol. 20, 1941, pp. 224–30). Find all solutions.

A2. The Sparkle Plenty Diamond Company has orders from three cities, as follows: Kuwait, 3500 carats; Paris, 8000 carats; New York, 7000 carats. Its diamond-cutting centers can produce jewels in the following quantities: Antwerp, 6500 carats; Rotterdam, 5000 carats; New York, 4500 carats. The shipping charges from these centers (in dollars per carat) are shown in Table 89. The company insists on running all of its cutting centers at maximum rate, but it does have another source of jewels—the great AKV diamond production center at Johannesburg, South Africa. There are, however, some restrictions on this source. First, AKV always sets its shipping rate to a given city equal to the highest rate charged by any of the cutting centers for that city. Also, AKV will only produce quantities of cut diamonds in multiplies of 1000 carats. Diamonds may be stored in special guarded vaults at each center at the following rates per carat: Antwerp, $40; Rotterdam, $50; New York, $70; Johannesburg, $20; all charges are paid by the company. The problem is, of course, to ship the jewels so as to minimize total shipping and storage costs, but there is one other condition. In an attempt to improve Arab-Dutch relations, the company wishes as much of the Kuwait order as possible to be filled from Rotterdam, without increasing total cost. Please include the total shipping and storage costs in the final answer.

Chart 56

10	5	6	7	25
8	2	7	6	25
9	3	4	8	50
15	20	30	35	

Table 89

From	To		
	Kuwait	Paris	New York
Antwerp	90	30	90
Rotterdam	80	20	80
New York	120	100	20

A3. Sparkle Plenty produces five types of cut diamonds: round, square, emerald, marquis, and tear. In the past the five cutters at the Antwerp center have each cut all five shapes, but it has beome apparent that some cutters are better at certain cuts than others. It has been decided to assign each cutter to do just one type of cut. Over the past year each cutter's work has been examined in London by the governing board of the Royal Jewellers' Guild. The cutters have been graded as shown in Table 90. Here, of course, A+ is best, D− worst. How should the assignment best be made? (For a hint on how to set up the problem, see the first sentence of the answer in Appendix B.)

Table 90

	Round	Square	Emerald	Marquis	Tear
Gretchen	A+	C	A−	B−	B−
Marie	C+	B+	B+	A	D+
Rolf	C−	C	C−	B	B
Lousia	B−	A−	A	B+	D
Anna	A	A−	D−	D	C

Supplementary Exercises 1. Solve the following transportation problems:

(a)

1	2	2	10
2	1	3	20
3	4	2	30
20	15	25	

A(b)

1	1	4	20
3	2	2	10
2	5	3	50
7	10	9	30
30	70	10	

(c)

0	4	3	2	25
1	2	3	1	60
2	1	4	2	40
8	10	10	20	40
60	50	50	15	

Chart 57

C2. Finish the transportation problem shown in Chart 57.

3 (1)	1 (4)	1 (5)	2	5	10
2 (3)	3	3	1 (2)	5	5
4	4	1	4	4 (4)	4
4	3	4	2	4 (2)	2
4	4	3	3 (3)	1 (1)	4
4	4	5	5	7	

3. Follow the advice on p. 195 to write the first chart needed to solve the transportation problem shown in Table 91.

Table 91

Factory	Store			Production
	A	B	C	
I	7	8	9	52
II	3	4	5	26
III	6	4	2	26
Demand	10	13	81	

Chart 58

4. So, you think that in a transportation problem you always ship a lot along the cheapest route and none, or very little, along the most expensive route? Then solve the problem presented in Chart 58.

5. Solve the following transportation problems. In each case, find minimal transportation costs and shipping schedules.

A(a)

Factory	Supply
I	30
II	55
III	15

Location	Demand
A	20
B	35
C	40
D	5

Shipping Costs

From	To			
	A	B	C	D
I	3	1	2	3
II	2	3	1	2
III	1	3	4	1

(b)

Factory	Supply
I	65
II	20
III	30

Location	Demand
A	40
B	20
C	55

Shipping Costs

From	To		
	A	B	C
I	1	2	1
II	2	3	4
III	1	4	5

(c) Factory	Supply
1	10
2	14
3	16

Location	Demand
1	10
2	12
3	8
4	10

Shipping Costs

From	To			
	1	2	3	4
1	3	2	4	5
2	4	5	2	2
3	1	2	3	4

6. Three factories make sewing machines (Table 92). They are sold to three stores (Table 93). A new factory, factory IV, is opened. It produces 20 machines (Table 94). Warehousing is required as shown in Table 95. Minimize total costs.

Table 92

Factory	Production
I	30
II	40
III	50

Table 93

Store	Demand
A	40
B	50
C	40

Table 94
Shipping Costs per Machine

Factory	Store		
	A	B	C
I	0	0	4
II	0	0	5
III	2	2	1
IV	4	5	7

Table 95

Factory	Warehousing cost per machine
I	3
II	3
III	3
IV	5

7. Seven persons apply for five jobs. Their scores on the aptitude test are shown in Table 96. Which person should be assigned to each job?

Table 96

Person	Job				
	I	II	III	IV	V
A	30	20	90	70	80
B	95	80	10	20	40
C	90	30	80	70	85
D	75	80	95	90	30
E	90	70	20	40	50
F	80	50	90	30	85
G	80	30	40	70	90

^8. PHE has three high-level openings at its San Francisco headquarters: financial analyst, acquisitions evaluator, and long-range planner. Four good candidates, A, B, C, and D (not their real names), have applied and have been rated for each job on a 1 (poor) to 10 (perfect) scale, as shown in Table 97.

Table 97

	Analyst	Evaluator	Planner
A	9	8	3
B	9	5	7
C	8	9	9
D	7	8	7

(a) Who should be hired for which job and who should not be hired?
(b) At the last minute, candidates B and C drop out. Who should be hired for which job and which job should be left unfilled? (Assume each job is equally important.)

Summary This chapter presents a method of solution for two types of problems: the transportation problem and the assignment problem. Three stages are involved:

Setup:

Transportation problems (supply = demand)   p. 183

Transportation problems (supply ≠ demand)   Sec. 4-5

Assignment problems   p. 211

Northwest corner method: p. 172

Patterns of change: p. 177

Rules for patterns of change:

1   p. 174

2   p. 175

3   p. 179

4   p. 179

5   p. 182

(The final chart gives the answer.)

Chapter 5
The
Simplex
Method

Section 5-1
Introduction

When the management at Appalachian Creations chose Burlington, Vermont, as the site for their new factory for the production of high-quality down jackets and sleeping bags, they first asked a team of investigators to file a preliminary report on the proposed operation. This report revealed the following four facts:

(1) While rip-stop nylon, zippers, snaps, and thread were readily available, high-quality goose down was not. It seemed best to order down through a Norwegian distributor, who could guarantee delivery of 600 pounds a week.

(2) The planned size of the planned factory limited the staff to 30 workers. There would be no problem in hiring 30 skilled people; however, they would provide only $30 \times 40 = 1200$ person-hours of work each week.

(3) The proposed jacket required 2 pounds of down and 6 hours of work, while the sleeping bags required 3 pounds of down and 4 hours of individual work.

(4) The president of AC decided that in order to satisfy the stockholders the company should make $3 net profit on each jacket and $4 net profit on each sleeping bag produced.

The problem for management was to determine a "program" for production of jackets and sleeping bags at the greatest possible profit. This problem exemplifies a typical **linear programming** (LP) situation: how should one utilize limited resources (in this case down and time) to produce an optimum situation (in this case the greatest total profit)? It is obvious that this type of problem will arise in essentially every production situation. Of course, in most applications to real life there will be more than two resources and two commodities. The interested reader may wish to glance ahead to the examples in Section 5-6.

Returning to the AC linear programming problem, we tabulate the facts in the preliminary report, as shown in Table 1. This table is unlike any other we have considered in that the numbers in it represent three different units: pounds, hours, and dollars. However, such tables will prove to be useful as an intermediate step between a practical problem and its mathematical formulation.

Table 1

	Per jacket	Per sleeping bag	Total available
Down	2 lbs.	3 lbs.	600 lbs.
Time	6 hrs.	4 hr.	1200 hrs.
Profit	$3	$4	

Since we are asked to find the number of jackets and the number of sleeping bags to be produced, it would seem natural to let x_1 represent the number of jackets and x_2 represent the number of sleeping bags. However, in view of the large numbers in the last column of the table, we choose instead to let

x_1 = the number of hundreds of jackets produced each week

x_2 = the number of hundreds of sleeping bags produced each week

An answer like $x_1 = 3$ and $x_2 = 2$ will then mean: make 300 jackets and 200 sleeping bags.

Notice that since each jacket requires 2 pounds of down, x_1 hundred jackets will use $2x_1$ hundred pounds of down. Similarly, x_2 hundred sleeping bags will use $3x_2$ hundred pounds of down. Thus, the total down consumption will be $2x_1 + 3x_2$ hundred pounds per week. Since there are only 6 hundred pounds of down available, $2x_1 + 3x_2$ must be no more than 6, i.e.,

$$2x_1 + 3x_2 \leq 6 \qquad \qquad \textbf{1}$$

The symbols $\leq$ and $\geq$ are discussed in Appendix A, Section 8.

Likewise, to make x_1 hundred jackets uses $6x_1$ hundred hours, and $4x_2$ hundred hours will be used to make x_2 hundred sleeping bags. Since only 12 hundred hours of time are available each week, the total weekly time consumption $6x_1 + 4x_2$, must not exceed 12:

$$6x_1 + 4x_2 \leq 12 \qquad \qquad \textbf{2}$$

The profit on x_1 hundred jackets will amount to $3x_1$ hundred dollars. On x_2 hundred sleeping bags it will be $4x_2$ hundred dollars. The total profit will be the sum $3x_1 + 4x_2$, which we denote by y.

A quantity such as $y = 3x_1 + 4x_2$ is called a *dependent variable,* since its value varies depending upon the choice of x_1 and x_2. For instance, choosing $x_1 = 1$ and $x_2 = 2$ results in the value $y = 3 \cdot 1 + 4 \cdot 2 = 11$ hundred dollars. Our aim is to find x_1 and x_2 so that $2x_1 + 3x_2 \leq 6$, $6x_1 + 4x_2 \leq 12$ and, most importantly, so that y has the largest value possible. Thus, choosing $x_1 = 1$ and $x_2 = 2$ is unacceptable since $2 \cdot 1 + 3 \cdot 2 = 8 > 6$, which violates inequality **1**. An acceptable choice would be $x_1 = 1$ and $x_2 = 1$. With this choice we have $2 \cdot 1 + 3 \cdot 1 = 5 < 6$ and $6 \cdot 1 + 4 \cdot 1 = 10 < 12$, so that inequalities **1** and **2** are satisfied with a profit of $3 \cdot 1 + 4 \cdot 1 = 7$ hundred dollars. (Other acceptable choices may result in a greater profit.)

We are now ready to illustrate the standard format for

presenting a mathematically formulated LP problem:

MAXIMIZE

$$y = 3x_1 + 4x_2 \qquad\qquad\qquad \textbf{3}$$

SUBJECT TO

$$x_1 \geq 0, \quad x_2 \geq 0 \qquad\qquad\qquad \textbf{4}$$
$$2x_1 + 3x_2 \leq 6 \qquad\qquad\qquad \textbf{5}$$
$$6x_1 + 4x_2 \leq 12 \qquad\qquad\qquad \textbf{6}$$

Given a practical problem, it is necessary to present its information in this form before our methods can be applied to find a solution. The instruction "Maximize $y = 3x_1 + 4x_2$ subject to ..." is an abbreviation for the phrase "Find x_1 and x_2 such that $3x_1 + 4x_2$ is as large as possible and such that ...."

The dependent variable y is called the **objective variable,** since the object of the problem is to maximize y. While it is obvious in our case that x_1 and x_2 cannot be negative, this fact must be expressed formally as in **4** above. The inequalities **5** and **6** are called **constraints**.

You may wonder why we do not treat this problem like the ones presented in Chapter 2. Can we not determine how many jackets and sleeping bags we should make to use up all the available down and all the available time? Why not simply solve the system

$$\begin{bmatrix} 2x_1 + 3x_2 = 6 \\ 6x_1 + 4x_2 = 12 \end{bmatrix}$$

obtaining $x_1 = \frac{6}{5}$ hundred $= 120$ and $x_2 = \frac{6}{5}$ hundred $= 120$? At this point it is difficult to raise an objection to this approach, since we will soon discover that problems of this size can often be solved in this way. As we will verify later (p. 233), the solution $x_1 = \frac{6}{5}$, $x_2 = \frac{6}{5}$ does indeed produce the maximum profit $y = 3 \cdot \frac{6}{5} + 4 \cdot \frac{6}{5} = \frac{42}{5} = 8.4$, that is, \$840. This method of solution is, however, unlikely to work for any but the simplest of LP problems.

Suppose, for instance, that in our problem there had also been a constraint on the supply of rip-stop nylon, say $3x_1 + 5x_2 \leq 7$. The resulting system

$$\begin{bmatrix} 2x_1 + 3x_2 = 6 \\ 6x_1 + 4x_2 = 12 \\ 3x_1 + 5x_2 = 7 \end{bmatrix}$$

is inconsistent, in the sense that it has no solutions at all, and hence the above "method" fails. In this extended problem, as in some practical LP problems, it is impossible to use up all of the time and material available. We are then in a new mathematical situation, more complicated than that of Chapter 2.

In the next section we will develop a geometrical method, using some of the ideas presented in Chapter 2, to solve LP problems that involve only two variables, such as the above problem at AC. In the first part of Section 5-3 we extend the discussion to problems involving three variables, for which the method presented in Section 5-2 is usually inadequate. This leads to the introduction of the **simplex method**, which was invented by George Dantzig in 1947. The matrix form of the simplex method, the form in which practical calculations are done (even for two variable LP problems), is introduced in the second part of Section 5-3. For interested readers we provide in Section 5-4 an explanation of how the simplex method works. In Section 5-5, the simplex method is adapted to solve minimization problems. Section 5-6 contains examples and applications.

In Section 6-4, we extend the simplex method, discuss a new minimization technique, and give a method for recognizing when an LP problem has multiple solutions. While the simplex method is suitable for hand calculation in relatively small problems, it is also well suited for use with a computer. In real life, LP problems involving several dozen commodities and resources often arise. These can be solved by machine computations using the simplex method. The last section of Chapter 6 discusses the use of computers in the solution of LP problems.

Why are problems of this sort called "linear programming" problems? First, the goal of such problems is to obtain a good *program* of production, that is, a set of instructions such as "make 120 jackets and 120 sleeping bags per week." Second, we will confine ourselves to problems that can be expressed mathematically by using only *linear* forms, such as those encountered in Chapter 2. The word "programming" in LP does not have to do with computer programming.

It may be of interest to point out that we have already used a special arrangement of the simplex method to solve the transportation problems of Chapter 4. A transportation problem with 3 factories and 3 stores, for example, can be rewritten as an LP problem involving 9 variables and 6 constraints. However, the transportation problem is so special that the form of the simplex method to be presented here bears little resemblance to the patterns of change technique used in the previous chapter.

At this point you may wish to reread the introduction to Chapter 2.

Exercises 5.1

1. In a well run PHE store a cheeseburger requires 5 oz. of meat, 0.7 oz. of cheese, and three person-minutes of preparation time, while a superburger requires 7 oz. of meat, 0.6 oz. of cheese, and 11 person-minutes time. Suppose the store has 21 lbs. 14 oz. of meat, 2 lbs. 10 oz. of cheese, and 5 hours 30 person-minutes of available prepara-

tion time. Each cheeseburger sold produces a profit of 10 cents, and each superburger a profit of 40 cents. It is desired to maximize profit. Introduce appropriate variables and units of measure, then write the problem as an LP problem in standard format.

**Section 5-2
The Geometry of Linear
Programming in Two
Variables**

Consider again the LP problem of Section 5-1:

MAXIMIZE

$$y = 3x_1 + 4x_2 \qquad \qquad \textbf{1}$$

SUBJECT TO

$$x_1 \geq 0, \quad x_2 \geq 0 \qquad \qquad \textbf{2}$$
$$2x_1 + 3x_2 \leq 6 \qquad \qquad \textbf{3}$$
$$6x_1 + 4x_2 \leq 12 \qquad \qquad \textbf{4}$$

Figure 1

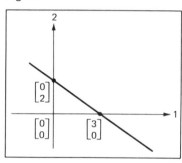

In seeking a new approach to this problem, we note that since two variables (x_1 and x_2) are involved, it might be possible to use our knowledge of coordinates (Section 2-1) to interpret the problem in the plane. Our plan is first to make a graph showing all points $\begin{bmatrix} x_1 \\ x_2 \end{bmatrix}$ that satisfy inequalities **2**, **3**, and **4**. We will then develop a technique for quickly determining which of these points yields the maximum value for y.

Since the problem involves several inequalities, we must learn how to graph an inequality.

Graphing Inequalities

To find all points $\begin{bmatrix} x_1 \\ x_2 \end{bmatrix}$ that satisfy $2x_1 + 3x_2 \leq 6$, we begin by graphing the *equation* $2x_1 + 3x_2 = 6$. From Section 2-1 we know that this graph is a line (Figure 1). Now it can be shown that the graph of the inequality $2x_1 + 3x_2 \leq 6$ will be one of the two **half-planes** bounded by this line. It will either be the **lower half-plane,** consisting of all points on or below this line, or else the **upper half-plane,** consisting of all points on or above this line. Obviously the origin $\begin{bmatrix} 0 \\ 0 \end{bmatrix}$ satisfies our inequality, since $2 \cdot 0 + 3 \cdot 0 = 0 < 6$, so the origin is in the graph of our inequality. We see then from Figure 1 that the graph is the lower half-plane (Figure 2).

Figure 2

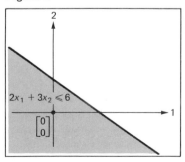

This same technique is used to graph any inequality of the form $ax_1 + bx_2 \leq c$ or the form $ax_1 + bx_2 \geq c$, where a and b are not both zero, as we see in the following examples.

Problem 1   Graph the inequality $3x_1 - 2x_2 \leq 6$.

Solution   First we graph the line $-3x_1 + 2x_2 = -6$. Since $\begin{bmatrix} 0 \\ 0 \end{bmatrix}$ satisfies the inequality, and the origin is *above* the line, the graph is the upper half-plane, Figure 3.

Problem 2   Graph the inequality $2x_1 + \frac{7}{2}x_2 \geq 14$.

Solution   First we graph the line $2x_1 + \frac{7}{2}x_2 = 14$. Since the origin does *not* satisfy the inequality ($2 \cdot 0 + \frac{7}{2} \cdot 0$ is not ≥ 14), the graph must be the half-plane that does *not* contain the origin, that is, the upper half-plane (see Figure 4).

Problem 3   Graph the inequality $4x_1 - 5x_2 \geq 0$.

Solution   First we graph the line $4x_1 - 5x_2 = 0$. Since the line passes through the origin, we cannot use $\begin{bmatrix} 0 \\ 0 \end{bmatrix}$ to help us determine which half-plane is the answer; instead, we may use any point that is not on this line. For easy computation we choose $\begin{bmatrix} 0 \\ 1 \end{bmatrix}$. This point does not satisfy the inequality ($4 \cdot 0 - 5 \cdot 1 = -5$, which is not ≥ 0); hence, the graph is the lower half-plane (see Figure 5).

There are two linear inequalities that occur in every standard format LP problem: $x_1 \geq 0$ and $x_2 \geq 0$. These simple inequalities are graphed in Figures 6 and 7, respectively. In Figure 6 the terms "upper" or "lower" are inappropriate; the graph in Figure 6 is a **right half-plane**.

Figure 3

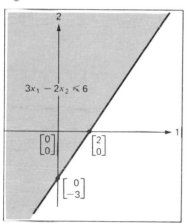

Figure 4

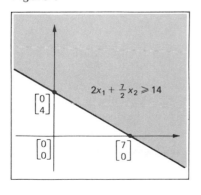

Figure 5

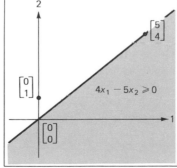

Figure 6

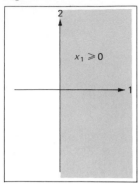

Figure 7

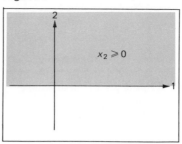

Graphing a Region of Feasibility

Figure 8

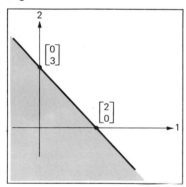

See Section 2-2 if necessary.

Figure 9

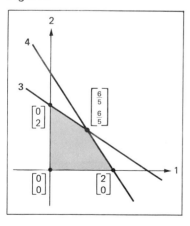

A point $\begin{bmatrix} x_2 \\ x_2 \end{bmatrix}$ that satisfies all the conditions of an LP problem is called a **feasible solution** of the problem (in our case, a feasible solution would satisfy inequalities **2**, **3**, and **4**). The **region of feasibility** of our problem consists of all points that are feasible solutions. To graph this region, we first use the technique described above to graph each of the inequalities **2**, **3**, and **4**. We have already graphed inequality **2** (Figures 6 and 7) and inequality **3** (Figure 2). The graph of inequality **4** appears in Figure 8. The graph of the region of feasibility consists of the points that these graphs have in common, as shown in Figure 9. Notice that we have labeled the constraint boundary lines **3** and **4**, after the inequalities from which they arose.

Our region of feasibility has four **corner points:** $\begin{bmatrix} 0 \\ 0 \end{bmatrix}$, $\begin{bmatrix} 0 \\ 2 \end{bmatrix}$, $\begin{bmatrix} 2 \\ 0 \end{bmatrix}$, and $\begin{bmatrix} 6/5 \\ 6/5 \end{bmatrix}$. The first three of these points were found when we graphed the inequalities. The fourth point is the intersection of lines **3** and **4**, found by solving the system of equations for these lines,

$$\begin{bmatrix} 2x_1 + 3x_2 = 6 \\ 6x_1 + 4x_2 = 12 \end{bmatrix}$$

We will see that determining the corners is important, since an optimal solution to the problem will always be found at some corner of the region of feasibility. Before we see why this is the case, we will investigate a few other examples. Note that while some linear programming problems in two variables have more complicated regions of feasibility, the process of graphing the region and locating the corners is always essentially the same.

Sometimes the positive quadrant is
referred to as the *first quadrant*.

Figure 10

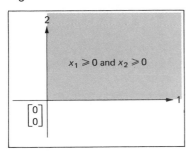

The inequalities $x_1 \geq 0$ and $x_2 \geq 0$ restrict us to the **positive quadrant** of the plane (Figure 10).

In the following problems, we draw each constraint graph in the positive quadrant and locate the corner points by solving the appropriate systems.

Problem 4   Draw the region of feasibility determined by the inequalities

$$x_1 \geq 0, x_2 \geq 0$$
$$-2x_2 + 3x_2 \leq 6 \qquad\qquad 5$$
$$x_1 - 4x_2 \leq 4 \qquad\qquad 6$$

Figure 11   Solution

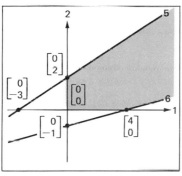

See Figure 11, where the lines with equations $-2x_1 + 3x_2 = 6$ and $x_1 - 4x_2 = 4$ have been labeled **5** and **6**, respectively. This region is *unbounded*, since the lines **5** and **6** never cross as one moves to the right in the positive quadrant.

Problem 5   Draw the region of feasibility determined by the inequalities

$$x_1 \geq 0, \ x_2 \geq 0$$
$$-2x_1 + 3x_2 \geq 6$$
$$x_1 - 4x_2 \geq 4$$

Solution   The lines we use here are the same as in the previous problem (Figure 11). However, in this case we are interested in the points that are simultaneously above line **5**, below line **6**, and in the first quadrant. That there are no such points is apparent from Figure 11. Thus, there is no such region or, as mathematicians would say, the region is **empty**. A situation like this occurs in practice whenever conditions of such stringency are placed on a problem that all possible solutions are ruled out.

Problem 6 Draw the region of feasibility determined by the inequalities

$$x_1 \geq 0, \; x_2 \geq 0$$
$$x_1 - x_2 \leq 1 \qquad\qquad 7$$
$$x_1 + x_2 \leq 3 \qquad\qquad 8$$
$$3x_1 + x_2 \leq 6 \qquad\qquad 9$$

Answer See Figure 12. Since, as we will see, the solution of an LP problem will occur at a corner point, and since the corner points can be found by solving the given equations simultaneously in pairs, the necessity of graphing may not be apparent. Notice however that in this problem the intersection of lines **7** and **8** yields neither a corner point nor a feasible solution. It is not easy to notice that this intersection is irrelevant by inspecting the given inequalities. In general, one should graph the lines first, and then solve only those systems of equations that yield corner points of the region of feasibility.

Figure 12

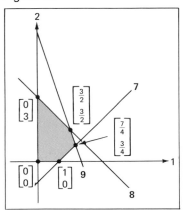

Problem 7 Draw the region of feasibility determined by the inequalities

$$x_1 \geq 0, \; x_2 \geq 0$$
$$2x_1 - x_2 \geq 1 \qquad\qquad 10$$
$$-x_1 + 2x_2 \geq 1 \qquad\qquad 11$$
$$x_1 + x_2 \leq 5 \qquad\qquad 12$$

Answer See Figure 13. Notice that the requirements $x_1 \geq 0$ and $x_2 \geq 0$ are unnecessary in view of constraints **10** and **11**.

Figure 13

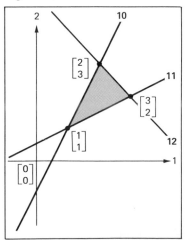

The Corner Point Theorem

We now return to our discussion of the AC problem. In Figure 9 we used all of the inequalities to determine the feasible region. The question remains as to how to interpret the objective variable $y = 3x_1 + 4x_2$ geometrically. Assigning different values to y yields different lines. For example, when $y = 4$ we obtain the line $4 = 3x_1 + 4x_2$. A different value, say $y = 5$, gives a different line: $5 = 3x_1 + 4x_2$. Since the system

$$\begin{bmatrix} 3x_1 + 4x_2 = 4 \\ 3x_1 + 4x_2 = 5 \end{bmatrix}$$

has no solution, these lines are parallel. In fact, no matter what value we assign to y we get a line that is parallel to these two. We see from Figure 14 that the larger the value we assign to y, the farther the line is from the origin. Every point in the plane is on exactly one of the lines $3x_1 + 4x_2 = y$, for some value of y. For example, the point $\begin{bmatrix} 2 \\ 3 \end{bmatrix}$ is on the line $y = 2 \cdot 2 + 4 \cdot 3 = 16$, that is, the line $3x_1 + 4x_2 = 16$.

Now let us return to the graph in Figure 9. Draw the "y-lines" that pass through the corners of the regions of feasibility (see Figure 15). Recall that we want y to be as large as is feasible. The largest of these y-values occurs at the corner $\begin{bmatrix} 6/5 \\ 6/5 \end{bmatrix}$, where $y = 3 \cdot 6/5 + 4 \cdot 6/5 = 42/5$. For any larger value of y, the corresponding line lies beyond the region of feasibility—no point on it satisfies the conditions of the problem. For example, the dotted line $y = 3x_1 + 4x_2 = 12$ in Figure 15 is such a line. Thus, the largest value of y that is yielded by a feasible solution occurs at the point with $x_1 = 6/5$ and $x_2 = 6/5$—which, as we noted before, corresponds to making $6/5$ hundred or 120 jackets and $6/5$ hundred or 120 sleeping bags. It can be shown that in all such problems the largest value of y occurs at a corner.

Figure 14

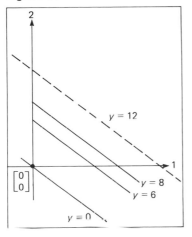

Figure 15

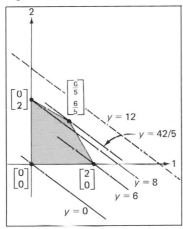

> **Corner Point Theorem: In any linear programming problem, an optimal value of the objective variable occurs at a corner point of the region of feasibility (unless, of course, there is no optimal value).**

This is true for all LP problems whether they are maximization or minimization problems (in the latter, which we will discuss in Section 5-5, the optimal value is the *smallest* value yielded by a feasible solution). This result also holds for problems involving more than two variables. The corner point theorem is the basis for the flow chart on p. 234, which shows how to solve a linear programming problem. We assume here that $x_1 \geq 0$ and $x_2 \geq 0$ are part of the input data and that all the numbers occurring in the formula for the objective variables are positive. These are the usual circumstances in practical applications, since the x's measure physical quantities. The technique of graphing particular lines corresponding to various values of y is slightly more difficult than the technique presented in the flow chart, but it will work in all cases.

Geometrical Method for Solving a Two-variable LP Problem in Standard Format

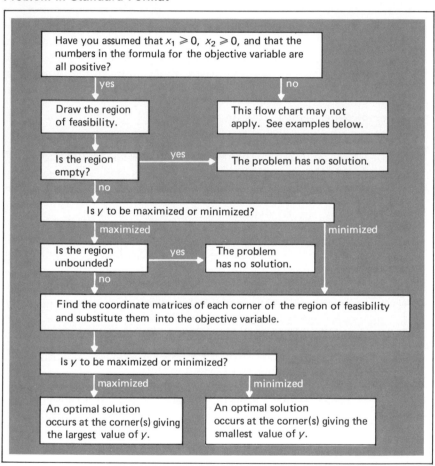

Problem 8   Maximize $y = 6x_1 + x_2$ subject to

$$x_1 \geq 0, \; x_2 \geq 0$$
$$x_1 - x_2 \leq 1$$
$$x_1 + x_2 \leq 3$$
$$3x_1 + x_2 \leq 6$$

Solution   We have already drawn the region of feasibility and found the corner points in Figure 12. Substituting these points into the formula for y results in Table 2.

Table 2

Corner point	y-value
$\begin{bmatrix} 0 \\ 0 \end{bmatrix}$	0
$\begin{bmatrix} 0 \\ 3 \end{bmatrix}$	3
$\begin{bmatrix} 3/2 \\ 3/2 \end{bmatrix}$	$21/2 = 10\frac{1}{2}$
$\begin{bmatrix} 7/4 \\ 3/4 \end{bmatrix}$	$45/4 = 11\frac{1}{4}$
$\begin{bmatrix} 1 \\ 0 \end{bmatrix}$	6

Sometimes it may be hard to tell which y-value is largest. For example, which is larger: $^{29}/_{12}$ or $^{12}/_{5}$? For help in making such determinations, see Appendix A, Section 8.

The largest of the y-values so obtained is $y = {}^{45}/_4$. Thus, the answer is $x_1 = {}^7/_4$, $x_2 = {}^3/_4$, and $y = {}^{45}/_4$.

Problem 9   Maximize $y = 2x_1 + 3x_2$ subject to

$$x_1 \geq 0, \ x_2 \geq 0$$
$$-2x_1 + 3x_2 \leq 6$$
$$x_1 - 4x_2 \leq 4$$

Solution   The region of feasibility (Figure 11) is unbounded; hence, the problem has no maximum solution.

Problem 10   Minimize (i.e., find the smallest value of) $y = 2x_1 + 5x_2$ subject to

$$x_1 \geq 0, \ x_2 \geq 0$$
$$-2x_1 + 3x_2 \geq 6$$
$$x_1 - 4x_2 \geq 4$$

Solution   By Problem 5, the region of feasibility is empty; hence, there is no solution.

Problem 11   Minimize $y = 2x_1 + 5x_2$ subject to

$$x_1 \geq 0, \ x_2 \geq 0$$
$$x_1 + x_2 \geq 3 \qquad \textbf{13}$$
$$2x_1 + x_2 \geq 4 \qquad \textbf{14}$$
$$2x_1 + 5x_2 \geq 10 \qquad \textbf{15}$$

Solution   We draw the region of feasibility and find its corner points (Figure 16). Substitution into the equation for the objective variable yields Table 3.

Table 3

Corner point	y-value
$\begin{bmatrix} 0 \\ 4 \end{bmatrix}$	20
$\begin{bmatrix} 1 \\ 2 \end{bmatrix}$	12
$\begin{bmatrix} 5/3 \\ 4/3 \end{bmatrix}$	10
$\begin{bmatrix} 5 \\ 0 \end{bmatrix}$	10

Figure 16

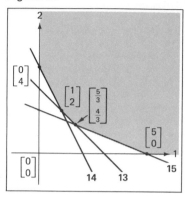

Thus, we find two minimum solutions: $x_1 = 5/3$, $x_2 = 4/3$ with $y = 10$ and $x_1 = 5$, $x_2 = 0$ with $y = 10$. (Actually there are even more than two minimum solutions in this case. We will learn how to determine all of them in Section 6-4.)

The following two problems are included so as to indicate techniques for handling cases to which the flow chart does not apply. Such cases are not likely to occur in practical applications.

Problem 12   Find the maximum and minimum values of $y = x_1 - x_2$ subject to

$$x_1 \geq 0, \quad x_2 \geq 0$$
$$-3x_1 + x_2 \leq 3 \qquad\qquad \textbf{16}$$
$$2x_1 - x_2 \geq 2 \qquad\qquad \textbf{17}$$

Figure 17

Solution   The flow chart does not apply, since the objective variable contains a negative number (the -1 before the x_2). The graph of the feasible region is shown in Figure 17. The line $y = 1$ passes through the corner point $\begin{bmatrix} 1 \\ 0 \end{bmatrix}$. For any larger value of y (such as $y = 2$), the y-line misses the region. Thus, the maximum solution is: $x_1 = 1$, $x_2 = 0$, $y = 1$. The line $y = -3$ passes through the corner point $\begin{bmatrix} 0 \\ 3 \end{bmatrix}$, but for every lower value of y (such as $y = -4$), the y-line also intersects the region. Thus, there is no minimum solution.

Problem 13   Find the maximum and minimum values of $y = x_1 + x_2$ subject to

$$-x_1 + 2x_2 \leq 2 \qquad \textbf{18}$$
$$2x_1 - x_2 \leq 2 \qquad \textbf{19}$$

Figure 18

Solution   We cannot use the flow chart because we are not told that $x_1 \geq 0$ and $x_2 \geq 0$. The feasible region is shown in Figure 18. The region has only one corner point: $\begin{bmatrix} 2 \\ 2 \end{bmatrix}$. It is clear from the two y-lines drawn above that the maximum solution is $x_1 = 2$, $x_2 = 2$, $y = 4$, and that there is no minimum solution.

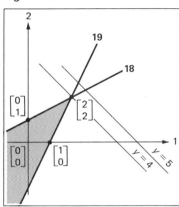

Exercises 5.2

A 1. (a) Maximize $y = 3x_1 + 2x_2$ subject to $x_1 \geq 0$, $x_2 \geq 0$, $x_1 + x_2 \leq 3$, $x_1 + 2x_2 \leq 4$.
   (b) Maximize $y = 2x_1 + 3x_2$ subject to the same conditions as in (a).
   (c) Maximize $y = x_1 + 4x_2$ subject to the same conditions as in (a).

2. (a) Minimize $y = 2x_1 + 3x_2$ subject to $x_1 \geq 0$, $x_2 \geq 0$, $x_1 + x_2 \geq 3$, $2x_1 + 5x_2 \geq 10$.
   (b) Minimize $y = x_1 + 6x_2$ subject to the same conditions as in (a).
   (c) Minimize $y = 3x_1 + x_2$ subject to the same conditions as in (a).

A 3. Minimize $y = 2x_1 + 3x_2$ subject to $x_1 \geq 0$, $x_2 \geq 0$, $4x_1 + x_2 \geq 4$, $x_1 + x_2 \geq 3$, $x_1 + 2x_2 \geq 4$.

4. Maximize $y = 4x_1 + 4x_2$ subject to $x_1 \geq 0$, $x_2 \geq 0$, $3x_1 + 8x_2 \leq 24$, $4x_1 + 5x_2 \leq 20$, $6x_1 + 4x_2 \leq 24$.

A 5. (a) Find the maximum and minimum values of $y = 6x_1 + x_2$ subject to $4x_1 + 3x_2 \geq 22$, $3x_1 - 2x_2 \leq 8$, $x_1 + 5x_2 \leq 31$.
   (b) Find the maximum and minimum values of $y = x_1 + 7x_2$ subject to the same conditions as in (a).

6. (a) Find the maximum and minimum values of $y = 3x_1 + 2x_2$ subject to $x_1 \geq 0$, $x_2 \geq 0$, $9x_1 + 5x_2 \leq 45$, $2x_1 + 6x_2 \leq 12$, $2x_1 + x_2 \geq 4$.
   (b) Find the maximum and minimum values of $y = -x_1 + x_2$ subject to the same conditions as in (a).

A 7. Find the maximum and minimum values of $y = 3x_1 + 2x_2$ subject to $x_1 \geq 0$, $x_2 \geq 0$, $x_1 + x_2 \leq 1$, $-x_1 + x_2 \geq 2$.

8. (a) Find the maximum and minimum values of $y = 3x_1 + 2x_2$ subject to $x_1 \geq 0$, $x_2 \geq 0$, $-x_1 + x_2 \leq 1$, $x_1 - x_2 \geq 1$.

(b) Find the maximum and minimum values of $y = 2x_1 - 3x_2$ subject to the same conditions as in (a).

A 9. Find the maximum and minimum values of $y = x_1 + x_2$ subject to $-x_1 \leq 2, -x_2 \leq 2, x_1 \leq 2, x_2 \leq 2$.

C 10. In a well-run PHE store, given the conditions of Exercise 1 in Section 5-1, how many cheeseburgers and how many superburgers should be made? What will the profit from this operation be? How much meat, cheese, and time will be left over?

Section 5.3
The Simplex Method for Maximization Problems

In this section we will describe the matrix form of the simplex method as it applies to a large class of LP maximization problems. It is important to be able to solve practical problems when you have been given the raw data, but in this section we will assume that the step from the raw data to their formulation as a system of inequalities has already been taken. In Section 5-6 we will discuss how to derive the appropriate system of inequalities from a given practical problem.

Geometric Introduction

Almost all linear programming (LP) problems in the real world involve more than two variables. If our solution technique for LP problems in two variables, presented in the previous section, is to have practical value, it must be adapted to LP problems in three, four, and more variables. To this end, consider the following rather simple LP problem in three variables:

MAXIMIZE

$$y = 5x_1 + 4x_2 + 8x_3 \qquad \mathbf{1}$$

SUBJECT TO

$$x_1 \geq 0, x_2 \geq 0, x_3 \geq 0 \qquad \mathbf{2}$$
$$2x_1 + 2x_2 + 3x_3 \leq 6 \qquad \mathbf{3}$$

Figure 19

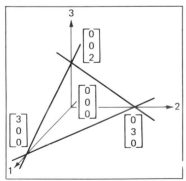

In order to approach this problem geometrically we must employ coordinates in three-dimensional space, as introduced in Section 2-6. The reader is not expected to become proficient in the use of the coordinates; this discussion is being presented as introduction to and motivation for the purely computational method that follows.

In Section 5-2 we learned that a linear inequality in two variables represents a half-plane. Similarly, it can be shown that a linear inequality in three variables represents a **half-space**. For example, inequality **3** above represents the lower half-space bounded by the plane $2x_1 + 2x_2 + 3x_3 = 6$ (Figure 19). This half-

Figure 20

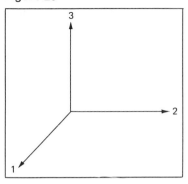

Figure 21

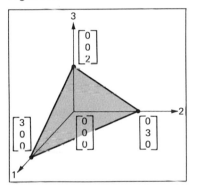

space consists of all points on or below this plane. Recall that in Section 5-2 we said that the "background inequalities" $x_1 \geq 0$, $x_2 \geq 0$ represent the positive quadrant (Figure 10). Similarly, the inequalities in condition **2** above represent in space the **positive octant** (Figure 20). The region of feasibility of our problem consists of all points common to the graphs in these two figures—that is, the solid pyramid of Figure 21. This region has four corner points: $\begin{bmatrix} 0 \\ 0 \\ 0 \end{bmatrix}$, $\begin{bmatrix} 3 \\ 0 \\ 0 \end{bmatrix}$, $\begin{bmatrix} 0 \\ 3 \\ 0 \end{bmatrix}$, and $\begin{bmatrix} 0 \\ 0 \\ 2 \end{bmatrix}$. According to the corner point theorem, p. 233, the maximum value of the objective variable (equation **1**) can be found by checking values at each of these points, which we do in Table 4.

Table 4

Corner point	y-value
$\begin{bmatrix} 0 \\ 0 \\ 0 \end{bmatrix}$	0
$\begin{bmatrix} 3 \\ 0 \\ 0 \end{bmatrix}$	15
$\begin{bmatrix} 0 \\ 3 \\ 0 \end{bmatrix}$	12
$\begin{bmatrix} 0 \\ 0 \\ 2 \end{bmatrix}$	16

The answer is $x_1 = 0$, $x_2 = 0$, $x_3 = 2$ and $y = 16$.

That was easy, but it was easy because there was only one constraint. In a three-variable LP problem involving more con-

See the discussion following Problem 6 in Section 5-2 above.

Figure 22

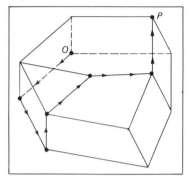

This remark holds for problems involving any number of variables.

straints, the region of feasibility may be more complicated than the region shown in Figure 21. It might, for example, look like the region in Figure 22. In a region as complex as that in Figure 22, there are many corner points, and they may be hard to find. Just solving the equations three at a time will not work, for the same reason that solving them two at a time did not work in the two-variable case. Also, even after the corner points are found, there may be a great number of them and the labor of substituting their coordinate matrices into the formula for y may be considerable. The simplex method was created to surmount all these difficulties.

In the region of feasibility shown in Figure 22, some pairs of corner points are joined by an *edge*, where two boundary planes intersect. The idea of the simplex method is to begin at a feasible corner and travel along these edges from corner to corner, always increasing the value of y, until we reach a corner at which y attains a maximum value. It can be shown that we will have reached a maximum corner when moving to any adjacent corner causes no increase in the value of y. Figure 22 shows an example of such a path (marked with arrows), leading from the origin O to a maximum point P.

The Matrix Form of the Simplex Method

However, before the simplex method can be applied in most practical situations, the problem must be subjected to Phase I, a technique described in Section 6-4. Unfortunately, Phase I is a special modification of the simplex method and is best understood after a discussion of the simplex method itself.

The method that we are about to describe is a compact matrix version of the geometric approach suggested above. It is remarkably simple and can be used effectively even if the user does not understand why it works. In the next section we will relate the algebraic and geometric approaches and briefly describe why the matrix technique works.

This technique applies to essentially all LP maximization problems that arise in applications. We illustrate it with the following example:

Problem 1   MAXIMIZE

$$y = 3x_1 + 4x_2 + 7x_3 \qquad \textbf{4}$$

SUBJECT TO

$$x_1 \geq 0, \quad x_2 \geq 0, \quad x_3 \geq 0 \qquad \textbf{5}$$
$$x_1 + 2x_2 + 4x_3 \leq 3 \qquad \textbf{6}$$
$$x_1 + \quad x_2 + 2x_3 \leq 2 \qquad \textbf{7}$$
$$2x_1 + 3x_2 + 8x_3 \leq 7 \qquad \textbf{8}$$

Before setting up the matrix that we will use in solving this problem, we introduce some convenient terminology. Recall that inequalities **6**, **7**, and **8** are called constraints. The numbers multiplying the variables in these constraints make up the **constraint matrix** of the problem:

$$\begin{bmatrix} 1 & 2 & 4 \\ 1 & 1 & 2 \\ 2 & 3 & 8 \end{bmatrix} \qquad \textbf{9}$$

The term "bound" is usually used in linear programming to refer to a bound on a variable. For example, in our case, 0 is a bound for x_1. A better (although more cumbersome) term would be *right-hand side*. When dealing with computers or persons familiar with LP, be careful of confusion which may result from our use of this word.

The **bounds** are the numbers on the right side of the inequality signs in inequalities **6**, **7**, and **8**. These numbers make up the **bound matrix** of the problem:

$$\begin{bmatrix} 3 \\ 2 \\ 7 \end{bmatrix} \qquad \textbf{10}$$

The numbers multiplying x_1, x_2, and x_3 in the formula for the objective variable (equation **4**) make up the **objective matrix** of the problem:

$$\begin{bmatrix} 3 & 4 & 7 \end{bmatrix} \qquad \textbf{11}$$

The matrix form of the simplex method as presented in this section does not apply to all maximization LP problems. Before setting up a problem, we must make sure that the following three conditions are satisfied:

Condition 1 The problem must stipulate all the x's that occur in it are greater than or equal to zero. That is, the inequalities $x_1 \geq 0$, $x_2 \geq 0$, etc., must appear.

Condition 2 All the constraints must contain $\leq$ signs rather than $\geq$ signs or $=$ signs. Constraints such as $3x_1 + 2x_2 \geq 2$ or $x_1 - x_2 = 5$ must not be among the conditions to be met.

Condition 3 All the entries in the bound matrix must be greater than or equal to zero. A problem with a constraint such as $3x_1 + 2x_2 + x_3 \leq -2$ cannot be considered.

If one or more of these conditions is not satisfied, then the matrix form of the simplex method as given in this section does not apply. Care must be taken, since most problems can be set up and "solved" whether or not they meet the conditions. The trouble is that when the conditions are not met the techniques will generally yield wrong answers. If not all the conditions are met, some other technique, such as the geometric method, must be employed. In Section 6-4 we will discuss modifications of the simplex method for handling problems in which Conditions 2 or

3 are not met. In Problem 1 above, however, all three conditions are met.

The matrix for the first step of the simplex method solution of this problem is set up as follows.

(1) Write the constraint matrix:

$$\begin{bmatrix} 1 & 2 & 4 \\ 1 & 1 & 2 \\ 2 & 3 & 8 \end{bmatrix}$$

(2) To its right, write the identity matrix that has the same number of rows as the constraint matrix (in this case, three rows):

$$\begin{bmatrix} 1 & 2 & 4 & 1 & 0 & 0 \\ 1 & 1 & 2 & 0 & 1 & 0 \\ 2 & 3 & 8 & 0 & 0 & 1 \end{bmatrix}$$

(3) To the right of the identity matrix, write the bound matrix:

$$\begin{bmatrix} 1 & 2 & 4 & 1 & 0 & 0 & 3 \\ 1 & 1 & 2 & 0 & 1 & 0 & 2 \\ 2 & 3 & 8 & 0 & 0 & 1 & 7 \end{bmatrix}$$

(4) Below the constraint matrix, write the objective matrix; fill out this row with zeros:

$$\begin{bmatrix} 1 & 2 & 4 & 1 & 0 & 0 & 3 \\ 1 & 1 & 2 & 0 & 1 & 0 & 2 \\ 2 & 3 & 8 & 0 & 0 & 1 & 7 \\ 3 & 4 & 7 & 0 & 0 & 0 & 0 \end{bmatrix} \qquad \textbf{12}$$

Matrix **12** is the matrix on which the simplex method will operate.

Except for the identity matrix and the bottom row, this matrix is much like the matrix of the system

$$\begin{bmatrix} x_1 + 2x_2 + 4x_3 = 3 \\ x_1 + x_2 + 2x_3 = 2 \\ 2x_1 + 3x_2 + 8x_3 = 7 \end{bmatrix}$$

The identity matrix is needed to account for the fact that we are dealing with inequalitities. Its presence is easily explained through the introduction of slack variables (see Section 5-4). Matrix **12** is, of course, the matrix for a system of equations that is closely related to our system of inequalities. The simplex method is essentially Gauss-Jordan reduction modified to account for the fact that our object is not just to solve a system of equations, but rather to solve a system of equations *and* maximize the objective variable.

The flow chart on p. 244 summarizes the above discussion and contains the necessary modifications of Gauss-Jordan reduction. We suggest that after glancing over this chart the reader work through the solution of Problem 1 presented below, referring to the chart as required.

As our first illustration of the flow chart, we complete the solution of Problem 1. We have already completed the setup and obtained the first matrix:

$$R \begin{bmatrix} 1 & 2 & 4 & 1 & 0 & 0 & 3 \\ 1 & 1 & 2 & 0 & 1 & 0 & 2 \\ 2 & 3 & 8 & 0 & 0 & 1 & 7 \\ 3 & 4 & 7 & 0 & 0 & 0 & 0 \end{bmatrix} \qquad \textbf{13}$$
$$ \quad C$$

The bottom row contains three positive entries, each of whose columns contain other positive entries; we arbitrarily choose the first such column (marked C). We divide all but the last positive entry in C into the last entry in its row: $3 \div 1 = 3$, $2 \div 1 = 2$, $7 \div 2 = 3\frac{1}{2}$. The smallest of these quotients is the second one; hence, we must choose R to be the second row. We use R to clear C and obtain:

$$\begin{array}{l} R_1 - R_2 \\[2pt] \\ R_3 - 2R_2 \\ R_4 - 3R_2 \end{array} \qquad R \begin{bmatrix} 0 & 1 & 2 & 1 & -1 & 0 & 1 \\ 1 & 1 & 2 & 0 & 1 & 0 & 2 \\ 0 & 1 & 4 & 0 & -2 & 1 & 3 \\ 0 & 1 & 1 & 0 & -3 & 0 & -6 \end{bmatrix} \qquad \textbf{14}$$
$$ \qquad\qquad C$$

The bottom row of this matrix contains two positive entries. We choose the first one (marked C) because it looks as if it will result in easier arithmetic when we clear. The quotients are $1 \div 1 = 1$, $2 \div 1 = 2$, and $3 \div 1 = 3$, so we must choose R to be R_1. Clearing, we obtain

$$\begin{bmatrix} 0 & 1 & 2 & 1 & -1 & 0 & 1 \\ 1 & 0 & 0 & -1 & 2 & 0 & 1 \\ 0 & 0 & 2 & -1 & -1 & 1 & 2 \\ 0 & 0 & -1 & -1 & -2 & 0 & -7 \end{bmatrix} \qquad \textbf{15}$$

Since the bottom row of this matrix contains no positive entries, we are ready to read the answer (p. 245).

The Simplex Method for an LP
Maximization Problem

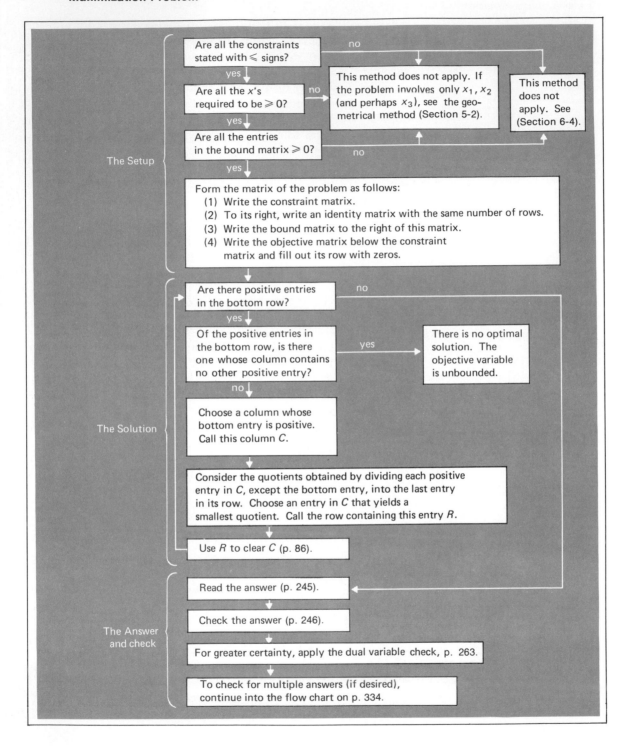

The Setup

Are all the constraints stated with ≤ signs? — no →

yes ↓

Are all the x's required to be ≥ 0? — no →

yes ↓

Are all the entries in the bound matrix ≥ 0? — no →

yes ↓

This method does not apply. If the problem involves only x_1, x_2 (and perhaps x_3), see the geometrical method (Section 5-2).

This method does not apply. See (Section 6-4).

Form the matrix of the problem as follows:
(1) Write the constraint matrix.
(2) To its right, write an identity matrix with the same number of rows.
(3) Write the bound matrix to the right of this matrix.
(4) Write the objective matrix below the constraint matrix and fill out its row with zeros.

The Solution

Are there positive entries in the bottom row? — no →

yes ↓

Of the positive entries in the bottom row, is there one whose column contains no other positive entry? — yes → There is no optimal solution. The objective variable is unbounded.

no ↓

Choose a column whose bottom entry is positive. Call this column C.

↓

Consider the quotients obtained by dividing each positive entry in C, except the bottom entry, into the last entry in its row. Choose an entry in C that yields a smallest quotient. Call the row containing this entry R.

↓

Use R to clear C (p. 86).

The Answer and check

Read the answer (p. 245).

↓

Check the answer (p. 246).

↓

For greater certainty, apply the dual variable check, p. 263.

↓

To check for multiple answers (if desired), continue into the flow chart on p. 334.

Reading the Answer   When we began, matrix **12** contained an identity matrix above
a row of zeros:

$$
\begin{bmatrix}
1 & 2 & 4 & 1 & 0 & 0 & 3 \\
1 & 1 & 2 & 0 & 1 & 0 & 2 \\
2 & 3 & 8 & 0 & 0 & 1 & 7 \\
3 & 4 & 7 & 0 & 0 & 0 & 0
\end{bmatrix}
$$

Each time a column was cleared, a column of this matrix was
relocated. In the final matrix, matrix **15**, we have

$$
\begin{bmatrix}
0 & 1 & 2 & 1 & -1 & 0 & 1 \\
1 & 0 & 0 & -1 & 2 & 0 & 1 \\
0 & 0 & 2 & -1 & -1 & 1 & 2 \\
0 & 0 & -1 & -1 & -2 & 0 & -7
\end{bmatrix}
$$

The first two columns of the identity matrix have been moved,
in opposite order, to the first two columns of the final matrix.
We now proceed as follows:

(1) Write x_1 above C_1, x_2 above C_2, and so on. In our case we
write

$$
\begin{array}{ccc}
x_1 & x_2 & x_3 \\
\end{array}
$$
$$
\begin{bmatrix}
0 & 1 & 2 & 1 & -1 & 0 & 1 \\
1 & 0 & 0 & -1 & 2 & 0 & 1 \\
0 & 0 & 2 & -1 & -1 & 1 & 2 \\
0 & 0 & -1 & -1 & -2 & 0 & -7
\end{bmatrix}
\qquad \textbf{16}
$$

(2) Variables that do not occur above columns of the relocated
identity matrix are zero. That is, variables that do not occur
above columns that you actually cleared are zero. In our
case: $x_3 = 0$.

(3) The value of each remaining variable occurs in the last column opposite the 1 in the column of that variable. In our
case:

$$
\begin{array}{ccc}
x_1 & x_2 & x_3 \\
\end{array}
$$
$$
\begin{bmatrix}
0 & 1 & 2 & 1 & -1 & 0 & 1 \\
1 & 0 & 0 & -1 & 2 & 0 & 1 \\
0 & 0 & 2 & -1 & -1 & 1 & 2 \\
0 & 0 & -1 & -1 & -2 & 0 & -7
\end{bmatrix}
\quad
\begin{array}{l}
x_2 = 1 \\
x_1 = 1
\end{array}
\qquad \textbf{17}
$$

(4) The maximum value of the objective variable y is the number
in the lower right corner (the southeast corner) of the final

matrix, *with the sign changed*. In our case, the maximum value is $y = 7$.

We have thus obtained an answer to our problem: $x_1 = 1$, $x_2 = 1$, $x_3 = 0$; maximum, $y = 7$. In this case, it happens that this is the only answer. But some problems have many answers and an answer found as above would be only one of them. We will consider this situation in Section 6-4.

The other columns of the final matrix contain some extra information about the solution, which may be useful. We will discuss this in Section 5-4.

One final remark: While this technique is correct and easily applied to all the problems and exercises in this chapter, if a problem has multiple solutions, there may be several choices of columns to be put into the identity matrix. In these cases the more general approach described in Section 6-4 (in the paragraph entitled "Locating the Cleared Columns," p. 333 is more appropriate.

Checking the Answer

Step 1   Substitute the solution into the equation for the objective variable. Does it yield the negative of the number in the southeast corner?

In our case, we have $y = 3 \cdot 1 + 4 \cdot 1 + 7 \cdot 0 = 7$, as it should.

Step 2   Substitute the solution into each constraint. Are they satisfied?

In our case we have

$$1 \cdot 1 + 2 \cdot 1 + 4 \cdot 0 = 3 \leq 3 \quad \checkmark$$
$$1 \cdot 1 + 1 \cdot 1 + 2 \cdot 0 = 2 \leq 2 \quad \checkmark$$
$$2 \cdot 1 + 3 \cdot 1 + 8 \cdot 0 = 5 \leq 7 \quad \checkmark$$

so our solution checks.

In Section 5-5, we present a more certain check: the dual variable check.

Warning: While each solution should certainly be subjected to the previous check, certain incorrect solutions will check. (See Problem 5.) This check is no guarantee that your answer is right. However, if it fails, your answer is certainly wrong.

Problem 2   Maximize $y = 3x_1 + 6x_2 + 4x_3 - 4x_4$ subject to

$$x_1 \geq 0, \ x_2 \geq 0, \ x_3 \geq 0, \ x_4 \geq 0$$
$$x_1 + 2x_2 + x_3 - 3x_4 \leq 2$$
$$x_1 - x_2 + 3x_3 + 5x_4 \leq 1$$
$$-x_1 + 4x_2 - 2x_3 + x_4 \leq 10$$

Solution   All constraints are with $\leq$ signs, all x's are ≥ 0, and all bounds are ≥ 0, so we set up the matrix:

$$\begin{bmatrix} 1 & 2^* & 1 & -3 & 1 & 0 & 0 & 2 \\ 1 & -1 & 3 & 5 & 0 & 1 & 0 & 1 \\ -1 & 4 & -2 & 1 & 0 & 0 & 1 & 10 \\ 3 & 6 & 4 & -4 & 0 & 0 & 0 & 0 \end{bmatrix}$$

Say we choose the second column to be C. Then the quotients $2 \div 2 = 1$, $10 \div 4 = \frac{5}{2}$ tell us to choose R to be R_1. This choice of C and R is indicated by the *. (Note that we did not consider the quotient $1 \div -1$, since -1 is not positive.) The result after clearing is:

$$\begin{bmatrix} \frac{1}{2} & 1 & \frac{1}{2} & -\frac{3}{2} & \frac{1}{2} & 0 & 0 & 1 \\ \frac{3}{2} & 0 & \frac{7}{2} & \frac{7}{2}^* & \frac{1}{2} & 1 & 0 & 2 \\ -3 & 0 & -4 & 7 & -2 & 0 & 1 & 6 \\ 0 & 0 & 1 & 5 & -3 & 0 & 0 & -6 \end{bmatrix}$$

Say we now take C to be C_4; then we must choose R to be R_2, as indicated by the *. Clearing yields:

$$\begin{array}{cccc} x_1 & x_2 & x_3 & x_4 \end{array}$$
$$\begin{bmatrix} \frac{8}{7} & 1 & 2 & 0 & \frac{5}{7} & \frac{3}{7} & 0 & \frac{13}{7} \\ \frac{3}{7} & 0 & 1 & 1 & \frac{1}{7} & \frac{2}{7} & 0 & \frac{4}{7} \\ -6 & 0 & -11 & 0 & -3 & -2 & 1 & 2 \\ -\frac{15}{7} & 0 & -4 & 0 & -\frac{26}{7} & -\frac{10}{7} & 0 & \frac{62}{7} \end{bmatrix}$$

Had we let C be C_3 it would have taken one more step to reach this answer.

The bottom row of this matrix contains no positive entries, so it is the final matrix. We label the first four columns as shown and read the answer: $x_1 = 0$, $x_2 = \frac{13}{7}$, $x_3 = 0$, $x_4 = \frac{4}{7}$, $y = \frac{62}{7}$.

Problem 3   Maximize $y = x_1 - x_2$ subject to

$$x_1 \geq 0, \; x_2 \geq 0, \; x_3 \geq 0$$
$$x_1 - 2x_2 \qquad \leq 2$$
$$2x_1 - 3x_2 - x_3 \leq 6$$
$$x_1 - 3x_2 + x_3 \leq 1$$

Solution   The first matrix is

$$\begin{bmatrix} 1 & -2 & 0 & 1 & 0 & 0 & 2 \\ 2 & -3 & -1 & 0 & 1 & 0 & 6 \\ 1^* & -3 & 1 & 0 & 0 & 1 & 1 \\ 1 & -1 & 0 & 0 & 0 & 0 & 0 \end{bmatrix}$$

Since the last row has only one positive entry there is only one choice for C: column 1. Checking the ratios, we let R be the third row. Clearing yields

$$\begin{bmatrix} 0 & 1^* & -1 & 1 & 0 & -1 & 1 \\ 0 & 3 & -3 & 0 & 1 & -2 & 4 \\ 1 & -3 & 1 & 0 & 0 & 1 & 1 \\ 0 & 2 & -1 & 0 & 0 & -1 & -1 \end{bmatrix}$$

Again there is only one choice for C. Checking quotients, we let R be the first row. Clearing yields

$$\begin{bmatrix} 0 & 1 & -1 & 1 & 0 & -1 & 1 \\ 0 & 0 & 0 & -3 & 1 & 1^* & 1 \\ 1 & 0 & -2 & 3 & 0 & -2 & 4 \\ 0 & 0 & 1 & -2 & 0 & 1 & -3 \end{bmatrix}$$

Column three of this matrix has a positive bottom entry but no other positive entries. Thus, according to the flow chart, there is no maximum.

The reason for the difficulty in Problem 3 is, of course, that the region of feasibility is unbounded. However, to further illustrate the relation between the geometric and algebraic approaches we will, in this one case, graph the region and indicate the path that the simplex solution followed (Figure 23). One can actually prove that the region is unbounded by noticing that any matrix of the form $\begin{bmatrix} x_1 \\ {}^2\!/_3 x_1 \\ 0 \end{bmatrix}$ is a solution if $x_1 \geq 0$. We show this by a direct check:

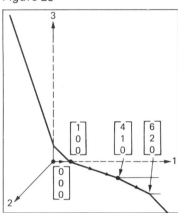

Figure 23

First constraint:

$$\begin{aligned} x_1 - 2x_2 &= x_1 - 2({}^2\!/_3 x_1) \\ &= x_1 - {}^4\!/_3 x_1 \\ &= -{}^1\!/_3 x_1 \leq 0 < 2 \end{aligned}$$

Second constraint:

$$\begin{aligned} 2x_1 - 3x_2 - x_3 &= 2x_1 - 3({}^2\!/_3 x_1) - 0 \\ &= 2x_1 - 2x_1 \\ &= 0 < 6 \end{aligned}$$

Third constraint:

$$\begin{aligned} x_1 - 3x_2 + x_3 &= x_1 - 3({}^2\!/_3 x_1) + 0 \\ &= x_1 - 2x_1 \\ &= -x_1 \leq 0 < 1 \end{aligned}$$

Finally, for this solution, $y = x_1 - x_2 = x_1 - {}^2\!/_3 x_1 = {}^1\!/_3 x_1$. Thus, we can make y as large as we want by taking x_1 to be a sufficiently large number. For instance, if we want y to be 10,000,000,000, we simply let $x_1 = 30,000,000,000$. The reader may check that

$$\begin{bmatrix} 30,000,000,000 \\ 20,000,000,000 \\ 0 \end{bmatrix} \text{ is a solution for which } y = 10,000,000,000.$$

Problem 4   Maximize $y = 3x_1 + 4x_2$ subject to

$$\begin{aligned} x_1 &\geq 0, \ x_2 \geq 0 \\ 2x_1 + 3x_2 &\leq 6 \\ 6x_1 + 4x_2 &\leq 12 \end{aligned}$$

Solution   This is the problem introduced in Section 5-1. Its matrix is

$$\begin{bmatrix} 2 & 3^* & 1 & 0 & 6 \\ 6 & 4 & 0 & 1 & 12 \\ 3 & 4 & 0 & 0 & 0 \end{bmatrix}$$

where we have indicated the entry that we will use to clear with a
*. The first application yields:

$$\begin{bmatrix} {}^2/_3 & 1 & {}^1/_3 & 0 & 2 \\ {}^{10}/_3{}^* & 0 & -{}^4/_3 & 1 & 4 \\ {}^1/_3 & 0 & -{}^4/_3 & 0 & -8 \end{bmatrix}$$

The second application yields:

$$\begin{bmatrix} 0 & 1 & {}^3/_5 & -{}^1/_5 & {}^6/_5 \\ 1 & 0 & -{}^2/_5 & {}^3/_{10} & {}^6/_5 \\ 0 & 0 & -{}^6/_5 & -{}^1/_{10} & -{}^{42}/_5 \end{bmatrix}$$

Again the answer is $x_1 = {}^6/_5$ and $x_2 = {}^6/_5$; the maximum is $y = {}^{42}/_5$.

Problem 5   Maximize $y = 2x_1 + x_2$ subject to

$$\begin{aligned} x_1 &\geq 0, \ x_2 \geq 0 \\ -x_1 + \ x_2 &\leq 2 \\ 3x_1 + 2x_2 &\geq 14 \\ x_1 - \ x_2 &\leq 3 \end{aligned}$$

Figure 24

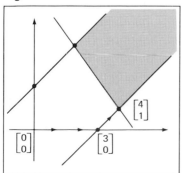

Solution   Since the second constraint violates Condition 2, the matrix
method does not apply and we must find some other way to
handle this problem. The fact that there are two variables
suggests a geometric approach. The graph in Figure 24 shows
that the region of feasibility is unbounded and that there is no
maximum. If we ignore the fact that Condition 2 is violated and
try to use the matrix method anyway by setting up the matrix

$$\begin{bmatrix} -1 & 1 & 1 & 0 & 0 & 2 \\ 3 & 2 & 0 & 1 & 0 & 14 \\ 1 & -1 & 0 & 0 & 1 & 3 \\ 2 & 1 & 0 & 0 & 0 & 0 \end{bmatrix}$$

then, in two steps, we obtain the final matrix

$$\begin{bmatrix} 0 & 0 & 1 & 0 & 1 & 5 \\ 0 & 1 & 0 & 1/5 & -3/5 & 1 \\ 1 & 0 & 0 & 1/5 & 2/5 & 4 \\ 0 & 0 & 0 & -3/5 & -1/5 & -9 \end{bmatrix}$$

and the apparent solution $x_1 = 4$, $x_2 = 1$, and $y = 9$. It is easy to verify that this answer "checks." However, as we have seen, it is incorrect.

Additional Checks and Shortcuts
The most common errors in the use of the simplex method will be avoided if the user will keep in mind the following simple facts:

(1) A column that is to be cleared must have a positive last entry.

(2) The entry that is used to clear a column must also be positive.

(3) Throughout the entire process, all the entries in the last column, except the bottom entry, are always ≥ 0.

(4) The number in the southeast corner is zero at the first step and becomes more negative (or stays the same) after each subsequent step.

The shortcuts of Section 2-5 do *not* apply here. However, you may if you wish multiply any row except the last by a positive number and not change the answer. The most effective shortcut is to decide which column to clear on the basis of a quick look ahead to the effect of various choices on the last row.

Exercises 5.3   C 1. Maximize $y = 2x_1 + 3x_2$ subject to $x_1 \geq 0$, $x_2 \geq 0$, $x_1 + x_2 \leq 2$, $x_1 + 4x_2 \leq 4$.

2. Maximize $y = x_1 + x_2$ subject to $x_1 \geq 0$, $x_2 \geq 0$, $-x_1 + x_2 \leq 1$, $3x_1 + x_2 \leq 9$.

A 3. Maximize $y = -2x_1 - 3x_2$ subject to $x_1 \geq 0$, $x_2 \geq 0$, $-x_1 + x_2 \leq 2$, $x_1 + x_2 \leq 2$.

4. Maximize $y = 3x_1 + 2x_2$ subject to $x_1 \geq 0$, $x_2 \geq 0$, $-x_1 + x_2 \leq 2$, $x_1 - 2x_2 \leq 4$.

C 5. Maximize $y = 2x_1 + x_2$ subject to $x_1 \geq 0$, $x_2 \geq 0$, $-x_1 + x_2 \leq 1$,
$x_1 + x_2 \geq 2$, $4x_1 - x_2 \leq 4$.

6. Maximize $y = x_1 + 2x_2$ subject to $x_1 \geq 0$, $x_2 \geq 0$, $-2x_1 + x_2 \leq 2$,
$x_1 - x_2 \leq 1$, $-x_1 + 2x_2 \leq 8$, $x_2 \leq 3$.

C 7. Maximize $y = 12x_1 + 16x_2 + 6x_3$ subject to $x_1 \geq 0$, $x_2 \geq 0$, $x_3 \geq 0$,
$2x_1 + x_2 + 2x_3 \leq 6$, $3x_1 + 8x_2 + 5x_3 \leq 22$, $2x_1 + 2x_2 + 3x_3 \leq 12$.

8. Maximize $y = 6x_1 + 7x_2 + 8x_3$ subject to $x_1 \geq 0$, $x_2 \geq 0$, $x_3 \geq 0$,
$3x_1 + 7x_2 + 5x_3 \leq 15$, $x_1 + x_2 + x_3 \leq 4$, $2x_1 + 4x_2 + 5x_3 \leq 11$.

9. Maximize $y = 7x_1 + 5x_2 + 6x_3$ subject to $x_1 \geq 0$, $x_2 \geq 0$, $x_3 \geq 0$,
$x_1 + x_2 - x_3 \leq 3$, $x_1 + 2x_2 + x_3 \leq 8$, $x_1 + x_2 \leq 5$.

A 10. Maximize $y = 5x_1 + 4x_2 + 2x_3 + 3x_4$ subject to $x_1 \geq 0$, $x_2 \geq 0$,
$x_3 \geq 0$, $x_4 \geq 0$, $2x_1 + x_2 + 3x_3 + 6x_4 \leq 5$, $3x_1 + 3x_2 + 2x_3 \leq 2$,
$4x_1 + 2x_2 + x_3 + 3x_4 \leq 9$.

C 11. In some of their early attempts to use the simplex method, the PHE
directors were rather unsophisticated. By failing to specify carefully
all constraints, they sometimes got correct but useless answers. In
planning the fishfeast dinner, for example, the first attempt to use LP
techniques was as follows: A fishfeast consists of fish, fries, and
slaw. Fish costs 21¢ per 3-oz. serving, fries cost 18¢ per 3-oz. serving, and slaw costs 14¢ per 3-oz. serving. If a fishfeast must cost no
more than $1.26, how much of each food should be used to produce
a feast of the *greatest weight*? (Use the geometric method.)

12. (a) Given an LP maximization problem with 5 variables (not counting the objective variable) and 6 constraints, what is the size of
the matrix set up in the simplex method?

(b) In general, if you know how many constraints and how many
variables (not including the objective variable) an LP maximization problem has, how do you determine the size of the matrix set up
in the simplex method?

C 13. Use the simplex method to solve the PHE problems of Exercise 1
in Section 5-1 and Exercise 11 above.

Section 5-4
The Algebra of the Simplex
Method

In this section we will attempt to remove some of the mystery
from the simplex method as it was presented in the previous section. First we return to Problem 1 of the previous section:

MAXIMIZE

$$y = 3x_1 + 4x_2 + 7x_3 \qquad\qquad 1$$

SUBJECT TO

$$x_1 \geq 0, \; x_2 \geq 0, \; x_3 \geq 0 \qquad\qquad 2$$
$$x_1 + 2x_2 + 4x_3 \leq 3 \qquad\qquad 3$$
$$x_1 + x_2 + 2x_3 \leq 2 \qquad\qquad 4$$
$$2x_1 + 3x_2 + 8x_3 \leq 7 \qquad\qquad 5$$

We will explain how we arrived at the initial matrix, (matrix **12** in Section 5-3) and we will show how to get more information from the final matrix (matrix **15** in Section 5-3) than we have obtained so far.

Since the algebraic techniques of Chapter 2 apply only to equalities and not to inequalities, we will need to reformulate this problem in terms of equalities in order to be able to use our previous work. We do this by introducing one new variable in each inequality. If we find a solution $\begin{bmatrix} x_1 \\ x_2 \\ x_3 \end{bmatrix}$ to the problem, it is likely that in constraint **3**, $x_1 + 2x_2 + 4x_3$ will not be equal to 3. For example, the solution $\begin{bmatrix} 1 \\ 1/2 \\ 0 \end{bmatrix}$ yields $x_1 + 2x_2 + 4x_3 = 2 < 3$. A new variable, x_4, is assigned the task of making up the difference between $x_1 + 2x_2 + 4x_3$ and 3. Thus, for the solution $\begin{bmatrix} 1 \\ 1/2 \\ 0 \end{bmatrix}$, x_4 would be $1 = 3 - 2$. In general, if we know the numerical values of x_1, x_2, and x_3, we will be able to determine x_4 as the quantity left over. That is,

$$x_4 = 3 - (x_1 + 2x_2 + 4x_3)$$

While the value of x_4 changes as we change the values of x_1, x_2, and x_3, we have arranged matters so that it is always true that

$$x_1 + 2x_2 + 4x_3 + x_4 = 3$$

A variable such as x_4 is often called a **slack variable**, since it takes up all of the slack in an inequality. While x_4 may be zero, it will never be negative, since the constraint $x_1 + 2x_2 + 4x_3 \leq 3$ guarantees that there will be some (or exactly no) slack.

We do the same thing for the other two inequalities, introducing new variables x_5 and x_6 to take up the slack in them. We now have the following system of linear equations:

$$\begin{bmatrix} x_1 + 2x_2 + 4x_3 + x_4 && = 3 \\ x_1 + x_2 + 2x_3 & + x_5 & = 2 \\ 2x_1 + 3x_2 + 8x_3 & + x_6 & = 7 \end{bmatrix} \qquad \textbf{6}$$

The new variables x_4, x_5, and x_6 are completely independent of one another. In practical applications they can often be given physical interpretations as representing leftover resources, unused machine time, etc. It will be convenient to have these new variables occur in the expression for the objective variable. However, since leftover resources contribute nothing to profit, the value of these new variables should not affect the value of the objective variable. We resolve this situation by rewriting the objective variable as

$$y = 3x_1 + 4x_2 + 7x_3 + 0x_4 + 0x_5 + 0x_6 \qquad \textbf{7}$$

While the new variables appear in the formula for y, the zeros multiplying them guarantee that they will not affect its value.

Matrix **12** in Section 5-3, which we used in solving the problem, is just the matrix of system **6** followed by the coefficients of the objective variable (equation **7**):

$$\begin{bmatrix} 1 & 2 & 4 & 1 & 0 & 0 & 3 \\ 1 & 1 & 2 & 0 & 1 & 0 & 2 \\ 2 & 3 & 8 & 0 & 0 & 1 & 7 \\ 3 & 4 & 7 & 0 & 0 & 0 & 0 \end{bmatrix} \qquad \textbf{8}$$

For reasons that will become clear in the second part of this section, we insert a 0 in the southeast corner, address (4,7).

Once the manipulations of the simplex method were completed and we had the final matrix (matrix **15** in Section 5-3), we began by writing x_1 above C_1, x_2 above C_2, and so on, to obtain

$$\begin{array}{cccccc} x_1 & x_2 & x_3 & x_4 & x_5 & x_6 \end{array}$$
$$\begin{bmatrix} 0 & 1 & 2 & 1 & -1 & 0 & 1 \\ 1 & 0 & 0 & -1 & 2 & 0 & 1 \\ 0 & 0 & 2 & -1 & -1 & ① \rightarrow 2 \\ 0 & 0 & -1 & -1 & -2 & 0 & -7 \end{bmatrix} \quad x_6 = 2 \qquad \textbf{9}$$

Now here, in contrast to system **17** in Section 5-3, we have also written the slack variables x_4, x_5, and x_6 above the appropriate columns. We can read not only the values of the original variables $x_1 = 1$, $x_2 = 1$, $x_3 = 0$, $y = 7$, as explained on p. 245, but also the values of the slack variables $x_4 = 0$, $x_5 = 0$, $x_6 = 2$. These values are borne out in the check of our solution made on p. 246; inequality **6** of Section 5-3 has slack of $x_4 = 0$; inequality **7** of Section 5-3 has slack of $x_5 = 0$; inequality **8** of that section has slack of $x_6 = 2$.

In order to illustrate more precisely the details of the simplex method, we will now solve one rather simple problem in accordance with the geometrical instructions of Section 5-3, and then translate each step of the solution into matrix methods. This illustration is provided for those readers who may be interested in the algebraic details behind the simplex method. Those who are interested only in practical methods may omit the rest of this section and go on to the exercises, skipping Exercise 1.

We will solve the following problem: Maximize $y = 2x_1 + x_2$ subject to

$$x_1 \geq 0, \, x_2 \geq 0$$
$$-x_1 + x_2 \leq 2$$
$$3x_1 + 2x_2 \leq 14$$
$$x_1 - x_2 \leq 3$$

We need three slack variables: x_3, x_4, x_5. These transform our constraint inequalities into the system of equalities

$$
\begin{bmatrix}
-x_1 + x_2 + x_3 & & = 2 \\
3x_1 + 2x_2 & + x_4 & = 14 \\
x_1 - x_2 & + x_5 = 3 \\
2x_1 + x_2 & = y
\end{bmatrix}
\quad
\begin{aligned}
&(1) \\
&(2) \\
&(3) \\
&(4)
\end{aligned}
$$

Here we have included the objective variable as well.

Instead of working with 2×1 matrices, $\begin{bmatrix} x_1 \\ x_2 \end{bmatrix}$, we must now

work with 5×1 matrices: $\begin{bmatrix} x_1 \\ x_2 \\ x_3 \\ x_4 \\ x_5 \end{bmatrix}$. The simplex method starts at the

origin $\begin{bmatrix} x_1 \\ x_2 \end{bmatrix} = \begin{bmatrix} 0 \\ 0 \end{bmatrix}$. When $x_1 = 0$ and $x_2 = 0$, equations 1, 2, and 3

tell us immediately that $x_3 = 2$, $x_4 = 14$, and $x_5 = 3$. Thus we begin with the solution

$$
\begin{bmatrix} x_1 \\ x_2 \\ x_3 \\ x_4 \\ x_5 \end{bmatrix} = \begin{bmatrix} 0 \\ 0 \\ 2 \\ 14 \\ 3 \end{bmatrix} \quad \text{with } y = 0
$$

In the formula for the objective variable, $y = 2x_1 + x_2$, we see that an increase in either x_1 or x_2 will cause y to increase. We arbitrarily decide to move along the 1-axis by increasing x_1 and holding $x_2 = 0$. As we do this, the variables x_3, x_4, and x_5 will also change. We can see exactly what happens by setting $x_2 = 0$ in equations 1, 2, and 3:

$$-x_1 + x_3 = 2$$
$$3x_1 + x_4 = 14$$
$$x_1 + x_5 = 3$$

and solving for x_3, x_4, and x_5:

$$x_3 = \ 2 + \ x_1 \qquad (1')$$
$$x_4 = 14 - 3x_1 \qquad (2')$$
$$x_5 = \ 3 - \ x_1 \qquad (3')$$

From these equations we see that as x_1 gets larger so will x_3. However, x_4 and x_5 will decrease as we increase x_1. Remember that to stay within the region of feasibility we must keep $x_3 \geq 0$, $x_4 \geq 0$, and $x_5 \geq 0$. There is no problem about x_3 since it is also increasing. But x_4 decreases to zero when $14 - 3x_1 = 0$; that is, when $x_1 = {}^{14}/_3 = 4{}^2/_3$. And x_5 decreases to zero when $3 - x_1 = 0$, that is, when $x_1 = 3$. Since $3 < 4{}^2/_3$, we must stop increasing x_1 when it hits 3. If we went beyond 3, x_5 would become negative and this would violate the constraint $x_5 \geq 0$. To calculate the effect of this increase, we simply set $x_1 = 3$, keep $x_2 = 0$, and obtain from equations $1'$, $2'$, and $3'$ the values $x_3 = 5$, $x_4 = 5$, and $x_5 = 0$. We now have the solution

$$\begin{bmatrix} x_1 \\ x_2 \\ x_3 \\ x_4 \\ x_5 \end{bmatrix} = \begin{bmatrix} 3 \\ 0 \\ 5 \\ 5 \\ 0 \end{bmatrix} \text{ with } y = 6$$

Figure 25

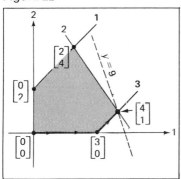

Geometrically, we have now moved along the edge of the region of feasibility (Figure 25) from the corner $\begin{bmatrix} 0 \\ 0 \end{bmatrix}$ to the corner $\begin{bmatrix} 3 \\ 0 \end{bmatrix}$.

At this corner we have $x_2 = 0$ and $x_5 = 0$. Moving away from this corner is more complicated, since the appropriate edge does not lie along the 1-axis or the 2-axis. We first rewrite the objective variable (equation 4) in terms of x_2 and x_5. To do this we subtract 2 times equation 3 from equation 4:

$$(2x_1 + x_2) - 2(x_1 - x_2 + x_5) = y - 2 \cdot 3$$
$$3x_2 - 2x_5 = y - 6 \qquad (4')$$

We see from this equation that y increases when x_2 increases, but that y decreases when x_5 increases. If we are to continue to increase y, we should therefore hold $x_5 = 0$ and increase x_2. How will x_1, x_3, and x_4 change as $x_5 = 0$ and x_2 increases? To answer this question, we must find expressions for x_1, x_3, and x_4 in terms of x_2. From equation 3 with $x_5 = 0$, we find $x_1 = 3 + x_2$. Substituting this expression for x_1 into equations 1 and 2 yields $-(3 + x_2) + x_2 + x_3 = 2$, or $x_3 = 5$, and $3(3 + x_2) + 2x_2 + x_4 = 14$, or $x_4 = 5 - 5x_2$. In summary:

$$x_3 = 5 \tag{1''}$$
$$x_4 = 5 - 5x_2 \tag{2''}$$
$$x_1 = 3 + x_2 \tag{3''}$$

From these equations we see that as x_2 increases only x_4 decreases and we must stop when $x_4 = 0$, that is, when $0 = 5 - 5x_2$, or $x_2 = 1$. We set $x_2 = 1$ in these equations, and recalling that $x_5 = 0$, we obtain the solution

$$\begin{bmatrix} 4 \\ 1 \\ 5 \\ 0 \\ 0 \end{bmatrix} \text{ with } y = 9 \tag{5}$$

In Figure 25, we have now moved along the edge from corner $\begin{bmatrix} 3 \\ 0 \end{bmatrix}$ to corner $\begin{bmatrix} 4 \\ 1 \end{bmatrix}$.

At this corner $x_4 = 0$ and $x_5 = 0$, so that to consider what to do next we must express y in terms of x_4 and x_5. Since we obtained the previous set of equations by assuming $x_5 = 0$ and since that is still the case, we can use equation $2''$ here as well. We first rewrite it as

$$5x_2 + x_4 = 5 \tag{2*}$$

and then, taking equation $4'$ minus $\frac{3}{5}$ times equation $2*$, we obtain $(3x_2 - 2x_5) - \frac{3}{5}(5x_2 + x_4) = y - 6 - \frac{3}{5} \cdot 5$ that is,

$$-\tfrac{3}{5}x_4 - 2x_5 = y - 9 \tag{4''}$$

From this equation we see that increasing either x_4 or x_5 would decrease y. We have arrived at the maximal corner! Since at this corner we have $x_4 = 0$ and $x_5 = 0$, we see from equation $4''$ that $0 = y - 9$ or $y = 9$. The final answer is matrix 5 above. The problem as originally stated involved only x_1 and x_2, and we are not really interested in the values of the slack variables. So our final answer is simply $x_1 = 4$, $x_2 = 1$, and $y = 9$.

The matrix form of the simplex method was invented to provide a convenient way to keep track of all the algebraic manipulations carried out above. The matrix for this problem is

$$\begin{bmatrix} -1 & 1 & 1 & 0 & 0 & 2 \\ 3 & 2 & 0 & 1 & 0 & 14 \\ 1 & -1 & 0 & 0 & 1 & 3 \\ 2 & 1 & 0 & 0 & 0 & 0 \end{bmatrix}$$

We choose C as the first column. The calculations involving equations 1, 2, and 3 determined R. Of course, an equivalent simpler technique is to divide the entries of C into the corresponding entry in the last column. The quotients are $^2/_{-1} = -2$, $^{14}/_3 = 4^2/_3$, $^3/_1 = 3$. Notice that these numbers are the same as those that occur in connection with equations 1′, 2′, 3′. As before, we choose the smallest non-negative ratio, and so the third row becomes R.

The algebraic manipulations by which we rewrote the formula for y and expressed the other variables in terms of x_2 and x_5 amount to using R to clear C. Doing this yields

$$\begin{bmatrix} 0 & 0 & 1 & 0 & 1 & 5 \\ 0 & 5 & 0 & 1 & -3 & 5 \\ 1 & -1 & 0 & 0 & 1 & 3 \\ 0 & 3 & 0 & 0 & -2 & -6 \end{bmatrix}$$

Note that the equations that this matrix represents are

$$\begin{bmatrix} x_3 & + & x_5 = 5 \\ 5x_2 & + x_4 - 3x_5 = 5 \\ x_1 - x_2 & + x_5 = 3 \\ 3x_2 & - 2x_5 = y - 6 \end{bmatrix}$$

We can obtain equations 1″, 2″, and 3″ from the first three equations by setting $x_5 = 0$. The fourth equation is just equation 4′.

The final step is just a repeat of the previous process for the new matrix; the result is

$$\begin{bmatrix} 0 & 0 & 1 & 0 & 1 & 5 \\ 0 & 1 & 0 & ^1/_5 & -^3/_5 & 1 \\ 1 & 0 & 0 & ^1/_5 & ^2/_5 & 4 \\ 0 & 0 & 0 & -^3/_5 & -^1/_5 & -9 \end{bmatrix}$$

which corresponds to the system

$$\begin{bmatrix} x_3 & + & x_5 = 5 \\ x_2 & + ^1/_5x_4 - ^3/_5x_5 = 1 \\ x_1 & + ^1/_5x_4 + ^2/_5x_5 = 4 \\ & - ^3/_5x_4 - ^1/_5x_5 = y - 9 \end{bmatrix}$$

We obtain the complete solution (matrix 5) from this system by setting $x_4 = x_5 = 0$.

Exercises 5.4   1. Carry through a discussion similar to the one above for the PHE burger problem set up in the solution to Exercise 1 in Section 5-1.

^2. (a) What are the values of the slack variables in the solution to Problem 2, page 246?

(b) What are the values of the slack variables in the solution to Problem 4, page 249?

(c) What does the answer to (b) mean in relation to the problem stated in Section 5-1?

Section 5-5
The Simplex Method for
Minimization Problems

We will present two methods for solving minimization problems. The technique presented in this section is easy, in that it requires only two minor modifications of the technique used in Section 5-3. It is particularly useful because it will enable us to provide a reliable check for both minimization and maximization problems. It is also the easiest way to solve matrix games (see Section 6-1). The other technique is presented in Section 6-4 along with a method for solving problems that do not satisfy Conditions 2 or 3. The technique is not much different from that of Section 5-3, and is the method computers use to solve minimization problems. You may choose to read Section 6-4 in place of or in addition to this section. The minimization problems presented in the next section are solved by the technique of this section. You may wish to rework them later using the method of Section 6-4.

An economic interpretation of duality, that is, the relation between a given LP problem and its dual, is given in the last part of this section. It is presented for those readers who seek a more complete understanding of duality.

We give no detailed explanation of how the technique of this section works in solving minimization problems. It can be shown that, given any LP minimization problem, there is a related LP maximization problem, called the *dual* problem. Either both or neither of these problems have a solution and, furthermore, the objective variable for each problem will have the same value when optimized. It also happens that the answer to the minimization problem can be determined from the final matrix in the solution of the dual maximization problem.

The method involves the following three steps:

Step 1   Set up a certain matrix as described below. Once set up, this matrix will look exactly like the maximization problem matrices of Section 5-3.

Step 2   Forgetting for a moment that the problem is one of minimization, use the technique of Section 5-3 to solve the problem as if the matrix had arisen from a maximization problem.

Step 3　Read the answer in the manner described below. The reading procedure here is different from (and easier than) the maximization case.

　　　　Thus, the only difference between maximization and minimization with the simplex method occurs in setting up the problem and reading the answer. The actual computation uses a technique we already know.

　　　　In describing how to set up the minimization problem, it will be convenient to introduce a new matrix operation: Given a matrix, one finds its **transpose** by switching its rows and columns, as in the following example.

Problem 1　Find the transpose of the matrix $\begin{bmatrix} 2 & 1 \\ 3 & 4 \\ 5 & 7 \end{bmatrix}$.

Solution　The matrix has three rows: 2　1; 3　4; and 5　7. We form a new matrix which has these rows as columns:

$$\begin{bmatrix} 2 & 3 & 5 \\ 1 & 4 & 7 \end{bmatrix}$$

Thus, the first row becomes the first column, and so forth. This new matrix is the transpose of the original matrix. The original matrix has two columns: $\begin{matrix} 2 \\ 3 \\ 5 \end{matrix}$ and $\begin{matrix} 1 \\ 4 \\ 7 \end{matrix}$. We could also have formed the transposed matrix by writing these columns as rows:

$$\begin{bmatrix} 2 & 3 & 5 \\ 1 & 4 & 7 \end{bmatrix}$$

Problem 2　Find the transposes of the following matrices:

$$\begin{bmatrix} 1 & 1 & 2 \\ 2 & 1 & 3 \\ 4 & 2 & 8 \end{bmatrix}, \; [3 \quad 2 \quad 7], \; \text{and} \; \begin{bmatrix} 3 \\ 4 \\ 7 \end{bmatrix}$$

Answer　The respective transposes are $\begin{bmatrix} 1 & 2 & 4 \\ 1 & 1 & 2 \\ 2 & 3 & 8 \end{bmatrix}, \begin{bmatrix} 3 \\ 2 \\ 7 \end{bmatrix}$, and $[3 \quad 4 \quad 7]$

　　　　The need for constraint, bound, and objective matrices arises in minimization problems just as in maximization problems. These concepts are the same as before (see p. 241).

Problem 3   What are the constraint, bound, and objective matrices for the following problem?

Minimize $y = 2x_1 + x_2 + 10x_3$ subject to

$$x_1 \geq 0, \; x_2 \geq 0, \; x_3 \geq 0$$
$$x_1 + x_2 - x_3 \geq 3$$
$$2x_1 - x_2 + 4x_3 \geq 6$$
$$x_1 + 3x_2 - 2x_3 \geq 4$$
$$-3x_1 + 5x_2 + x_3 \geq -4$$

Solution   The constraint matrix is

$$\begin{bmatrix} 1 & 1 & -1 \\ 2 & -1 & 4 \\ 1 & 3 & -2 \\ -3 & 5 & 1 \end{bmatrix}$$

The bound matrix is $\begin{bmatrix} 3 \\ 6 \\ 4 \\ -4 \end{bmatrix}$. The objective matrix is $\begin{bmatrix} 2 & 1 & 10 \end{bmatrix}$.

Since minimization problems actually involve using the simplex method to solve a related maximization problem, certain conditions must be met, as follows:

Condition 1   The problem must stipulate that all the x's that occur in it are greater than or equal to zero. That is, it must be assumed that $x_1 \geq 0, \; x_2 \geq 0$, etc.

Condition 2   All the constraints must contain $\geq$ signs as opposed to $\leq$ signs or $=$ signs. Constraints such as $3x_1 + 2x_2 + x_3 \leq 2$ or $4x_1 - 2x_2 + 3x_3 = 5$ must not be among the conditions to be met.

Condition 3   All the entries in the objective matrix must be greater than or equal to zero. Thus, an objective variable such as $y = 3x_1 - 2x_2 + x_3$ is not allowed.

The same cautions that were made at this point in the discussion of maximization problems (p. 241) apply here. We now present the method of solution, when the conditions are met, in flow-chart form (p. 261). The instructions for reading the answer follow the presentation of the first example, Problem 4, below. Note that the bound and objective matrices in the minimization matrix are in opposite locations from where they were in the maximization matrix. The "solution" section of this flow chart is exactly the same as in the flow chart on p. 244.

The Simplex Method for a Minimization LP Problem

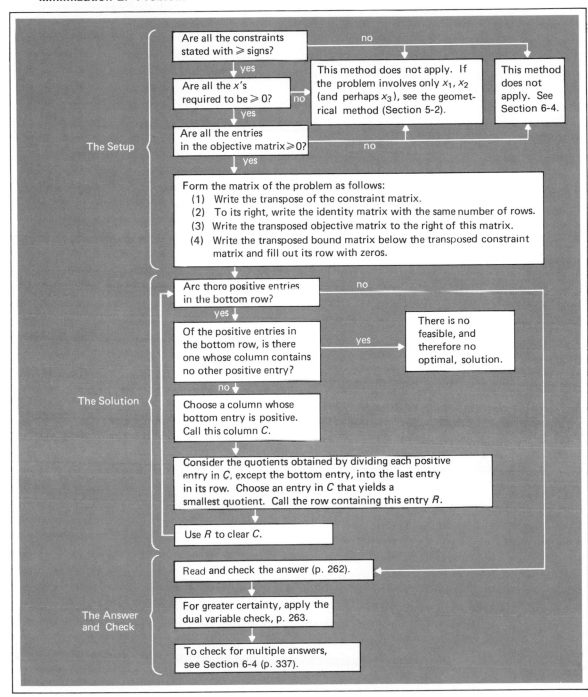

The Setup

Are all the constraints stated with ⩾ signs?

no

yes

Are all the x's required to be ⩾ 0?

no

yes

Are all the entries in the objective matrix ⩾ 0?

no

yes

This method does not apply. If the problem involves only x_1, x_2 (and perhaps x_3), see the geometrical method (Section 5-2).

This method does not apply. See Section 6-4.

Form the matrix of the problem as follows:
(1)   Write the transpose of the constraint matrix.
(2)   To its right, write the identity matrix with the same number of rows.
(3)   Write the transposed objective matrix to the right of this matrix.
(4)   Write the transposed bound matrix below the transposed constraint matrix and fill out its row with zeros.

The Solution

Are there positive entries in the bottom row?

no

yes

Of the positive entries in the bottom row, is there one whose column contains no other positive entry?

yes

There is no feasible, and therefore no optimal, solution.

no

Choose a column whose bottom entry is positive. Call this column C.

Consider the quotients obtained by dividing each positive entry in C, except the bottom entry, into the last entry in its row. Choose an entry in C that yields a smallest quotient. Call the row containing this entry R.

Use R to clear C.

The Answer and Check

Read and check the answer (p. 262).

For greater certainty, apply the dual variable check, p. 263.

To check for multiple answers, see Section 6-4 (p. 337).

We use the following example to illustrate how the answer is read once the matrix operations are completed.

Problem 4   Minimize $y = 3x_1 + 2x_2 + 7x_3$ subject to

$$x_1 \geq 0, \ x_2 \geq 0, \ x_3 \geq 0$$
$$x_1 + \ x_2 + 2x_3 \geq 3$$
$$2x_1 + 1x_2 + 3x_3 \geq 4$$
$$4x_1 + 2x_2 + 8x_3 \geq 7$$

Solution   The constraint matrix is $\begin{bmatrix} 1 & 1 & 2 \\ 2 & 1 & 3 \\ 4 & 2 & 8 \end{bmatrix}$, the bound matrix is $\begin{bmatrix} 3 \\ 4 \\ 7 \end{bmatrix}$, and the objective matrix is $\begin{bmatrix} 3 & 2 & 7 \end{bmatrix}$. We found the transposes of these matrices in Problem 2. The matrix for the minimization problem is

$$\begin{bmatrix} 1 & 2 & 4 & 1 & 0 & 0 & 3 \\ 1 & 1 & 2 & 0 & 1 & 0 & 2 \\ 2 & 3 & 8 & 0 & 0 & 1 & 7 \\ 3 & 4 & 7 & 0 & 0 & 0 & 0 \end{bmatrix}$$

We have already encountered this matrix in the first example in Section 5-3 (matrix **12** of that section). There, we applied the simplex method to this matrix, with the result

$$\begin{bmatrix} 0 & 1 & 2 & 1 & -1 & 0 & 1 \\ 1 & 0 & 0 & -1 & 2 & 0 & 1 \\ 0 & 0 & 2 & -1 & -1 & 1 & 2 \\ 0 & 0 & -1 & -1 & -2 & 0 & -7 \end{bmatrix}$$

Reading the Answer   One obtains the answer to the minimization problem as follows:

(1) Write the letters x_1, x_2, etc., below the columns where the identity matrix *used* to be. In our case we have

$$\begin{bmatrix} 0 & 1 & 2 & 1 & -1 & 0 & 1 \\ 1 & 0 & 0 & -1 & 2 & 0 & 1 \\ 0 & 0 & 2 & -1 & -1 & 1 & 2 \\ 0 & 0 & -1 & -1 & -2 & 0 & -7 \end{bmatrix}$$
$$\quad x_1 \qquad x_2 \ x_3$$

(2) The value of a particular x is the entry directly above it with the sign changed. Thus, in our case, we have $x_1 = 1$, $x_2 = 2$, and $x_3 = 0$.

(3) The value of the optimized objective variable is the entry in the southeast corner with the sign changed, just as before. In this case it is $y = 7$.

We check the answer in exactly the same way as before:

$$y = 3 \cdot 1 + 2 \cdot 2 + 7 \cdot 0 = 7$$
$$x_1 + x_2 + 2x_3 = 1 \cdot 1 + 1 \cdot 2 + 2 \cdot 0 = 3 \geq 3$$
$$2x_1 + 1x_2 + 3x_3 = 2 \cdot 1 + 1 \cdot 2 + 3 \cdot 0 = 4 \geq 4$$
$$4x_1 + 2x_2 + 8x_3 = 4 \cdot 1 + 2 \cdot 2 + 8 \cdot 0 = 8 \geq 7$$

For greater certainty, we may apply the following technique.

The Dual Variable Check The dual variable check is a means of checking an answer obtained through the simplex method. It is as certain a check as substitution is for Gauss-Jordan reduction. That is, if you do not make arithmetic errors you can be sure you have the right answer if it checks, provided the variables are nonnegative.

We illustrate the technique by using it to check a problem which we have already worked. Problem 2 on p. 246 asked that we maximize $y = 3x_1 + 6x_2 + 4x_3 - 4x_4$ subject to

$$x_1 \geq 0,\ x_2 \geq 0,\ x_3 \geq 0,\ x_4 \geq 0$$
$$x_1 + 2x_2 + x_3 - 3x_4 \leq 2$$
$$x_1 - x_2 + 3x_3 + 5x_4 \leq 1$$
$$-x_1 + 4x_2 - 2x_3 + x_4 \leq 10$$

The first matrix we set up was

$$\begin{bmatrix} 1 & 2^* & 1 & -3 & 1 & 0 & 0 & 2 \\ 1 & -1 & 3 & 5 & 0 & 1 & 0 & 1 \\ -1 & 4 & -2 & 1 & 0 & 0 & 1 & 10 \\ 3 & 6 & 4 & -4 & 0 & 0 & 0 & 0 \end{bmatrix}$$

From this we can read the dual minimization problem, namely: minimize $y = 2x_1 + x_2 + 10x_3$ subject to

$$x_1 \geq 0,\ x_2 \geq 0,\ x_3 \geq 0$$
$$x_1 + x_2 - x_3 \geq 3$$
$$2x_1 - x_2 + 4x_3 \geq 6$$
$$x_1 + 3x_2 - 2x_3 \geq 4$$
$$-3x_1 + 5x_2 + x_3 \geq -4$$

This is just Problem 3. The final matrix was found on p. 247 to be

$$\begin{bmatrix} 8/7 & 1 & 2 & 0 & 5/7 & 3/7 & 0 & 13/7 \\ 3/7 & 0 & 1 & 1 & 1/7 & 2/7 & 0 & 4/7 \\ -6 & 0 & -11 & 0 & -3 & -2 & 1 & 2 \\ -15/7 & 0 & -4 & 0 & -26/7 & -10/7 & 0 & -62/7 \end{bmatrix}$$

From this matrix we read two answers:

MAXIMIZATION ANSWER:

$$x_1 = 0, \; x_2 = {}^{13}/_7, \; x_3 = 0, \; x_4 = {}^4/_7, \; y = {}^{62}/_7$$

MINIMIZATION ANSWER:

$$x_1 = {}^{26}/_7, \; x_2 = {}^{10}/_7, \; x_3 = 0, \; y = {}^{62}/_7$$

We check that each of these solutions is feasible by substitution into the appropriate inequalities.

MAXIMIZATION PROBLEM:

$$0 + {}^{26}/_7 + 0 - {}^{12}/_7 = {}^{14}/_7 = 2 \leq 2$$
$$0 - {}^{13}/_7 + 0 + {}^{20}/_7 = {}^7/_7 = 1 \leq 1$$
$$0 + {}^{52}/_7 - 0 + {}^4/_7 = {}^{56}/_7 = 8 \leq 10$$

MINIMIZATION PROBLEM:

$$
\begin{aligned}
{}^{26}/_7 + {}^{10}/_7 - 0 &= {}^{36}/_7 = 5{}^1/_7 \geq 3 \\
{}^{52}/_7 - {}^{10}/_7 + 0 &= {}^{42}/_7 = 6 \geq 6 \\
{}^{26}/_7 + {}^{30}/_7 - 0 &= {}^{56}/_7 = 8 \geq 4 \\
-{}^{78}/_7 + {}^{50}/_7 + 0 &= -{}^{28}/_7 = -4 \geq -4
\end{aligned}
$$

Finally, we note that these feasible solutions give the same value to their respective objective variables.

MAXIMIZATION PROBLEM:

$$0 + {}^{78}/_7 + 0 - {}^{16}/_7 = {}^{62}/_7$$

MINIMIZATION PROBLEM:

$$ {}^{52}/_7 + {}^{10}/_7 + 0 = {}^{62}/_7$$

This provides a certain check for both problems. We know that the optimal values of the objective variables for dual problems are equal from the fact that the matrices are identical for a maximization problem and its dual minimization problem. The following can also be shown to be true:

> **If feasible solutions to a maximization problem and its dual minimization problem give the same value to the objective variable, then each feasible solution is optimal.**

As another example, we apply this additional check to Problem 4, by substituting $x_1 = 1$, $x_2 = 1$, $x_3 = 0$ into the dual maximization problem:

$$1 + 2 + 0 = 3 \le 3 \quad ✔$$
$$1 + 1 + 0 = 2 \le 2 \quad ✔$$
$$2 + 3 + 0 = 5 \le 7 \quad ✔$$
$$y = 3 + 4 + 0 = 7 \quad ✔$$

To help further the reader's understanding of the dual variable check and of duality in general, we offer as an optimal topic the following example:

An Economic Interpretation of Duality

The Almine Coal Company has four mines. Each mine produces three types of coal, as shown in Table 5. Its problem is to fill its

Table 5

Mine	Daily coal production (in hundreds of tons)			Cost of operation (in thousands of dollars)
	Type			
	A	B	C	
I	4	5	3	4
II	2	1	0	2
III	3	2	5	3
IV	4	2	6	3
Orders	20	30	40	

orders at minimum cost. If we let x_1 be the number of days to operate mine I, x_2 the number of days to operate mine II, and so forth, we are led immediately to the following problem:

Minimize $4x_1 + 2x_2 + 3x_3 + 3x_4$ subject to

$$x_1 \ge 0, \ x_2 \ge 0, \ x_3 \ge 0, \ x_4 \ge 0$$
$$4x_1 + 2x_2 + 3x_3 + 4x_4 \ge 20$$
$$5x_1 + \ x_2 + 2x_3 + 2x_4 \ge 30$$
$$3x_1 \qquad + 5x_3 + 6x_4 \ge 40$$

The setup for the simplex method is

$$\begin{bmatrix} 4 & 5 & 3 & 1 & 0 & 0 & 0 & 4 \\ 2 & 1 & 0 & 0 & 1 & 0 & 0 & 2 \\ 3 & 2 & 5 & 0 & 0 & 1 & 0 & 3 \\ 4 & 2 & 6 & 0 & 0 & 0 & 1 & 3 \\ 20 & 30 & 40 & 0 & 0 & 0 & 0 & 0 \end{bmatrix}$$

The managers of a private company, Kraye & Sons, have access to this information and feel they can mine coal more cheaply. However, they do not have the distribution capability of Almine. They decide to offer to operate Almine's facilities and then sell the coal, already sorted by type, to Almine. Kraye & Sons' problem is to find an offer, attractive to Almine, which produces the greatest profit for them. Suppose we let x_1 be the price to be charged for 100 tons of type A coal, x_2 the price for the same quantity of type B, and x_3 the price for type C. Since Almine needs 2000 tons of type A, 3000 tons of type B, and 4000 tons of type C, Kraye's sales will generate a daily income of

$$20x_1 + 30x_2 + 40x_3$$

hundred dollars. This is to be maximized.

Mine I produces 4 hundred tons of type A, 5 hundred tons of type B, and 3 hundred tons of type C. At prices x_1, x_2, and x_3 it would cost Almine $4x_1 + 5x_2 + 3x_3$ to buy this much coal from Kraye. Almine can obtain this much coal for \$4000 by operating the mine itself. Kraye cannot expect Almine to pay more than this for coal. Thus, they must set their prices so that

$$4x_1 + 5x_2 + 3x_3 \le 4$$

A similar argument for each of the other mines forces Kraye to set prices so that

$$\begin{aligned}
2x_1 + \ x_2 \quad\ \ &\le 2 \\
3x_1 + 2x_2 + 5x_3 &\le 3 \\
4x_1 + 2x_2 + 6x_3 &\le 3
\end{aligned}$$

In summary, their problem is to maximize $20x_1 + 30x_2 + 40x_3$ subject to

$$\begin{aligned}
x_1 \ge 0, \ x_2 \ge 0, \ x_3 &\ge 0 \\
4x_1 + 5x_2 + 3x_3 &\le 4 \\
2x_1 + \ x_2 \quad\ \ &\le 2 \\
3x_1 + 2x_2 + 5x_3 &\le 3 \\
4x_1 + 2x_2 + 6x_3 &\le 3
\end{aligned}$$

The setup for the simplex method is

$$\begin{bmatrix}
4 & 5 & 3 & 1 & 0 & 0 & 0 & 4 \\
2 & 1 & 0 & 0 & 1 & 0 & 0 & 2 \\
3 & 2 & 5 & 0 & 0 & 1 & 0 & 3 \\
4 & 2 & 6 & 0 & 0 & 0 & 1 & 3 \\
20 & 30 & 40 & 0 & 0 & 0 & 0 & 0
\end{bmatrix}$$

Note that this is precisely the same matrix we set up for the minimization problem at Almine. Recall that after the simplex method is completed, the negative of the number in the lower right corner is *both* the minimal cost and maximal profit. The best offer from Kraye's viewpoint will cost Almine exactly the same as mining the coal themselves. Whether or not to accept the offer will be a matter of complete indifference, mathematically speaking.

Exercises 5.5

1. Minimize $y = 4x_1 + 2x_2$ subject to $x_1 \geq 0$, $x_2 \geq 0$, $x_1 + x_2 \geq 2$, $4x_1 + x_2 \geq 2$.

A 2. Minimize $y = 9x_1 + x_2$ subject to $x_1 \geq 0$, $x_2 \geq 0$, $x_1 + x_2 \geq 1$, $3x_1 - x_2 \geq 1$.

C 3. Minimize $y = 30x_1 + 51x_2$ subject to $x_1 \geq 0$, $x_2 \geq 0$, $2x_1 + x_2 \geq 16$, $x_1 + 2x_2 \geq 11$, $x_1 + x_2 \leq 9$.

A 4. Minimize $y = 2x_1 + x_2$ subject to $x_1 \geq 0$, $x_2 \geq 0$, $-x_1 + x_2 \geq 1$, $x_1 + x_2 \leq 2$, $16x_1 - x_2 \geq 4$.

5. Minimize $y = 2x_1 + x_2 + 8x_3 + 3x_4$ subject to $x_1 \geq 0$, $x_2 \geq 0$, $x_3 \geq 0$, $x_4 \geq 0$, $-2x_1 + x_2 - x_3 \geq 1$, $x_1 - x_2 + 2x_3 + x_4 \geq 2$.

6. Minimize $y = 10x_1 + 12x_2 + 12x_3$ subject to $x_1 \geq 0$, $x_2 \geq 0$, $x_3 \geq 0$, $5x_1 + 2x_2 + x_3 \geq 3$, $x_1 + 2x_2 + x_3 \geq 2$.

C 7. Minimize $y = 3x_1 + 8x_2 + 5x_3$ subject to $x_1 \geq 0$, $x_2 \geq 0$, $x_3 \geq 0$, $x_1 + x_2 + x_3 \geq 7$, $x_1 + 2x_2 + x_3 \geq 5$, $-x_1 + x_2 \geq 6$.

8. Minimize $y = 5x_1 + 6x_2 + 2x_3$ subject to $x_1 \geq 0$, $x_2 \geq 0$, $x_3 \geq 0$, $x_1 + 2x_2 + 3x_3 \geq 6$, $3x_1 + 2x_2 + x_3 \geq 6$, $-6x_1 - 4x_2 - 3x_3 > -11$.

9. Minimize $2x_1 + 3x_2 + 7x_3$ subject to $x_1 \geq 0$, $x_2 \geq 0$, $x_3 \geq 0$, $x_1 + 2x_2 + 3x_3 \geq 4$, $2x_1 + 4x_2 + 8x_3 \geq 7$, $x_1 + x_2 + 2x_3 \geq 3$.

A 10. Minimize $y = 8x_1 + 4x_2 + 9x_3$ subject to $x_1 \geq 0$, $x_2 \geq 0$, $x_3 \geq 0$, $3x_1 - 2x_2 + 2x_3 \geq 4$, $-2x_1 + x_2 + 2x_3 \geq 5$, $2x_1 + 4x_2 + 3x_3 \geq 5$.

*11. A complicated pilot study involving trading of certain frozen food items between stores in PHEM, the model PHE company, has led to the necessity to minimize $y = 6x_1 + 3x_2 + 2x_3 + 4x_4$, where x_1, x_2, x_3, and x_4 must not be negative, subject to the conditions $2x_2 + x_3 \geq x_1 + x_4 + 3$, $x_2 + x_3 - 4 \geq 2x_1 + 3x_4$, $2(x_1 + x_2) \geq 3x_3 - 2x_4 + 2$, and $4 - 3x_1 - 2x_3 \leq x_4 - 3x_2$. Do it.

**Section 5-6
Linear Programming
at Appalachian
Creations**

In this section we present and solve two problems that occurred recently at Appalachian Creations, Inc. While our discussion and analysis of these problems is rather complete, it should be pointed out that this is a mathematics book and not a management text. What we are really presenting here is the "middle part" of these problems. The first part, concerned with the recognition of the problem and the collection of data, and the last part, concerned

(Another Author's message.)

with the presentation and implementation of the solution, are best discussed in courses in management techniques. These "outer parts" of linear programming obviously vary greatly from company to company, and in the final analysis they can only be learned through actual experience. The "middle part" however is a piece of

ABSTRACT MATHEMATICS

and as such it is essentially the same for *all* types of problems. Hence, it can really be learned from this book!

Problem 1

The technical climbing division of AC believes they can sell the following quantities of the following items: pitons, 800; carabiners, 600; nuts, 1000; ascenders, 90. These beliefs may be somewhat optimistic; at any rate the company has decided as a matter of policy to order no more than these quantities. The company has four possible sources for these items, small factories in Switzerland that operate up to 32 hours a week. In one full week, the factory at Visp produces 300 pitons, 200 carabiners, and 40 ascenders; the factory at Brig produces 200 pitons, 200 carabiners, 400 nuts, and 40 ascenders; the factory at Ayer produces 100 pitons, 400 carabiners, 300 nuts, and 20 ascenders; and the factory at Binn produces 400 pitons, 400 nuts, and 20 ascenders. The price structure at the factories is such that sale of one week's production from Visp results in a profit of $300 for AC, and sale of one week's production from Brig, Ayer, and Binn results in a profit of $600, $800, and $400, respectively. How long should each factory work so as to produce the maximum profit?

Solution

Let us agree to measure pitons, carabiners, and nuts in units of 100 each, ascenders in units of 10, and profit in units of $100. The data above may then be tabulated as shown in Table 6.

Table 6

Product	Factory				Constraint
	Visp	Brig	Ayer	Binn	
Pitons	3	2	1	4	at most 8
Carabiners	2	2	4	0	at most 6
Nuts	0	4	3	4	at most 10
Ascenders	4	4	2	2	at most 9
Profit	3	6	8	4	maximize

Let x_1 be the number of weeks the Visp factory operates and let x_2, x_3, x_4 be the number of weeks the factories operate at Brig, Ayer, and Binn, respectively. Let y be the total profit. Then we are to maximize

$$y = 3x_1 + 6x_2 + 8x_3 + 4x_4 \qquad \textbf{1}$$

subject to

$$
\begin{aligned}
&x_1 \geq 0,\ x_2 \geq 0,\ x_3 \geq 0,\ x_4 \geq 0 \\
&3x_1 + 2x_2 +\ x_3 + 4x_4 \leq\ 8 &\textbf{2}\\
&2x_1 + 2x_2 + 4x_3 \qquad\ \leq\ 6 &\textbf{3}\\
&\qquad\quad 4x_2 + 3x_3 + 4x_4 \leq 10 &\textbf{4}\\
&4x_1 + 4x_2 + 2x_3 + 2x_4 \leq\ 9 &\textbf{5}
\end{aligned}
$$

We apply the simplex method, which results in the following sequence of matrices:

$$
\begin{bmatrix}
3 & 2 & 1 & 4 & 1 & 0 & 0 & 0 & 8\\
2 & 2 & 4^* & 0 & 0 & 1 & 0 & 0 & 6\\
0 & 4 & 3 & 4 & 0 & 0 & 1 & 0 & 10\\
4 & 4 & 2 & 2 & 0 & 0 & 0 & 1 & 9\\
3 & 6 & 8 & 4 & 0 & 0 & 0 & 0 & 0
\end{bmatrix}
$$

$$
\begin{array}{l}
R_1 - \tfrac{1}{4}R_2 \\
\tfrac{1}{4}R_2 \\
R_3 - \tfrac{3}{4}R_2 \\
R_4 - \tfrac{1}{2}R_2 \\
R_5 - 2R_2
\end{array}
\begin{bmatrix}
\tfrac{3}{2} & \tfrac{3}{2} & 0 & 4 & 1 & -\tfrac{1}{4} & 0 & 0 & \tfrac{13}{2}\\
\tfrac{1}{2} & \tfrac{1}{2} & 1 & 0 & 0 & \tfrac{1}{4} & 0 & 0 & \tfrac{3}{2}\\
-\tfrac{3}{2} & \tfrac{5}{2} & 0 & 4^* & 0 & -\tfrac{3}{4} & 1 & 0 & \tfrac{11}{2}\\
3 & 3 & 0 & 2 & 0 & -\tfrac{1}{2} & 0 & 1 & 6\\
-1 & 2 & 0 & 4 & 0 & -2 & 0 & 0 & -12
\end{bmatrix}
$$

$$
\begin{array}{l}
R_1 -\ \ R_3 \\
\\
\tfrac{1}{4}R_3 \\
R_4 - \tfrac{1}{2}R_3 \\
R_5 -\ \ R_3
\end{array}
\begin{bmatrix}
4^* & -1 & 0 & 0 & 1 & \tfrac{1}{2} & -1 & 0 & 1\\
\tfrac{1}{2} & \tfrac{1}{2} & 1 & 0 & 0 & \tfrac{1}{4} & 0 & 0 & \tfrac{3}{2}\\
-\tfrac{3}{8} & \tfrac{5}{8} & 0 & 1 & 0 & -\tfrac{3}{16} & \tfrac{1}{4} & 0 & \tfrac{11}{8}\\
\tfrac{15}{4} & \tfrac{7}{4} & 0 & 0 & 0 & \tfrac{1}{8} & -\tfrac{1}{2} & 1 & \tfrac{13}{4}\\
\tfrac{1}{2} & -\tfrac{1}{2} & 0 & 0 & 0 & -\tfrac{5}{4} & -1 & 0 & -\tfrac{35}{2}
\end{bmatrix}
$$

$$
\begin{array}{l}
\tfrac{1}{4}R_1 \\
R_2 - \tfrac{1}{8}R_1 \\
R_3 + \tfrac{3}{32}R_1 \\
R_4 - \tfrac{15}{16}R_1 \\
R_5 - \tfrac{1}{8}R_1
\end{array}
\begin{array}{ccccccccc}
X_1 & X_2 & X_3 & X_4 & X_5 & X_6 & X_7 & X_8 & \\
\begin{bmatrix}
1 & — & 0 & 0 & — & — & — & 0 & \tfrac{1}{4}\\
0 & — & 1 & 0 & — & — & — & 0 & \tfrac{11}{8}\\
0 & — & 0 & 1 & — & — & — & 0 & \tfrac{47}{32}\\
0 & — & 0 & 0 & — & — & — & 1 & \tfrac{37}{16}\\
0 & -\tfrac{3}{8} & 0 & 0 & -\tfrac{1}{8} & -\tfrac{21}{16} & -\tfrac{7}{8} & 0 & -\tfrac{141}{8}
\end{bmatrix}
\end{array}
$$

In this last matrix, we have illustrated a way to save computational labor. We calculated the bottom row first. Since the result indicated that this was the final matrix, we calculated only the cleared columns and the last column. The answer can be read from this incomplete final matrix: $y = {}^{141}/_8$, $x_1 = {}^1/_4$, $x_2 = 0$, $x_3 = {}^{11}/_8$, $x_4 = {}^{47}/_{32}$, and the slack variable values, $x_5 = x_6 = x_7 = 0$, $x_8 = {}^{37}/_{16}$. We substitute these values into **1** and constraints **2** through **5** above, they check. For example, in constraint **5**: $4({}^1/_4) + 4(0) + 2({}^{11}/_8) + 2({}^{47}/_{32}) = {}^{107}/_{16}$; ${}^{107}/_{16} +$ the slack $= {}^{107}/_{16} + {}^{37}/_{16} = 9$:✔. The total number of ascenders is $({}^{107}/_{16})(10) = 66{}^7/_8$; since a partially completed device will not be shipped, the company might receive only 66 ascenders. Since Binn, for example, works on a 32-hour week, it will work ${}^{47}/_{32}$ week $= 1$ week, 15 hours to fill the order. Recalling the other units in which we counted, we may express the final answer as follows: Order 8 hours production from Visp, none from Brig, one week, 12 hours from Ayer, and one week, 15 hours from Binn. The company will receive 800 pitons, 600 carabiners, 1000 nuts, and 66 ascenders, which can be sold for a profit of a little less than $1762.50 (a little less because of the uncompleted ascender).

The dual variable check, p. 263, could also be applied here.

Problem 2

The managers at AC wish to blend their own brand of instant breakfast cereal for backpackers by mixing various quantities of the prepared mixes they have on hand. They have consulted a nutritionist, who has recommended that a six-person portion of their cereal should contain at least 3 oz. raisins, 2 oz. currants, and 2 oz. nuts. A ten-oz. package of each of the five kinds of prepared mix they have on hand contains the proportions of these ingredients listed in Table 7.

Table 7

Type	Mix (oz. per package)			Cost
	Raisins	Nuts	Currants	(cents per package)
A	1	2	$1/_2$	30
B	2	1	2	90
C	3	0	2	120
D	1	1	3	70
E	2	3	2	90

The last column of this table gives the cost to the company of each type of mix. How should they prepare the most economical cereal mix, how much of each food will be in each six-person portion, how much will the portion weigh, and how much will the ingredients cost the company?

Solution   Let x_1, x_2, x_3, x_4, x_5 be the number of 10-oz. packages of mix of type A, B, C, D, E, respectively, used to make one six-person portion. Measure the cost y in units of 10 cents. We are to minimize

$$y = 3x_1 + 9x_2 + 12x_3 + 7x_4 + 9x_5 \qquad \textbf{6}$$

subject to

$$x_1 \geq 0,\ x_2 \geq 0,\ x_3 \geq 0,\ x_4 \geq 0,\ x_5 \geq 0$$
$$x_1 + 2x_2 + 3x_3 + x_4 + 2x_5 \geq 3 \qquad \textbf{7}$$
$$2x_1 + x_2 + x_4 + 3x_5 \geq 2 \qquad \textbf{8}$$
$$\tfrac{1}{2}x_1 + 2x_2 + 2x_3 + 3x_4 + 2x_5 \geq 2 \qquad \textbf{9}$$

The simplex method yields the following sequence of matrices:

$$\begin{bmatrix}
1^* & 2 & \tfrac{1}{2} & 1 & 0 & 0 & 0 & 0 & 3 \\
2 & 1 & 2 & 0 & 1 & 0 & 0 & 0 & 9 \\
3 & 0 & 2 & 0 & 0 & 1 & 0 & 0 & 12 \\
1 & 1 & 3 & 0 & 0 & 0 & 1 & 0 & 7 \\
2 & 3 & 2 & 0 & 0 & 0 & 0 & 1 & 9 \\
3 & 2 & 2 & 0 & 0 & 0 & 0 & 0 & 0
\end{bmatrix}$$

$$\begin{matrix}
R_2 - 2R_1 \\
R_3 - 3R_1 \\
R_4 - R_1 \\
R_5 - 2R_1 \\
R_6 - 3R_1
\end{matrix}
\Longrightarrow
\begin{bmatrix}
1 & 2 & \tfrac{1}{2} & 1 & 0 & 0 & 0 & 0 & 3 \\
0 & -3 & 1 & -2 & 1 & 0 & 0 & 0 & 3 \\
0 & -6 & \tfrac{1}{2} & -3 & 0 & 1 & 0 & 0 & 3 \\
0 & -1 & \tfrac{5}{2}^* & -1 & 0 & 0 & 1 & 0 & 4 \\
0 & -1 & 1 & -2 & 0 & 0 & 0 & 1 & 3 \\
0 & -4 & \tfrac{1}{2} & -3 & 0 & 0 & 0 & 0 & -9
\end{bmatrix}$$

$$R_6 - \tfrac{1}{5}R_4 \Longrightarrow
\begin{bmatrix}
 & & & & & & & & \\
0 & -\tfrac{19}{5} & 0 & -\tfrac{14}{5} & 0 & 0 & -\tfrac{1}{5} & 0 & -\tfrac{49}{5}
\end{bmatrix}$$

$$\begin{matrix} & x_1 & x_2 & x_3 & x_4 & x_5 \end{matrix}$$

There is no need to compute more than the last row of the final matrix to get the answer: $x_1 = {}^{14}/_5$, $x_4 = {}^{1}/_5$, $x_2 = x_3 = x_5 = 0$, $y = {}^{49}/_5$. The answer checks when substituted into **6** and constraints **7, 8** and **9** above. The AC cereal contains, in one six-person portion, 3 oz. raisins, ${}^{29}/_5 = 5.2$ oz. nuts, and 2 oz. currants, which explains why they called it "the nuttiest breakfast on the trail." Such a portion is prepared from $({}^{14}/_5)(10) = 28$ oz. type A mix and $({}^{1}/_5)(10) = 2$ oz, type D; it weighs 30 oz. and costs the company $({}^{49}/_5)(10) = \$.98$.

Note that if we wish to apply the dual variable check, p. 263, to this problem, then we must fill in more of the last matrix.

Exercises 5.6 ᴬ 1. AC wishes to make a high-calorie candy by hardening a mixture of two syrups into a pellet. The syrups contain small quantities of two undesirable chemicals, call them A and B. Three grams (g) of syrup I contain 10 milligrams (mg) of A, 10 of B; 3 g of syrup II contain 10 mg of A and 40 of B. Three g of I and II will deliver 20 and 30 calories, respectively. How should the syrups be mixed to produce a pellet with the greatest caloric content but with no more than 20 mg of A and 40 of B?

2. My doctor says I should take at least 20 mg of phlorotaste and 30 mg of lypollogin a day for the next two months. While the prescription he writes is for each of these in pure form, my druggist tells me that getting them already mixed together is much cheaper. In fact, 8 oz. of hyposin contain 1 mg of phlorotaste and 5 mg of lypollogin and cost $7. Eight oz. of eneron contain 7 mg of phlorate and 2 mg of lypollogin and cost $10. How many ounces of each of these should I buy to fill my needs at minimum cost? (I have been assured, by the way, that there is no danger of overdosage with these substances.)

3. Our local toy company makes trucks, beach balls, music boxes, and tricycles from limited amounts of wood, vinyl plastic, and steel, according to Table 8. How many of each toy should they make to maximize their profit?

Table 8

	Wood	Vinyl plastic	Steel	Profit
Trucks	1	0	1	2
Beach balls	0	3	0	1
Music boxes	4	4	2	3
Tricycles	0	0	2	2
Supply	21	15	14	

ᶜ4. "I don't know how much these two temporary secretarial services charge," said the AC personnel director, "but they charge the same rate per hour. We need to get 60 letters typed, 700 circulars copied, and 60 stencils cut by tomorrow. I am told that in 5 hours the Jiffy Service can type 30 letters, copy 100 circulars, and cut 20 stencils; the Rapid Service can type 20 letters, copy 700 circulars, and cut 30 stencils. What should I do?"

ᴬ5. The advertising managers at AC are ready to mail out some circulars. For ecological reasons, they want no more than 600 circulars to be sent to New York, 400 to Boston, and 600 to Chicago. For a certain fee the Pelham Distributing Company will mail out 3000 circulars, of which 30 will go to New York, 10 to Boston, and 40 to Chicago. For the same fee the Quinlon Company will mail out 5000 circulars, of which 40 will go to New York, 20 to Boston, 30 to Chicago, and the

Rex Company will mail 2000 circulars, of which 50 will go to New York, 10 to Boston, and 50 to Chicago. How many circulars can AC distribute while remaining ecologically sound and which companies should do how much?

A 6. "In testing boots," explained the AC research director, "it isn't the number of miles walked that matters, but the type of footing encountered. We have spoken with our chief tester, who has described three different routes she knows in this area. Route A is 6 miles long. Three miles of this have a rocky footing, 4 miles are steep, and 5 miles are wet. Route B is 4 miles long with 1 mile rocky, 2 steep, and 1 wet. Route C is 6 miles long with 4 miles rocky, 3 steep, and 5 wet. We want to test these boots over at least 12 miles of rocky terrain, 20 of steep, and 8 of wet, while keeping the total mileage walked as low as we can. How can we do it?"

* 7. Petaluma Hamburger Emporium is designing a fruit compote for its dessert offering, to be made with strawberries, apples, and peaches. One ladleful of uncooked strawberries costs 40¢, contains 2 milligrams (mg) salt and, when cooked, has a volume of 3 cc and a weight of 1 oz. One ladleful of uncooked apples costs 30¢, contains no salt and cooks down to 2 cc and 2 oz. The peaches are precooked; one ladleful costs 50¢, weighs 3 oz., has 3 mg of salt, and has a volume of 2 cc. A serving of compote should contain at least 4 mg. salt and weigh at least 3 oz. For the sake of taste and appearance there is one more restriction: when one ladleful of strawberries is cooked and mixed with one ladleful of peaches, the resulting mix should have a volume at least 5 cc more than the volume of one ladleful of cooked apples. In what proportions should the fruits be mixed to make the cheapest compote, and how much, per serving, will the compote cost?

Review Exercises

Look back over the chapter; then do the following exercises without looking back again. Answers to all of these exercises are in Appendix B.

A 1. Find the maximum and the minimum values of $y = -3x_1 + 5x_2$ subject to $x_1 + x_2 \leq 2$, $3x_1 - x_2 \leq -6$, $x_1 - 7x_2 \leq 18$.

A 2. (a) Find the maximum and minimum values of $y = x_1 + 2x_2$ subject to $2x_1 - x_2 \geq 2$, $2x_1 - 3x_2 \leq 6$.

(b) Find the maximum and minimum values of $y = x_1 + 2x_2$ subject to $2x_1 - x_2 \leq 2$, $2x_1 - 3x_2 \geq 6$.

A 3. Use the simplex method to solve the following problems:

(a) Maximize $y = x_1 + 2x_2$ subject to $x_1 \geq 0$, $x_2 \geq 0$, and $3x_1 + 2x_2 \leq 6$, $x_1 - 3x_2 \leq 7$.

(b) Minimize $y = 4x_1 + 7x_2$ subject to $x_1 \geq 0$, $x_2 \geq 0$, $2x_1 + 3x_2 \geq 4$, $3x_1 + 4x_2 \geq 6$, $-x_1 - x_2 \geq 2$.

A 4. Maximize $y = 3x_1 + 3x_2 + 4x_3 + 2x_4$ subject to $x_1 \geq 0$, $x_2 \geq 0$, $x_3 \geq 0$, $x_4 \geq 0$, $2x_1 + 3x_2 + 4x_3 + 2x_4 \leq 4$, $3x_1 + 2x_2 + x_3 \leq 5$, $2x_1 + 2x_2 + 3x_4 \leq 3$, $x_1 + x_2 + 2x_3 + 2x_4 \leq 4$.

A 5. Minimize $y = 4x_1 + 5x_2 + 3x_3 + 4x_4$ subject to $x_1 \geq 0$, $x_2 \geq 0$, $x_3 \geq 0$, $x_4 \geq 0$, $2x_1 + 3x_2 + 2x_3 + x_4 \geq 3$, $3x_1 + 2x_2 + 2x_3 + x_4 \geq 3$, $4x_1 + x_2 + 2x_4 \geq 4$, $2x_1 + 3x_3 + 2x_4 \geq 2$.

A 6. A candy wholesaler has a supply of 6,000 rolls of mints, 10,000 gumballs, and 8,000 candy bars. She has 4 types of boxes. Type A can accommodate 2 rolls of mints, 3 gumballs, and 2 candy bars; B holds 3 rolls of mints and 4 gumballs; C holds 2 rolls of mints, 2 gumballs, and 3 candy bars; D holds 6 rolls of mints only. The filled boxes can be wholesaled at a per-box profit (in cents) of 50, 30, 20, and 60, respectively. How should she maximize profit?

C 7. The Titan Conglomerate needs 9 million gallons of fuel oil and 7 million gallons of gasoline. The Supergas refinery will deliver 40,000 gallons of oil and 80,000 gallons of gasoline every day for $110,000; the Exxtrabig refinery will deliver 60,000 gallons of oil and 30,000 of gasoline every day for $70,000. How does Titan get the best deal? How much do they pay?

Supplementary Exercises

A 1. Maximize $y = 7x_1 + 5x_2$ subject to $x_1 \geq 0$, $x_2 \geq 0$, $7x_1 + 8x_2 \leq 9$, $6x_1 + 7x_2 \leq 15$.

2. Minimize $y = 3x_1 + 5x_2$ subject to $x_1 \geq 0$, $x_2 \geq 0$, $7x_1 + 6x_2 \geq 7$, $8x_1 + 7x_2 \geq 5$.

A 3. Maximize $y = 5x_1 + 5x_2 + 7x_3$ subject to $x_1 \geq 0$, $x_2 \geq 0$, $x_3 \geq 0$, $-2x_1 + x_2 + 3x_3 \leq 9$, $3x_1 + 4x_3 \leq 6$, $2x_1 + 4x_2 + 2x_3 \leq 2$.

4. Minimize $y = 9x_1 + 6x_2 + 2x_3$ subject to $x_1 \geq 0$, $x_2 \geq 0$, $x_3 \geq 0$, $-2x_1 + 3x_2 + 2x_3 \geq 5$, $x_1 + 4x_3 \geq 5$, $3x_1 + 4x_2 + 2x_3 \geq 7$.

A 5. Maximize $y = x_1 + 3x_2 + 6x_3 + 2x_4 + 2x_5$ subject to $x_1 \geq 0$, $x_2 \geq 0$, $x_3 \geq 0$, $x_4 \geq 0$, $x_5 \geq 0$, $3x_1 + 2x_2 - 6x_3 + 4x_4 + 5x_5 \leq 10$, $2x_1 + 3x_3 + 5x_4 + 2x_5 \leq 8$.

6. Minimize $y = 10x_1 + 8x_2$ subject to $x_1 \geq 0$, $x_2 \geq 0$, $3x_1 + 2x_2 \geq 1$, $2x_1 \geq 0$, $-6x_1 + 3x_2 \geq 6$, $4x_1 + 5x_2 \geq 2$, $5x_1 + 2x_2 \geq 2$.

A 7. Maximize $y = 4x_1 + 8x_2 + 12x_3$ subject to $x_1 \geq 0$, $x_2 \geq 0$, $x_3 \geq 0$, $x_1 + 5x_2 + 9x_3 \leq 13$, $2x_1 + 6x_2 + 10x_3 \leq 14$, $3x_1 + 7x_2 + 11x_3 \leq 15$.

8. Minimize $y = 13x_1 + 14x_2 + 15x_3$ subject to $x_1 \geq 0$, $x_2 \geq 0$, $x_3 \geq 0$, $x_1 + 2x_2 + 3x_3 \geq 4$, $5x_1 + 6x_2 + 7x_3 \geq 8$, $9x_1 + 10x_2 + 11x_3 \geq 12$.

A 9. Maximize $y = 6x_1 + 12x_2$ subject to $x_1 \geq 0$, $x_2 \geq 0$, $2x_1 + 8x_2 \leq 14$, $x_1 + 7x_2 \leq 13$, $4x_1 + 10x_2 \leq 16$, $3x_1 + 9x_2 \leq 15$, $5x_1 + 11x_2 \leq 17$.

10. Minimize $y = 13x_1 + 14x_2 + 15x_3 + 16x_4 + 17x_5$ subject to $x_1 \geq 0$, $x_2 \geq 0$, $x_3 \geq 0$, $x_4 \geq 0$, $x_5 \geq 0$, $x_1 + 2x_2 + 3x_3 + 4x_4 + 5x_5 \geq 6$, $7x_1 + 8x_2 + 9x_3 + 10x_4 + 11x_5 \geq 12$.

A 11. Maximize $y = 3x_1 + 4x_2 + 5x_3 + 6x_4$ subject to $x_1 \geq 0$, $x_2 \geq 0$, $x_3 \geq 0$, $x_4 \geq 0$, $x_1 \leq 6$, $2x_1 + x_2 \leq 5$, $2x_2 + x_3 \leq 4$, $2x_3 + x_4 \leq 3$, $2x_4 \leq 2$.

12. Minimize $y = 6x_1 + 5x_2 + 4x_3 + 3x_4 + 2x_5$ subject to $x_1 \geq 0$, $x_2 \geq 0$, $x_3 \geq 0$, $x_4 \geq 0$, $x_5 \geq 0$, $x_1 + 2x_2 \geq 3$, $x_2 + 2x_3 \geq 4$, $x_3 + 2x_4 \geq 5$, $x_4 + 2x_5 \geq 6$.

*13. Maximize $y = 4x_1 + 5x_2 + 6x_3 + 7x_4$ subject to $x_1 \geq 0$, $x_2 \geq 0$, $x_3 \geq 0$, $x_4 \geq 0$, $3x_1 \leq 7$, $2x_1 + 3x_2 \leq 6$, $x_1 + 2x_2 + 3x_3 \leq 5$, $x_2 + 2x_3 + 3x_4 \leq 4$.

14. Minimize $y = 7x_1 + 6x_2 + 5x_3 + 4x_4$ subject to $x_1 \geq 0$, $x_2 \geq 0$, $x_3 \geq 0$, $x_4 \geq 0$, $3x_1 + 2x_2 + x_3 \geq 4$, $3x_2 + 2x_3 + x_4 \geq 5$, $3x_3 + 2x_4 \geq 6$, $x_4 \geq 7$.

A 15. There are 2 mines. One operation of mine I produces 1 ton of high-grade ore, 3 tons of medium-grade and 5 tons of low-grade. One operation of mine II produces 2 tons high-grade, 2 tons medium-grade, and 2 tons low-grade. The company makes $200 profit in one operation of mine I and $208 profit in one operation of mine II. Storage restrictions require that the company make no more than 88 tons of high-grade ore, 160 tons of medium-grade ore, and 200 tons of low-grade ore. How many operations of each mine should it make to maximize its profit and not exceed storage? What will the maximum profit be?

A 16. Two liquid vitamin supplements are available. Supplement I contains 2 units of vitamin A, 1 unit of vitamin B, 1 unit of vitamin C per tablespoon. Supplement II contains 1 unit of vitamin A, 2 units of vitamin B, and 3 units of vitamin C per tablespoon. Each costs 30¢ per tablespoon. I need 3 units of vitamin A, 5 units of vitamin B, and 2 units of vitamin C. How can I meet my needs most economically, and what will be the minimum cost?

17. A manufacturer produces 4 products, each of which requires wood, plastic, and steel in its construction. A requires 2 units of wood, 1 unit of plastic, and 4 units of steel. B requires 1 unit of wood, 3 units of plastic, and 2 units of steel. C requires 3 units of wood, 2 units of plastic, and 1 unit of steel. D requires 6 units of wood, no units of plastic, and 3 units of steel. The manufacturer has only 5 units of wood, 2 units of plastic, and 9 units of steel. His profit is $5 on each A, $4 on each B, $2 on each C, and $3 on each D. How many of each item should he produce to maximize his profit, and what will his maximum profit be?

18. A company produces 3 items: A, B, and C. The main ingredients are materials I, II, and III. The company has only 7 units of I, 3 units of II, and 7 units of III. Each A requires 3 units of I, 1 unit of II, and 1 unit of III. Each B requires 5 units of I, 2 units of II, and 6 units of III. Each C requires 1 unit of I, 1 unit of II, and 3 units of III. The profit is: $9 for each A, $4 for each B, and $6 for each C. How many of each item should they produce to maximize their profit?

19. A manufacturer has orders for 3 items. These can be produced by 3 methods. In 1 hour, method 1 produces 2 A's, 1 B, and 2 C's. In 1 hour, method 2 produces 3 A's, 8 B's, and 5 C's. In 1 hour, method 3 produces 2 A's, 2 B's, and 3 C's. It costs $6 per hour to operate

method 1. It costs $22 per hour to operate method 2. It costs $12 per hour to operate method 3. There are orders for 12 A's, 16 B's, and 6 C's. While the orders must be met, overproduction is no concern. How many hours should he operate each method to fill the orders at minimum cost? What will this cost be?

20. Minimize $y = 2x_1 + 2x_2$ subject to $x_1 \geq 0$, $x_2 \geq 0$, $-x_1 + x_2 \geq -2$, $x_1 + x_2 \leq -3$.

A 21. Maximize $y = 3x_1 + 2x_2$ subject to $x_1 \geq 0$, $x_2 \geq 0$, $5x_1 + x_2 \leq 10$, $2x_1 + 2x_2 \leq 12$, $x_1 + x_2 \leq 12$.

A 22. Minimize $y = 2x_1 + x_2 + 8x_3 + 3x_4$ subject to $x_1 \geq 0$, $x_2 \geq 0$, $x_3 \geq 0$, $x_4 \geq 0$, $-2x_1 + x_2 - x_3 \geq 1$, $x_1 - x_2 + 2x_3 + x_4 \geq 2$.

23. Maximize $y = 2x_1 + 6x_2 + 5x_3$ subject to $x_1 \geq 0$, $x_2 \geq 0$, $x_3 \geq 0$, $5x_1 + 4x_2 + 12x_3 \leq 9$, $x_1 + 2x_2 + 4x_3 \leq 4$, $4x_1 + x_2 + 5x_3 \leq 3$.

24. Minimize $y = 6x_1 + 22x_2 + 12x_3$ subject to $x_1 \geq 0$, $x_2 \geq 0$, $x_3 \geq 0$, $2x_1 + 3x_2 + 2x_3 \geq 12$, $x_1 + 8x_2 + 2x_3 \geq 16$, $2x_1 + 5x_2 + 3x_3 \geq 6$.

A 25. Maximize $y = 2x_1 + x_2$ subject to $-x_1 + x_2 \leq 1$, $x_1 - x_2 \leq 1$, $x_1 + x_2 \geq 4$, $x_1 + x_2 \leq 6$.

A 26. Minimize $y = 15x_1 + 4x_2 + 11x_3$ subject to $x_1 \geq 0$, $x_2 \geq 0$, $x_3 \geq 0$, $3x_1 + x_2 + 2x_3 \geq 6$, $7x_1 + x_2 + 4x_3 \geq 7$, $5x_1 + x_2 + 5x_3 \geq 8$.

27. Maximize $y = 16x_1 + 11x_2 + 12x_3$ subject to $x_1 \geq 0$, $x_2 \geq 0$, $x_3 \geq 0$, $2x_1 + x_2 + x_3 \leq 30$, $x_1 + 2x_2 + 3x_3 \leq 51$.

28. Minimize $y = 3x_1 + 8x_2 + 5x_3$ subject to $x_1 \geq 0$, $x_2 \geq 0$, $x_3 \geq 0$, $x_1 + x_2 + x_3 \geq 7$, $x_1 + 2x_2 + x_3 \geq 5$, $-x_1 + x_2 \geq 6$.

A 29. Maximize $y = 3x_1 + 6x_2 + 4x_3 - 4x_4$ subject to $x_1 \geq 0$, $x_2 \geq 0$, $x_3 \geq 0$, $x_4 \geq 0$, $x_1 + 2x_2 + x_3 - 3x_4 \leq 2$, $x_1 - x_2 + 3x_3 + 5x_4 \leq 1$, $-x_1 + 4x_2 - 2x_3 + x_4 \leq 10$.

A 30. Minimize $y = 5x_1 + 2x_2 + 9x_3$ subject to $x_1 \geq 0$, $x_2 \geq 0$, $x_3 \geq 0$, $2x_1 + 3x_2 + 4x_3 \geq 5$, $x_1 + 3x_2 + 2x_3 \geq 4$, $3x_1 + 2x_2 + x_3 \geq 2$, $6x_1 + 3x_3 \geq 3$.

31. (See Problem 3 of Section 2-3, p. 81.) Suppose each hour of operation of the Springfield factory costs $20 and each hour of operation of the Danbury factory costs $30. What is the cheapest way to fill the Falmouth order?

Summary	objective variable in an LP problem	p. 226
Terms	constraint in an LP problem	p. 226
	region of feasibility of an LP problem	p. 230
	constraint matrix of an LP problem	p. 241
	bound matrix of an LP problem	p. 241
	objective matrix of an LP problem	p. 241
	transpose of a matrix	p. 259
Computational Techniques	simplex method for maximization problems	p. 244
	simplex method for minimization problems	p. 261

Chapter 6
Additional
Topics

**Section 6-1
Game Theory**

So far all the applications that we have developed have involved only internal matters at AC. For example, we've used systems of equations to help us fill orders; patterns of change to minimize transportation costs; the simplex method to maximize profits. All these techniques are very important. However, it would also be useful to be able to analyze situations involving competition between companies—situations in which a course of action that is optimal for one company is detrimental to another. Game theory can be used to study situations involving conflict. The subject was invented by John von Neumann in 1928. He and Oskar Morgenstern subsequently developed the topic extensively in their book, *Theory of Games and Economic Behavior* (Princeton University Press, 1944).

This section contains a very brief introduction to game theory. A more detailed treatment may be found in *The Compleat Strategyst* (McGraw-Hill, 1966) by J. D. Williams. The latter work presupposes less mathematical background than the present book provides. For this reason, the method of solving problems in game theory presented there is less general than ours. However, the Williams book does provide many interesting and amusing examples.

While most of the real business world is too complex for direct application of the existing theory of games, studying relatively simple games can give some insights into the more complicated situations of business. We will limit our considerations rather severely in that we will consider only games with two players in which one player's gain is always exactly offset by the other player's loss. That is, if one player should win $5, the other player must lose $5. If we agree that "losing" is the same as "winning a negative amount," then at the end of the game, adding the winnings of one player to the winnings of the other player must always result in zero. For this reason such games are called **two-person zero-sum games.**

A game involves strategies. A **strategy** is a complete plan of action. It may be as simple as calling heads in a game of flipping coins, or it may include elaborate sequences of moves, each dependent upon all the previous moves in the game. For example, a strategy for the player moving first in checkers would include: (1) a first move, (2) a list of second moves indicating which move to employ depending upon her first move and the opponent's first move, (3) a list of third moves, and so forth. Once both players have chosen such strategies they might simply hand their decisions over to an intermediary who could actually play the game and then announce the result. The point is that once the two

We remark that there are computer programs that can play very good games of chess and checkers, but this is a long way from finding an optimal strategy. These programs use more sophisticated techniques than we discuss in this book.

chosen strategies are made known, the outcome of the game is completely determined. Actually, to list even one strategy for checkers would be a lifetime task. To list them all and then to use the technique of this section to determine an optimal procedure would be beyond the capability of even the largest computer in existence. We pass on to games simpler than checkers.

Example 1

Player A gives player B $3 before the start of the following game. Each player writes 1, 2, or 3 on a slip of paper. If A writes 1, then B pays A $1 if B has also written 1, $2 if B has written 2, and $4 if B has written 3. If A writes 2, then B pays A $4 if B has written 1, $3 if B has written 2, and $4 if B has written 3. If A writes 3, then B pays A $0 if B has written 1, $1 if B has written 2, and $5 if B has written 3. We tabulate these data in Table 1.

Table 1
Amount Player B Pays Player A

A writes	B writes		
	1	2	3
1	1	2	4
2	4	3	4
3	0	1	5

The associated matrix is called the **payoff matrix:**

$$\begin{bmatrix} 1 & 2 & 4 \\ 4 & 3 & 4 \\ 0 & 1 & 5 \end{bmatrix}$$

Every two-person zero-sum game is completely determined by its payoff matrix. We will therefore consider only the matrices and not the games from which they arose. The matrix

$$\begin{bmatrix} -1 & 2 & 0 & 3 \\ 2 & 1 & 1 & 3 \\ 4 & 2 & -1 & 2 \end{bmatrix}$$

represents a game in which A has three strategies and B has four. If A and B both select strategy 3, for example, then this matrix informs us that B pays A -1 dollars; i.e., A pays B $1. Given any game with relatively few strategies, its matrix can be determined by simply checking the result as A applies each of her strategies against every strategy of B.

We are concerned with two questions. Given a payoff matrix how does one determine (a) the optimal strategies for each player, and (b) the amount each player can expect to win? The simplex method can be used to answer both these questions at once. However, some games admit a simpler approach.

Returning to the payoff matrix of Example 1, we consider each player's situation separately. Since player B wants to pay A as little as possible, she might be tempted to try strategy 1, since it contains the entry 0. However, if she selects strategy 1 and A happens to select strategy 2, then she ends up paying $4. Rather than take this chance, she decides to give up working for herself and starts working against A. How can she guarantee that no matter what happens, A will win as little as possible? If she chooses strategy 1, then the worst that can happen is for A to choose strategy 2, in which case B loses $4. If she chooses strategy 2, then A can win at most $3 (by selecting strategy 2). Finally, if she chooses strategy 3, the worst that can happen is a loss of $5. Thus, she selects strategy 2 and minimizes the worst that can happen.

Payoff matrix:
$$\begin{bmatrix} 1 & 2 & 4 \\ 4 & 3 & 4 \\ 0 & 1 & 5 \end{bmatrix}$$

The situation for A is similar except that she wants to win as much as possible. Again, the greedy temptation to select strategy 3 in hopes of a $5 win must be avoided. The best bet is to work against B and select a strategy that results in the maximum loss for B. Strategy 1 would yield a loss of at least $1. Strategy 2 would result in a loss of at least $3, and strategy 3 in a loss of at least $0. Thus, A selects strategy 2 to obtain the maximum assured loss for B.

Notice that each player's best strategy involves the (2,2) entry 3. Thus, the minimum assured win for A in B's best strategy equals the maximum assured loss for B in A's best strategy. When this happens, the game is said to have a **saddle point**. It can be shown that if a game has a saddle point, then the best plan for each player is to use the same strategy constantly. The most A can hope to win equals the least B can expect to lose. In Example 1, A and B should both always use strategy 2. Each game will then result in B's paying A $3. The number 3 is called the **value** of the game. It is the amount that A should pay B to make the game fair.

Most games do not have saddle points. However, it is worth applying the simple check shown in the flow chart on p. 281 before turning to the more difficult technique that follows.

Problem 1 Use the flow chart to find the saddle point (if any) of the following payoff matrices:

(a) $\begin{bmatrix} 4 & -1 & 1 & 0 \\ -2 & -1 & 2 & 0 \\ 3 & 0 & 1 & 1 \\ 5 & -1 & -3 & 1 \end{bmatrix}$

(b) $\begin{bmatrix} 0 & 2 & 3 & 2 & 1 \\ -1 & 1 & -1 & -3 & 4 \\ 3 & 4 & 0 & -2 & 3 \end{bmatrix}$

Solution   (a) The smallest entry in R_1 is -1. We write a -1 to the right of this row, and proceed similarly for each of the other rows. The largest entry in C_1 is 5. We write 5 below this column and proceed similarly for each of the other columns. Here is the result:

$$
\begin{bmatrix}
4 & -1 & 1 & 0 \\
-2 & -1 & 2 & 0 \\
3 & 0 & 1 & 1 \\
5 & -1 & -3 & 1
\end{bmatrix}
\begin{matrix}
-1 \\
-2 \\
0^* \\
-3
\end{matrix}
$$
$$
\begin{matrix}
5 & 0^* & 2 & 1
\end{matrix}
$$

The zero in the added column is starred since it is the largest of the entries in this column. The zero in the added row is starred since it is the smallest of the entries in this row. Since the starred numbers are equal, there is a saddle point. Player A should always use strategy 3 and player B should always use strategy 2. The game is *fair*, that is, its value is zero, so that neither player should expect to gain or lose by playing it.

To Determine a Saddle Point

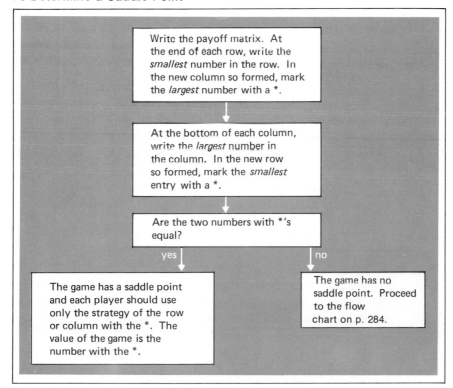

Write the payoff matrix. At the end of each row, write the *smallest* number in the row. In the new column so formed, mark the *largest* number with a *.

At the bottom of each column, write the *largest* number in the column. In the new row so formed, mark the *smallest* entry with a *.

Are the two numbers with *'s equal?

yes

The game has a saddle point and each player should use only the strategy of the row or column with the *. The value of the game is the number with the *.

no

The game has no saddle point. Proceed to the flow chart on p. 284.

(b) We proceed as before:

$$\begin{bmatrix} 0 & 2 & 3 & 2 & 1 \\ -1 & 1 & -1 & -3 & 4 \\ 3 & 4 & 0 & -2 & 3 \end{bmatrix} \begin{matrix} 0* \\ -3 \\ -2 \end{matrix}$$
$$\begin{matrix} \;3 & 4 & 3 & 2* & 4 \end{matrix}$$

Since the starred numbers are not equal, the game has no saddle point and must be approached through other techniques.

Example 2   Player A and player B each write the number 1 or the number 2 on a slip of paper. If the numbers turn out to be the same, B pays A \$1. If the numbers turn out to be different, A pays B \$1. The payoff matrix is

$$\begin{bmatrix} 1 & -1 \\ -1 & 1 \end{bmatrix}$$

We check for saddle points—

$$\begin{bmatrix} 1 & -1 \\ -1 & 1 \end{bmatrix} \begin{matrix} -1 \\ -1 \end{matrix}$$
$$\begin{matrix} \;\;1 & \;\;1 \end{matrix}$$

—and find that there are none.

While in a game with a saddle point it is best for each player to use the same strategy continually, such a policy in this game could have disastrous consequences. Suppose player B chooses strategy 1 and decides to stick to it. After a while, player A is likely to notice this, and has merely to select strategy 1 to drive B into bankruptcy. Similarly, player A should not employ only one strategy. The best choice for each player is to select a strategy at random. In this way A and B will win equally often and neither will experience any gain or loss over the long run. Actually, suppose A decides to follow our advice but B elects to use only strategy 1. Since A will be selecting each strategy at random, she will win half of the time anyway. Thus, in the long run A will not lose anything by using this technique. On the other hand, A can gain considerably if she discovers what B is doing.

The procedure of using each of several strategies a certain percentage of the time is called a **mixed strategy**. In the above example, the mixed strategy for each player would be the same: use the first strategy half the time and use the second strategy half the time. To actually implement such a mixed strategy one might flip a coin: heads would mean choose strategy 1, tails would mean choose strategy 2.

It is important that the choice of strategy be left up to chance, since otherwise the risk is run that the other player will discover the pattern and take advantage of it.

In most games the proper mixture of strategies is difficult to determine by just looking at the matrix. It turns out that the simplex method can be used to determine this mixture, as explained in the flow chart on p. 284.

Dominant Strategies Before starting to solve a game, it is often worth comparing strategies, since it can happen that some strategies are so bad that they should never be used. For example, consider B's first two strategies in the following game:

$$\begin{bmatrix} 7 & 5 & 2 \\ 8 & -3 & 1 \\ -1 & -4 & 3 \end{bmatrix} \qquad \textbf{1}$$

In every case strategy 1 is worse than strategy 2. That is, each entry in column 1 is larger than the corresponding entry in column 2. We say that strategy 2 *dominates* strategy 1, and we may reduce the payoff matrix by eliminating column 1, since B would never use it.

$$\begin{bmatrix} 5 & 2 \\ -3 & 1 \\ -4 & 3 \end{bmatrix}$$

Now consider this reduced game from the standpoint of A. Player A would never use strategy 2, since strategy 1 is always better—that is, every entry in row 2 is less than the corresponding entry in row 1. We reduce the game again by eliminating this useless row:

$$\begin{bmatrix} 5 & 2 \\ 4 & 3 \end{bmatrix} \qquad \textbf{2}$$

This game cannot be reduced further.

Problem 2 Solve the game with payoff matrix **1**.

Solution We first note that the game has no saddle point. Next we eliminate dominated strategies, as explained above, to reduce the game to one with payoff matrix **2**. It is simpler to apply the simplex method to this game than to the original 3×3 game. The flow chart instructs us to set up the matrix

$$\begin{bmatrix} 5 & 2 & 1 & 0 & 1 \\ -4 & 3 & 0 & 1 & 1 \\ 1 & 1 & 0 & 0 & 0 \end{bmatrix}$$

To Determine a Best Mixed Strategy

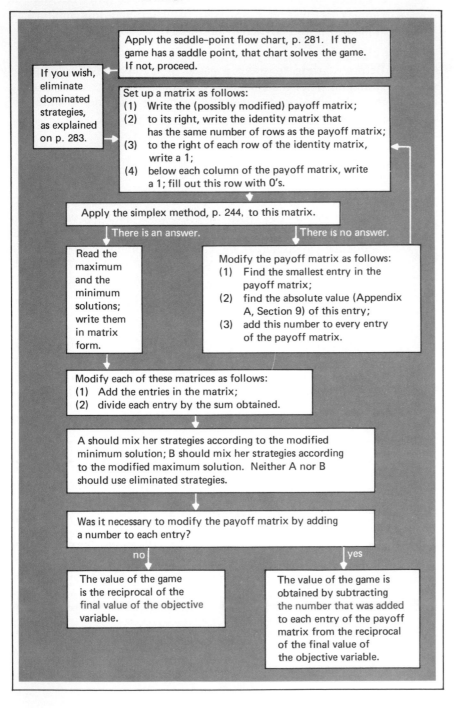

Apply the saddle-point flow chart, p. 281. If the game has a saddle point, that chart solves the game. If not, proceed.

If you wish, eliminate dominated strategies, as explained on p. 283.

Set up a matrix as follows:
(1) Write the (possibly modified) payoff matrix;
(2) to its right, write the identity matrix that has the same number of rows as the payoff matrix;
(3) to the right of each row of the identity matrix, write a 1;
(4) below each column of the payoff matrix, write a 1; fill out this row with 0's.

Apply the simplex method, p. 244, to this matrix.

There is an answer.

There is no answer.

Read the maximum and the minimum solutions; write them in matrix form.

Modify the payoff matrix as follows:
(1) Find the smallest entry in the payoff matrix;
(2) find the absolute value (Appendix A, Section 9) of this entry;
(3) add this number to every entry of the payoff matrix.

Modify each of these matrices as follows:
(1) Add the entries in the matrix;
(2) divide each entry by the sum obtained.

A should mix her strategies according to the modified minimum solution; B should mix her strategies according to the modified maximum solution. Neither A nor B should use eliminated strategies.

Was it necessary to modify the payoff matrix by adding a number to each entry?

no

yes

The value of the game is the reciprocal of the final value of the objective variable.

The value of the game is obtained by subtracting the number that was added to each entry of the payoff matrix from the reciprocal of the final value of the objective variable.

The simplex method leads to the final matrix

$$\begin{bmatrix} 1 & 0 & 3/23 & -2/23 & 1/23 \\ 0 & 1 & 4/23 & 5/23 & 9/23 \\ 0 & 0 & -7/23 & -3/23 & -10/23 \end{bmatrix}$$

from which we read the minimum solution $[7/23 \quad 3/23]$ and the maximum solution $\begin{bmatrix} 1/23 \\ 9/23 \end{bmatrix}$. The sum of the entries of the first matrix is $10/23$. Dividing each entry by $10/23$, we obtain the modified minimum solution $[7/10 \quad 3/10]$. Similarly, we obtain the modified maximum solution $\begin{bmatrix} 1/10 \\ 9/10 \end{bmatrix}$. Recall that for A strategy 2 (that is, R_2) was eliminated. Thus, A should use her first strategy $7/10$ of the time, never use her second strategy and use her third strategy $3/10$ of the time. Recall that for B strategy 1 (that is, C_1) was eliminated. Thus, B should never use strategy 1, use her second strategy $1/10$ of the time, and use her third strategy $9/10$ of the time. According to the flow chart, the value of the game is $1/(10/23) = 23/10$.

Note: If you fail to eliminate a dominated strategy you will still get the right answer, but with a little more effort than necessary. The simplex method will also locate a saddle point if you should fail to notice it yourself.

Problem 3   Solve the game with the following payoff matrix:

$$\begin{bmatrix} 1 & 2 & 1 \\ 3 & 1 & 2 \\ 1 & 3 & 2 \end{bmatrix}$$

Solution   Following the instructions of the flow chart, we find that there are no saddle points and no dominated strategies. We then set up the following matrix:

$$\begin{bmatrix} 1 & 2 & 1 & 1 & 0 & 0 & 1 \\ 3 & 1 & 2 & 0 & 1 & 0 & 1 \\ 1 & 3 & 2 & 0 & 0 & 1 & 1 \\ 1 & 1 & 1 & 0 & 0 & 0 & 0 \end{bmatrix}$$

The simplex method yields the final matrix:

$$\begin{bmatrix} 0 & 0 & -1/2 & 1 & -1/8 & -5/8 & 1/4 \\ 1 & 0 & 1/2 & 0 & 3/8 & -1/8 & 1/4 \\ 0 & 1 & 1/2 & 0 & -1/8 & 3/8 & 1/4 \\ 0 & 0 & 0 & 0 & -1/4 & -1/4 & -1/2 \end{bmatrix}$$

The solution to the minimization problem is

$$[0 \quad 1/4 \quad 1/4]$$

which we modify to

$$[0 \quad 1/2 \quad 1/2]$$

Thus, player A should completely avoid strategy 1 and flip a coin to decide which of strategies 2 or 3 she should use. The solution to the maximization problem is $\begin{bmatrix} 1/4 \\ 1/4 \\ 0 \end{bmatrix}$, which we modify to $\begin{bmatrix} 1/2 \\ 1/2 \\ 0 \end{bmatrix}$.

Player B should mix her strategies as follows: use strategy 1 half the time, use strategy 2 half the time, and avoid strategy 3 altogether. The value of the optimized objective variable is $1/2$. Thus, the value of the game is 2. In the long run player A can expect to win \$2 for each time the game is played. Thus, player A can reasonably be expected to pay \$2 each time she plays the game.

Problem 4 Solve the game

$$\begin{bmatrix} 1 & -1 \\ -1 & 1 \end{bmatrix}$$

Solution In Example 2, we found that this game has no saddle point. The matrix to set up for the simplex method is

$$\begin{bmatrix} 1^* & -1 & 1 & 0 & 1 \\ -1 & 1 & 0 & 1 & 1 \\ 1 & 1 & 0 & 0 & 0 \end{bmatrix}$$

If we apply the simplex method to this matrix, the following matrix results:

$$\begin{bmatrix} 1 & -1 & 1 & 0 & 1 \\ 0 & 0 & 1 & 1 & 2 \\ 0 & 2 & -1 & 0 & -1 \end{bmatrix}$$

The second column has a positive last entry, but it has no other positive entries. Thus, the simplex method fails to provide a solution. The smallest entry in the original payoff matrix is -1. The flow chart instructs us to add 1, the absolute value of -1, to each entry in the payoff matrix and try again. The matrix for the simplex method becomes

$$\begin{bmatrix} 2 & 0 & 1 & 0 & 1 \\ 0 & 2 & 0 & 1 & 1 \\ 1 & 1 & 0 & 0 & 0 \end{bmatrix}$$

Two operations yield the solution

$$\begin{bmatrix} 1 & 0 & 1/2 & 0 & 1/2 \\ 0 & 1 & 0 & 1/2 & 1/2 \\ 0 & 0 & -1/2 & -1/2 & -1 \end{bmatrix}$$

The answer for the minimization problem is

$$[\tfrac{1}{2} \quad \tfrac{1}{2}]$$

Since $\tfrac{1}{2} + \tfrac{1}{2} = 1$, the required modification does not affect this solution. The answer for the maximization problem is $\begin{bmatrix} \tfrac{1}{2} \\ \tfrac{1}{2} \end{bmatrix}$. The value of the optimized objective variable is 1. Since we added 1 to the payoff matrix, to find the value of the game we must subtract 1 from the reciprocal of this value: $\tfrac{1}{1} - 1 = 0$. Thus, each player should use each of her available strategies at random and equally often.

It can happen that the simplex method results in multiple solutions (see Section 6-4). In this case, the players will have a whole range of possible mixed strategies to choose from. We will not discuss this matter here.

See exercise 16, p. 345.

Exercises 6.1

Solve the following games:

A1. $\begin{bmatrix} 1 & 2 \\ 3 & 4 \end{bmatrix}$

2. $\begin{bmatrix} -1 & 3 \\ 2 & -2 \end{bmatrix}$

A3. $\begin{bmatrix} -1 & 4 \\ 5 & 2 \end{bmatrix}$

C4. $\begin{bmatrix} 2 & 3 \\ 4 & 4 \end{bmatrix}$

C5. $\begin{bmatrix} -1 & -2 \\ -3 & 1 \end{bmatrix}$

6. $\begin{bmatrix} 3 & 7 \\ 4 & 13 \end{bmatrix}$

C7. Player A and player B each write the number 1, 2, or 3 on a slip of paper. If the numbers turn out the same, the game is called a draw. If the numbers are different but their sum is even, B pays A $2. If the sum is odd, A pays B $1. How should each player proceed? What amount should B pay A before the game to make the game fair?

Solve the following games:

A8. $\begin{bmatrix} -1 & 2 & 0 & 3 \\ 2 & 1 & 1 & 3 \\ 4 & 2 & -1 & 2 \end{bmatrix}$

9. $\begin{bmatrix} 2 & 1 & 0 \\ 1 & 2 & 2 \\ 0 & 3 & 3 \end{bmatrix}$

10. $\begin{bmatrix} 1 & 1 & 2 & 2 \\ -1 & 2 & 3 & 2 \\ 2 & 0 & 1 & 0 \\ 0 & 3 & 3 & 3 \end{bmatrix}$

C11. $\begin{bmatrix} 1 & 3 & -4 & -1 & 1 & 0 \\ 2 & 3 & -3 & 2 & 1 & 2 \\ 4 & 3 & -1 & 3 & 2 & 3 \\ 2 & 1 & 0 & 5 & 4 & 4 \\ 2 & 1 & 4 & 4 & 3 & 5 \\ 5 & 4 & 0 & 4 & 3 & 4 \end{bmatrix}$

A 12. PHE is trying to assess the effects of two types of TV commercials in which they compare their products to those of their archcompetitor, MacDougal's (McD). They know that MacDougal's is also considering two types of commercials: the hard sell and the soft sell. The PHE advertising agency estimates that if PHE and McD both use hard sell, then PHE will get $6 million of McD business; if they both use soft sell, PHE will get $5 million; if PHE uses hard and McD uses soft, PHE will get $3 million; if PHE uses soft and McD uses hard, PHE will get $2 million. Both types of commercials are in the can for both companies. How should each company air the commercials, and what will be the result?

**Section 6-2
Introduction to Markov
Chains**

Appalachian Creations is not the only sporting goods company on the East Coast: the other big suppliers there are Adirondack Knapsacks (AK) and Pocono Productions (PP). These rival companies vie for dominance of the market. Thus, even though sales of the AC brand of freeze-dried backpacker foods were up in July, the managent was distressed by the report summarized in Table 2.

**Table 2
Fractions of Consumers'
Brand Switching (June–July)**

We read from Table 2, for example, that $2/3$ of AC's June customers remained with them in July, while $1/6$ switched to AK and $1/6$ switched to PP.

From	To		
	AC	AK	PP
AC	$2/3$	$1/6$	$1/6$
AK	$1/4$	$1/2$	$1/4$
PP	$1/8$	$7/8$	0

The management at AC was interested in the long-range effects of this pattern of brand switching. Knowing where current practices were leading would be helpful in directing sales campaigns. In the next section, we will learn how to compute the ultimate effect of such brand switching. In order for us to be able to solve this problem, we will have to assume that future tables of fractions of consumers' brand switching will be exactly the same as Table 2 in all of the months to come. This is not likely to happen. In fact, the VP in charge of sales at AC spent her entire July advertising budget trying to convince consumers not to buy AK foods. She was fairly successful. However, purchasers of PP foods were turned off by her approach and in the ensuing backlash, they purchased only AK foods in August (see Table 3).

Table 3
Fractions of Consumers'
Brand Switching (July–August)

From	To		
	AC	AK	PP
AC	$2/3$	0	$1/3$
AK	$1/2$	0	$1/2$
PP	0	1	0

In the next section we will learn how to compute the long-range effects of this buying trend as well, but as before we will have to assume that these consumer buying practices will persist forever. Although this is still not likely to happen, and even though buying practices will probably never be anything like those predicted, it is still useful to know the direction in which things are headed. Comparing the long-term effects of the practices of Tables 2 and 3 can give a more subtle evaluation of the advertising policies than would be given by merely checking gross sales. Is the advertising policy that resulted in Table 3 steering the company in a better direction than the policy that resulted in Table 2? If the practices of Table 2 persist, will it mean ruin for the company and, if so, did the VP help avert financial disaster or did she move the company closer to bankruptcy? We will return to these questions later.

As a second example of the kind of situation with which we will be dealing in these sections, consider what happened when AC decided to invest in a rather sophisticated machine for the automatic production of packframes. Such machines were available from two different companies: American Aluminum Tubing (AAT) and International Interstate Tubing (IIT). These machines were so complicated that breakdowns were common. The experience of several West Coast manufacturers revealed the following information: If the AAT machine made it through a week of operation without any breakdowns, then 75% of the time it would make it through the next week without any breakdowns. If it broke down during one week, it would be skillfully repaired and would break down again only 12.5% of the time the next week. The corresponding percentages for the IIT machine were 62.5% and 10%, respectively. Tables 4 and 5 summarize these data. Before purchasing a machine, AC wants to know how frequently each machine will break down in the long run.

Table 4
AAT Breakdown Rate (%)

From	To	
	Not broken	Broken
Not broken	75	25
Broken	87.5	12.5

Table 5
IIT Breakdown Rate (%)

From	To	
	Not broken	Broken
Not broken	62.5	37.5
Broken	90	10

Actually, AC plans to buy only one machine. To say that the AAT machine breaks down 25% of the time means that if AC had 100 working machines in one week, 25 of them would be expected to break down during the next week.

The mathematical model for situations like these is called a **Markov chain**, after the mathematician A. A. Markov, who invented the concept in 1907.

The examples above illustrate the two fundamental properties of a Markov chain:

(1) A Markov chain involves certain *states* (or positions) of individuals or things. Everyone or everything under consideration is always in one of these states.

(2) At regular intervals, transition occurs, that is, certain percentages of the individuals in each state move or change to another state. These *transition percentages* persist—that is, the percentage that moves from one given state to another state is the same each time transition occurs.

In our first example, there were three states and each backpacker was in exactly one of them: he or she was a consumer of AC, AK, or PP foods. In the second example, the machine was either broken or not broken.

In the first example, brand switching was checked regularly each month. In the second example, the machines were surveyed weekly. In applications to science, the time interval of a Markov chain can range anywhere from much less than a second to more than a thousand years.

While the assumption that the transition percentages persist limits the usefulness of Markov chains in business situations, we have already pointed out that they can at least give an idea of where things are headed. The model has found many more applications in the natural sciences, where the assumption of persistence is often quite realsitic.

Our two assumptions have the following consequences: If a matrix is made from the table of transition percentages (e.g., Tables 2,3, 4, or 5), then this matrix will be square and each of its rows will sum to 1 (or 100%)—that is, the entries in each row will add up to 1 (or 100%). It will be square because we are concerned with the transition percentage for moving from any state into any state. Each row sums to 1 (or 100%) because all the individuals in any given state move from that state into some state. A square matrix each of whose entries are between 0 and 1 and whose rows sum to 1 is sometimes called a **Markov matrix**. We are now ready to begin solving problems involving Markov chains.

In the next section we will complete the solution of the problems posed above. Our concern for the rest of this section will be with the computation of short-range effects of Markov chains.

Problem 1   Rather than be forced to resign, an unusually sneaky government employee decides to look busy. Since the warehouse he supervises stocks only obsolete equipment, he is never bothered by requisitions. He received his last shipment for storage 17 years ago. This particular warehouse consists of two large rooms and contains 14,400 items. Each week he labels $1/3$ of the stock in room A "For Shipment" and $1/4$ of the stock in room B "For Shipment." He then ships the indicated items to the other room. Supposing the material to be evenly divided when he begins this project, what will be the distribution after one week, two weeks, and three weeks?

Solution   One way to solve this problem is by use of a "tree" diagram (Figure 1). The fractions on each arrow indicate what part of the contents of the room at the tail of the arrow is shipped to the room at the head of the arrow. The letter O at the bottom refers to the "outside": when the material arrived at the warehouse, one half

Figure 1

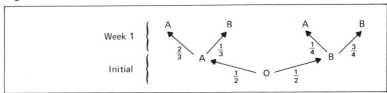

was sent to each room. Since the problem stipulates that we have 14,400 items altogether, it is possible to compute the numbers involved in each transfer. These numbers have been included in Figure 2.

By adding the numbers at the top of the tree in Figure 2, we see that after one week there are $4800 + 1800 = 6600$ items in room A and $2400 + 5400 = 7800$ items in room B. We can determine the distribution of goods after two weeks by simply extending the tree, as shown in Figure 3. Room A now contains $3200 + 600 + 1200 + 1350 = 6350$ items, and room B contains the other 8050 items.

Figure 2

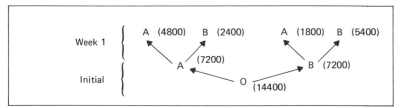

Figure 3

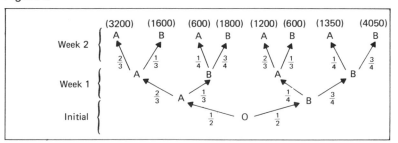

If we are interested only in the fraction of items in each room at the end of two weeks we could simply divide: $^{6350}/_{14400} = {}^{127}/_{288}$ in room A and $^{8050}/_{14400} = {}^{161}/_{288}$ in room B. However, there is a more direct way to obtain these fractions: For each letter at the top of the tree, multiply the fractions on each branch leading down from that letter to O. Figure 4 shows the result. The sum of the fractions at the A's is $^{2}/_{9} + {}^{1}/_{24} + {}^{1}/_{12} + {}^{3}/_{32} = {}^{127}/_{288}$, and the sum of the fractions at the B's is $^{1}/_{9} + {}^{1}/_{8} + {}^{1}/_{24} + {}^{9}/_{32} = {}^{161}/_{288}$. To obtain the actual number of items, we simply multiply:

$$\text{For A: } {}^{127}/_{288} \times 14,400 = 6350$$
$$\text{For B: } {}^{161}/_{288} \times 14,400 = 8050$$

Figure 4

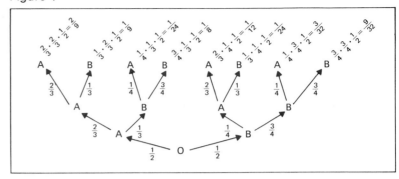

These computations were not too bad, but to extend the tree for a third week involves sixteen more branches. If we were interested in four or five weeks, these techniques would be very tedious. Fortunately, it is possible to use matrices to condense and simplify our work. We first write the transition-fraction table (Table 6) as a matrix:

Table 6
Transition Fractions

From	To	
	Room A	Room B
Room A	$2/_9$	$1/_3$
Room B	$1/_4$	$3/_4$

TRANSITION MATRIX

$$\begin{bmatrix} 2/_3 & 1/_3 \\ 1/_4 & 3/_4 \end{bmatrix}$$

Next, we write the initial-state table (Table 7) as a row matrix:

Table 7
Initial-state

	Room	
	A	B
Amount	7200	7200

INITIAL-STATE MATRIX

$$[7200 \quad 7200]$$

We now multiply the **intial-state matrix** and the **transition matrix**.

$$[7200 \quad 7200] \cdot \begin{bmatrix} {}^2\!/_3 & {}^1\!/_3 \\ {}^1\!/_4 & {}^3\!/_4 \end{bmatrix}$$

$$= [{}^2\!/_3 \times 7200 + {}^1\!/_4 \times 7200 \quad {}^1\!/_3 \times 7200 + {}^3\!/_4 \times 7200]$$
$$= [4800 + 1800 \quad 2400 + 5400]$$
$$= [6600 \quad 7800]$$

This answer is the same as that obtained above. Even the arithmetic is identical. We call the row matrix so obtained a **distribution matrix**, since it indicates the distribution of items, in this case at the end of the first week. Here is the general rule:

> **To obtain the distribution matrix for a certain period, multiply the distribution matrix for the previous period by the transition matrix.**

Thus, to obtain the distribution figures for the second week we multiply the Week 1 distribution matrix and the transition matrix:

$$[6600 \quad 7800] \begin{bmatrix} {}^2\!/_3 & {}^1\!/_3 \\ {}^1\!/_4 & {}^3\!/_4 \end{bmatrix} = [6350 \quad 8050]$$

Our answer matches that found above: at the end of the second week, there are 6350 items in room A and 8050 items in room B. It is now relatively simple to compute the distribution for the third week:

$$[6350 \quad 8050] \begin{bmatrix} {}^2\!/_3 & {}^1\!/_3 \\ {}^1\!/_4 & {}^3\!/_4 \end{bmatrix} = [6245{}^5\!/_6 \quad 8154{}^1\!/_6] \qquad \mathbf{1}$$

Fortunately, some of the items in the warehouse are liquid and can be divided to meet the conditions of this answer: $6245{}^5\!/_6$ items in room A and $8154{}^1\!/_6$ items in room B.

Suppose we were asked to compute the distribution of items in the sixth week. We could perform three more multiplications of the various distribution matrices. However, there is a slightly more direct way to obtain this answer. We first reconsider our procedure for obtaining the third week's distribution. For this we multiplied:

$$[6350 \quad 8050] \begin{bmatrix} {}^2\!/_3 & {}^1\!/_3 \\ {}^1\!/_4 & {}^3\!/_4 \end{bmatrix}$$

But [6350 8050] was itself obtained by a multiplication. Substituting, we obtain

$$[6350 \quad 8050]\begin{bmatrix} {}^2/_3 & {}^1/_3 \\ {}^1/_4 & {}^3/_4 \end{bmatrix} = \left([6600 \quad 7800]\begin{bmatrix} {}^2/_3 & {}^1/_3 \\ {}^1/_4 & {}^3/_4 \end{bmatrix}\right)\begin{bmatrix} {}^2/_3 & {}^1/_3 \\ {}^1/_4 & {}^3/_4 \end{bmatrix}$$

Finally, $[6600 \quad 7800]$ was obtained through a multiplication. If we substitute again, we obtain

$$[6350 \quad 8050]\begin{bmatrix} {}^2/_3 & {}^1/_3 \\ {}^1/_4 & {}^3/_4 \end{bmatrix}$$

$$= \left(\left([7200 \quad 7200]\begin{bmatrix} {}^2/_3 & {}^1/_3 \\ {}^1/_4 & {}^3/_4 \end{bmatrix}\right)\begin{bmatrix} {}^2/_3 & {}^1/_3 \\ {}^1/_4 & {}^3/_4 \end{bmatrix}\right)\begin{bmatrix} {}^2/_3 & {}^1/_3 \\ {}^1/_4 & {}^3/_4 \end{bmatrix}$$

Recall from Section 1-6 that the manner in which we insert parentheses is irrelevant. We can just as correctly write

$$[6350 \quad 8050]\begin{bmatrix} {}^2/_3 & {}^1/_3 \\ {}^1/_4 & {}^3/_4 \end{bmatrix}$$

$$= [7200 \quad 7200]\left(\begin{bmatrix} {}^2/_3 & {}^1/_3 \\ {}^1/_4 & {}^3/_4 \end{bmatrix}\begin{bmatrix} {}^2/_3 & {}^1/_3 \\ {}^1/_4 & {}^3/_4 \end{bmatrix}\begin{bmatrix} {}^2/_3 & {}^1/_3 \\ {}^1/_4 & {}^3/_4 \end{bmatrix}\right)$$

If we agree to call the product in parentheses the *third power* of the matrix, and write

$$\begin{bmatrix} {}^2/_3 & {}^1/_3 \\ {}^1/_4 & {}^3/_4 \end{bmatrix}\begin{bmatrix} {}^2/_3 & {}^1/_3 \\ {}^1/_4 & {}^3/_4 \end{bmatrix}\begin{bmatrix} {}^2/_3 & {}^1/_3 \\ {}^1/_4 & {}^3/_4 \end{bmatrix} = \begin{bmatrix} {}^2/_3 & {}^1/_3 \\ {}^1/_4 & {}^3/_4 \end{bmatrix}^3$$

just as we would for numbers, then the third week's distribution is just the product

$$[7200 \quad 7200]\begin{bmatrix} {}^2/_3 & {}^1/_3 \\ {}^1/_4 & {}^3/_4 \end{bmatrix}^3 = [7200 \quad 7200]\begin{bmatrix} {}^{203}/_{432} & {}^{229}/_{432} \\ {}^{229}/_{576} & {}^{347}/_{576} \end{bmatrix}$$

In other words, to obtain the distribution matrix for the third week we multiply the initial-state matrix by the third power of the transition matrix. Now if we want to know the distribution matrix for the sixth week, we need only multiply the initial-state matrix by the sixth power of the transition matrix:

$$[7200 \quad 7200]\begin{bmatrix} {}^2/_3 & {}^1/_3 \\ {}^1/_4 & {}^3/_4 \end{bmatrix}^6$$

To find the sixth power of our matrix, we simply multiply the third power by itself, that is,

$$\begin{bmatrix} {}^2/_3 & {}^1/_3 \\ {}^1/_4 & {}^3/_4 \end{bmatrix}^6$$

$$= \left(\begin{bmatrix} {}^2/_3 & {}^1/_3 \\ {}^1/_4 & {}^3/_4 \end{bmatrix}\begin{bmatrix} {}^2/_3 & {}^1/_3 \\ {}^1/_4 & {}^3/_4 \end{bmatrix}\begin{bmatrix} {}^2/_3 & {}^1/_3 \\ {}^1/_4 & {}^3/_4 \end{bmatrix}\right)\left(\begin{bmatrix} {}^2/_3 & {}^1/_3 \\ {}^1/_4 & {}^3/_4 \end{bmatrix}\begin{bmatrix} {}^2/_3 & {}^1/_3 \\ {}^1/_4 & {}^3/_4 \end{bmatrix}\begin{bmatrix} {}^2/_3 & {}^1/_3 \\ {}^1/_4 & {}^3/_4 \end{bmatrix}\right)$$

$$= \begin{bmatrix} {}^2\!/_3 & {}^1\!/_2 \\ {}^1\!/_4 & {}^3\!/_4 \end{bmatrix}^3 \begin{bmatrix} {}^2\!/_3 & {}^1\!/_2 \\ {}^1\!/_4 & {}^3\!/_4 \end{bmatrix}^3$$

$$= \begin{bmatrix} \dfrac{203}{432} & \dfrac{229}{432} \\[2mm] \dfrac{229}{576} & \dfrac{347}{576} \end{bmatrix} \begin{bmatrix} \dfrac{203}{432} & \dfrac{229}{432} \\[2mm] \dfrac{229}{576} & \dfrac{347}{576} \end{bmatrix}$$

$$= \begin{bmatrix} \dfrac{812}{1728} & \dfrac{916}{1728} \\[2mm] \dfrac{687}{1728} & \dfrac{1041}{1728} \end{bmatrix} \begin{bmatrix} \dfrac{812}{1728} & \dfrac{916}{1728} \\[2mm] \dfrac{687}{1728} & \dfrac{1041}{1728} \end{bmatrix}$$

$$= \dfrac{1}{1728} \, \dfrac{1}{1728} \begin{bmatrix} 812 & 916 \\ 687 & 1041 \end{bmatrix} \begin{bmatrix} 812 & 916 \\ 687 & 1041 \end{bmatrix}$$

$$= \dfrac{1}{2{,}985{,}984} \begin{bmatrix} 1{,}288{,}636 & 1{,}697{,}348 \\ 1{,}273{,}011 & 1{,}712{,}973 \end{bmatrix}$$

The final answer is then

$$[7200 \quad 7200] \left(\dfrac{1}{2{,}985{,}984} \begin{bmatrix} 1{,}288{,}636 & 1{,}697{,}348 \\ 1{,}273{,}011 & 1{,}712{,}973 \end{bmatrix} \right)$$

$$= \dfrac{1}{2{,}985{,}984} [7200 \quad 7200] \begin{bmatrix} 1{,}288{,}636 & 1{,}697{,}348 \\ 1{,}273{,}011 & 1{,}712{,}973 \end{bmatrix}$$

$$= \dfrac{7{,}200}{2{,}985{,}984} [1 \quad 1] \begin{bmatrix} 1{,}288{,}636 & 1{,}697{,}348 \\ 1{,}273{,}011 & 1{,}712{,}973 \end{bmatrix}$$

$$= \dfrac{25}{10368} [2{,}561{,}647 \quad 3{,}410{,}321]$$

$$= \left[6176\dfrac{8407}{10368} \quad 8223\dfrac{1961}{10368} \right]$$

Notice that the distribution has not changed much since the third week (shown in equation **1**). In the next section, we will learn that making more and more shifts moves the distribution closer and closer to the theoretical long-range distribution

$$[6171{}^3\!/_7 \quad 8228{}^4\!/_7]$$

We could just as easily have worked this problem under the assumption that the stock was initially divided in some other proportion. Suppose for example that ${}^1\!/_3$ of the stock was initially in room A and ${}^2\!/_3$ in room B. In this case, the initial-state matrix would be

$$[4800 \quad 9600]$$

We would use this matrix in place of the matrix $[7200 \quad 7200]$.

Problem 2   Mr. Olney has taught his second-grade class a sand-passing game to help them learn about fractions. The game is for three players, whom we will refer to as A, B, and C. Each of the players is to sit before a large box of sand with a measuring scoop. Every five minutes Mr. Olney rings his bell. When he does, each player transfers a certain fraction of his or her sand to each of the other two players. Player A gives $1/6$ of his sand to player B, $1/6$ of his sand to player C, and keeps the other $2/3$ for himself. Player B gives $1/4$ of his sand to A, $1/4$ to C, and keeps $1/2$ of it for himself. Player C gives $1/8$ of her sand to A and $7/8$ to B. She keeps none for herself. The passing of the sand constitutes one round. The game consists of three 5-minute rounds. Before the start of the game, Mr. Olney pours unequal amounts of sand into the three boxes. The object of the game is to figure out where to sit so as to have the most sand at the end of the three rounds. Suppose that for the first four games Mr. Olney distributes the sand and the players sit as shown in Table 8. Who will win each of the games?

Table 8
Fractional Distribution of Sand

Game	Player		
	A	B	C
1	$1/3$	$1/3$	$1/3$
2	$1/2$	0	$1/2$
3	$3/4$	0	$1/4$
4	0	0	1

Solution   The transition matrix for the game is

$$\begin{bmatrix} 2/3 & 1/6 & 1/6 \\ 1/4 & 1/2 & 1/4 \\ 1/8 & 7/8 & 0 \end{bmatrix}$$

Since the game lasts for three rounds, we are interested in the third power of this matrix:

$$\begin{bmatrix} 2/3 & 1/6 & 1/6 \\ 1/4 & 1/2 & 1/4 \\ 1/8 & 7/8 & 0 \end{bmatrix}^3 = \begin{bmatrix} 2/3 & 1/6 & 1/6 \\ 1/4 & 1/2 & 1/4 \\ 1/8 & 7/8 & 0 \end{bmatrix}\begin{bmatrix} 2/3 & 1/6 & 1/6 \\ 1/4 & 1/2 & 1/4 \\ 1/8 & 7/8 & 0 \end{bmatrix}\begin{bmatrix} 2/3 & 1/6 & 1/6 \\ 1/4 & 1/2 & 1/4 \\ 1/8 & 7/8 & 0 \end{bmatrix}$$

$$= \begin{bmatrix} 73/144 & 49/144 & 22/144 \\ 31/96 & 49/96 & 16/96 \\ 29/96 & 44/96 & 23/96 \end{bmatrix}\begin{bmatrix} 2/3 & 1/6 & 1/6 \\ 1/4 & 1/2 & 1/4 \\ 1/8 & 7/8 & 0 \end{bmatrix}$$

$$= \frac{1}{16} \begin{bmatrix} \frac{73}{9} & \frac{49}{9} & \frac{22}{9} \\ \frac{31}{6} & \frac{49}{6} & \frac{16}{6} \\ \frac{29}{6} & \frac{44}{6} & \frac{23}{6} \end{bmatrix} \begin{bmatrix} \frac{2}{3} & \frac{1}{6} & \frac{1}{6} \\ \frac{1}{4} & \frac{1}{2} & \frac{1}{4} \\ \frac{1}{8} & \frac{7}{8} & 0 \end{bmatrix}$$

$$= \frac{1}{16} \begin{bmatrix} \frac{1528}{216} & \frac{1342}{216} & \frac{586}{216} \\ \frac{838}{144} & \frac{1048}{144} & \frac{418}{144} \\ \frac{797}{144} & \frac{1127}{144} & \frac{380}{144} \end{bmatrix}$$

$$= \frac{1}{16 \cdot 72 \cdot 6} \begin{bmatrix} 3056 & 2684 & 1172 \\ 2514 & 3144 & 1254 \\ 2391 & 3381 & 1140 \end{bmatrix}$$

In this problem we are not actually given the quantity of sand involved. Instead, we use the fractions in the four distribution matrices:

$[\frac{1}{3}$	$\frac{1}{3}$	$\frac{1}{3}]$	**2**
$[\frac{1}{2}$	0	$\frac{1}{2}]$	**3**
$[\frac{3}{4}$	0	$\frac{1}{4}]$	**4**
$[0$	0	$1]$	**5**

We multiply each of these distribution matrices by the third power of the transition matrix:

FOR MATRIX 2:

$$[\tfrac{1}{3} \quad \tfrac{1}{3} \quad \tfrac{1}{3}] \left(\frac{1}{16 \cdot 72 \cdot 6} \right) \begin{bmatrix} 3056 & 2684 & 1172 \\ 2514 & 3144 & 1254 \\ 2391 & 3381 & 1140 \end{bmatrix}$$

$$= \frac{1}{3} \cdot \frac{1}{16 \cdot 72 \cdot 6} [1 \quad 1 \quad 1] \begin{bmatrix} 3056 & 2684 & 1172 \\ 2514 & 3144 & 1254 \\ 2391 & 3381 & 1140 \end{bmatrix}$$

$$= \frac{1}{3} \cdot \frac{1}{16 \cdot 72 \cdot 6} [7961 \quad 9209 \quad 3566]$$

Without multiplying out the denominator of the fraction, we see that after three transitions B is ahead, A is in second place, and C is last.

FOR MATRIX 3:

$$[\tfrac{1}{2} \quad 0 \quad \tfrac{1}{2}] \left(\frac{1}{16\cdot72\cdot6}\right) \begin{bmatrix} 3056 & 2684 & 1172 \\ 2514 & 3144 & 1254 \\ 2391 & 3381 & 1140 \end{bmatrix}$$

$$= \frac{1}{2}\cdot\frac{1}{16\cdot72\cdot6} \ [1 \quad 0 \quad 1] \begin{bmatrix} 3056 & 2684 & 1172 \\ 2514 & 3144 & 1254 \\ 2391 & 3381 & 1140 \end{bmatrix}$$

$$= \frac{1}{16\cdot72\cdot6} \ [5447 \quad 6065 \quad 2312]$$

The order of finish is the same here as in the previous case.

FOR MATRIX 4:

$$[\tfrac{3}{4} \quad 0 \quad \tfrac{1}{4}] \left(\frac{1}{16\cdot72\cdot6}\right) \begin{bmatrix} 3056 & 2684 & 1172 \\ 2514 & 3144 & 1254 \\ 2391 & 3381 & 1140 \end{bmatrix}$$

$$= \frac{1}{4}\cdot\frac{1}{16\cdot72\cdot6} \ [3 \quad 0 \quad 1] \begin{bmatrix} 3056 & 2684 & 1172 \\ 2514 & 3144 & 1254 \\ 2391 & 3381 & 1140 \end{bmatrix}$$

$$= \frac{1}{4}\cdot\frac{1}{16\cdot72\cdot6} \ [11559 \quad 11443 \quad 4656]$$

In this case A wins, B finishes second, and C is last.

FOR MATRIX 5:

$$[0 \quad 0 \quad 1] \left(\frac{1}{16\cdot72\cdot6}\right) \begin{bmatrix} 3056 & 2684 & 1172 \\ 2514 & 3144 & 1254 \\ 2391 & 3381 & 1140 \end{bmatrix}$$

$$= \frac{1}{16\cdot72\cdot6} \ [2391 \quad 3381 \quad 1140]$$

This results in the same order of finish as with matrices **2** and **3**.

We remarked early in this section that in order to be classified as a Markov chain, a process must have a persistent transition matrix. While this hypothsis is essential in the analysis of long-range effects, we can also handle situations in which the transition matrix varies, provided we are only interested in short-term results.

Problem 3 I am the TV program manager for Channel 2 in a small Midwestern town. Our only competition is Channel 7. I decide to survey the Tuesday-night viewing habits of everyone who owns a set. Here are the results:

(1) At the start of the evening, $3/8$ of the potential viewers watch our 6:00–7:00 local news and $4/8$ watch Channel 7's news.

(2) At 7:00, $6/8$ of our viewers stay with us and $1/8$ change channels. Of Channel 7's viewers, $3/4$ stay with them, and the rest switch their sets off. Of the people who were not watching, $3/4$ switch on our national news and $1/4$ switch on Channel 7's.

(3) At 7:30, $3/4$ of our viewers remain faithful and the rest switch to Channel 7; $1/4$ of their viewers remain faithful and $1/2$ switch to us; anyone with his set off keeps it off.

What fraction of viewers will be watching our 7:30 show?

Solution This process has three states: a viewer either watches Channel 2, watches Channel 7, or has his set off. The initial-state matrix is

$$[\begin{matrix} 3/8 & 4/8 & 1/8 \end{matrix}]$$

The first transition matrix is:

$$\begin{bmatrix} 6/8 & 1/8 & 1/8 \\ 0 & 3/4 & 1/4 \\ 3/4 & 1/4 & 0 \end{bmatrix}$$

The second transition matrix is:

$$\begin{bmatrix} 3/4 & 1/4 & 0 \\ 1/2 & 1/4 & 1/4 \\ 0 & 0 & 1 \end{bmatrix}$$

To answer the question, we simply multiply:

$$[\begin{matrix} 3/8 & 4/8 & 1/8 \end{matrix}] \begin{bmatrix} 6/8 & 1/8 & 1/8 \\ 0 & 3/4 & 1/4 \\ 3/4 & 1/4 & 0 \end{bmatrix} \begin{bmatrix} 3/4 & 1/4 & 0 \\ 1/2 & 1/4 & 1/4 \\ 0 & 0 & 1 \end{bmatrix}$$

$$= (1/8)(1/8)(1/4) \; [\begin{matrix} 3 & 4 & 1 \end{matrix}] \begin{bmatrix} 6 & 1 & 1 \\ 0 & 6 & 2 \\ 6 & 2 & 0 \end{bmatrix} \begin{bmatrix} 3 & 1 & 0 \\ 2 & 1 & 1 \\ 0 & 0 & 4 \end{bmatrix}$$

$$= 1/256 \; [\begin{matrix} 3 & 4 & 1 \end{matrix}] \begin{bmatrix} 20 & 7 & 5 \\ 12 & 6 & 14 \\ 22 & 8 & 2 \end{bmatrix}$$

$$= {}^{1}\!/_{256} \begin{bmatrix} 130 & 53 & 73 \end{bmatrix}$$

We see that $^{130}\!/_{256}$ of the viewers will be watching our show; $^{53}\!/_{256}$ of the viewers will be watching Channel 7's show; and $^{73}\!/_{256}$ of the viewers will have their sets off.

> **The general procedure is this: to determine a subsequent distribution, multiply the initial state matrix by the appropriate product of transition matrices.**

Tree diagrams can handle even more complicated situations such as a game where a player drops out after losing three times in a row. In this case the number of states does not remain constant. We will not consider such matters in this book.

Exercises 6.2 C 1. Each week a dispatcher sends a rather large group of traveling salespeople to two towns: town A and town B. Where a salesperson is sent is determined by where he or she went in the previous week. Half the people who went to town A are sent back there and the rest are sent to town B. Everyone who went to town B is sent back to town B. Suppose the dispatcher begins by sending $^{1}\!/_{3}$ of all his salespeople to town A and $^{2}\!/_{3}$ to town B. What is the situation after six weeks?

A 2. In Exercise 1, what do you think will be the long-range effect of the dispatcher's policy?

3. A product is purchased by 128,000 consumers. Initially $^{1}\!/_{4}$ of them buy it from us and $^{3}\!/_{4}$ buy it from them. Each month we get $^{1}\!/_{2}$ of their customers and they get $^{1}\!/_{4}$ of ours. How many customers will we have after three months?

A 4. A charitable organization accepts donations from a special group of potential contributors in the amount of $2 and $5. Of those who did not contribute this month, $^{3}\!/_{4}$ will contribute $5 next month and the rest will contribute $2. Of these who contributed $2 this month, $^{1}\!/_{2}$ will contribute $5 next month and the rest will contribute nothing. Everyone who contributed $5 last month will contribute nothing this month. Assuming this trend continues, how much money can the charity expect to receive in the third month if their 768,000 potential contributors are initially evenly divided among the three possible states?

5. Market research predicts the following monthly brand-switching trends by consumers (Tables 9, 10, and 11). If $^{2}\!/_{5}$ of the consumers bought Brand X in June, what fraction (according to these transition matrices) will buy Brand X in September?

<table>
<tr><td rowspan="2">Table 9
June–July</td><td colspan="3">**From**   **To**</td></tr>
</table>

Table 9
June–July

From	To	
	Brand X	Brand Z
Brand X	$1/2$	$1/2$
Brand Z	$1/4$	$3/4$

Table 10
July–August

From	To	
	Brand X	Brand Z
Brand X	1	0
Brand Z	$1/3$	$2/3$

Table 11
August–September

From	To	
	Brand X	Brand Z
Brand X	$3/4$	$1/4$
Brand Z	0	1

c6. In its continuing competition with MacDougal's, Petaluma Hamburger Emporium is planning four consecutive weeks of discount sales of certain products. There are two different discount plans: (1) half-price beverages, and (2) free fries with any $1.50 purchase. It is believed that one week of plan 1 will result in PHE's retaining $2/3$ of its customers and McD's getting $1/3$ while McD retains $3/4$ of its customers and PHE gets $1/4$. One week of plan 2 is expected to result in PHE's retaining $1/2$ of its customers and McD's getting $1/2$ while McD retains $1/4$ of its customers and PHE gets $3/4$. Just before the four-week sale, PHE has twice as many customers as McD. What will the distribution be at the end of the sale under each of the following schemes of operation?

(a) first week, plan 1; second, plan 2; third, 1; fourth, 2

(b) first, 2; second, 1; third, 2; fourth, 1

(c) first and second, 1; third and fourth, 2

Comment on these results.

**Section 6-3
Regular and Absorbing
Markov Chains**

In this section, we will give procedures for detemining the long-range effects of two commonly occuring types of Markov chains: the regular chain and the absorbing chain. These are most easily defined in terms of their transition matrices. A chain is **regular** if some power of its transition matrix has no zero entries. For example, the transition matrix

$$\begin{bmatrix} 1/2 & 1/2 & 0 \\ 0 & 1/2 & 1/2 \\ 1/2 & 0 & 1/2 \end{bmatrix}$$

yields a regular chain, since its second power,

$$\begin{bmatrix} 1/2 & 1/2 & 0 \\ 0 & 1/2 & 1/2 \\ 1/2 & 0 & 1/2 \end{bmatrix}\begin{bmatrix} 1/2 & 1/2 & 0 \\ 0 & 1/2 & 1/2 \\ 1/2 & 0 & 1/2 \end{bmatrix} = \begin{bmatrix} 1/4 & 1/2 & 1/4 \\ 1/4 & 1/4 & 1/2 \\ 1/2 & 1/4 & 1/4 \end{bmatrix}$$

has no zero entries. The transition matrix

$$\begin{bmatrix} 1/4 & 3/4 \\ 1/2 & 1/2 \end{bmatrix}$$

also provides a regular chain, since its first power, namely itself, has no zero entries.

A state in a Markov chain is an **absorbing state** if its row in the transition matrix has a 1 on the main diagonal. For example, the first, third, and fourth states in the following transition matrix are absorbing states.

$$\begin{bmatrix} 1 & 0 & 0 & 0 \\ 1/4 & 1/2 & 0 & 1/4 \\ 0 & 0 & 1 & 0 \\ 0 & 0 & 0 & 1 \end{bmatrix} \qquad \textbf{1}$$

In an absorbing state, the 1 on the main diagonal indicates that at each transition everything in the absorbing state stays there. Once an item enters an absorbing state, it can never leave. A Markov chain is called an **absorbing chain** if some of the material in each state will eventually end up in an absorbing state. In the matrix

$$\begin{bmatrix} 0 & 1/2 & 1/2 \\ 0 & 1 & 0 \\ 1/2 & 0 & 1/2 \end{bmatrix} \qquad \textbf{2}$$

state 2 is an absorbing state. At each transition, $1/2$ of the material in state 1 moves into the absorbing state. While nothing moves directly from state 3 into state 2, in one transition $1/2$ of the material in state 3 moves into state 1 and from there $1/2$ of it moves into the absorbing state 2 on the next transition. Thus, this matrix defines an absorbing chain.

One way to determine whether or not a matrix yields an absorbing chain is by making a tree diagram, as follows:

(1) List all the states. In the Markov chain with transition matrix

Figure 5

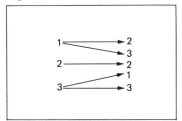

Figure 6

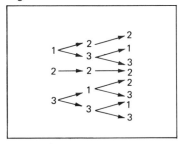

Figure 7

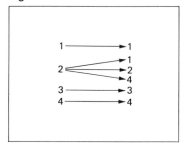

2 we have three:

$$
\begin{array}{c}
1 \\
2 \\
3
\end{array}
$$

(2) For each state listed, draw an arrow for each nonzero entry in its row. At the end of the arrow, place the number of the column of the nonzero entry. In our case, the first row has nonzero entries in columns 2 and 3; the second row has a nonzero entry only in the second column; the third row has nonzero entries in columns 1 and 3. We extend the tree (Figure 5).

(3) Repeat step 2 with the new list of states. Our case is shown in Figure 6. Continue until it is clear that either (a) there is a sequence of branches leading from each state on the left into an absorbing state, or (b) for some state on the left, it becomes apparent that there will never be a sequence of branches leading to an absorbing state.

In Figure 7, we carry out this procedure for the Markov chain with transition matrix **1**. In one step we move from each state into some absorbing state.

As a final example, consider the matrix

$$
\begin{bmatrix}
{}^{1}\!/_{2} & {}^{1}\!/_{4} & 0 & {}^{1}\!/_{4} & 0 \\
{}^{1}\!/_{2} & {}^{1}\!/_{4} & 0 & {}^{1}\!/_{8} & {}^{1}\!/_{8} \\
0 & 0 & 1 & 0 & 0 \\
0 & 0 & 0 & {}^{7}\!/_{8} & {}^{1}\!/_{8} \\
0 & 1 & 0 & 0 & 0
\end{bmatrix}
$$

While state 3 is an absorbing state, the fact that its column is cleared indicates that it is impossible to ever move into this state from another state. Since this is the only absorbing state, this matrix does not define an absorbing chain. We leave to the reader the task of drawing the appropriate tree diagram.

Our first problem involves a regular Markov chain:

Problem 1   At a party I offer peanuts and cashews in two huge bowls at opposite ends of my apartment. The bowl at the north end contains a mixture of $^1/_4$ peanuts and $^3/_4$ cashews. The bowl at the south end has a mixture of $^1/_2$ peanuts and $^1/_2$ cashews. My friend Luis loves peanuts but hates cashews. If he takes a nut from a bowl and it turns out to be a peanut, he selects another nut from the same bowl. Otherwise, he moves across the room to the other bowl. Luis eats nuts at the rate of one per minute. My party lasts for four hours. Where will he most likely be at the end of the party?

Solution   We label the bowls N and S. If Luis is at bowl N, then $1/4$ of the time he will get a peanut and remain there. Otherwise, he will move to bowl S. At Bowl S he will get a peanut about $1/2$ of the time. Otherwise, he will move to bowl N. Table 12 is the transition table for this problem.

Table 12
Transition Table

From	To	
	N	S
N	$1/4$	$3/4$
S	$1/2$	$1/2$

The associated matrix is

$$\begin{bmatrix} 1/4 & 3/4 \\ 1/2 & 1/2 \end{bmatrix}$$

The problem not only asks us to find the 240th power of this matrix, but has even neglected to supply us with an initial state matrix—that is, it has not mentioned how likely it is that Luis would begin at bowl N.

We could make a valiant attempt to compute

$$\begin{bmatrix} 1/4 & 3/4 \\ 1/2 & 1/2 \end{bmatrix}^{240}$$

$$\begin{bmatrix} 1/4 & 3/4 \\ 1/2 & 1/2 \end{bmatrix} \begin{bmatrix} 1/4 & 3/4 \\ 1/2 & 1/2 \end{bmatrix} = \begin{bmatrix} 7/16 & 9/16 \\ 3/8 & 5/8 \end{bmatrix} \quad \text{(the 2nd power)}$$

$$\begin{bmatrix} 7/16 & 9/16 \\ 3/8 & 5/8 \end{bmatrix} \begin{bmatrix} 7/16 & 9/16 \\ 3/8 & 5/8 \end{bmatrix} = \begin{bmatrix} 103/256 & 153/256 \\ 51/128 & 77/128 \end{bmatrix} \quad \text{(the 4th power)}$$

$$\begin{bmatrix} \dfrac{103}{256} & \dfrac{153}{256} \\ \dfrac{51}{128} & \dfrac{77}{128} \end{bmatrix} \begin{bmatrix} \dfrac{103}{256} & \dfrac{153}{256} \\ \dfrac{51}{128} & \dfrac{77}{128} \end{bmatrix} = \begin{bmatrix} \dfrac{26215}{65536} & \dfrac{39321}{65536} \\ \dfrac{13107}{32768} & \dfrac{19661}{32768} \end{bmatrix} \quad \text{(the 8th power)}$$

Maybe looking at decimals instead of fractions would simplify matters. We compute

The symbol $\doteq$ means "is approximately equal to."

$$\begin{bmatrix} 7/16 & 9/16 \\ 3/8 & 5/8 \end{bmatrix} \doteq \begin{bmatrix} .43 & .57 \\ .38 & .62 \end{bmatrix} \quad \text{(the 2nd power)}$$

$$\begin{bmatrix} \dfrac{103}{256} & \dfrac{153}{256} \\[2mm] \dfrac{51}{128} & \dfrac{77}{128} \end{bmatrix} \doteq \begin{bmatrix} .402 & .598 \\[2mm] .398 & .602 \end{bmatrix} \quad \text{(the 4th power)}$$

$$\begin{bmatrix} \dfrac{26215}{65536} & \dfrac{39321}{65536} \\[2mm] \dfrac{13107}{32768} & \dfrac{19661}{32768} \end{bmatrix} \doteq \begin{bmatrix} .400009 & .599991 \\[2mm] .399994 & .600006 \end{bmatrix} \quad \text{(the 8th power)}$$

As we raise the transition matrix to higher and higher powers, we seem to be getting closer and closer to the matrix

$$\begin{bmatrix} .4 & .6 \\ .4 & .6 \end{bmatrix} = \begin{bmatrix} {}^{2}\!/_{5} & {}^{3}\!/_{5} \\ {}^{2}\!/_{5} & {}^{3}\!/_{5} \end{bmatrix}$$

In fact, we get so close to this "limit" matrix that for all practical purposes we can assume

$$\begin{bmatrix} {}^{1}\!/_{4} & {}^{3}\!/_{4} \\ {}^{1}\!/_{2} & {}^{1}\!/_{2} \end{bmatrix}^{240} = \begin{bmatrix} {}^{2}\!/_{5} & {}^{3}\!/_{5} \\ {}^{2}\!/_{5} & {}^{3}\!/_{5} \end{bmatrix}$$

For a regular matrix, that is, a transition matrix of a regular Markov chain (p. 302), it turns out to be fairly easy to compute the limit without doing any matrix multiplications at all. We will return to this in a moment. Right now there is that other problem, the lack of an initial-state matrix. Let us try some particular examples. Suppose that at $^{1}\!/_{3}$ of my parties Luis begins his nut eating at the north end. The initial-state matrix is then

$$\begin{bmatrix} {}^{1}\!/_{3} & {}^{2}\!/_{3} \end{bmatrix}$$

and the final distribution matrix is

$$\begin{bmatrix} {}^{1}\!/_{3} & {}^{2}\!/_{3} \end{bmatrix} \begin{bmatrix} {}^{2}\!/_{5} & {}^{3}\!/_{5} \\ {}^{2}\!/_{5} & {}^{3}\!/_{5} \end{bmatrix} = \begin{bmatrix} {}^{2}\!/_{5} & {}^{3}\!/_{5} \end{bmatrix}$$

Thus, he is more likely to be at the south end. That is, he will end up at the south end at $^{3}\!/_{5}$ of my parties. Suppose we assume that initially he always goes to the north end. In this case the initial state matrix is

$$\begin{bmatrix} 1 & 0 \end{bmatrix}$$

and the final distribution is

$$\begin{bmatrix} 1 & 0 \end{bmatrix} \begin{bmatrix} {}^{2}\!/_{5} & {}^{3}\!/_{5} \\ {}^{2}\!/_{5} & {}^{3}\!/_{5} \end{bmatrix} = \begin{bmatrix} {}^{2}\!/_{5} & {}^{3}\!/_{5} \end{bmatrix}$$

which is the same as before. Here are a few other cases:

$$\begin{bmatrix} 1/8 & 7/8 \end{bmatrix} \begin{bmatrix} 2/5 & 3/5 \\ 2/5 & 3/5 \end{bmatrix} = \begin{bmatrix} 2/5 & 3/5 \end{bmatrix}$$

$$\begin{bmatrix} .0713 & .9297 \end{bmatrix} \begin{bmatrix} 2/5 & 3/5 \\ 2/5 & 3/5 \end{bmatrix} = \begin{bmatrix} 2/5 & 3/5 \end{bmatrix}$$

$$\begin{bmatrix} 0 & 1 \end{bmatrix} \begin{bmatrix} 2/5 & 3/5 \\ 2/5 & 3/5 \end{bmatrix} = \begin{bmatrix} 2/5 & 3/5 \end{bmatrix}$$

In fact, the final distribution will be the same, *no matter what the initial-state matrix is.* The limit matrix will always have identical rows and the final distribution will be the same as each of these rows.

This property of Markov chains is called *washing out.* In a regular Markov chain, the effect of the initial condition eventually disappears. We now turn to the question of computing the final distribution associated with a regular Markov chain. The flow chart for this procedure is given on p. 308. We will illustrate its use with the problem following.

This flow chart actually applies to a larger class of Markov chains than just the regular chains. In 1936 A. Kolmogorov showed that any *ergodic* Markov chain approaches a steady state or final distribution which can be found using this technique. The definition of an ergodic chain is too complicated to be discussed in detail here. The main condition is that each entry of the transition matrix must become nonzero in some power. However, not all entries need necessarily be nonzero in the *same* power, as is See Exercise 9.   the requirement for regular chains. This condition alone is not enough to ensure that the chain is ergodic.

Problem 2   What will be the long-range effect of the brand-switching trend described in Table 2 of the previous section?

Solution   The transition matrix from that table is

$$\begin{bmatrix} 2/3 & 1/6 & 1/6 \\ 1/4 & 1/2 & 1/4 \\ 1/8 & 7/8 & 0 \end{bmatrix}$$

This is a regular transition matrix, since the product

$$\begin{bmatrix} 2/3 & 1/6 & 1/6 \\ 1/4 & 1/2 & 1/4 \\ 1/8 & 7/8 & 0 \end{bmatrix} \begin{bmatrix} 2/3 & 1/6 & 1/6 \\ 1/4 & 1/2 & 1/4 \\ 1/8 & 7/8 & 0 \end{bmatrix} = \begin{bmatrix} 73/144 & 49/144 & 22/144 \\ 31/96 & 49/96 & 16/96 \\ 29/96 & 44/96 & 23/96 \end{bmatrix}$$

is a matrix with no zero entries. Following the instructions of the flow chart, we write the transpose of the transition matrix

$$\begin{bmatrix} {}^2\!/_3 & {}^1\!/_4 & {}^1\!/_8 \\ {}^1\!/_6 & {}^1\!/_2 & {}^7\!/_8 \\ {}^1\!/_6 & {}^1\!/_4 & 0 \end{bmatrix}$$

subtract 1 from each entry on its main diagonal,

$$\begin{bmatrix} -{}^1\!/_3 & {}^1\!/_4 & {}^1\!/_8 \\ {}^1\!/_6 & -{}^1\!/_2 & {}^7\!/_8 \\ {}^1\!/_6 & {}^1\!/_4 & -1 \end{bmatrix}$$

**Computing the Final
Distribution of a Regular
Markov Chain**

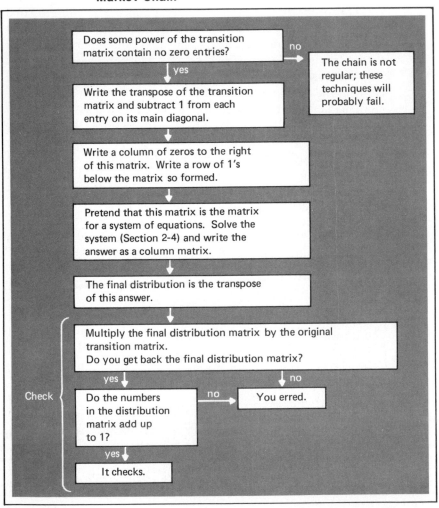

add a column of zeros

$$\begin{bmatrix} -1/3 & 1/4 & 1/8 & 0 \\ 1/6 & -1/2 & 7/8 & 0 \\ 1/6 & 1/4 & -1 & 0 \end{bmatrix}$$

and finally, insert a row of 1's beneath:

$$\begin{bmatrix} -1/3 & 1/4 & 1/8 & 0 \\ 1/6 & -1/2 & 7/8 & 0 \\ 1/6 & 1/4 & -1 & 0 \\ 1 & 1 & 1 & 1 \end{bmatrix}$$

Treating this as a system of four equations in three unknowns, we solve by Gauss-Jordan reduction:

$$\begin{bmatrix} 1 & -3/4 & -3/8 & 0 \\ 0 & -3/8 & 15/16 & 0 \\ 0 & 3/8 & -15/16 & 0 \\ 0 & 7/4 & 11/8 & 1 \end{bmatrix}$$

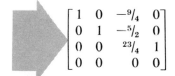

$$\begin{bmatrix} 1 & 0 & -9/4 & 0 \\ 0 & 1 & -5/2 & 0 \\ 0 & 0 & 23/4 & 1 \\ 0 & 0 & 0 & 0 \end{bmatrix}$$

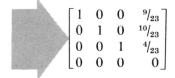

$$\begin{bmatrix} 1 & 0 & 0 & 9/23 \\ 0 & 1 & 0 & 10/23 \\ 0 & 0 & 1 & 4/23 \\ 0 & 0 & 0 & 0 \end{bmatrix}$$

The answer is $\begin{bmatrix} 9/23 \\ 10/23 \\ 4/23 \end{bmatrix}$. Thus, the long-range distribution will be

$$\begin{bmatrix} 9/23 & 10/23 & 4/23 \end{bmatrix}$$

Applying this to the original problem, we see that if consumer trends continue, AC eventually will control $9/23$ of the market, AK will control $10/23$, and PP will control $4/23$.

We check this answer by computing

$$\begin{bmatrix} 9/23 & 10/23 & 4/23 \end{bmatrix} \begin{bmatrix} 2/3 & 1/6 & 1/6 \\ 1/4 & 1/2 & 1/4 \\ 1/8 & 7/8 & 0 \end{bmatrix} = \begin{bmatrix} 9/23 & 10/23 & 4/23 \end{bmatrix}$$

Analyzing this checking procedure should remove some of the mystery from the flow chart. All that the procedure involves is solving the matrix equation

$$[x_1 \quad x_2 \quad x_3] \begin{bmatrix} 2/3 & 1/6 & 1/6 \\ 1/4 & 1/2 & 1/4 \\ 1/8 & 7/8 & 0 \end{bmatrix} = [x_1 \quad x_2 \quad x_3]$$

with the side condition that $x_1 + x_2 + x_3 = 1$.

Our answer also tells us that as we raise the matrix

$$\begin{bmatrix} 2/3 & 1/6 & 1/6 \\ 1/4 & 1/2 & 1/4 \\ 1/8 & 7/8 & 0 \end{bmatrix}$$

to successively higher powers, we move closer and closer to the limit matrix

$$\begin{bmatrix} 9/23 & 10/23 & 4/23 \\ 9/23 & 10/23 & 4/23 \\ 9/23 & 10/23 & 4/23 \end{bmatrix}$$

all of whose rows are the same as the final distribution matrix.

Problem 3   What will be the long-range effect of the brand-switching trend described in Table 3 of the previous section?

Solution   The transition matrix for this problem is given by

$$\begin{bmatrix} 2/3 & 0 & 1/3 \\ 1/2 & 0 & 1/2 \\ 0 & 1 & 0 \end{bmatrix}$$

This transition matrix is regular since, as you may check, its third power has no zero entries. The system to be solved is

$$\begin{bmatrix} -1/3 & 1/2 & 0 & 0 \\ 0 & -1 & 1 & 0 \\ 1/3 & 1/2 & -1 & 0 \\ 1 & 1 & 1 & 1 \end{bmatrix}$$

and the final distribution matrix is

$$[3/7 \quad 2/7 \quad 2/7]$$

Now AC will control $3/7$ of the market and the other two companies will share the rest evenly. Note that since $9/23 < 3/7$, the sales VP did indeed move the company in a better direction.

We now turn our attention to absorbing chains.

Problem 4   In 1960, when AC and AK were the only major Eastern sporting goods companies, a newcomer, Sporting and Recreation of Boston (SARB), caused quite a stir. Market research indicated that while hardly anyone had heard of SARB, practically everyone who bought their products was so impressed that they never again bought anything from AC or AK. From the brand-switching (transition) table for May–June 1960 (Table 13), we see that $1/2$ of AC's customers remained loyal, $1/4$ went to AK, and $1/4$ went to SARB. All of AK's customers went to AC and, as we have said, all of SARB's customers stayed with that company.

Table 13
Fractions of Consumer Brand
Switching (May–June 1960)

From	To		
	AC	AK	SARB
AC	$1/2$	$1/4$	$1/4$
AK	1	0	0
SARB	0	0	1

In May, the customers were divided thus: $1/2$ bought from AC, $3/8$ from AK, and $1/8$ from SARB. Assuming that this buying trend persists, what will be the distribution of consumers after sixteen months? What will happen in the long run?

Solution

From the table, we immediately obtain the transition matrix

$$\begin{bmatrix} 1/2 & 1/4 & 1/4 \\ 1 & 0 & 0 \\ 0 & 0 & 1 \end{bmatrix}$$

Since the problem asks us to determine the situation in sixteen months, we raise this matrix to the 16th power, as follows:

$$\begin{bmatrix} 1/2 & 1/4 & 1/4 \\ 1 & 0 & 0 \\ 0 & 0 & 1 \end{bmatrix}\begin{bmatrix} 1/2 & 1/4 & 1/4 \\ 1 & 0 & 0 \\ 0 & 0 & 1 \end{bmatrix} = \begin{bmatrix} 1/2 & 1/8 & 3/8 \\ 1/2 & 1/4 & 1/4 \\ 0 & 0 & 1 \end{bmatrix} \quad \text{(the 2nd power)}$$

$$\begin{bmatrix} 1/2 & 1/8 & 3/8 \\ 1/2 & 1/4 & 1/4 \\ 0 & 0 & 1 \end{bmatrix}\begin{bmatrix} 1/2 & 1/8 & 3/8 \\ 1/2 & 1/4 & 1/4 \\ 0 & 0 & 1 \end{bmatrix} = \begin{bmatrix} 10/32 & 3/32 & 19/32 \\ 3/8 & 1/8 & 1/2 \\ 0 & 0 & 1 \end{bmatrix} \quad \text{(the 4th power)}$$

$$\begin{bmatrix} 10/32 & 3/32 & 19/32 \\ 3/8 & 1/8 & 1/2 \\ 0 & 0 & 1 \end{bmatrix}\begin{bmatrix} 10/32 & 3/32 & 19/32 \\ 3/8 & 1/8 & 1/2 \\ 0 & 0 & 1 \end{bmatrix}$$

$$= \begin{bmatrix} 68/512 & 21/512 & 423/512 \\ 42/256 & 13/256 & 201/256 \\ 0 & 0 & 1 \end{bmatrix} \quad \text{(the 8th power)}$$

$$\begin{bmatrix} \dfrac{68}{512} & \dfrac{21}{512} & \dfrac{423}{512} \\[2mm] \dfrac{42}{256} & \dfrac{13}{256} & \dfrac{201}{256} \\[2mm] 0 & 0 & 1 \end{bmatrix} \begin{bmatrix} \dfrac{68}{512} & \dfrac{21}{512} & \dfrac{423}{512} \\[2mm] \dfrac{42}{256} & \dfrac{13}{256} & \dfrac{201}{256} \\[2mm] 0 & 0 & 1 \end{bmatrix}$$

$$= \begin{bmatrix} \dfrac{6388}{262144} & \dfrac{1974}{262144} & \dfrac{253782}{262144} \\[2mm] \dfrac{7896}{262144} & \dfrac{2440}{262144} & \dfrac{251808}{262144} \\[2mm] 0 & 0 & 1 \end{bmatrix} \quad \text{(the 16th power)}$$

If we replace the fractions in this matrix by decimal approximations we obtain the matrix

$$\begin{bmatrix} 0.0244 & 0.0075 & 0.9681 \\ 0.0301 & 0.0093 & 0.9606 \\ 0.0000 & 0.0000 & 1.0000 \end{bmatrix}$$

Note that after 16 months, 96.81% of AC's customers and 96.06% of AK's customers have switched to SARB. The distribution we are looking for is obtained by multiplying the initial-state matrix

$$\begin{bmatrix} \frac{1}{2} & \frac{3}{8} & \frac{1}{8} \end{bmatrix}$$

by the transition matrix:

$$\begin{bmatrix} \frac{1}{2} & \frac{3}{8} & \frac{1}{8} \end{bmatrix} \begin{bmatrix} 0.0244 & 0.0075 & 0.9681 \\ 0.0301 & 0.0093 & 0.9606 \\ 0.0000 & 0.0000 & 1.0000 \end{bmatrix}$$

$$= \begin{bmatrix} .0235 & .0073 & .9692 \end{bmatrix}$$
$$\doteq \begin{bmatrix} 2.3\% & 0.7\% & 96.92\% \end{bmatrix}$$

Thus, after 16 months, approximately 2.3% of the backpackers will be buying AC products, 0.7% will be buying AK products, and approximately 96.92% will be buying SARB products, SARB will have essentially cornered the market, and AC and AK will both be forced out of business. In the long run the distribution will be

$$\begin{bmatrix} 0 & 0 & 100\% \end{bmatrix}$$

Fortunately for the other companies, SARB began to have management problems and a little price war forced them to merge with AC before things got out of hand.

Another point to be made here is that the transition matrices gave AC an idea of how much time they had to act. For instance, in just two months the distribution of consumers would be

$$\begin{bmatrix} \frac{1}{2} & \frac{3}{8} & \frac{1}{8} \end{bmatrix} \begin{bmatrix} \frac{1}{2} & \frac{1}{8} & \frac{3}{8} \\ \frac{1}{2} & \frac{1}{4} & \frac{1}{4} \\ 0 & 0 & 1 \end{bmatrix} = \begin{bmatrix} \frac{14}{32} & \frac{5}{32} & \frac{13}{32} \end{bmatrix}$$

which indicates that SARB would already have had much of the market.

In the previous problem, the state represented by SARB was an absorbing state. While not many individuals switched to SARB at any one transition (only $\frac{1}{4}$ of AC's customers), once at SARB, they all stayed; that is, they were absorbed. Thus, SARB would slowly drain off all of the business. Actually, there are two kinds of absorbing states. One, like SARB's, has disastrous consequences for the competition. But the other is not so bad:

Problem 5   In August 1967, the management of AC nearly panicked when they heard that another company, Elite Klimbing Gear (EKG), had achieved 100% consumer loyalty. However, a short investigation disclosed that EKG's equipment was for show and was never intended for the rigors of the mountains. The consumers all understood this. None of AC's customers bought EKG's equipment, although $\frac{1}{2}$ did switch to AK. AK's customers reacted in the same way. What will happen as time goes on?

Solution   The transition table is Table 14.

Table 14
Fractions of Consumer Brand
Switching (August–September
1967)

From	To		
	AC	AK	EKG
AC	$\frac{1}{2}$	$\frac{1}{2}$	0
AK	$\frac{1}{2}$	$\frac{1}{2}$	0
EKG	0	0	1

The transition matrix is

$$\begin{bmatrix} \frac{1}{2} & \frac{1}{2} & 0 \\ \frac{1}{2} & \frac{1}{2} & 0 \\ 0 & 0 & 1 \end{bmatrix}$$

Squaring this matrix yields

$$\begin{bmatrix} \frac{1}{2} & \frac{1}{2} & 0 \\ \frac{1}{2} & \frac{1}{2} & 0 \\ 0 & 0 & 1 \end{bmatrix} \begin{bmatrix} \frac{1}{2} & \frac{1}{2} & 0 \\ \frac{1}{2} & \frac{1}{2} & 0 \\ 0 & 0 & 1 \end{bmatrix} = \begin{bmatrix} \frac{1}{2} & \frac{1}{2} & 0 \\ \frac{1}{2} & \frac{1}{2} & 0 \\ 0 & 0 & 1 \end{bmatrix}$$

Any higher power of this matrix will therefore just equal the original matrix. Thus, after the first month, no further change will occur. Suppose that AC begins with a customers, AK with b customers, and EKG with c customers. We compute:

$$[a \quad b \quad c] \begin{bmatrix} \frac{1}{2} & \frac{1}{2} & 0 \\ \frac{1}{2} & \frac{1}{2} & 0 \\ 0 & 0 & 1 \end{bmatrix} = [\frac{1}{2}(a + b) \quad \frac{1}{2}(a + b) \quad c]$$

Thus, after one month, and from then on, AC and AK will have the same number of customers while EKG will have only its original number of customers.

While this problem is similar to the previous one in that no one leaves the state EKG, it differs in that no one ever enters this state from another state. Note that the 1 on the main diagonal indicates that everyone in that state returns to it, and that whether or not its column is cleared indicates whether or not it can be entered from another state.

For absorbing chains, there are two cases to be considered. If there is only one absorbing state, the initial conditions wash out and eventually all of the distribution ends up being absorbed into that state. If there is more than one absorbing state, then while everything is absorbed into one of these states, nonabsorbing states will send a certain fraction to each absorbing state. For example, in the absorbing transition matrix

$$\begin{bmatrix} 1 & 0 & 0 & 0 \\ 0 & 1 & 0 & 0 \\ 0 & \frac{1}{2} & \frac{1}{4} & \frac{1}{4} \\ 0 & 0 & 0 & 1 \end{bmatrix}$$

everything in states 1, 2, or 4 stays there and some of the distribution initially in state 3 ends up being absorbed by one of these states. Since state 1 is isolated, nothing that did not start there will end up there. The only problem is to determine how states 2 and 4 share in the final distribution. The flow chart on p. 315 shows how to do this.

Problem 6 Given the initial distribution

$$[\frac{1}{8} \quad \frac{1}{2} \quad \frac{1}{4} \quad \frac{1}{8}]$$

what will be the long-run distribution of the Markov chain with
transition matrix

$$\begin{bmatrix} 1 & 0 & 0 & 0 \\ 0 & 1 & 0 & 0 \\ 0 & \frac{1}{2} & \frac{1}{4} & \frac{1}{4} \\ 0 & 0 & 0 & 1 \end{bmatrix}$$

Solution   There are three absorbing states: states 1, 2, and 4. Choosing
state 1, so that C is column 1, the flow chart instructs us to write

**To Determine the Fraction of
Each State in a Given Initial
Distribution that Finally Enters
a Given Absorbing State**

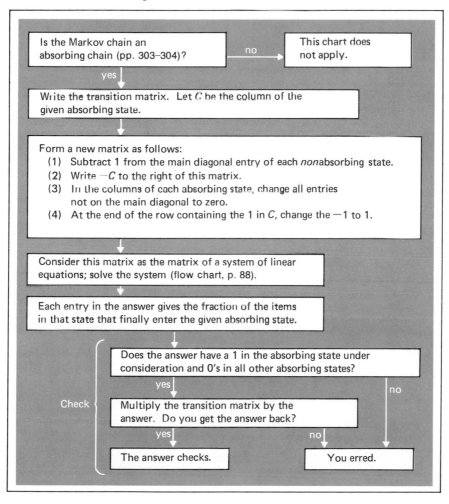

the transition matrix

$$
\begin{array}{c}
 \quad C \\
\begin{bmatrix}
1 & 0 & 0 & 0 \\
0 & 1 & 0 & 0 \\
0 & \frac{1}{2} & \frac{1}{4} & \frac{1}{4} \\
0 & 0 & 0 & 1
\end{bmatrix}
\end{array}
$$

subtract 1 from the main diagonal entry of each nonabsorbing state,

$$
\begin{bmatrix}
1 & 0 & 0 & 0 \\
0 & 1 & 0 & 0 \\
0 & \frac{1}{2} & -\frac{3}{4} & \frac{1}{4} \\
0 & 0 & 0 & 1
\end{bmatrix}
$$

write the negative of column C to the right,

$$
\begin{bmatrix}
1 & 0 & 0 & 0 & -1 \\
0 & 1 & 0 & 0 & 0 \\
0 & \frac{1}{2} & -\frac{3}{4} & \frac{1}{4} & 0 \\
0 & 0 & 0 & 1 & 0
\end{bmatrix}
$$

change all entries in C_1, C_2, and C_4 not on the main diagonal to zero,

$$
\begin{bmatrix}
1 & 0 & 0 & 0 & -1 \\
0 & 1 & 0 & 0 & 0 \\
0 & 0 & -\frac{3}{4} & 0 & 0 \\
0 & 0 & 0 & 1 & 0
\end{bmatrix}
$$

and change the -1 at the end of R_1 to 1:

$$
\begin{bmatrix}
1 & 0 & 0 & 0 & 1 \\
0 & 1 & 0 & 0 & 0 \\
0 & 0 & -\frac{3}{4} & 0 & 0 \\
0 & 0 & 0 & 1 & 0
\end{bmatrix}
$$

Next, we solve the system having this matrix. Obviously, the answer is

$$
\begin{bmatrix}
1 \\
0 \\
0 \\
0
\end{bmatrix}
$$

This indicates that all of the initial distribution in state 1 stays there and none enters from other states. We knew this was going to happen, since state 1 is an isolated absorbing state. Choosing state 2 results in the following more interesting sequence of matrices:

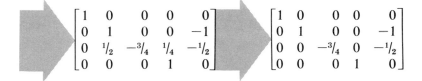

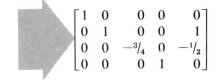

$$\begin{bmatrix} 1 & 0 & & 0 & 0 & & 0 \\ 0 & 1 & & 0 & 0 & & 1 \\ 0 & 0 & -3/4 & 0 & -1/2 \\ 0 & 0 & & 0 & 1 & & 0 \end{bmatrix}$$

We check this result by multiplying the original transition matrix by this column vector:

$$\begin{bmatrix} 1 & 0 & 0 & 0 \\ 0 & 1 & 0 & 0 \\ 0 & 1/2 & 1/4 & 1/4 \\ 0 & 0 & 0 & 1 \end{bmatrix} \begin{bmatrix} 0 \\ 1 \\ 2/3 \\ 0 \end{bmatrix} = \begin{bmatrix} 0 \\ 1 \\ 2/3 \\ 0 \end{bmatrix}$$

From this check we see that all the flow chart involves is the solution of the matrix equation

$$\begin{bmatrix} 1 & 0 & 0 & 0 \\ 0 & 1 & 0 & 0 \\ 0 & 1/2 & 1/4 & 1/4 \\ 0 & 0 & 0 & 1 \end{bmatrix} \begin{bmatrix} x_1 \\ x_2 \\ x_3 \\ x_4 \end{bmatrix} = \begin{bmatrix} x_1 \\ x_2 \\ x_3 \\ x_4 \end{bmatrix}$$

with the side conditions that $x_1 = 0$, $x_2 = 1$, and $x_4 = 0$.

Solving this system yields the answer

$$\begin{bmatrix} 0 \\ 1 \\ 2/3 \\ 0 \end{bmatrix}$$

Thus, nothing goes from states 1 or 4 into state 2. Everything in state 2 stays there, and two-thirds of the initial contents of state 3 moves to state 2.

Finally, the sequence for state 4 ends with the matrix

$$\begin{bmatrix} 1 & 0 & & 0 & 0 & & 0 \\ 0 & 1 & & 0 & 0 & & 0 \\ 0 & 0 & -3/4 & 0 & -1/4 \\ 0 & 0 & & 0 & 1 & & 1 \end{bmatrix}$$

with solution

$$\begin{bmatrix} 0 \\ 0 \\ 1/3 \\ 1 \end{bmatrix}$$

Thus, nothing goes from states 1 or 2 into state 4, one-third of the initial contents of state 3 goes to state 4, and, of course, all of the contents of state 4 remain there.

In conclusion, we see that $2/3$ of the initial contents of state 3 goes into state 2 and the rest goes into state 4. The initial distribution

$$[\tfrac{1}{8} \quad \tfrac{1}{2} \quad \tfrac{1}{4} \quad \tfrac{1}{8}]$$

indicates that $1/4$ of the distribution begins in state 3. Since $2/3 \times 1/4 = 1/6$ and $1/3 \times 1/4 = 1/12$, we must add $1/6$ to the contents of state 2 and $1/12$ to the contents of state 4. The final long-range distribution will then be

$$[\tfrac{1}{8} \quad \tfrac{2}{3} \quad 0 \quad \tfrac{5}{24}]$$

Exercises 6.3

^{c}1. Compute the fourth power of the matrix $\begin{bmatrix} \tfrac{2}{3} & 0 & \tfrac{1}{3} \\ \tfrac{1}{2} & 0 & \tfrac{1}{2} \\ 0 & 1 & 0 \end{bmatrix}$ and compute the product

$$[\tfrac{1}{8} \quad \tfrac{1}{2} \quad \tfrac{1}{4} \quad \tfrac{1}{8}] \begin{bmatrix} 1 & 0 & 0 & 0 \\ 0 & 1 & 0 & 0 \\ 0 & \tfrac{1}{2} & \tfrac{1}{4} & \tfrac{1}{4} \\ 0 & 0 & 0 & 1 \end{bmatrix}^4$$

Then compare your answers with the results predicted in the text. (To compare, change all the entries to decimals.)

^{c}2. Compute the long-range distribution associated with the transition matrix $\begin{bmatrix} \tfrac{1}{3} & \tfrac{1}{3} & \tfrac{1}{3} \\ \tfrac{1}{4} & \tfrac{1}{2} & \tfrac{1}{4} \\ 0 & 1 & 0 \end{bmatrix}$.

^{c}3. What long-range effect will the transition matrix

$$\begin{bmatrix} 1 & 0 & 0 & 0 & 0 \\ \tfrac{1}{4} & 0 & \tfrac{1}{2} & \tfrac{1}{4} & 0 \\ 0 & 0 & 1 & 0 & 0 \\ 0 & 0 & 0 & 1 & 0 \\ 0 & \tfrac{1}{3} & \tfrac{1}{3} & 0 & \tfrac{1}{3} \end{bmatrix}$$

have on the initial distribution $[\tfrac{1}{10} \quad \tfrac{1}{5} \quad \tfrac{3}{10} \quad \tfrac{3}{10} \quad \tfrac{1}{10}]$?

^{c}4. With reference to the second example of Section 6-2, which machine should AC buy?

5. What is the long-range distribution for the transition matrix $\begin{bmatrix} 0 & \tfrac{3}{4} & 0 & \tfrac{1}{4} \\ 0 & 0 & 1 & 0 \\ \tfrac{1}{2} & \tfrac{1}{2} & 0 & 0 \\ 0 & 0 & 0 & 1 \end{bmatrix}$?

6. What is the long-range distribution for the transition matrix $\begin{bmatrix} \tfrac{1}{3} & \tfrac{1}{3} & \tfrac{1}{3} \\ \tfrac{1}{2} & \tfrac{1}{2} & 0 \\ 0 & \tfrac{1}{2} & \tfrac{1}{2} \end{bmatrix}$?

7. What is the long-range effect of the transition matrix

$$\begin{bmatrix} 1/4 & 0 & 1/2 & 0 & 0 & 1/4 \\ 0 & 1 & 0 & 0 & 0 & 0 \\ 1/2 & 1/4 & 0 & 1/4 & 0 & 0 \\ 0 & 0 & 0 & 1 & 0 & 0 \\ 0 & 0 & 0 & 0 & 1 & 0 \\ 0 & 0 & 0 & 1/2 & 1/2 & 0 \end{bmatrix}$$

on the initial-distribution matrix $[\,1/12 \quad 2/12 \quad 3/12 \quad 4/12 \quad 1/12 \quad 1/12\,]$?

^8. What is the long-range effect of the transition matrix

$$\begin{bmatrix} 7/8 & 1/8 & 0 \\ 0 & 15/16 & 1/16 \\ 7/8 & 0 & 1/8 \end{bmatrix}$$

on any given initial-distribution matrix?

9. Compute the first five powers of the transition matrix $\begin{bmatrix} 0 & 1 & 0 & 0 \\ 0 & 0 & 1 & 0 \\ 0 & 0 & 0 & 1 \\ 1 & 0 & 0 & 0 \end{bmatrix}$.

What can you conclude about the long-range effect of such a transition matrix? (Note: this matrix satisfies the main condition for
ergodic matrices. It is, however, not ergodic.)

*10. What is the long-range effect of the transition matrix

$$\begin{bmatrix} 1 & 0 & 0 & 0 \\ 0 & 1/3 & 1/3 & 1/3 \\ 0 & 1/4 & 1/2 & 1/4 \\ 0 & 0 & 1 & 0 \end{bmatrix}$$

on the initial distribution $[\,1/4 \quad 1/8 \quad 1/2 \quad 1/8\,]$?

^11. The three PHE stores in Little Rock are keeping track of their customers. Every week, $1/2$ of the customers in store 1 switch to store 2
and $1/3$ to store 3; $5/6$ of the customers in store 2 switch to store 1; $1/5$
of the customers in store 3 switch to store 1 and $3/5$ to store 2. The
total pool of customers stays the same: 1200 kids, 50 parents, and 1
retired dentist. In the long run, how many customers will each store
have?

Section 6-4
More on the Simplex
Method

In this section we consider two ways in which the simplex
method can be extended to a wider class of linear programming
problems, namely: (1) problems that were ruled out by Conditions 2 or 3 in Sections 5-3 and 5-5, and (2) problems that have
more than one answer. We also explain another technique for
solving LP minimization problems using the simplex method.

Introduction to Mixed Constraints

The operations research group at Appalachian Creations was
about to tackle the following simple LP problem: Fluids A and

B are to be mixed to make a new type of sunscreen. Sale of 1 oz. of A, when included in the mixture, yields a \$3 profit, and sale of 1 oz. of B in the mixture yields a \$4 profit. Each oz. of A contains 3 milligrams (mg) of a harmful chemical; each oz. of B contains 5 mg of the chemical. If a tube of the mixture can weigh at most 4 oz. and contain at most 15 mg of the bad chemical, how does AC maximize profit?

It was routine for the staff to set up this problem. Let x_1, x_2 be the number of ounces of A, B, respectively, in a tube. The staff were required to

MAXIMIZE

$$y = 3x_1 + 4x_2$$

SUBJECT TO

$$x_1 \geq 0, \; x_2 \geq 0$$
$$x_1 + \; x_2 \leq \; 4 \qquad\qquad \mathbf{1}$$
$$3x_1 + 5x_2 \leq 15 \qquad\qquad \mathbf{2}$$

Just as they were about to start using the simplex method, the following memo arrived: "One oz. fluid A contains 1 g (gram) of a very beneficial chemical, and each oz. of B contains 2 g of the chemical. Every tube must contain at least 2 g of this stuff in order for us to mount a good ad campaign." In terms of the above notation, this meant that there was now an additional constraint:

$$x_1 + 2x_2 \geq 2 \qquad\qquad \mathbf{3}$$

Figure 8

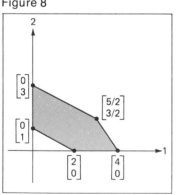

This shook them up a little. The flow chart in Section 5-3 does not allow a constraint like constraint **3** (with a $\geq$ sign) in an LP maximization problem. Fortunately this problem involves only two variables. Thus, the staff was able to graph the region of feasibility (Figure 8), check the corner points, and find the answer: $x_1 = {}^5\!/_2$, $x_2 = {}^3\!/_2$, $y = {}^{27}\!/_2$. So each tube should contain 2.5 oz. fluid A and 1.5 oz. fluid B for a profit of \$13.50.

But the staff continued to worry. What if they were given an LP maximization problem with more than two variables that had *mixed constraints*, that is, where some constraints had $\leq$ signs, some had $\geq$ signs, and maybe even some had $=$ signs?

A Simple Technique

See Appendix A, Sections 8 and 10.

First we must learn how to reverse inequality signs by multiplying by -1. The effect of such a multiplication on an inequality is to change $\leq$ to $\geq$ and $\geq$ to $\leq$.

Problem 1   Rewrite the constraint

$$4x_1 - 3x_2 + \tfrac{2}{3}x_3 \geq \tfrac{5}{2} \qquad\qquad \textbf{4}$$

as an equivalent constraint involving a $\leq$ sign.

Solution   Multiply both sides of constraint **4** by -1; then $\geq$ must be changed to $\leq$:

$$-4x_1 + 3x_2 - \tfrac{2}{3}x_3 \leq -\tfrac{5}{2} \qquad\qquad \textbf{5}$$

Phase I   We are now able to rewrite the sunscreen problem so that all constraints conform to Condition 2 of Section 5-3.

MAXIMIZE

$$y = 3x_1 + 4x_2$$

SUBJECT TO

$$x_1 \geq 0,\ x_2 \geq 0$$
$$x_1 + x_2 \leq 4$$
$$3x_1 + 5x_2 \leq 15$$
$$-x_1 - 2x_2 \leq -2$$

However, the third constraint fails to satisfy Condition 3. Condition 3 is important because it guarantees that the origin is a feasible point. That is, since 0 is less than any positive number, substituting 0 for each variable will make the left side of each constraint 0, satisfying the constraint, unless, of course, the constraint bound is negative.

The simplex method begins with a feasible point and moves along the edges of the feasible region until it finds the corner at which the objective variable is optimized. In Chapter 5, we always set out from the origin. In the original AC example in this section, the first matrix for the simplex method is

$$\begin{bmatrix} 1 & 1 & 1 & 0 & 4 \\ 3 & 5 & 0 & 1 & 15 \\ 3 & 4 & 0 & 0 & 0 \end{bmatrix}$$

If we read the answer from this matrix, we see immediately that $x_1 = 0$, $x_2 = 0$, since their columns are uncleared. For the modified problem, the matrix is

$$\begin{bmatrix} 1 & 1 & 1 & 0 & 0 & 4 \\ 3 & 5 & 0 & 1 & 0 & 15 \\ -1 & -2 & 0 & 0 & 1 & -2 \\ 3 & 4 & 0 & 0 & 0 & 0 \end{bmatrix}$$

The answer still reads $x_1 = 0$, $x_2 = 0$, but this time such a result is unacceptable since it is not feasible.

Checking the graph (Figure 8), we notice that there is an edge leading from the origin to a corner of the feasible region. In fact, we could travel away from $\begin{bmatrix} 0 \\ 0 \end{bmatrix}$ along either axis to reach such a corner. Since the simplex method moves along edges, is it possible to modify the technique so that it can find a first feasible solution? Yes, it is. The new technique is called Phase I (see the flow chart on p. 323.

Problem 2 Maximize $y = 3x_1 + 4x_2$ subject to

$$x_1 \geq 0, \ x_2 \geq 0$$
$$x_1 + \ x_2 \leq \ 4$$
$$3x_1 + 5x_2 \leq 15$$
$$x_1 + 2x_2 \geq \ 2$$

Solution 1 We have already set up the first matrix:

$$\begin{array}{c} \ \ \ \ \ C \\ R \begin{bmatrix} 1 & 1 & 1 & 0 & 0 & 4 \\ 3 & 5 & 0 & 1 & 0 & 15 \\ -1^* & -2 & 0 & 0 & 1 & -2 \\ 3 & 4 & 0 & 0 & 0 & 0 \end{bmatrix} \end{array}$$

We see that R_3 is the only row with a negative last entry. Thus, R_3 becomes R. There are two choices for C. We pick $C = C_1$. We note that there is no row except the last below R_3. We use R to clear C:

$$\begin{bmatrix} 0 & -1 & 1 & 0 & 1^* & 2 \\ 0 & -1 & 0 & 1 & 3 & 9 \\ 1 & 2 & 0 & 0 & -1 & 2 \\ 0 & -2 & 0 & 0 & 3 & -6 \end{bmatrix}$$

Now there are no negative entries (except the last entry) in the last column. We apply the familiar simplex method

$$\begin{bmatrix} 0 & -1 & 1 & 0 & 1 & 2 \\ 0 & 2^* & -3 & 1 & 0 & 3 \\ 1 & 1 & 1 & 0 & 0 & 4 \\ 0 & 1 & -3 & 0 & 0 & -12 \end{bmatrix} \quad \mathbf{1}$$

$$\begin{bmatrix} 0 & 0 & -1/2 & 1/2 & 1 & 7/2 \\ 0 & 1 & -3/2 & 1/2 & 0 & 3/2 \\ 1 & 0 & 5/2 & -1/2 & 0 & 5/2 \\ 0 & 0 & -3/2 & -1/2 & 0 & -27/2 \end{bmatrix}$$

from which we read the answer

$$x_1 = {}^5/_2, \; x_2 = {}^3/_2, \; \text{and} \; y = {}^{27}/_2$$

as before.

Solution 2

To further illustrate the flow chart, we interchange the first and last constraints and solve this problem again. Now the first matrix

Phase I: Finding an Initial Feasible Solution for a Maximization Problem

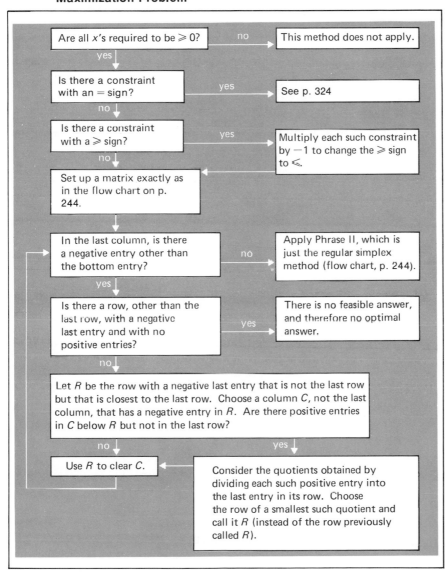

is

$$\begin{array}{c} \quad\;\; C \\ R\begin{bmatrix} -1 & -2 & 1 & 0 & 0 & -2 \\ 3 & 5 & 0 & 1 & 0 & 15 \\ 1^* & 1 & 0 & 0 & 1 & 4 \\ 3 & 4 & 0 & 0 & 0 & 0 \end{bmatrix} \end{array}$$

Again, there is only one row with a negative entry. We let R be R_1. If we again pick C_1 to be C, we now have a situation in which there are positive entries in C below R. The ratios are $^{15}/_3 = 5$ and $^4/_1 = 4$. We must choose the 1. Clearing yields the matrix

$$\begin{bmatrix} 0 & -1 & 1 & 0 & 1 & 2 \\ 0 & 2^* & 0 & 1 & -3 & 3 \\ 1 & 1 & 0 & 0 & 1 & 4 \\ 0 & 1 & 0 & 0 & -3 & -12 \end{bmatrix}$$

This is matrix **1** with C_3 and C_5 interchanged. The problem will be finished after one more clearing.

Equality Constraints If one or more of the constraints is an equation rather than an inequality, there are several ways to proceed. One way is simply to omit the appropriate slack variables. Recall from Section 5-4 that we used slack variables to change each inequality into an equality. If we already have an equality, we merely omit the slack variable by deleting a column from the identity matrix. For example, look at this problem:

Problem 3 Maximize $y = 7x_1 + 2x_2$ subject to

$$x_1 \geq 0, x_2 \geq 0$$
$$8x_1 + 3x_2 \leq 24$$
$$x_1 + x_2 \leq 4$$
$$-x_1 + x_2 = 2$$

Solution We need only two slack variables, x_3 and x_4:

$$\begin{bmatrix} 8x_1 + 3x_2 + x_3 & & = 24 \\ x_1 + x_2 & + x_4 = & 4 \\ -x_1 + x_2 & & = 2 \end{bmatrix}$$

The initial matrix is

$$\begin{bmatrix} 8 & 3 & 1 & 0 & 24 \\ 1 & 1 & 0 & 1 & 4 \\ -1 & 1^* & 0 & 0 & 2 \\ 7 & 2 & 0 & 0 & 0 \end{bmatrix}$$

Be careful. Direct application of the simplex method here will generally fail, because while this problem does not require three slack variables, it can be shown that the simplex method requires the presence of all three columns of the identity matrix. (They may appear in any order.)

**Rules for Problems
Involving Equations**

(1) Omit the columns of the identity matrix that corresponds to the rows of the equations.
(2) Use Gauss-Jordan reduction to clear columns to restore the missing columns of the identity matrix.
(3) Use Phase I, if necessary, and then complete the problem using the simplex method.

In our case we need a third column for the identity matrix. We arbitrarily choose the second column and clear it with the third row:

$$\begin{bmatrix} 11 & 0 & 1 & 0 & 18 \\ 2^* & 0 & 0 & 1 & 2 \\ -1 & 1 & 0 & 0 & 2 \\ 9 & 0 & 0 & 0 & -4 \end{bmatrix}$$

There is no need for Phase I. We merely apply the simplex method:

$$\begin{bmatrix} 0 & 0 & 1 & -11/2 & 7 \\ 1 & 0 & 0 & 1/2 & 1 \\ 0 & 1 & 0 & 1/2 & 3 \\ 0 & 0 & 0 & -9/2 & -13 \end{bmatrix}$$

We obtain $x_1 = 1$, $x_2 = 3$, $y = 13$.

Problem 4 Maximize $y = 2x_1 + x_2$ subject to

$$x_1 \geq 0, \ x_2 \geq 0$$
$$3x_1 + 2x_2 \leq 6$$
$$x_1 - \ x_2 = 4$$

Solution The initial matrix is

$$\begin{bmatrix} 3 & 2 & 1 & 6 \\ 1 & -1 & 0 & 4 \\ 2 & 1 & 0 & 0 \end{bmatrix}$$

Note: the simplex method would not allow us to use R_2 to clear C_1, since of the ratios $6/3$ and $4/1$ the one obtained from R_2 is not the smallest. Nevertheless, we do use R_2 to clear C_1, and then we repair the damage to the simplex method by going through Phase I.

We obtain the second column for the identity matrix by using the second row to clear the first column:

$$\begin{bmatrix} 0 & 5 & 1 & -6 \\ 1 & -1 & 0 & 4 \\ 0 & 3 & 0 & -8 \end{bmatrix}$$

The -6 tells us to use Phase I. The fact that there are no other negative entries in its row tells us that there is no feasible, and therefore, no optimal solution. This is easy to see geometrically (Figure 9). There are no points that are in the positive quadrant, below line **a**: $3x_1 + 2x_2 = 6$, and on line **b**: $x_1 - x_2 = 4$.

A New Minimization Technique

Phase I can be used to allow a more straightforward solution of minimization problems. Let us reconsider Problem 4 in Section 5-5.

Problem 5

Minimize $y = 2x_1 + 3x_2 + 7x_3$ subject to

$$x_1 \geq 0, \ x_2 \geq 0, \ x_3 \geq 0$$
$$x_1 + \ x_2 + 2x_3 \geq 3$$
$$x_1 + 2x_2 + 3x_3 \geq 4$$
$$2x_1 + 4x_2 + 8x_3 \geq 7$$

Solution

Consider the variable y' which is the negative of the objective variable:

$$y' = -2x_1 - 3x_2 - 7x_3$$

The larger we make this variable, the smaller we make y. For example, if we increase y' from -20 to -10, we decrease y from 20 to 10. Thus, if we maximize y' we will have minimized y. Therefore, corresponding to the above minimization problem there is a maximization problem, namely

MAXIMIZE

$$y' = -2x_1 - 3x_2 - 7x_3$$

SUBJECT TO

$$x_1 \geq 0, \ x_2 \geq 0, \ x_3 \geq 0$$
$$x_1 + \ x_2 + 2x_3 \geq 3$$
$$x_1 + 2x_2 + 3x_3 \geq 4$$
$$2x_1 + 4x_2 + 8x_3 \geq 7$$

Now every constraint violates Condition 2. We multiply -1 through each constraint and set up the initial matrix:

$$\begin{bmatrix} -1 & -1 & -2 & 1 & 0 & 0 & -3 \\ -1 & -2 & -3 & 0 & 1 & 0 & -4 \\ -2^* & -4 & -8 & 0 & 0 & 1 & -7 \\ -2 & -3 & -7 & 0 & 0 & 0 & 0 \end{bmatrix}$$

Figure 9

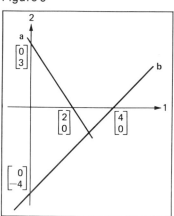

There are three negative numbers in the last column. The flow chart for Phase I instructs us to use the -7, since it is farthest down in the matrix. We have a choice of any of the first three columns. We arbitrarily select C_1. Clearing yields the matrix

$$\begin{bmatrix} 0 & 1 & 2 & 1 & 0 & -\frac{1}{2} & \frac{1}{2} \\ 0 & 0 & 1 & 0 & 1 & -\frac{1}{2}* & -\frac{1}{2} \\ 1 & 2 & 4 & 0 & 0 & -\frac{1}{2} & \frac{7}{2} \\ 0 & 1 & 1 & 0 & 0 & -1 & 7 \end{bmatrix}$$

We must use Phase I again. This time we have no choice. We must let $R_2 = R$ and $C_6 = C$. We then obtain

$$\begin{bmatrix} 0 & 1* & 1 & 1 & -1 & 0 & 1 \\ 0 & 0 & -2 & 0 & -2 & 1 & 1 \\ 1 & 2 & 3 & 0 & 1 & 0 & 4 \\ 0 & 1 & -1 & 0 & -2 & 0 & 8 \end{bmatrix}$$

Now we are ready to apply the ordinary Simplex Method.

$$\begin{bmatrix} 0 & 1 & 1 & 1 & -1 & 0 & 1 \\ 0 & 0 & -2 & 0 & -1 & -2 & 1 \\ 1 & 0 & 1 & -2 & 1 & 0 & 2 \\ 0 & 0 & -2 & -1 & -1 & 0 & 7 \end{bmatrix}$$

Since all the entries (except the last) in the last row are zero or negative, we can read the answer.

Warning: We have just solved a maximization problem. We read the *maximization* answer:

$$x_1 = 2, \ x_2 = 1, \ x_3 = 0, \ y' = -7$$

To relate this to our minimization problem, we simply change the sign of y': The answer to the minimization problem is then

$$x_1 = 2, \ x_2 = 1, \ x_3 = 0, \ y = 7$$

Minimization with mixed constraints is now easily accomplished by maximizing the negative of the objective variable.

Problem 6 Minimize $y = 5x_1 + 6x_2$ subject to

$$\begin{aligned} x_1 \geq 0, \ x_2 &\geq 0 \\ -2x_1 + x_2 &= -2 \\ 2x_1 + 3x_2 &\geq 6 \\ 3x_1 + 5x_2 &\leq 15 \end{aligned}$$

Solution We convert to the equivalent maximization problem: maximize

$y' = -5x_1 - 6x_2$ subject to

$$
\begin{aligned}
x_1 \geq 0,\ x_2 &\geq 0 \\
-2x_1 + x_2 &= -2 \\
2x_1 + 3x_2 &\geq 6 \\
3x_1 + 5x_2 &\leq 15
\end{aligned}
$$

Next, we multiply -1 through the second constraint so that it satisfies Condition 2. The matrix is

$$
\begin{bmatrix}
-2 & 1^* & 0 & 0 & -2 \\
-2 & -3 & 1 & 0 & -6 \\
3 & 5 & 0 & 1 & 15 \\
-5 & -6 & 0 & 0 & 0
\end{bmatrix}
$$

First, we must generate the missing column of the identity matrix. We could clear either of the first two columns. However, it looks as if it will be easier to clear the second column:

$$
\begin{bmatrix}
-2 & 1 & 0 & 0 & -2 \\
-8 & 0 & 1 & 0 & -12 \\
13^* & 0 & 0 & 1 & 25 \\
-17 & 0 & 0 & 0 & -12
\end{bmatrix}
$$

We must now use Phase I. The instructions lead us first to the -12, then to the -8, and finally to the 13. Clearing yields

$$
\begin{bmatrix}
0 & 1 & 0 & {}^{2}/_{13} & {}^{24}/_{13} \\
0 & 0 & 1 & {}^{8}/_{13}{}^* & {}^{44}/_{13} \\
1 & 0 & 0 & {}^{1}/_{13} & {}^{25}/_{13} \\
0 & 0 & 0 & {}^{17}/_{13} & {}^{269}/_{13}
\end{bmatrix}
$$

Applying the ordinary simplex method, we are led to the fourth column and the ${}^{8}/_{13}$.

$$
\begin{bmatrix}
0 & 1 & -{}^{1}/_{4} & 0 & 1 \\
0 & 0 & 1 & {}^{8}/_{13} & {}^{44}/_{13} \\
1 & 0 & -{}^{1}/_{8} & 0 & {}^{3}/_{2} \\
0 & 0 & -{}^{17}/_{8} & 0 & {}^{27}/_{2}
\end{bmatrix}
$$

The answer to this maximization problem is

$$x_1 = {}^{3}/_{2},\ x_2 = 1,\ y' = -{}^{27}/_{2}$$

Thus, the answer to the minimization problem is

$$x_1 = {}^{3}/_{2},\ x_2 = 1,\ y = {}^{27}/_{2}$$

This minimization technique is important because usually there are many more variables than constraints in problems arising in practice. It is much easier for a machine to use the simplex method on a problem with many columns than one with many rows. The technique of Section 5-5 converts the column of each variable into a row. Therefore, computers are usually programmed to solve minimization problems by maximizing the negative of the objective variable.

Geometric Introduction to Multiple Answers

Figure 10

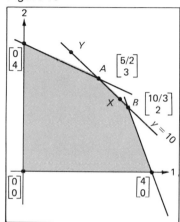

It can happen that more than one corner point of the region of feasibility of an LP problem is an optimal solution. Consider, for example, this problem:

MAXIMIZE

$$y = 2x_1 + \tfrac{5}{3}x_2$$

SUBJECT TO

$$x_1 \geq 0,\ x_2 \geq 0$$
$$2x_1 + 5x_2 \leq 20$$
$$6x_1 + 5x_2 \leq 30$$
$$3x_1 + x_2 \leq 12$$

The point X in Figure 10, for instance, is a solution. On the other hand, the point Y in Figure 10 is *not* a solution because, while it is on line AB, it is not between A and B, and hence it is not in the region of feasibility of the problem.

The region of feasibility is shown in Figure 10. Substituting the five corner points into y, we obtain the information in Table 15. A maximum value of y occurs at point A *and* at point B. Thus, the line y = 10 is, in fact, just the line AB determined by the points A and B. It follows that all of the points on line AB that lie *between* A and B are also solutions of this problem.

Table 15

Corner point	y-value
$\begin{bmatrix} 0 \\ 0 \end{bmatrix}$	0
$\begin{bmatrix} 0 \\ 4 \end{bmatrix}$	$\tfrac{20}{3}$
$\begin{bmatrix} \tfrac{5}{2} \\ 3 \end{bmatrix}$	10
$\begin{bmatrix} \tfrac{10}{3} \\ 2 \end{bmatrix}$	10
$\begin{bmatrix} 4 \\ 0 \end{bmatrix}$	8

Convex Combinations

Using analytic geometry, it can be shown that the points between A and B are all points $\begin{bmatrix} x_1 \\ x_2 \end{bmatrix}$ given by the formula

$$\begin{bmatrix} x_1 \\ x_2 \end{bmatrix} = t \begin{bmatrix} 5/2 \\ 3 \end{bmatrix} + (1-t) \begin{bmatrix} 10/3 \\ 2 \end{bmatrix} \qquad \textbf{2}$$

where t may be any number between 0 and 1, including 0 and 1 themselves. Now $t \begin{bmatrix} 5/2 \\ 3 \end{bmatrix} = \begin{bmatrix} 5/2 t \\ 3t \end{bmatrix}$ and $(1-t) \begin{bmatrix} 10/3 \\ 2 \end{bmatrix} = \begin{bmatrix} 10/3(1-t) \\ 2(1-t) \end{bmatrix} = \begin{bmatrix} 10/3 - 10/3 t \\ 2 - 2t \end{bmatrix}$. Thus, $\begin{bmatrix} x_1 \\ x_2 \end{bmatrix} = \begin{bmatrix} 5/2 t \\ 3t \end{bmatrix} + \begin{bmatrix} 10/3 - 10/3 t \\ 2 - 2t \end{bmatrix} = \begin{bmatrix} 10/3 - 5/6 t \\ 2 + t \end{bmatrix}$. The solutions of the problem are, then,

$$x_1 = {}^{10}\!/_3 - {}^5\!/_6 t, \ x_2 = 2 + t, \ y = 10,$$

where $\qquad\qquad\qquad\qquad\qquad\qquad\qquad\qquad\qquad$ **3**

$$0 \le t \le 1$$

For example, if $t = 0$ we have, from solution **3**, $x_1 = {}^{10}\!/_3$ and $x_2 = 2$, which is corner point B. If $t = 1$ we have $x_1 = {}^5\!/_2$ and $x_2 = 3$, which is corner point A. If say, $t = {}^1\!/_2$, we have $x_1 = {}^{35}\!/_{12}, \ x_2 = {}^5\!/_2$.

The point $\begin{bmatrix} 35/12 \\ 5/2 \end{bmatrix}$ is the midpoint of the line segment AB. If $t = {}^2\!/_3$ we obtain the point $\begin{bmatrix} 25/9 \\ 8/3 \end{bmatrix}$, which is ${}^2\!/_3$ of the way from B to A.

It is possible for a problem to have more than two corner point solutions. Suppose for example that a three-variable LP problem has solutions $\begin{bmatrix} 4 \\ 2 \\ 1 \end{bmatrix}$, $\begin{bmatrix} 3 \\ 1 \\ 4 \end{bmatrix}$, and $\begin{bmatrix} 5 \\ 0 \\ 3 \end{bmatrix}$; then, all points in the triangle determined by these points (Figure 11) are also solutions. How do we describe these points? The description is analogous to result **2**; they are all points given by the formula

$$\begin{bmatrix} x_1 \\ x_2 \\ x_3 \end{bmatrix} = t_1 \begin{bmatrix} 4 \\ 2 \\ 1 \end{bmatrix} + t_2 \begin{bmatrix} 3 \\ 1 \\ 4 \end{bmatrix} + t_3 \begin{bmatrix} 5 \\ 0 \\ 3 \end{bmatrix} \qquad \textbf{4}$$

where t_1, t_2, t_3 are any numbers such that
$$t_1 \ge 0, \ t_2 \ge 0, \ t_3 \ge 0, \text{ and } t_1 + t_2 + t_3 = 1$$

Let us spend some time illustrating formula **4**.

Suppose we take $t_1 = {}^1\!/_2$ and $t_2 = {}^1\!/_3$ in formula **4**. Then we

Figure 11

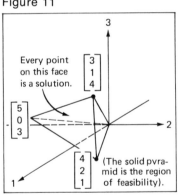

Every point on this face is a solution. $\begin{bmatrix} 3 \\ 1 \\ 4 \end{bmatrix}$

$\begin{bmatrix} 5 \\ 0 \\ 3 \end{bmatrix}$

$\begin{bmatrix} 4 \\ 2 \\ 1 \end{bmatrix}$ (The solid pyramid is the region of feasibility).

To see the analogy between formula **4** and formula **2**, you must in the latter let $t = t_1$ and $1 - t = t_2$.

This point, which is called the *centroid* of the triangle, is the point of intersection of the medians, and the point at which a triangle cut out of cardboard will balance.

See Appendix A, Sections 8 and 10 for a review of inequalities, if needed.

must take $t_3 = \frac{1}{6}$ in order to have $t_1 + t_2 + t_3 = 1$. Then
$$\begin{bmatrix} x_1 \\ x_2 \\ x_3 \end{bmatrix} = \frac{1}{2}\begin{bmatrix} 4 \\ 2 \\ 1 \end{bmatrix} + \frac{1}{3}\begin{bmatrix} 3 \\ 1 \\ 4 \end{bmatrix} + \frac{1}{6}\begin{bmatrix} 5 \\ 0 \\ 3 \end{bmatrix} = \begin{bmatrix} 3\frac{5}{6} \\ 1\frac{1}{3} \\ 2\frac{1}{3} \end{bmatrix}$$ is a solution.

If we take $t_1 = t_2 = t_3$, then we must have $t_1 = t_2 = t_3 = \frac{1}{3}$, and we obtain the solution $\begin{bmatrix} x_1 \\ x_2 \\ x_3 \end{bmatrix} = \frac{1}{3}\begin{bmatrix} 4 \\ 2 \\ 1 \end{bmatrix} + \frac{1}{3}\begin{bmatrix} 3 \\ 1 \\ 4 \end{bmatrix} + \frac{1}{3}\begin{bmatrix} 5 \\ 0 \\ 3 \end{bmatrix} = \begin{bmatrix} 4 \\ 1 \\ 8/3 \end{bmatrix}$.

If we take $t_1 = t_2 = 0$, then we must take $t_3 = 1$, and formula **4** gives us back the corner point $\begin{bmatrix} 5 \\ 0 \\ 3 \end{bmatrix}$.

Suppose we set $t_1 = 0$ and allow t_2 and t_3 to vary. Then we must have $t_2 + t_3 = 1$, $t_2 \geq 0$, and $t_3 \geq 0$. If we let $t_2 = t$, then $t_3 = 1 - t$. The condition $t_2 \geq 0$ becomes $t \geq 0$, that is $0 \leq t$; the condition $t_3 \geq 0$ becomes $1 - t \geq 0$, that is, $t \leq 1$. Thus, formula **4** gives us the points
$$\begin{bmatrix} x_1 \\ x_2 \\ x_3 \end{bmatrix} = t\begin{bmatrix} 3 \\ 1 \\ 4 \end{bmatrix} + (1 - t)\begin{bmatrix} 5 \\ 0 \\ 3 \end{bmatrix}, \text{ where } 0 \leq t \leq 1,$$
which are just the points on the line segment between the corners $\begin{bmatrix} 3 \\ 1 \\ 4 \end{bmatrix}$ and $\begin{bmatrix} 5 \\ 0 \\ 3 \end{bmatrix}$.

If we use values t_1, t_2, t_3 such that $t_1 + t_2 + t_3 = 1$ but one or more of the conditions $t_1 \geq 0$, $t_2 \geq 0$, or $t_3 \geq 0$, is not satisfied then the point obtained will lie on the plane determined by the three points but it will *not* lie in the triangular region determined by them.

Suppose $t_1 = \frac{3}{4}$ and $t_2 = \frac{1}{3}$. Then in order to satisfy $t_1 + t_2 + t_3 = 1$, we must have $\frac{3}{4} + \frac{1}{3} + t_3 = 1$, $\frac{13}{12} + t_3 = 1$, $t_3 = -\frac{1}{12}$. But then t_3 is not ≥ 0; so the point $\frac{3}{4}\begin{bmatrix} 4 \\ 2 \\ 1 \end{bmatrix} + \frac{1}{3}\begin{bmatrix} 3 \\ 1 \\ 4 \end{bmatrix} - \frac{1}{12}\begin{bmatrix} 5 \\ 0 \\ 3 \end{bmatrix}$ is not in the triangle, though it is in the plane determined by these points.

Mathematicians call formulas **2** and **4 convex combinations** of the corner points. We may by analogy form a convex combination of any number of points. Two examples will suffice:

Example 1   The convex combinations of the points $\begin{bmatrix} 1 \\ -1 \\ 2 \end{bmatrix}$, $\begin{bmatrix} 2 \\ 1 \\ -1 \end{bmatrix}$, $\begin{bmatrix} -1 \\ -2 \\ 2 \end{bmatrix}$, and

$\begin{bmatrix} 1 \\ 2 \\ 1 \end{bmatrix}$ are all points given by

$$\begin{bmatrix} x_1 \\ x_2 \\ x_3 \end{bmatrix} = t_1 \begin{bmatrix} 1 \\ -1 \\ 2 \end{bmatrix} + t_2 \begin{bmatrix} 2 \\ 1 \\ -1 \end{bmatrix} + t_3 \begin{bmatrix} -1 \\ -2 \\ 2 \end{bmatrix} + t_4 \begin{bmatrix} 1 \\ 2 \\ 1 \end{bmatrix}$$

where t_1, t_2, t_3, and t_4 are any numbers such that $t_1 \geq 0$, $t_2 \geq 0$, $t_3 \geq 0$, $t_4 \geq 0$, and $t_1 + t_2 + t_3 + t_4 = 1$.

Example 2   The convex combinations of the points $\begin{bmatrix} 0.3 \\ 7.2 \\ -1.5 \\ 6.0 \\ 3.9 \end{bmatrix}$ $\begin{bmatrix} -4.1 \\ 0.0 \\ 3.7 \\ -8.8 \\ 2.1 \end{bmatrix}$ and

$\begin{bmatrix} 0.0 \\ 6.6 \\ -7.3 \\ -0.9 \\ -1.2 \end{bmatrix}$ are all points given by

$$\begin{bmatrix} x_1 \\ x_2 \\ x_3 \\ x_4 \\ x_5 \end{bmatrix} = t_1 \begin{bmatrix} 0.3 \\ 7.2 \\ -1.5 \\ 6.0 \\ 3.9 \end{bmatrix} + t_2 \begin{bmatrix} -4.1 \\ 0.0 \\ 3.7 \\ -8.8 \\ 2.1 \end{bmatrix} + t_3 \begin{bmatrix} 0.0 \\ 6.6 \\ -7.3 \\ -0.9 \\ -1.2 \end{bmatrix}$$

where t_1, t_2, and t_3 are any numbers such that $t_1 \geq 0$, $t_2 \geq 0$, $t_3 \geq 0$, and $t_1 + t_2 + t_3 = 1$.

Maximization Problems with Multiple Answers   How can the simplex method be used to determine whether an LP problem has multiple answers, and if there are, how can the method be used to find all corner-point solutions? Consider first a maximization problem. The simplex method will reach a corner point of the region of feasibility at which the objective variable is maximized. We will then check whether there is any edge leading away from this corner along which we can travel *without changing the value of the objective variable*. If one or more such edges exist, we will travel along each of them to reach other maximal corner points, at which we repeat the search. In terms of the matrix version of the simplex method, all this amounts to is working with uncleared columns that end in zero. Such columns are called **special**.

It is sometimes difficult to determine exactly which columns are cleared, especially since in a problem with multiple answers there may be many choices for columns to designate as cleared columns. The following technique can help eliminate some of this confusion.

Locating the Cleared Columns   In the final matrix, place a * below each column that ends in zero, contains a 1, and has all other entries zero.

For each column of the original identity matrix, choose a starred column that has its 1 in the same row. The columns so chosen are the cleared columns.

Notice that clearing a special column (or any other column for that matter) results in the unclearing of a column that had been designated cleared. In fact, if you clear a special column the column that becomes uncleared actually becomes a special column. This interchanging of special and cleared columns is sometimes merely a matter of redesignation. For example in the final matrix

$$\begin{bmatrix} 1 & 1 & 0 & 3 \\ 0 & 0 & 1 & 2 \\ 0 & 0 & 0 & -6 \end{bmatrix}$$

we could designate C_1 and C_3 to be cleared and C_2 to be special (whence $x_1 = 3$, $x_2 = 0$, $x_3 = 2$, $y = 6$). Clearing C_2, which requires no work in this case, results in unclearing C_1. Now the cleared columns are C_2 and C_3 and the special column is C_1 (whence $x_1 = 0$, $x_2 = 3$, $x_3 = 2$, and $y = 6$). Applying the simplex method involved no work, although it did change the answer (but not the value of the objective variable).

The idea of the flow chart on p. 334 is to clear all special columns and try all designations for cleared columns. In a large matrix this can be complicated. The flow chart attempts to eliminate some needless backtracking and duplication effort. It is not necessary to use the flow chart if you are sure you have all answers that can be obtained from the final matrix by clearing special columns.

The statement in the last box of this flow chart implies that this method will not necessarily find all the answers to a problem. This is in fact the case. There is one further possibility, which we will discuss at the end of this section.

To Find Multiple Answers to an LP Maximization Problem

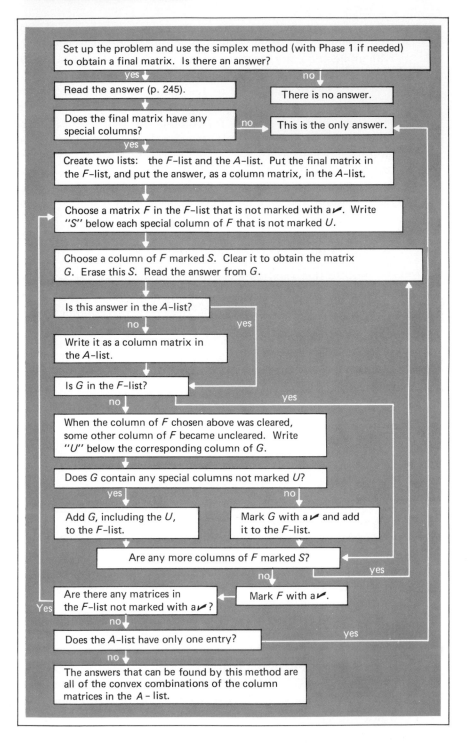

Set up the problem and use the simplex method (with Phase 1 if needed) to obtain a final matrix. Is there an answer?

yes ↓   no ↓

Read the answer (p. 245).

There is no answer.

Does the final matrix have any special columns?   no →   This is the only answer.

yes ↓

Create two lists: the F-list and the A-list. Put the final matrix in the F-list, and put the answer, as a column matrix, in the A-list.

Choose a matrix F in the F-list that is not marked with a ✔. Write "S" below each special column of F that is not marked U.

Choose a column of F marked S. Clear it to obtain the matrix G. Erase this S. Read the answer from G.

Is this answer in the A-list?

no ↓   yes

Write it as a column matrix in the A-list.

Is G in the F-list?

no ↓   yes

When the column of F chosen above was cleared, some other column of F became uncleared. Write "U" below the corresponding column of G.

Does G contain any special columns not marked U?

yes ↓   no ↓

Add G, including the U, to the F-list.

Mark G with a ✔ and add it to the F-list.

Are any more columns of F marked S?

no ↓   yes

Mark F with a ✔.

Are there any matrices in the F-list not marked with a ✔?   Yes

no ↓

Does the A-list have only one entry?   yes

no ↓

The answers that can be found by this method are all of the convex combinations of the column matrices in the A-list.

Problem 7 Suppose the final matrix in the simplex method solution of a certain LP maximization problem is

$$x_1 \quad x_2 \quad x_3 \quad x_4$$

$$\begin{bmatrix} 0 & 1 & 0 & 0 & 0 & 1 & 2 & 0 & 1 \\ 0 & 0 & 2 & 1 & 0 & 0 & 1 & 0 & 2 \\ 1 & 0 & -1 & 0 & 0 & 0 & 0 & 1 & 3 \\ 0 & 0 & 0 & 0 & 1 & 0 & 0 & 1 & 4 \\ 0 & 0 & -3 & 0 & 0 & 0 & 0 & -1 & -5 \end{bmatrix}$$

Find all answers to the problem.

Solution In this solution we indicate a few shortcuts and memory aids not in the flow chart. None of the operations of the flow chart will alter the last row; hence, columns 3 and 8 will never be cleared. We eliminate C_3, C_8, and R_5 from the final matrix, taking care to remember that all solutions will have $x_3 = 0$, and obtain

$$x_1 \quad x_2 \quad x_4$$

$$\begin{bmatrix} 0 & 1 & 0 & 0 & 1^* & 2^* & 1 \\ 0 & 0 & 1 & 0 & 0 & 1 & 2 \\ 1 & 0 & 0 & 0 & 0 & 0 & 3 \\ 0 & 0 & 0 & 1 & 0 & 0 & 4 \end{bmatrix}, \begin{bmatrix} 3 \\ 1 \\ 0 \\ 2 \end{bmatrix} \qquad \textbf{5}$$

$$K \quad K \quad K \quad K \quad S_1 \quad S_2$$

We have labeled cleared columns K and special columns with a subscripted S. We write the answer read from the matrix **5** next to the matrix separated by a comma. We have indicated (by*) the choice of R for both S_1 and S_2. In matrix **5** clear S_1:

$$x_1 \quad x_2 \quad x_4$$

$$\begin{bmatrix} 0 & 1 & 0 & 0 & 1 & 2^* & 1 \\ 0 & 0 & 1 & 0 & 0 & 1 & 2 \\ 1 & 0 & 0 & 0 & 0 & 0 & 3 \\ 0 & 0 & 0 & 1 & 0 & 0 & 4 \end{bmatrix}, \begin{bmatrix} 3 \\ 0 \\ 0 \\ 2 \end{bmatrix} \qquad \textbf{6}$$

$$K \quad U \quad K \quad K \quad K \quad S$$

In matrix **5**, clear S_2:

$$x_1 \quad x_2 \quad x_4$$

$$\begin{bmatrix} 0 & \tfrac{1}{2} & 0 & 0 & \tfrac{1}{2}^* & 1 & \tfrac{1}{2} \\ 0 & -\tfrac{1}{2} & 1 & 0 & -\tfrac{1}{2} & 0 & \tfrac{3}{2} \\ 1 & 0 & 0 & 0 & 0 & 0 & 3 \\ 0 & 0 & 0 & 1 & 0 & 0 & 4 \end{bmatrix}, \begin{bmatrix} 3 \\ 0 \\ 0 \\ \tfrac{3}{2} \end{bmatrix} \qquad \textbf{7}$$

$$K \quad U \quad K \quad K \quad S \quad K$$

Check off matrix **5**. In matrix **6**, clear S: we obtain matrix **7** again.

Check off matrix **6**. In matrix **7**, clear S: we obtain matrix **6**. Check off matrix **7**. Everything is checked off. The answer:

$$\begin{bmatrix} x_1 \\ x_2 \\ x_3 \\ x_4 \end{bmatrix} = t_1 \begin{bmatrix} 3 \\ 1 \\ 0 \\ 2 \end{bmatrix} + t_2 \begin{bmatrix} 3 \\ 0 \\ 0 \\ 2 \end{bmatrix} + t_3 \begin{bmatrix} 3 \\ 0 \\ 0 \\ 3/2 \end{bmatrix}$$

where $t_1 \geq 0$, $t_2 \geq 0$, $t_3 \geq 0$, and $t_1 + t_2 + t_3 = 1$.

In the following problem, we abandon the flow chart to illustrate the alternative technique of clearing columns that end in zero until all possible answers have been obtained.

Problem 8 Find all solutions of the following problem: maximize $y = 2x_1 + \frac{5}{3}x_2$ subject to

$$x_1 \geq 0,\ x_2 \geq 0$$
$$2x_1 + 5x_2 \leq 20$$
$$6x_1 + 5x_2 \leq 30$$
$$3x_1 + \ x_2 \leq 12$$

Solution We have already solved this problem in this section by the geometrical method. Now we proceed as follows:

$$\begin{bmatrix} 2 & 5 & 1 & 0 & 0 & 20 \\ 6 & 5 & 0 & 1 & 0 & 30 \\ 3^* & 1 & 0 & 0 & 1 & 12 \\ 2 & 5/3 & 0 & 0 & 0 & 0 \end{bmatrix} \Rightarrow \begin{bmatrix} 0 & 13/3 & 1 & 0 & -2/3 & 12 \\ 0 & 3^* & 0 & 1 & -2 & 6 \\ 1 & 1/3 & 0 & 0 & 1/3 & 4 \\ 0 & 1 & 0 & 0 & -2/3 & -8 \end{bmatrix}$$

$$\Rightarrow \begin{bmatrix} 0 & 0 & 1 & -13/9 & 20/9^* & 10/3 \\ 0 & 1 & 0 & 1/3 & -2/3 & 2 \\ 1 & 0 & 0 & -1/9 & 5/9 & 10/3 \\ 0 & 0 & 0 & -1/3 & 0 & -10 \end{bmatrix} \text{ (put } \begin{bmatrix} 10/3 \\ 2 \end{bmatrix} \text{ in a list)}$$

$$\Rightarrow \begin{bmatrix} 0 & 0 & 9/20 & -13/20 & 1 & 3/2 \\ 0 & 1 & 3/10 & -1/10 & 0 & 3 \\ 1 & 0 & -1/4 & 1/4 & 0 & 5/2 \\ 0 & 0 & 0 & -1/3 & 0 & -10 \end{bmatrix} \text{ (put } \begin{bmatrix} 5/2 \\ 3 \end{bmatrix} \text{ in a list)}$$

Column 3 was previously cleared, so our list is complete. The answer: $\begin{bmatrix} x_1 \\ x_2 \end{bmatrix} = t \begin{bmatrix} 10/3 \\ 2 \end{bmatrix} + (1 - t) \begin{bmatrix} 5/2 \\ 3 \end{bmatrix}$, where $0 \leq t \leq 1$.

Our present formula looks different from equation **2**, but it gives the same answers. To obtain equation **2**, replace t in the present formula by $1 - t$.

Minimization Problems with Multiple Answers

We can find multiple answers to minimization problems by maximizing the negative of the objective variable and locating the multiple answers for this maximization problem as described above.

If you originally used the technique of Section 5-5, you will be able to tell that the problem has multiple answers by the presence of zeros in the last column. If zeros occur here, then solve the associated maximization problem to find all the answers.

Problem 9

Minimize $y = 2x_1 + x_2 + 3x_3$ subject to

$$x_1 \geq 0, \; x_2 \geq 0, \; x_3 \geq 0$$
$$x_1 + 2x_2 + \; x_3 \geq 3$$
$$4x_1 + 2x_2 + 6x_3 \geq 4$$

Solution

(Using the technique of Section 5-5):

$$
\begin{bmatrix}
1 & 4^* & 1 & 0 & 0 & 2 \\
2 & 2 & 0 & 1 & 0 & 1 \\
1 & 6 & 0 & 0 & 1 & 3 \\
3 & 4 & 0 & 0 & 0 & 0
\end{bmatrix}
\Rightarrow
\begin{bmatrix}
1/4 & 1 & 1/4 & 0 & 0 & 1/2 \\
3/2^* & 0 & -1/2 & 1 & 0 & 0 \\
-1/2 & 0 & -3/2 & 0 & 1 & 0 \\
2 & 0 & -1 & 0 & 0 & -2
\end{bmatrix}
$$

$$
\Rightarrow
\begin{bmatrix}
0 & 1 & 1/3 & -1/6 & 0 & 1/2 \\
1 & 0 & -1/3 & 2/3 & 0 & 0 \\
0 & 0 & -5/3 & 1/3 & 1 & 0 \\
0 & 0 & -1/3 & -4/3 & 0 & -2
\end{bmatrix}
$$

The zeros in the last column indicate that the problem has multiple answers. We set up and solve the associated maximization problem:

$$
\begin{bmatrix}
-1 & -2 & -1 & 1 & 0 & -3 \\
-4 & -2^* & -6 & 0 & 1 & -4 \\
-2 & -1 & -3 & 0 & 0 & 0
\end{bmatrix}
$$

$$
\Rightarrow
\begin{bmatrix}
3^* & 0 & 5 & 1 & -1 & 1 \\
2 & 1 & 3 & 0 & -1/2 & 2 \\
0 & 0 & 0 & 0 & -1/2 & 2
\end{bmatrix}
,
\begin{bmatrix}
0 \\
2 \\
0
\end{bmatrix}
$$

$$
\Rightarrow
\begin{bmatrix}
3 & 0 & 5^* & 1 & -1 & 1 \\
0 & 1 & -1/3 & -2/3 & 1/6 & 4/3 \\
0 & 0 & 0 & 0 & -1/2 & 2
\end{bmatrix}
,
\begin{bmatrix}
1/3 \\
4/3 \\
0
\end{bmatrix}
$$

$$\begin{bmatrix} 3 & 0 & 5 & 1 & -1 & 1 \\ \frac{1}{5} & 1 & 0 & -\frac{3}{5} & \frac{1}{10} & \frac{7}{5} \\ 0 & 0 & 0 & 0 & -\frac{1}{2} & 2 \end{bmatrix}, \begin{bmatrix} 0 \\ \frac{7}{5} \\ \frac{1}{5} \end{bmatrix}$$

We check that clearing any of the special columns will not yield new corner points. Answer: $\begin{bmatrix} x_1 \\ x_2 \\ x_3 \end{bmatrix} = t_1 \begin{bmatrix} 0 \\ 2 \\ 0 \end{bmatrix} + t_2 \begin{bmatrix} \frac{1}{3} \\ \frac{4}{3} \\ 0 \end{bmatrix} + t_3 \begin{bmatrix} 0 \\ \frac{7}{5} \\ \frac{1}{5} \end{bmatrix}$,

where $t_1 \geq 0$, $t_2 \geq 0$, $t_3 \geq 0$, and $t_1 + t_2 + t_3 = 1$; $y = 2$.

The next problem shows a situation in which we are motivated to look for multiple answers and in which our search proves fruitful.

Problem 10

In making up their special insect spray, AC insists that each bottle contain at least 6 g (grams) hydrone and 12 g pyrene. Aleph Chemical Company offers them a substance, costing $6.00 per oz., that contain 3 g hydrone and 3 g pyrene per oz. Beth Pharmaceuticals, Inc., offers a substance at $8.00 per oz. containing 1 g hydrone and 4 g pyrene per oz. AC wants to make up 7000 bottles of spray at minimum cost. At the same time they do not, if possible, want to antagonize either of these important suppliers. What should they do?

Solution

Let x_1 be the number of ounces of Aleph substance and x_2 the number of ounces of Beth substance ordered for one bottle. We are to minimize $y = 6x_1 + 8x_2$ subject to

$$x_1 \geq 0, x_2 \geq 0$$
$$3x_1 + x_2 \geq 6$$
$$3x_1 + 4x_2 \geq 12$$

$$\begin{bmatrix} -3 & -1 & 1 & 0 & -6 \\ -3^* & -4 & 0 & 1 & -12 \\ -6 & -8 & 0 & 0 & 0 \end{bmatrix} \Rightarrow \begin{bmatrix} 0 & 3^* & 1 & -1 & 6 \\ 1 & \frac{4}{3} & 0 & -\frac{1}{3} & 4 \\ 0 & 0 & 0 & -2 & 24 \end{bmatrix}, \begin{bmatrix} 4 \\ 0 \end{bmatrix}$$

While $x_1 = 4$, $x_2 = 0$ is an answer, it is unacceptable because it completely cuts out Beth. We look for other answers:

$$\begin{bmatrix} 0 & 1 & \frac{1}{3} & -\frac{1}{3} & 2 \\ 1 & 0 & -\frac{4}{9} & \frac{1}{9} & \frac{4}{3} \\ 0 & 0 & 0 & -2 & 24 \end{bmatrix}, \begin{bmatrix} \frac{4}{3} \\ 2 \end{bmatrix}$$

We see that $\begin{bmatrix} \frac{4}{3} \\ 2 \end{bmatrix}$ is also an answer. This is better, but not perfect: Beth would get a larger order than Aleph. The *general* answer to this problem is $\begin{bmatrix} x_1 \\ x_2 \end{bmatrix} = t \begin{bmatrix} 4 \\ 0 \end{bmatrix} + (1-t) \begin{bmatrix} 4/3 \\ 2 \end{bmatrix}$, where $0 \le t \le 1$, that is, $x_1 = 8/3t + 4/3$, $x_2 = 2 - 2t$, $0 \le t \le 1$. The ideal would be to have $x_1 = x_2$. This would require $8/3t + 4/3 = 2 - 2t$, that is, $14/3t = 2/3$, that is, $t = 1/7$. Since $t = 1/7$ satisfies $0 \le t \le 1$, this really is an acceptable answer. We have, for 1 bottle of spray, $x_1 = (8/3)(1/7) + 4/3 = 12/7$, $x_2 = 2 - 2(1/7) = 12/7$. Thus, for the full supply of 7000 bottles, the best order is 12,000 oz. from Aleph, 12,000 oz. from Beth; total cost, $168,000.

Alternate Solution

Another approach is to try to guarantee that $x_1 = x_2$ right from the start. We use the word "try" because stipulating $x_1 = x_2$ might rule out all solutions, in which case this approach would fail. The problem now reads:

MINIMIZE

$$y = 6x_1 + 8x_2$$

SUBJECT TO

$$x_1 \ge 0, \ x_2 \ge 0$$
$$3x_1 + x_2 \ge 6$$
$$3x_1 + 4x_2 \ge 12$$
$$x_1 - x_2 = 0$$

$$\begin{bmatrix} -3 & -1 & 1 & 0 & -6 \\ -3 & -4 & 0 & 1 & -12 \\ 1^* & -1 & 0 & 0 & 0 \\ -6 & -8 & 0 & 0 & 0 \end{bmatrix}$$

$$\begin{bmatrix} 0 & -4 & 1 & 0 & -6 \\ 0 & -7^* & 0 & 1 & -12 \\ 1 & -1 & 0 & 0 & 0 \\ 0 & -14 & 0 & 0 & 0 \end{bmatrix}$$

$$\begin{bmatrix} 0 & 0 & 1 & -4/7 & 6/7 \\ 0 & 1 & 0 & -1/7 & 12/7 \\ 1 & 0 & 0 & -1/7 & 12/7 \\ 0 & 0 & 0 & -2 & 24 \end{bmatrix}$$

Answer: $x_1 = 12/7$, $x_2 = 12/7$, $y = 24$. Only Phase I was used in the solution process.

One Other Possibility

There is one case in which the method described above will fail to find all the answers. Recall that the simplex method is a compact matrix technique that is equivalent to a geometric search for optimum values of the objective variable at corner points of the region of feasibility. If the optimum value occurs at two different corner points, then each point on the edge between

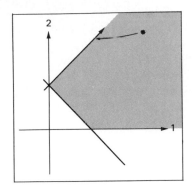

the corners will be optimal as well. We have noted before that the region of feasibility may be unbounded. In this case some of its edges may have infinite length; that is, the region of feasibility may contain an edge that does not lie between two corner points. For example, the inequalities $x_1 \geq 0$, $x_2 \geq 0$,

$$-x_1 + x_2 \leq 1, \qquad\qquad\qquad \mathbf{1}$$
$$x_1 + x_2 \geq 1,$$

define the region shown in Figure 12. The line determined by equation **1** involves only one corner point of the region of feasibility. If we now specify the objective variable to be

$$y = -x_1 + x_2$$

we have a situation in which each point on line **1** necessarily gives the same value to y, namely the value 1. In fact, 1 is the maximum value for y in this region. Thus the following problem has multiple solutions.

Problem 11   Maximize $y = -x_1 + x_2$ subject to

$$x_1 \geq 0,\ x_2 \geq 0$$
$$-x_1 + x_2 \leq 1$$
$$x_1 + x_2 \geq 1$$

Solution

$$\begin{bmatrix} -1 & 1 & 1 & 0 & 1 \\ -1 & -1^* & 0 & 1 & -1 \\ -1 & 1 & 0 & 0 & 0 \end{bmatrix}$$

$$\begin{bmatrix} -2 & 0 & 1 & 1^* & 0 \\ 1 & 1 & 0 & -1 & 1 \\ -2 & 0 & 0 & 1 & -1 \end{bmatrix}$$

$$\begin{bmatrix} -2 & 0 & 1 & 1 & 0 \\ -1 & 1 & 1 & 0 & 1 \\ 0 & 0 & -1 & 0 & -1 \end{bmatrix}$$

The simplex method finds only the answer $x_1 = 0$, $x_2 = 1$, $y = 1$. When it arrived at the corner $\begin{bmatrix} 0 \\ 1 \end{bmatrix}$, it noted that at no neighboring corner was the value of y higher, and quit.

This situation occurs when, and only when, one of the matrices in the F-list has a special column that *cannot be cleared*. In Problem 12 the only matrix in the F-list is the last one. In this matrix, column 1 is a special column but it cannot be cleared since it has no positive entries. If the solution to a problem yields special columns which cannot be cleared, the answer to the problem has an **unbounded component**.

Recall that the *F*-list includes the original final matrix and all other final matrices (if any) encountered in the search for multiple solutions.

Before describing the procedure for finding the unbounded components of an answer, we introduce two new terms and a new procedure. An **extra-special column** of a final matrix is a special column with no positive entries. A **non-negative combination** of column matrices is obtained by multiplying the matrices by non-negative numbers and adding the resulting matrices. For example, a non-negative combination of the matrices $\begin{bmatrix} -1 \\ 2 \\ 1 \end{bmatrix}, \begin{bmatrix} 0 \\ 1 \\ 4 \end{bmatrix}$, and

Note that a non-negative combination is not the same as a convex combination, since we do not also require that the r's add up to 1.

$\begin{bmatrix} -6 \\ -2 \\ 7 \end{bmatrix}$ is any expression of the form

$$r_1 \begin{bmatrix} -1 \\ 2 \\ 1 \end{bmatrix} + r_2 \begin{bmatrix} 0 \\ 1 \\ 4 \end{bmatrix} + r_3 \begin{bmatrix} -6 \\ -2 \\ 7 \end{bmatrix},$$

where $r_1 \geq 0, r_2 \geq 0, r_3 \geq 0$

Finding an Unbounded Component

Now suppose a final matrix F of a maximization LP problem has an extra-special column E. Here is the procedure for obtaining the unbounded component of E and F.

Step 1   Read the answer from F; write it as a column matrix A.

Step 2   Write the system of equations whose matrix is the matrix F with its last row deleted.

Step 3   In the system, set the variable corresponding to column E equal to one. Set the variables not corresponding to column E or to cleared columns equal to zero.

Step 4   Using the information in Step 3, make a column matrix B of the values of the original variables in the problem (not the slack variables).

Step 5   The unbounded component of E and F is the matrix B − A.

We illustrate this procedure with Problem 12 above, where

$$F = \begin{bmatrix} -2 & 0 & 1 & 1 & 0 \\ -1 & 1 & 1 & 0 & 1 \\ 0 & 0 & -1 & 0 & -1 \end{bmatrix} \text{ and } E = C_1.$$

Step 1 $A = \begin{bmatrix} 0 \\ 1 \end{bmatrix}$

Step 2
$$\begin{bmatrix} -2x_1 & & + x_3 + x_4 = 0 \\ -\ x_1 + x_2 + x_3 & & = 1 \end{bmatrix}$$

Step 3
$$x_1 = 1, \ x_3 = 0; \quad \begin{bmatrix} -2 + x_4 = 0 \\ -1 + x_2 = 1 \end{bmatrix}$$

Step 4 The original variables are x_1, x_2. Step 3 tells us $x_1 = 1$. In the second equation in the system in Step 3, we solve for x_2 and obtain $x_2 = 2$. Thus, $B = \begin{bmatrix} 1 \\ 2 \end{bmatrix}$.

Step 5 The unbounded component of E and F is $\begin{bmatrix} 1 \\ 2 \end{bmatrix} - \begin{bmatrix} 0 \\ 1 \end{bmatrix} = \begin{bmatrix} 1 \\ 1 \end{bmatrix}$.

For each matrix in the *F*-list we use the preceeding five steps to determine the unbounded component of each of its extra-special columns. Once we have found all the unbounded components the flow chart on p. 343 tells us how to write out the complete answer.

Applying this flow chart to problem 12, we discover that the unbounded component $\begin{bmatrix} 1 \\ 1 \end{bmatrix}$ found above is the only member of the *U*-list. The complete answer is then $\begin{bmatrix} 0 \\ 1 \end{bmatrix} + r\begin{bmatrix} 1 \\ 1 \end{bmatrix}$, where $r \geq 0$, that is, $\begin{bmatrix} r \\ r + 1 \end{bmatrix}$. Note that any point of the form $\begin{bmatrix} r \\ r + 1 \end{bmatrix}$ satisifes the equation

$$-x_1 + x_2 = 1$$

$(-r + r + 1 = 1 \quad \checkmark)$. Such points also optimize $y = -x_1 + x_2$ in the same manner.

Problem 12 Maximize $y = 9x_1 - 18x_2 + 12x_3$ subject to $x_1 \geq 0$, $x_2 \geq 0$, $x_3 \geq 0$, and $3x_1 - 6x_2 + 4x_3 \leq 12$.

Solution

Since there is only one constraint, the identity matrix used in the setup is $[1]$.

$$\begin{bmatrix} 3^* & -6 & 4 & 1 & 12 \\ 9 & -18 & 12 & 0 & 0 \end{bmatrix}$$

$$\begin{bmatrix} 1 & -2 & {}^4/_3{}^* & {}^1/_3 & 4 \\ 0 & 0 & 0 & -3 & -36 \end{bmatrix} = F_1, A_1 = \begin{bmatrix} 4 \\ 0 \\ 0 \end{bmatrix}$$

$$\begin{bmatrix} {}^3/_4 & -{}^3/_2 & 1 & {}^1/_4 & 3 \\ 0 & 0 & 0 & -3 & -36 \end{bmatrix} = F_2, A_2 = \begin{bmatrix} 0 \\ 0 \\ 3 \end{bmatrix}$$

The F-list contains F_1 and F_2, each of which have the extra-special column C_2. The five-step procedure applied to F_1 and $E = C_2$ yields the unbounded component $\begin{bmatrix} 6 \\ 1 \\ 0 \end{bmatrix} - \begin{bmatrix} 4 \\ 0 \\ 0 \end{bmatrix} = \begin{bmatrix} 2 \\ 1 \\ 0 \end{bmatrix} = U_1$. The

five-step procedure applied to F_2 and $E = C_2$ yields $\begin{bmatrix} 0 \\ 1 \\ 9/2 \end{bmatrix} - \begin{bmatrix} 0 \\ 0 \\ 3 \end{bmatrix} = \begin{bmatrix} 0 \\ 1 \\ 3/2 \end{bmatrix} = U_2$.

**To Find the Complete Solution
of an LP Maximization
Problem**

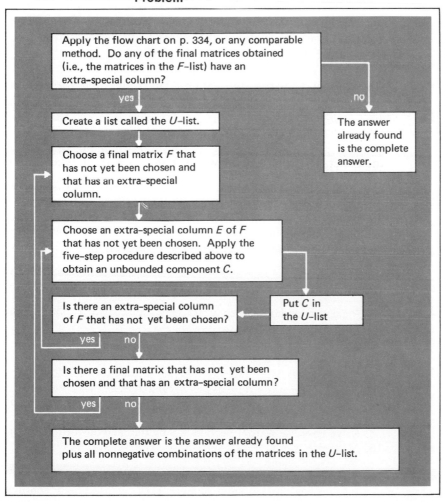

This completes the U-list; the complete answer to this problem is, then,

$$\begin{bmatrix} x_1 \\ x_2 \\ x_3 \end{bmatrix} = t_1 \begin{bmatrix} 4 \\ 0 \\ 0 \end{bmatrix} + t_2 \begin{bmatrix} 0 \\ 0 \\ 3 \end{bmatrix} + r_1 \begin{bmatrix} 2 \\ 1 \\ 0 \end{bmatrix} + r_2 \begin{bmatrix} 0 \\ 1 \\ ^3/_2 \end{bmatrix} = \begin{bmatrix} 4t_1 + 2r_1 \\ r_1 + r_2 \\ 3t_2 + ^3/_2 r_2 \end{bmatrix}, \ y = 36,$$

where $t_1 \geq 0$, $t_2 \geq 0$, $r_1 \geq 0$, $r_2 \geq 0$ and $t_1 + t_2 = 1$. Check: $y = 9(4t_1 + 2r_1) - 18(r_1 + r_2) + 12(3t_2 + ^3/_2 r_2) = 36t_1 + 18r_1 - 18r_1 - 18r_2 + 36t_2 + 18r_2 = 36t_1 + 36t_2 = 36(t_1 + t_2) = 36(1) = 36$ ✔; $3x_1 - 6x_2 + 4x_3 = 3(4t_1 + 2r_1) - 6(r_1 + r_2) + 4(3t_2 + ^3/_2 r_2) = 12t_1 + 6r_1 - 6r_1 - 6r_2 + 12t_2 + 6r_2 = 12(t_1 + t_2) = 12 \leq 12$ ✔.

In order to use this technique on an LP minimization problem, simply rewrite the problem as a maximization problem, as explained earlier in this section. If you use duality (Section 5-5) to solve minimization problems, the existence of multiple answers will be announced by the presence of one or more zeros as any but the bottom entry in the last column.

One Final Remark   Although it is extremely unlikely, it is possible for the simplex method, as presented in this book, to cycle, that is, to go on forever. This can only happen if a zero occurs in the last column (but not the last entry) at some point. There are ways to modify the simplex method to guarantee that cycling will not occur. However, cycling is so rare that machines are generally not programmed to avoid it. In hand computation you will notice that you are cycling if you keep returning to the same matrix. If this happens, try clearing a different column. It is difficult to find problems that cycle. Even if you find one you will have to cycle willfully. No problem ever forces you to cycle.

Exercises 6.4   A 1. Maximize and minimize $y = x_1 + x_2$ subject to $1 \leq x_1 \leq 2$ and $1 \leq x_2 \leq 2$.

2. Maximize $y = -2x_1 + x_2$ subject to $x_1 \geq 0$, $x_2 \geq 0$, $4x_1 + 2x_2 \geq -2$, and $-4x_1 + 2x_3 \leq 3$.

A 3. Maximize $y = 3x_1 + 4x_2 + x_3 + 2x_4$ subject to $x_1 \geq 0$, $x_2 \geq 0$, $x_3 \geq 0$, $x_4 \geq 0$, $7x_1 + 3x_2 + x_3 + 2x_4 \leq 14$, and $x_1 + 2x_2 + x_3 + 2x_4 = 4$.

4. Minimize $y = x_1 + 2x_2$ subject to $x_1 \geq 0$, $x_2 \geq 0$, $x_1 + x_2 \geq 2$, and $x_1 + x_2 \leq 3$.

5. Solve Exercise 4 from Section 5-5 by the methods of this section.

A 6. Maximize $y = x_1 + 3x_2$ subject to $x_1 \geq 0$, $x_2 \geq 0$, $x_1 + x_2 \geq 2$, $2x_1 + x_2 \leq 2$, and $-x_1 \leq 1$.

C 7. The PHE managers want to sell a lot of their large family-size Petalunches through store number 3749, just to check consumer

reaction. They want to make at least $15 profit. The store has plenty
of everything except buns: it has only 800 hamburger buns and 1000
hot-dog buns. There are three types of lunches:

Type	Hamburgers	Hot dogs	Profit (in cents)
I	8	3	15
II	4	4	10
III	5	12	20

Assuming they can sell whatever they make, what do they do?

8. Find all solutions of the following: maximize $y = \frac{7}{4}x_1 + 2x_2$ subject
to $x_1 \geq 0$, $x_2 \geq 0$, $x_1 + 4x_2 \leq 12$, $7x_1 + 8x_2 \leq 56$, $5x_1 - 3x_2 \leq 30$.

9. Maximize $y = 3x_1 + x_2 + 2x_3$ subject to $x_1 \geq 0$, $x_2 \geq 0$, $x_3 \geq 0$,
$x_1 + 2x_2 + x_3 \leq 4$, $6x_1 + 2x_2 + 4x_3 \leq 5$.

10. Minimize $y = 6x_1 + 4x_2 + 12x_3$ subject to $x_1 \geq 0$, $x_2 \geq 0$, $x_3 \geq 0$,
$x_1 + 2x_2 + 2x_3 \geq 3$, $3x_1 + 2x_2 + 6x_3 \geq 4$.

11. In Problem 9 of this section, suppose x_1, x_2, and x_3 represent the
number of parkas, tents, and stoves, respectively, to be produced in
a season (counted in hundreds). Find a solution to the problem in
which 10,101 parkas and 6,300 tents are made.

*12. PHE blends its hamburger meat from two types of ground beef. 100
g type A costs 60 ¢ and contains 40 g fat and 20 g protein; 100 g type
B costs $1.50 and contains 30 g fat and 50 g protein.

(a) What are the cheapest ways to prepare a batch of meat containing at least 1200 kilograms of fat and 1000 of protein (1 kilogram
= 1000 grams)?

(b) Is there a cheapest way, as found in part a, that would use up all
of PHE's supply of type A meat, 4000 kilograms?

(c) Is there a cheapest way, as found in part a, that would use up all
of PHE's supply of type B meat, 1200 kilograms?

13. Maximize $y = 3x_1 + 4x_2 + x_3 + 2x_4$ subject to $x_1 \geq 0$, $x_2 \geq 0$, $x_3 \geq 0$,
$x_4 \geq 0$, $7x_1 + 3x_2 + x_3 + 2x_4 \leq 14$, and $x_1 + 2x_2 + x_3 + 2x_4 = 6$.

14. Minimize $y = 8x_1 + 4x_2 + 6x_3$ subject to $x_1 \geq 0$, $x_2 \geq 0$, $x_3 \geq 0$,
$4x_1 + 2x_2 + 3x_3 \geq 12$, and $20x_1 - 12x_2 + 15x_3 \leq 60$.

15. Maximize $y = 24x_1 + x_2 + 4x_3$ subject to $x_1 \geq 0$, $x_2 \geq 0$, $x_3 \geq 0$, and
$24x_1 + 3x_2 + 4x_3 \leq 48$.

*16. (For those who know Section 6-1.) Find all solutions of the game
$$\begin{bmatrix} 1 & 2 & 3 \\ 2 & 1 & 0 \\ -1 & 2 & 5 \end{bmatrix}.$$

17. Minimize $y = x_2$ subject to $x_1 \geq 0$, $x_2 \geq 0$, $x_1 - x_2 \geq 0$.

^18. A final matrix for a certain maximization problem was
$$\begin{bmatrix} 0 & -1 & 1 & -1 & 0 & -1 & 3 \\ 1 & -2 & 0 & 0 & 0 & 2 & 6 \\ 0 & -1 & 0 & -1 & 1 & 1 & 2 \\ 0 & 0 & 0 & 0 & 0 & -3 & -12 \end{bmatrix}.$$ Find all the answers to the problem.

Section 6-5
Determinants and their
Applications

While the topic of determinants has little direct application to business problems, it is of value in connection with the further study of some of the topics discussed earlier. Moreover, the mathematical background necessary to learn how to calculate and use determinants has already been supplied in this book as part of other work. Thus, the present section makes a suitable extra topic for interested students.

Definition and Calculation

In Section 3-6 we defined the determinant of a general 2×2 matrix to be

$$\det \begin{bmatrix} a & b \\ c & d \end{bmatrix} = ad - bc \qquad \qquad \textbf{1}$$

The notation det, which is read "the determinant of," is introduced here for the first time. We learned that a 2×2 matrix has an inverse if its determinant is different from zero, but that it does not have an inverse if its determinant equals zero. The goal of this section is to define determinants for larger square matrices: 3×3, 4×4, etc. These determinants are harder to calculate than determinants of 2×2 matrices, unless the larger matrix is of a special type. If the matrix is of this special type, then its determinant is very easy to calculate. We will show how to use this fact to calculate these larger determinants.

We begin by investigating the effect of each of the three types of row equations on the determinant of the general 2×2 matrix in equation **1**. Suppose we add kR_1 to R_2, where k is some number:

$$\begin{bmatrix} a & b \\ c & d \end{bmatrix} \xrightarrow{R_2 + kR_1} \begin{bmatrix} a & b \\ c + ka & d + kb \end{bmatrix}.$$ The determinant of

this matrix is $\det \begin{bmatrix} a & b \\ c + ka & d + kb \end{bmatrix} = a(d + kb) - b(c + ka)$

$= ad + kab - bc - kab = ad - bc = \det \begin{bmatrix} a & b \\ c & d \end{bmatrix}$. So the determinant is not changed by this operation. Also, if we try

$$\begin{bmatrix} a & b \\ c & d \end{bmatrix} \xrightarrow{R_1 + kR_2} \begin{bmatrix} a + kc & b + kd \\ c & d \end{bmatrix},$$ we again have that

$\det \begin{bmatrix} a + kc & b + kd \\ c & d \end{bmatrix} = \det \begin{bmatrix} a & b \\ c & d \end{bmatrix}$, as the reader should check.

Thus, all row operations of this type leave the 2×2 determinant unchanged. Now suppose we multiply R_1 of matrix **1** by k:

$$\begin{bmatrix} a & b \\ c & d \end{bmatrix} \xrightarrow{\;kR_1\;} \begin{bmatrix} ka & kb \\ c & d \end{bmatrix}. \text{ Then } \det \begin{bmatrix} ka & kb \\ c & d \end{bmatrix} = (ka)d - (kb)c$$

$$= \; k(ad - bd) \;=\; k \det \begin{bmatrix} a & b \\ c & d \end{bmatrix}. \quad \text{Similarly,} \quad \text{we} \quad \text{have}$$

$\det \begin{bmatrix} a & b \\ kc & kd \end{bmatrix} = k \det \begin{bmatrix} a & b \\ c & d \end{bmatrix}.$ Thus, if a row of a 2×2 matrix is multiplied by a number, then the determinant is also multiplied by that number. We may restate this result as follows. Any number may be factored out of any row and multiplied by the determinant.

Examples

$$\det \begin{bmatrix} 6 & 8 \\ -3 & 2 \end{bmatrix} = \det \begin{bmatrix} 2 \cdot 3 & 2 \cdot 4 \\ -3 & 2 \end{bmatrix} = 2 \det \begin{bmatrix} 3 & 4 \\ -3 & 2 \end{bmatrix}.$$

$$\det \begin{bmatrix} 1 & -4 \\ -{}^2\!/_3 & {}^5\!/_2 \end{bmatrix} = \det \begin{bmatrix} 1 & -4 \\ -{}^1\!/_6(4) & -{}^1\!/_6(-15) \end{bmatrix} = -{}^1\!/_6 \det \begin{bmatrix} 1 & -4 \\ 4 & -15 \end{bmatrix}.$$

$$\det \begin{bmatrix} {}^1\!/_3 & {}^1\!/_4 \\ {}^1\!/_5 & {}^1\!/_6 \end{bmatrix} = \det \begin{bmatrix} {}^1\!/_{12}(4) & {}^1\!/_{12}(3) \\ {}^1\!/_5 & {}^1\!/_6 \end{bmatrix} = {}^1\!/_{12} \det \begin{bmatrix} 4 & 3 \\ {}^1\!/_5 & {}^1\!/_6 \end{bmatrix}$$

$$= {}^1\!/_{12} \det \begin{bmatrix} 4 & 3 \\ {}^1\!/_{30}(6) & {}^1\!/_{30}(5) \end{bmatrix} = ({}^1\!/_{12})({}^1\!/_{30}) \det \begin{bmatrix} 4 & 3 \\ 6 & 5 \end{bmatrix}.$$

Now, suppose we switch rows: $\begin{bmatrix} a & b \\ c & d \end{bmatrix} \bowtie \begin{bmatrix} c & d \\ a & b \end{bmatrix}.$ Then $\det \begin{bmatrix} c & d \\ a & b \end{bmatrix} = cb - da = -(ad - bc) = -\det \begin{bmatrix} a & b \\ c & d \end{bmatrix}.$ Thus, if the rows of a 2×2 matrix are switched, the determinant changes sign.

See p. 259 for the definition of transpose.

Finally, suppose we transpose the matrix $\begin{bmatrix} a & b \\ c & d \end{bmatrix}$ to obtain $\begin{bmatrix} a & c \\ b & d \end{bmatrix}.$

Then $\det \begin{bmatrix} a & c \\ b & d \end{bmatrix} = ad - cb = ad - bc = \det \begin{bmatrix} a & b \\ c & d \end{bmatrix}.$ Thus, the determinant is not changed by transposing the matrix. The row operations of the transposed matrix correspond to **column operations** (never studied before in this book) of the given matrix. Thus, in view of our above results for row operations, we may state the following additional results for the determinant of any 2×2 matrix:

(1) If any multiple of any column is added to another column, the determinant does not change.
(2) Any number may be factored out of any column and multiplied by the determinant.
(3) If any two columns are switched, the determinant changes sign.

Example

$$\det \begin{bmatrix} 4 & -\frac{1}{3} \\ 6 & \frac{2}{3} \end{bmatrix} = \det \begin{bmatrix} 4 & \frac{1}{3}(-1) \\ 6 & \frac{1}{3}(2) \end{bmatrix} = \frac{1}{3} \det \begin{bmatrix} 4 & -1 \\ 6 & 2 \end{bmatrix}$$

$$\begin{array}{c} C_1 - 3C_2 \\ = \det \begin{bmatrix} 7 & -1 \\ 0 & 2 \end{bmatrix} \end{array}$$

We can even do both row and column operations in the same calculation:

Example

$$\det \begin{bmatrix} 1 & 2 \\ 3 & 4 \end{bmatrix} = 2 \det \begin{bmatrix} 1 & 1 \\ 3 & 2 \end{bmatrix} = 2 \det \begin{bmatrix} 1 & 1 \\ 0 & -1 \end{bmatrix} \quad R_2 - 3R_1$$

$$\begin{array}{c} C_2 - C_1 \\ = 2 \det \begin{bmatrix} 1 & 0 \\ 0 & -1 \end{bmatrix} = -2 \det \begin{bmatrix} 0 & 1 \\ -1 & 0 \end{bmatrix} = 2 \det \begin{bmatrix} -1 & 0 \\ 0 & 1 \end{bmatrix} \end{array}$$

The operations in the last two steps were: switch C_1, C_2 and switch R_1, R_2.

Now it can be shown that determinants of square matrices of *any* size have all of the pleasant properties discovered above for determinants of 2×2 matrices. This fact provides the basis for the most efficient way to calculate the determinant of a large square matrix. Consider first the following determinants of

The main diagonal of a square matrix, first defined in Section 3-1, is the diagonal from the upper left corner to the lower right corner.

2×2 matrices: $\det \begin{bmatrix} a & b \\ 0 & d \end{bmatrix} = ad - b \cdot 0 = ad$; $\det \begin{bmatrix} a & 0 \\ c & d \end{bmatrix}$
$= ad - 0 \cdot c = ad$. In words: if either the entry above the main diagonal or the entry below the main diagonal is zero, then the determinant is simply the product of the entries on the main diagonal. In general it can be shown that if a square matrix has either all zero entries above the main diagonal or all zero entries below the main diagonal, then the determinant of the matrix is just the product of the entries on the main diagonal. This is the motivation for the flow chart on p. 349, which is both our definition of the determinant and the best method of calculation.

To Find the Determinant of a Matrix

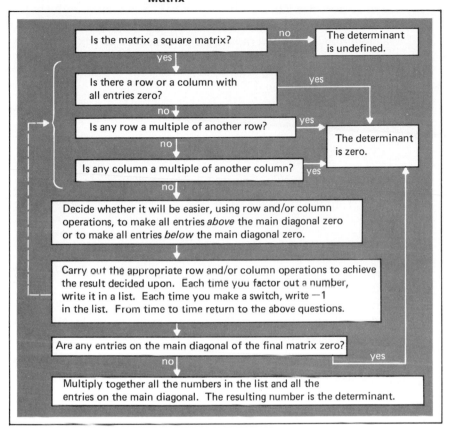

This flow chart raises a veritable catechism of questions, which we settle before giving examples. How do we know that we can always succeed in getting all those entries to be zero? Well, Gauss-Jordan reduction (Section 2-4) is one way that will always make all the entries below the main diagonal zero, though there may often be easier ways. How do we know that all the different possible ways to do these calculations for a given matrix will lead to the same determinant? This can be shown, but it would take us too far afield to explain how. What if we fail to notice at some point that one row or column is a multiple of another? We will still get the right answer (zero) eventually. How do we decide whether to eliminate the upper or lower part of the matrix? If one of the parts already has a number of zeros, it may be easier to clear it out. Experience will help in this choice. Sometimes both choices are equally easy (or equally hard). In higher mathematics, is there a definition for the determinant of a nonsquare matrix? Not really.

Problem 1   Find $\det \begin{bmatrix} \frac{2}{3} & -1 & 2 \\ 3 & 0 & \frac{1}{2} \\ 4 & 3 & \frac{2}{3} \end{bmatrix}$.

Solution $\begin{bmatrix} \frac{2}{3} & -1 & 2 \\ 3 & 0 & \frac{1}{2} \\ 4 & 3 & \frac{2}{3} \end{bmatrix}$ ➡ $\begin{bmatrix} -1 & \frac{2}{3} & 2 \\ 0 & 3 & \frac{1}{2} \\ 3 & 4 & \frac{2}{3} \end{bmatrix}$ (we switched C_1 and C_2;

put -1 in list)

➡ $R_3 + 3R_1$ $\begin{bmatrix} -1 & \frac{2}{3} & 2 \\ 0 & 3 & \frac{1}{2} \\ 0 & 6 & \frac{20}{3} \end{bmatrix}$ $R_3 - 2R_2$ ➡ $\begin{bmatrix} -1 & \frac{2}{3} & 2 \\ 0 & 3 & \frac{1}{2} \\ 0 & 0 & \frac{17}{3} \end{bmatrix}$

We now have all entries below the main diagonal zero. Our list contains -1; the entries on the main diagonal are -1, 3, and $\frac{17}{3}$. Then by the flow chart the answer is

$$\det \begin{bmatrix} \frac{2}{3} & -1 & 2 \\ 3 & 0 & \frac{1}{2} \\ 4 & 3 & \frac{2}{3} \end{bmatrix} = (-1)(-1)(3)(\tfrac{17}{3}) = 17.$$

Problem 2   Find $\det \begin{bmatrix} 1 & 2 & 3 \\ 4 & 5 & 6 \\ 7 & 8 & 9 \end{bmatrix}$.

Solution $C_2 - 2C_1$ $C_3 - 3C_1$

$\begin{bmatrix} 1 & 2 & 3 \\ 4 & 5 & 6 \\ 7 & 8 & 9 \end{bmatrix}$ ➡ $\begin{bmatrix} 1 & 0 & 0 \\ 4 & -3 & -6 \\ 7 & -6 & -12 \end{bmatrix}$

We now have $C_3 = 2C_2$. Answer: 0.

There are millions of ways to calculate the determinant of a given matrix. Here for instance is another way to do this one:

$\begin{bmatrix} 1 & 2 & 3 \\ 4 & 5 & 6 \\ 7 & 8 & 9 \end{bmatrix}$ $\begin{matrix} R_2 - 4R_1 \\ R_3 - 7R_1 \end{matrix}$ ➡ $\begin{bmatrix} 1 & 2 & 3 \\ 0 & -3 & -6 \\ 0 & -6 & -12 \end{bmatrix}$

$$R_3 - 2R_2 \quad \begin{bmatrix} 1 & 2 & 3 \\ 0 & -3 & -6 \\ 0 & 0 & 0 \end{bmatrix}$$

All entries below the main diagonal are now zero. The main diagonal itself contains a zero; hence, the determinant of the matrix is 0.

Problem 3   Find $\det \begin{bmatrix} 7 & 6 & 2 & 0 \\ -3 & 9 & 6 & 12 \\ 1 & -3 & 6 & 2 \\ 0 & 3 & -4 & -6 \end{bmatrix}$.

Solution

$$\begin{bmatrix} 7 & 6 & 2 & 0 \\ -3 & 9 & 6 & 12 \\ 1 & -3 & 6 & 2 \\ 0 & 3 & -4 & -6 \end{bmatrix} \xrightarrow{\frac{1}{3}R_2} \begin{bmatrix} 7 & 6 & 2 & 0 \\ -1 & 3 & 2 & 4 \\ 1 & -3 & 6 & 2 \\ 0 & 3 & -4 & -6 \end{bmatrix}$$

$$\xrightarrow{\frac{1}{3}C_2 \; \frac{1}{2}C_4 \; \frac{1}{2}C_5} \begin{bmatrix} 7 & 2 & 1 & 0 \\ -1 & 1 & 1 & 2 \\ 1 & -1 & 3 & 1 \\ 0 & 1 & -2 & -3 \end{bmatrix}$$

(put 3, 3, 2, and 2 in the list)

$$\xrightarrow{C_2 - 2C_3} \begin{bmatrix} 7 & 0 & 1 & 0 \\ -1 & -1 & 1 & 2 \\ 1 & -7 & 3 & 1 \\ 0 & 5 & -2 & -3 \end{bmatrix} \xrightarrow{R_2 - R_1} \begin{bmatrix} 7 & 0 & 1 & 0 \\ 8 & -1 & 0 & 2 \\ 0 & -7 & 3 & 1 \\ 0 & 5 & -2 & -3 \end{bmatrix}$$

$$\xrightarrow{C_4 + 2C_2} \begin{bmatrix} 7 & 0 & 1 & 0 \\ -8 & -1 & 0 & 0 \\ 1 & -7 & 3 & -13 \\ 0 & 5 & -2 & 7 \end{bmatrix} \xrightarrow{C_1 - 7C_3} \begin{bmatrix} 0 & 0 & 1 & 0 \\ -8 & -1 & 0 & 0 \\ -20 & 7 & 3 & -13 \\ 14 & 5 & -2 & 7 \end{bmatrix}$$

We don't calculate entries we will never need.

$$\xrightarrow{C_1 - 8C_2} \begin{bmatrix} 0 & 0 & 1 & 0 \\ 0 & -1 & 0 & 0 \\ 36 & 7 & 3 & -13 \\ -26 & 5 & -2 & 7 \end{bmatrix} \xrightarrow{R_3 + \frac{13}{7}R_4} \begin{bmatrix} 0 & 0 & 1 & 0 \\ 0 & -1 & 0 & 0 \\ -\frac{86}{7} & - & - & 0 \\ -26 & 5 & -2 & 7 \end{bmatrix}$$

$$\begin{bmatrix} 1 & 0 & 0 & 0 \\ 0 & -1 & 0 & 0 \\ - & - & -86/7 & 0 \\ -2 & 5 & -26 & 7 \end{bmatrix}$$ (put -1 in list)

Answer: $(3)(3)(2)(2)(-1)(1)(-1)(-86/7)(7) = -3096$.

Problem 4   Find det $\begin{bmatrix} 1/2 & -1/4 & 3/4 & 1/2 \\ 3/4 & 0 & 0 & 1/4 \\ -1/4 & 1/2 & 1/2 & 1/4 \\ 1 & 1/4 & 5/4 & 1 \end{bmatrix}$.

Solution

$$\begin{bmatrix} 2 & -1 & 3 & 2 \\ 3 & 0 & 0 & 1 \\ -1 & 2 & 2 & 1 \\ 4 & 1 & 5 & 4 \end{bmatrix}$$ ($1/4$ in list four times)

$$\begin{bmatrix} 2 & -1 & 3 & 2 \\ 3 & 0 & 0 & 1 \\ 3 & 0 & 8 & 5 \\ 6 & 0 & 8 & 6 \end{bmatrix} \quad \begin{bmatrix} 0 & -1 & 0 & 0 \\ 3 & 0 & 0 & 1 \\ 3 & 0 & 8 & 5 \\ 6 & 0 & 8 & 6 \end{bmatrix}$$

$$\begin{bmatrix} 0 & -1 & 0 & 0 \\ 3 & 0 & 0 & 1 \\ 3 & 0 & 8 & 5 \\ 3 & 0 & 0 & 1 \end{bmatrix}$$

Now $R_2 = R_4$. Answer: 0.

Now that we have learned how to calculate determinants, we consider a number of practical applications of them.

Inverses   In Section 3-6, we learned that a 2×2 matrix has an inverse if, and only if, its determinant is not zero. The same is true for larger matrices.

If the determinant of a square matrix is not zero, then the matrix has an inverse. If the determinant is zero, the matrix has no inverse.

2

Problem 5   Does $\begin{bmatrix} 1 & 1 & -1 & 0 & -1 \\ -1 & 1 & 0 & 1 & 1 \\ 1 & 1 & 1 & -1 & -1 \\ -1 & -1 & 1 & 0 & -1 \\ 0 & 0 & 1 & 1 & 1 \end{bmatrix}$ have an inverse?

Solution

$$\begin{bmatrix} 1 & 1 & -1 & 0 & -1 \\ -1 & 1 & 0 & 1 & 1 \\ 1 & 1 & 1 & -1 & -1 \\ -1 & -1 & 1 & 0 & -1 \\ 0 & 0 & 1 & 1 & 1 \end{bmatrix} \Rightarrow \begin{bmatrix} 1 & 1 & -1 & 0 & -1 \\ 0 & 2 & -1 & 1 & 0 \\ 0 & 0 & 2 & -1 & 0 \\ 0 & 0 & 0 & 0 & -2 \\ 0 & 0 & 1 & 1 & 1 \end{bmatrix}$$

$$\Rightarrow \begin{bmatrix} 1 & 1 & -1 & 0 & -1 \\ 0 & 2 & -1 & 1 & 0 \\ 0 & 0 & 0 & -3 & -2 \\ 0 & 0 & 0 & 0 & -2 \\ 0 & 0 & 1 & 1 & 1 \end{bmatrix} \Rightarrow \begin{bmatrix} 1 & 1 & -1 & 0 & -1 \\ 0 & 2 & -1 & 1 & 0 \\ 0 & 0 & 1 & 1 & 1 \\ 0 & 0 & 0 & -3 & -2 \\ 0 & 0 & 0 & 0 & -2 \end{bmatrix}$$

There is a formula that uses determinants to find the inverse of a matrix, but it requires much more computational labor than does the method of Section 3-3.

(put   1, −1 in list)
We multiply: $(-1)(-1)(1)(2)(1)(-3)(-2) = 12 \neq 0$; so the matrix does have an inverse.

Systems of Linear Equations

Consider a system of linear equations in which the number of equations equals the number of different variables, so that the coefficient matrix of the system is a square matrix. If the determinant of this matrix is not zero, then by result **2** the matrix has an inverse, and by our work in Section 3-4, the system has a unique solution. It can be shown that the converse of this result is also the case; we summarize these facts in the following:

> **Consider a system of linear equations in which the number of equations equals the number of different variables. If the determinant of the coefficient matrix of the system is not zero, then the system has a unique solution. If this determinant is zero, then either the system has no solutions or else it has an infinite number of solutions.**

3

Problem 6   Does the system

$$\begin{bmatrix} 0.50x_1 - 0.25x_2 + 0.75x_3 + 0.50x_4 = 9.85 \\ 0.75x_1 \qquad\qquad\qquad + 0.25x_4 = -10.60 \\ -0.25x_1 + 0.50x_2 + 0.50x_3 + 0.25x_4 = 8.35 \\ 1.00x_1 + 0.25x_2 + 1.25x_3 + 1.00x_4 = -7.65 \end{bmatrix}$$

have a unique solution?

Solution   By the result of Problem 4, the coefficient matrix has determinant zero. Hence, the system does not have a unique solution.

Cramer's Rule   This is a rule for finding the value of any one variable in a system of linear equations in which the number of different variables equals the number of equations, provided the system has a unique solution. The rule, attributed to G. Cramer (1704–1752), can be used repeatedly to find the value of all the variables, but the amount of calculation involved is much greater than in Gauss-Jordan reduction. We present the rule in flow-chart form below.

Problem 7   For the system

$$\begin{bmatrix} ^2/_3x_1 - x_2 + 2x_3 = 4 \\ 3x_1 \qquad\quad + ^1/_2x_3 = 0 \\ 4x_1 + 3x_2 + ^2/_3x_3 = 0 \end{bmatrix}$$

Cramer's Rule: To Find the Value of Any One of the Variables in a System of Linear Equations

what is the value of x_3?

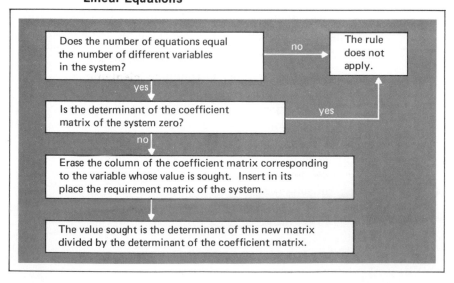

Solution   By Problem 1, the determinant of the coefficient matrix is 17. We form a new matrix by replacing C_3 of the coefficient matrix by the requirement matrix $\begin{bmatrix} 4 \\ 0 \\ 0 \end{bmatrix}$, and we calculate the determinant:

$$\begin{bmatrix} 2/3 & -1 & 4 \\ 3 & 0 & 0 \\ 4 & 3 & 0 \end{bmatrix} \implies \begin{bmatrix} 0 & 0 & 4 \\ 3 & 0 & 0 \\ 4 & 3 & 0 \end{bmatrix} \implies \begin{bmatrix} 4 & 0 & 0 \\ 0 & 0 & 3 \\ 0 & 3 & 4 \end{bmatrix} \quad \text{(put } -1 \text{ in list)}$$

$$\begin{bmatrix} 4 & 0 & 0 \\ 0 & 3 & 4 \\ 0 & 0 & 3 \end{bmatrix} \quad \text{(put another } -1 \text{ in list)}$$

Answer: $(-1)(-1)(4)(3)(3) = 36$. So $x_3 = {}^{36}/_{17}$.

Collinearity of Points in a Plane   Suppose we have three points in a coordinate plane and we wish to know if they are *collinear,* that is, if they all lie on one line. Let the coordinates of the points be $\begin{bmatrix} p_1 \\ p_2 \end{bmatrix}$, $\begin{bmatrix} q_1 \\ q_2 \end{bmatrix}$, and $\begin{bmatrix} r_1 \\ r_2 \end{bmatrix}$. If the points are collinear, then there is a linear equation, say $ax_1 + bx_2 = c$, with a and b not both zero, that is satisfied by all three points. Rewrite this equation $ax_1 + bx_2 - c = 0$. Then we have the system

$$\begin{bmatrix} ap_1 + bp_2 - c = 0 \\ aq_1 + bq_2 - c = 0 \\ ar_1 + br_2 - c = 0 \end{bmatrix} \qquad \textbf{4}$$

In this system it is a, b, and c that are the unknown variables! One solution of system **4** is, obviously, $a = b = c = 0$, but this is not a suitable solution since we must not have both a and b zero. Thus, we require that system **4** have another solution, that is, that system **4** *not* have a unique solution. The coefficient matrix of system **4** is

$$\begin{bmatrix} p_1 & p_2 & -1 \\ q_1 & q_2 & -1 \\ r_1 & r_2 & -1 \end{bmatrix} \qquad \textbf{5}$$

System **4** does not have a unique solution provided the determinant of matrix **4** is zero (see result **3**). In the determinant of matrix **5**, we may factor out -1 from C_3 without affecting whether or not the determinant is zero. We may also transpose the matrix without altering the determinant. The result is the flow chart on p. 356.

To Determine if the Points
$\begin{bmatrix} p_1 \\ p_2 \end{bmatrix}$, $\begin{bmatrix} q_1 \\ q_2 \end{bmatrix}$ and $\begin{bmatrix} r_1 \\ r_2 \end{bmatrix}$ are
Collinear

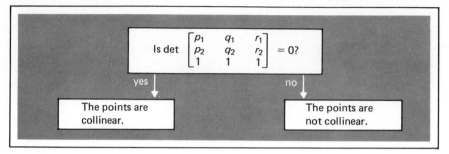

Is det $\begin{bmatrix} p_1 & q_1 & r_1 \\ p_2 & q_2 & r_2 \\ 1 & 1 & 1 \end{bmatrix} = 0$?

yes → The points are collinear.

no → The points are not collinear.

Problem 8   Are the points $\begin{bmatrix} 5 \\ -6 \end{bmatrix}$, $\begin{bmatrix} 29 \\ -38 \end{bmatrix}$, and $\begin{bmatrix} -37 \\ 50 \end{bmatrix}$ collinear?

Solution

$$\begin{bmatrix} 5 & 29 & -37 \\ -6 & -38 & 50 \\ 1 & 1 & 1 \end{bmatrix} \Rightarrow \begin{bmatrix} 0 & 24 & -42 \\ 0 & -32 & 56 \\ 1 & 1 & 1 \end{bmatrix}$$

$$\Rightarrow \begin{bmatrix} 0 & 4 & -7 \\ 0 & 4 & -7 \\ 1 & 1 & 1 \end{bmatrix} \quad \text{(list 6 and } -8\text{)}$$

We see that $R_1 = R_2$, so the determinant is zero. The points are collinear.

Equation for a Line   Given two distinct points $\begin{bmatrix} p_1 \\ p_2 \end{bmatrix}$ and $\begin{bmatrix} q_1 \\ q_2 \end{bmatrix}$, a variable point $\begin{bmatrix} x_1 \\ x_2 \end{bmatrix}$ is on the line determined by them provided it is collinear with them. By the flow chart above, these points are collinear if, and only if,

$$\det \begin{bmatrix} p_1 & q_1 & x_1 \\ p_2 & q_2 & x_2 \\ 1 & 1 & 1 \end{bmatrix} = 0 \qquad\qquad \textbf{6}$$

When specific numbers are given for p_1, p_2, q_1, q_2, this determinant may be found; it will be an expression involving x_1 and x_2. Then *equation* **6** *will be an equation for the line determined by* $\begin{bmatrix} p_1 \\ p_2 \end{bmatrix}$ *and* $\begin{bmatrix} q_1 \\ q_2 \end{bmatrix}$. This equation can be written in the form of either equation **5** in Section 2-1 or equation **6** in that section (see p. 64).

Problem 9 Find an equation for the line through $\begin{bmatrix} 3 \\ 4 \end{bmatrix}$ and $\begin{bmatrix} 5 \\ 6 \end{bmatrix}$.

Solution $\begin{bmatrix} 3 & 5 & x_1 \\ 4 & 6 & x_2 \\ 1 & 1 & 1 \end{bmatrix} \Rightarrow \begin{bmatrix} 1 & 1 & 1 \\ 4 & 6 & x_2 \\ 3 & 5 & x_1 \end{bmatrix}$ (put -1 in list) $\Rightarrow \begin{bmatrix} 1 & 1 & 1 \\ 0 & 2 & x_2 - 4 \\ 0 & 2 & x_1 - 3 \end{bmatrix}$

$$\Rightarrow \begin{bmatrix} 1 & 1 & 1 \\ 0 & 2 & x_2 - 4 \\ 0 & 0 & (x_1 - 3) - (x_2 - 4) \end{bmatrix} = \begin{bmatrix} 1 & 1 & 1 \\ 0 & 2 & x_2 - 4 \\ 0 & 0 & x_1 - x_2 + 1 \end{bmatrix}$$

We then have $\det \begin{bmatrix} 3 & 5 & x_1 \\ 4 & 6 & x_2 \\ 1 & 1 & 1 \end{bmatrix} = (-1)(1)(2)(x_1 - x_2 + 1) =$

$-2(x_1 - x_2 + 1) = -2x_1 + 2x_2 - 2$, and equation **6** gives

$$-2x_1 + 2x_2 - 2 = 0$$

as an equation for the line. This may be written

$$-2x_1 + 2x_2 = 2$$

which is of the form of equation **5** in Section 2-1. A "neater" form may be had by multiplying this equation through by $\frac{1}{2}$:

$$-x_1 + x_2 = 1,$$

in agreement with the answer to Exercise 9(i) of Section 2-1. This equation when written in the form of equation **6** in Section 2-1 is

$$[-1 \quad 1] \begin{bmatrix} x_1 \\ x_2 \end{bmatrix} = 1$$

Area of a Triangle Three points are on a line if, and only if, the "triangle" they form has area zero. Could it be that the determinant in the previous flow chart has something to do with the area of the triangle formed by the points? Yes, it has. It can be shown that this determinant is just twice the area of the triangle, except for a possible minus sign caused by the order in which the points are written in the determinant. We present this fact as follows.

The *absolute value* of a number is defined as the distance from the number to zero. See Appendix A, Section 9, if needed.

The area of the triangle formed by the points $\begin{bmatrix} p_1 \\ p_2 \end{bmatrix}$, $\begin{bmatrix} q_1 \\ q_2 \end{bmatrix}$, and $\begin{bmatrix} r_1 \\ r_2 \end{bmatrix}$ is equal to the absolute value of   **7**

$$\frac{1}{2} \det \begin{bmatrix} p_1 & q_1 & r_1 \\ p_2 & q_2 & r_2 \\ 1 & 1 & 1 \end{bmatrix}$$

Problem 10   Find the area of the triangle formed by the points $\begin{bmatrix} -2 \\ -3 \end{bmatrix}$, $\begin{bmatrix} 3 \\ 4 \end{bmatrix}$, and $\begin{bmatrix} 5 \\ -7 \end{bmatrix}$.

Solution

$$\begin{bmatrix} -2 & 3 & 5 \\ -3 & 4 & -7 \\ 1 & 1 & 1 \end{bmatrix} \Rightarrow \begin{bmatrix} 0 & 5 & 7 \\ 0 & 7 & -4 \\ 1 & 1 & 1 \end{bmatrix} \Rightarrow \begin{bmatrix} 0 & 0 & {}^{69}/_{7} \\ 0 & 7 & -4 \\ 1 & 1 & 1 \end{bmatrix}$$

$$\Rightarrow \begin{bmatrix} 1 & 1 & 1 \\ 0 & 7 & -4 \\ 0 & 0 & {}^{69}/_{7} \end{bmatrix} \quad \text{(put } -1 \text{ in list)}$$

The determinant is $(-1)(1)(7)({}^{69}/_{7}) = -69$. Thus, the area is the absolute value of $\frac{1}{2}(-69)$, that is, ${}^{69}/_{2}$.

Volume of a Tetrahedron   A *tetrahedron* is a pyramid with a triangular base and three triangular faces. For example, a bounded region of feasibility in space with just four corner points, not all in one plane, is a tetrahedron. The following statement is closely analogous to result **7**.

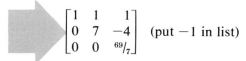

The volume of the tetrahedron with

corner points $\begin{bmatrix} p_1 \\ p_2 \\ p_3 \end{bmatrix}$, $\begin{bmatrix} q_1 \\ q_2 \\ q_3 \end{bmatrix}$, $\begin{bmatrix} r_1 \\ r_2 \\ r_3 \end{bmatrix}$, **and** $\begin{bmatrix} s_1 \\ s_2 \\ s_3 \end{bmatrix}$

is equal to the absolute value of   **8**

$$\frac{1}{6}\det \begin{bmatrix} p_1 & q_1 & r_1 & s_1 \\ p_2 & q_2 & r_2 & s_2 \\ p_3 & q_3 & r_3 & s_3 \\ 1 & 1 & 1 & 1 \end{bmatrix}$$

Problem 11   Find the volume of the tetrahedron of Figure 21 in Section 5-3.

Solution

$$\begin{bmatrix} 0 & 3 & 0 & 0 \\ 0 & 0 & 3 & 0 \\ 0 & 0 & 0 & 2 \\ 1 & 1 & 1 & 1 \end{bmatrix} \Rightarrow \begin{bmatrix} 3 & 0 & 0 & 0 \\ 0 & 3 & 0 & 0 \\ 0 & 0 & 2 & 0 \\ 1 & 1 & 1 & 1 \end{bmatrix} \quad \text{(put } -1, -1, -1 \text{ in list)}$$

Then $\frac{1}{6}$ times the determinant is $(\frac{1}{6})(-1)(-1)(-1)(3)(3)(2)(1) = -3$, and so the volume is 3.

Problem 12   Find the volume of the tetrahedron with corner points $\begin{bmatrix} 2 \\ 2 \\ 6 \end{bmatrix}$, $\begin{bmatrix} 1 \\ 0 \\ 4 \end{bmatrix}$,

$$\begin{bmatrix} 0 \\ 3 \\ 5 \end{bmatrix}, \text{ and } \begin{bmatrix} -1 \\ 1 \\ 3 \end{bmatrix}.$$

Solution

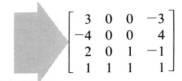

$$\begin{bmatrix} 2 & 1 & 0 & -1 \\ 2 & 0 & 3 & 1 \\ 6 & 4 & 5 & 3 \\ 1 & 1 & 1 & 1 \end{bmatrix} \Rightarrow \begin{bmatrix} 1 & 0 & -1 & -2 \\ 2 & 0 & 3 & 1 \\ 2 & 0 & 1 & -1 \\ 1 & 1 & 1 & 1 \end{bmatrix}$$

$$\Rightarrow \begin{bmatrix} 3 & 0 & 0 & -3 \\ -4 & 0 & 0 & 4 \\ 2 & 0 & 1 & -1 \\ 1 & 1 & 1 & 1 \end{bmatrix}$$

We see that $R_2 = -\frac{4}{3}R_1$; so the determinant is zero; so the volume
is zero. The four points are *coplanar*, that is, they all lie on the
same plane.

Exercises 6.5   1. Use the flow chart to find each of the following determinants.
(Check your answers to Parts a and b by equation **1**.)

(a) $\det \begin{bmatrix} \frac{2}{3} & -\frac{1}{4} \\ 3 & \frac{2}{5} \end{bmatrix}$

A(b) $\det \begin{bmatrix} -\frac{7}{4} & \frac{5}{6} \\ 21 & -10 \end{bmatrix}$

(c) $\det \begin{bmatrix} -1 & 2 & -3 \\ 4 & -5 & 6 \\ -7 & 8 & -9 \end{bmatrix}$

A(d) $\det \begin{bmatrix} 1 & 2 & 3 \\ 8 & 9 & 4 \\ 7 & 6 & 5 \end{bmatrix}$

(e) $\det \begin{bmatrix} 1 & \frac{1}{2} & \frac{1}{3} \\ \frac{1}{4} & \frac{1}{5} & \frac{1}{6} \\ \frac{1}{7} & \frac{1}{8} & \frac{1}{9} \end{bmatrix}$

A(f) $\det \begin{bmatrix} 0 & 2 & 3 & 0 \\ 2 & 0 & 0 & 3 \\ 0 & 3 & 2 & 0 \\ 3 & 0 & 0 & 2 \end{bmatrix}$

(g) $\det \begin{bmatrix} -2 & 5 & 3 & 4 \\ 3 & -4 & -2 & -5 \\ 4 & 9 & -4 & 9 \\ 7 & 1 & 6 & 12 \end{bmatrix}$

^A(h) $\det \begin{bmatrix} -\frac{1}{8} & -\frac{1}{9} & \frac{1}{7} & -\frac{1}{6} \\ 0 & 1 & 2 & 3 \\ 0 & 8 & 9 & 4 \\ 0 & 7 & 6 & 5 \end{bmatrix}$

(i) $\det \begin{bmatrix} -5 & 4 & -2 & 3 & -4 \\ -2 & 3 & 4 & -2 & 5 \\ 3 & -4 & 1 & 5 & -3 \\ -1 & 3 & -4 & -4 & -2 \\ 4 & -2 & 3 & -3 & 4 \end{bmatrix}$

^A(j) $\det \begin{bmatrix} -1 & -1 & 1 & 1 & -1 & -1 \\ 1 & 1 & 1 & -1 & -1 & -1 \\ -1 & -1 & -1 & -1 & 1 & -1 \\ 1 & -1 & 1 & -1 & 1 & -1 \\ 1 & -1 & -1 & 1 & -1 & -1 \\ 1 & 1 & -1 & -1 & -1 & 1 \end{bmatrix}$

*(k) $\det \begin{bmatrix} a_1 & a_2 & a_3 \\ b_1 & b_2 & b_3 \\ c_1 & c_2 & c_3 \end{bmatrix}$

2. There is a method for evaluating the determinant of a 3×3 matrix (but not for larger matrices) that is somewhat like formula **1** for 2×2 matrices. We illustrate it by evaluating $\det \begin{bmatrix} 3 & 0 & -2 \\ -4 & 2 & 3 \\ 2 & -1 & 4 \end{bmatrix}$.

(1) Form a 5×3 matrix consisting of the given matrix followed by R_1 and R_2 of the given matrix:

$$\begin{bmatrix} 3 & 0 & -2 \\ -4 & 2 & 3 \\ 2 & -1 & 4 \\ 3 & 0 & -2 \\ -4 & 2 & 3 \end{bmatrix}$$

(2) Multiply the numbers in the diagonals and list the products in columns on the left and right of the matrix.

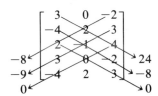

(3) Add up the columns:

$$\begin{bmatrix} -8 \\ -9 \\ 0 \\ \hline -17 \end{bmatrix} \qquad \begin{bmatrix} 24 \\ -8 \\ 0 \\ \hline 16 \end{bmatrix}$$

(4) The determinant equals the right-hand sum minus the left-hand sum:

$$\det \begin{bmatrix} 3 & 0 & -2 \\ -4 & 2 & 3 \\ 2 & -1 & 4 \end{bmatrix} = 16 - (-17) = 33$$

(a) As a check, calculate the above determinant by the flow-chart method.
Now use this method to rework:
(b) Exercise 1(c)
(c) Exercise 1(d)
(d) Exercise 1(e)
*(e) Exercise 1(k)
^{C}(f) Try the analogous method for the 4 × 4 matrix of exercise 1 (f). Does it give the right answer?

^{A}3. In connection with the design of mobile "burger vans" to serve the crews of the Alaska oil pipeline, PHE engineers came up with the following system of linear equations:

$$\begin{bmatrix} x_1 + 2x_2 - x_3 + 2x_4 - x_5 + 2x_6 = 18 \\ -x_1 - x_2 + 2x_3 \qquad + x_5 - x_6 = 0 \\ 3x_2 - x_3 + 2x_4 \qquad + 3x_6 = 9 \\ 2x_1 \qquad + 2x_3 \qquad + 3x_5 \qquad = 0 \\ 2x_2 \qquad - 3x_4 - 2x_5 + x_6 = -9 \\ x_1 \qquad + x_3 - x_4 + x_5 - 3x_6 = 0 \end{bmatrix}$$

What is the determinant of the coefficient matrix of this system?

4. Does the system $\begin{bmatrix} 8x_1 + 9x_2 + 4x_3 = 19 \\ x_1 + 2x_2 + 3x_3 = -47 \\ 7x_1 + 6x_2 + 5x_3 = 101 \end{bmatrix}$ have a unique solution?

^{A}5. Does the matrix $\begin{bmatrix} 4 & 9 & -4 & 9 \\ 3 & -4 & -2 & -5 \\ 7 & 1 & 6 & -12 \\ -2 & 5 & 3 & 4 \end{bmatrix}$ have an inverse?

6. In the system $\begin{bmatrix} x_1 + \frac{1}{2}x_2 + \frac{1}{3}x_3 = \frac{1}{2} \\ \frac{1}{4}x_1 + \frac{1}{5}x_2 + \frac{1}{6}x_3 = 0 \\ \frac{1}{7}x_1 + \frac{1}{8}x_2 + \frac{1}{9}x_3 = \frac{1}{3} \end{bmatrix}$, find x_2.

See Problem 1(j).   C7. In the system $\begin{bmatrix} -x_1 - x_2 + x_3 + x_4 - x_5 - x_6 = 0 \\ x_1 + x_2 + x_3 - x_4 - x_5 - x_6 = 0 \\ -x_1 - x_2 - x_3 - x_4 + x_5 - x_6 = 0 \\ x_1 - x_2 + x_3 - x_4 + x_5 - x_6 = 0 \\ x_1 - x_2 - x_3 + x_4 - x_5 - x_6 = 4 \\ x_1 + x_2 - x_3 - x_4 - x_5 + x_6 = 0 \end{bmatrix}$, find x_5.

8. Find the equations of the lines joining the following pairs of points:

(a) $\begin{bmatrix} -3 \\ 2 \end{bmatrix}, \begin{bmatrix} 6 \\ 4 \end{bmatrix}$ (b) $\begin{bmatrix} 5 \\ 0 \end{bmatrix}, \begin{bmatrix} 7 \\ 0 \end{bmatrix}$ (c) $\begin{bmatrix} 5 \\ 0 \end{bmatrix}, \begin{bmatrix} 0 \\ -3 \end{bmatrix}$ (d) $\begin{bmatrix} 2/3 \\ 3/4 \end{bmatrix}, \begin{bmatrix} 5/6 \\ 7/8 \end{bmatrix}$

A9. Are the points $\begin{bmatrix} 8 \\ 0 \end{bmatrix}, \begin{bmatrix} -13 \\ 14 \end{bmatrix}$, and $\begin{bmatrix} -1 \\ 6 \end{bmatrix}$ collinear?

10. What is the area of the triangle formed by the points $\begin{bmatrix} -1/2 \\ 2/3 \end{bmatrix}, \begin{bmatrix} 3/4 \\ -4/5 \end{bmatrix}$, and $\begin{bmatrix} -6/7 \\ 7/8 \end{bmatrix}$?

A11. Find the volume of the tetrahedron with corners $\begin{bmatrix} 1 \\ 2 \\ 3 \end{bmatrix}, \begin{bmatrix} 1 \\ 2 \\ 0 \end{bmatrix},$

$\begin{bmatrix} 1 \\ 0 \\ 0 \end{bmatrix}$ and $\begin{bmatrix} 0 \\ 0 \\ 1 \end{bmatrix}$.

12. Find the volume of the tetrahedron with corners $\begin{bmatrix} -1 \\ 2 \\ 1 \end{bmatrix}, \begin{bmatrix} 1 \\ 2 \\ 0 \end{bmatrix}, \begin{bmatrix} 1 \\ -1 \\ 3 \end{bmatrix}$

and $\begin{bmatrix} 4 \\ 3 \\ 10 \end{bmatrix}$.

*13. Proceeding by analogy with formula 6, develop a formula, using a determinant, for an equation for the plane through $\begin{bmatrix} p_1 \\ p_2 \\ p_3 \end{bmatrix}, \begin{bmatrix} q_1 \\ q_2 \\ q_3 \end{bmatrix}$, and $\begin{bmatrix} r_1 \\ r_2 \\ r_3 \end{bmatrix}$ assuming these points are not collinear.

14. Using the result of Exercise 13 (Appendix B), find equations for the planes through the following triples of points:

(a) $\begin{bmatrix} 1 \\ 0 \\ 0 \end{bmatrix}, \begin{bmatrix} 0 \\ 2 \\ 0 \end{bmatrix}, \begin{bmatrix} 0 \\ 0 \\ 3 \end{bmatrix}$ (b) $\begin{bmatrix} 1 \\ 2 \\ 3 \end{bmatrix}, \begin{bmatrix} 3 \\ 1 \\ 2 \end{bmatrix}, \begin{bmatrix} 2 \\ 3 \\ 1 \end{bmatrix}$ (c) $\begin{bmatrix} -1 \\ 2/3 \\ -4 \end{bmatrix}, \begin{bmatrix} 5/6 \\ -7 \\ 8/9 \end{bmatrix}, \begin{bmatrix} -10 \\ 11/12 \\ -13 \end{bmatrix}$

C15. The system of Exercise 3 resulted from comparing a new method of constructing PHE burger vans with the old method. Variable x_5 represents the gain in assembly time, in person-weeks of the new method. Assuming a person-week is 92 person-hours, how many person-hours are gained by the new method?

Section 6-6
Linear Programming and Computers

It has been the aim of this book to present, discuss, and explain various business applications of matrices. Although most of our examples have had a business flavor, this same mathematics has equally direct applications to economics, natural sciences, social sciences, and to other areas of mathematics. If we have seemed preoccupied with techniques it is because we feel that the only way to actually learn this kind of mathematics, with its applications and limitations, is through practical experience.

Once these methods are understood, computers can often be used as computational aids. In fact, most business applications involve problems so large that hand computation is out of the question. However, it must be kept in mind that a computer will not usually help you understand or interpret your problem. Before you submit a problem to a computer, you must have the problem properly formulated—and you must be sure that the mathematical expressions are actually relevant to the real-world situations being modeled. Furthermore, you must have some feeling for the answer. This is especially important because machines will use only the data you actually give them, not the data you intended to supply. It is easy to introduce errors into the input data. Whether these are merely typographical errors or actual errors of formulation, the machine will uselessly crank out an answer anyway. The only way to use a computer effectively is to understand the techniques it employs.

While computers can be programmed to perform all the computations we have presented, we will limit ourselves here to discussing the most complicated and perhaps most important technique, the simplex method. Our first few illustrations will involve problems we have already solved by hand. Later we will discuss some problems for which we are glad to have a computer available.

The particular computer program we will discuss is called LPS (Linear Programming System) and was developed by IBM (International Business Machines) for business use. We will relate LPS only to the types of problems that we have already treated. LPS itself is tremendously sophisticated and allows many variations. Our discussion will be vague and limited to an explanation of the computer output that follows. If you have LPS at your disposal, you will be able to learn the particulars by consulting the appropriate IBM manuals.

The first task is to get the data into the machine. Cards are punched giving each *nonzero* entry of the constraint and objective matrices. An unspecified entry is interpreted by the machine as zero. Next, the bound matrix is punched into cards. This time the zeros must be punched as well. To these cards are joined the title card, a card specifying the bound matrix, and a few other instruction cards, including the cards that call up the massive LPS program from storage. Next, a computer operator feeds the cards into the card reader, waits a few minutes, and hands you the output. This includes a listing of the input data for checking purposes. Finally, you must read and interpret the answer. Let us see how each of these steps is implemented in a few particular examples.

Example 1

(This is Problem 3, p. 247.)

Maximize $z = x_1 - x_2$ subject to $x_1 \geq 0$, $x_2 \geq 0$, $x_3 \geq 0$ and

$$y_1 = x_1 - 2x_2 \leq 2$$
$$y_2 = 2x_1 - 3x_2 - x_3 \leq 6$$
$$y_3 = x_1 - 3x_2 + x_3 \leq 1$$

Note the minor changes. We have given the names y_1, y_2, and y_3 to each constraint and denoted the objective variable by z instead of y.

First we punch the constraint matrix. Perversely, LPS requires us to mention columns before rows. We begin with x_1. This variable has the coefficient 1 in the z row, 1 in the y_1 row, 2 in the y_2 row, and 1 in the y_3 row. We punch four cards as follows:

```
X1      Z       1.0
X1      Y1      1.0
X1      Y2      2.0
X1      Y3      1.0
```

(The X is in card column 5; the Z is in card column 15 and the . (decimal point) is in card column 30.)

Next, we turn to x_2 and punch the following four cards:

```
X2      Z       -1.0
X2      Y1      -2.0
X2      Y2      -3.0
X2      Y3      -3.0
```

Finally, we do x_3. Since we do not have to include cards where the coefficients are zero, we merely punch

```
X3      Y2      -1.0
X3      Y3      1.0
```

Unless you specify otherwise, LPS assumes that $x_1 \geq 0$, $x_2 \geq 0$, $x_3 \geq 0$. However, we must tell it that y_1 is bounded above by 2, y_2 by 6 and y_3 by 1. Furthermore, z is not bounded, it is *free*. The possible specifications are

LB for a $\geq$ (lower bound)
UB for a $\leq$ (upper bound)
FX for an $=$ (fixed variable)
FR for no bound (free variable)

This is a peculiarity of the LPS program and not of computer programs in general.

Even though you must use these specifications in your input, the computer will use the abbreviations

LO for LB and UP for UB

in one place, while in another place it uses LL to indicate that a variable attains its lower bound and UL to indicate that a variable attains its upper bound in the answer.

Apart from all of this LPS requires that you give the bounds a name. We decided to use the word "bound." Here are the cards:

```
FR      BOUND   Z
UB      BOUND   Y1      2.0
UB      BOUND   Y2      6.0
UB      BOUND   Y3      1.0
```

(The F is in card column 2; the B of BOUND is in card column 5; the Z is in card column 15; the decimal point is in card column 30.)

After these cards, we insert a few machine instruction cards. The printed output begins by listing our input:

```
NAME              LP1


COLUMNS

        X1      Z               1.000000
        X1      Y1              1.000000
        X1      Y2              2.000000
        X1      Y3              1.000000
        X2      Z              -1.000000
        X2      Y1             -2.000000
        X2      Y2             -3.000000
        X2      Y3             -3.000000
        X3      Y2             -1.000000
        X3      Y3              1.000000

BOUNDS

    FR BOUND    Z
    UP BOUND    Y1              2.000000
    UP BOUND    Y2              6.000000
    UP BOUND    Y3              1.000000

ENDATA
```

Next follow some data about the solution process. These data will tell us if the objective variable was optimized and, if not, whether it was because the feasible region was unbounded or nonexistent, that is, empty.

```
LPSOLUTION
ITERATION      VALUE OF    NUMBER      SUM OF
  NUMBER        •Z       •  INFEAS     INFEAS


    0            0.000     0           0.000
SOLUTION UNBOUNDED
ERROR BELOW TOLERANCE     0.000000
```

Finally, we obtain more detailed information about the answer.

```
LP1


VARIABLE     ENTRIES    SOLUTION      UPPER          LOWER
              TYPE      ACTIVITY      BOUND          BOUND


X1         B*   4       4.001  *899000.000          0.000
Z          B*   0       3.000  *899000.000   *899000.000
Y1         UL   0       2.000        2.000          0.000

Y2         B*   0       5.000        6.000          0.000
Y3         UL   0       1.000        1.000          0.000
X2         B*   4       1.001  *899000.000          0.000

X3         LL   2       0.000  *899000.000          0.000
VARIABLE CAUSES UNBOUNDED SOLUTION
```

We will say more about reading this part of the output later. For the moment, we simply note that the computer found the problem to be unbounded—just as we observed before when we solved this problem by hand.

Our next example yields a more interesting answer.

Example 2
(This is Problem 1, p. 268.)

Maximize $z = 3x_1 + 6x_2 + 8x_3 + 4x_4$ subject to $x_1 \geq 0$, $x_2 \geq 0$, $x_3 \geq 0$, $x_4 \geq 0$ and

$$
\begin{aligned}
y_1 &= 3x_1 + 2x_2 + x_3 + 4x_4 \leq 8 \\
y_2 &= 2x_1 + 2x_2 + 4x_3 \leq 6 \\
y_3 &= 4x_2 + 3x_3 + 4x_4 \leq 10 \\
y_4 &= 4x_1 + 4x_2 + 2x_3 + 2x_4 \leq 9
\end{aligned}
$$

The cards are punched, the operator feeds them into the machine, and the following information is printed out.

```
NAME            LP2


COLUMNS

     X1        Z          3.000000
     X1        Y1         3.000000
```

```
            X1        Y2              2.000000
            X1        Y4              4.000000
            X2        Z               6.000000
            X2        Y1              2.000000
            X2        Y2              2.000000
            X2        Y3              4.000000
            X2        Y4              4.000000
            X3        Z               8.000000
            X3        Y1              1.000000
            X3        Y2              4.000000
            X3        Y3              3.000000
            X3        Y4              2.000000
            X4        Z               4.000000
            X4        Y1              4.000000
            X4        Y3              4.000000
            X4        Y4              2.000000

        BOUNDS

         FR  BOUND    Z
         UP  BOUND    Y1              8.000000
         UP  BOUND    Y2              6.000000
         UP  BOUND    Y3             10.000000
         UP  BOUND    Y4              9.000000

        ENDATA
```

```
    LPSOLUTION
    ITERATION     VALUE OF    NUMBER     SUM OF
      NUMBER       'Z      '  INFEAS     INFEAS


        0           0.000       0        0.000
        3          17.626       0        0.000
    SOLUTION OPTIMUM
    ERROR BELOW TOLERANCE     0.000000
```

```
    LP2


    VARIABLE    ENTRIES   SOLUTION      UPPER        LOWER
              TYPE        ACTIVITY      BOUND        BOUND


    X1        B*    4      0.251  *899000.000         0.000
    Z         B*    0     17.626  *899000.000  *899000.000
    Y1        UL    0      8.000       8.000          0.000

    Y2        UL    0      6.000       6.000          0.000
    Y4        B*    0      6.688       9.000          0.000
    X2        LL    5      0.000  *899000.000         0.000

    Y3        UL    0     10.000      10.000          0.000
    X3        B*    5      1.376  *899000.000         0.000
    X4        B*    4      1.469  *899000.000         0.000
```

After checking our input, we read that 3 *iterations* (or clearings of columns) were required to achieve the solution $z = 17.626$, which is optimal. The actual answer is 17.625, but the machine works with decimal arithmetic and hence there is often a slight round-off error.

The last nine lines contain the answer. We will interpret each of these lines.

Line 1   This line refers to x_1. We see that it is bounded (the B*) and are told that it occurred in 4 constraints (counting z). Under the heading SOLUTION ACTIVITY we find its value, .251, in the final answer. (Our hand calculations yielded .25.) Nest we see the number *899000.000. This is a puzzling and unexpected number, and besides, though we find it under the heading UPPER BOUND, we did not specify an upper bound for x_1. The explanation is simple: under upper bound and lower bound are listed our specifications. If you do not specify an upper bound, the program tries to insert infinity. But the machine prints *899000.000. Thus, all this number indicates is that no upper bound was specified. Under the heading LOWER BOUND, we find 0.000. This is just as we wished.

Line 2   Here we see that z, the objective variable, is bounded, occurs in no constraints, and has the value 17.626 in the final answer. The last columns indicate that neither an upper nor a lower bound was specified for z.

Line 3 y_1 achieves its upper bound, occurs in no constraints (it *is* a constraint), and takes on the value 8.000. Its upper bound was specified to be 8.000, and no lower bound was indicated, although the machine supplies the harmless lower bound zero.

Line 4 y_2 achieves its upper bound, occurs in no constraints, and takes on the value 6.000 with bounds as indicated.

Line 5 y_4 is bounded, occurs in no constraints, takes on the value 6.688, and has bounds as indicated.

Line 6 x_2 achieves its lower bound, occurs in five constraints (including the objective variable), takes on the value 0, and was unbounded above but bounded below by 0.

Line 7 y_3 achieves its upper bound, 10.

Line 8 x_3 is bounded and has the value 1.376. (We obtained 1.375 by hand computation.)

Line 9 x_4 is bounded and has the value 1.469. (We obtained 1.46875 by hand computation.)

Summary

$$x_1 = .251$$
$$x_2 = 0.000$$
$$x_3 = 1.376$$
$$x_4 = 1.469$$

Constraints 1, 2, and 4 are met exactly, while constraint 3 is not met but has slack of $9.000 - 6.688 = 2.312$.

The value of the objective variable is 17.626.

Example 3   Minimize $z = 3x_1 + 9x_2 + 12x_3 + 7x_4 + 9x_5$ subject to $x_1 \geq 0$,
(This is Problem 2. p. 270.) $x_2 \geq 0$, $x_3 \geq 0$, $x_4 \geq 0$, $x_5 \geq 0$, and

$$y_1 = \quad x_1 + 2x_2 + 3x_3 + \quad x_4 + 2x_5 \geq 3$$
$$y_2 = \quad 2x_1 + \quad x_2 \qquad\quad + \quad x_4 + 3x_5 \geq 2$$
$$y_3 = \tfrac{1}{2}x_1 + 2x_2 + 2x_3 + 3x_4 + 2x_5 \geq 2$$

The punched cards are fed in and the following is printed out.

```
NAME              LP3


COLUMNS

      X1        Z              3.000000
      X1        Y1             1.000000
      X1        Y2             2.000000
      X1        Y3             0.500000
      X2        Z              9.000000
      X2        Y1             2.000000
      X2        Y2             1.000000
      X2        Y3             2.000000
      X3        Z             12.000000
      X3        Y1             3.000000
      X3        Y3             2.000000
      X4        Z              7.000000
      X4        Y1             1.000000
      X4        Y2             1.000000
      X4        Y3             3.000000
      X5        Z              9.000000
      X5        Y1             2.000000
      X5        Y2             3.000000
      X5        Y3             2.000000


  BOUNDS

   FR BOUND     Z
   LO BOUND     Y1             3.000000
   LO BOUND     Y2             2.000000
   LO BOUND     Y3             2.000000

  ENDATA
```

```
LPSOLUTION
ITERATION    VALUE OF    NUMBER      SUM OF
  NUMBER       *Z      *  INFEAS     INFEAS


      0              0.000      3         4.221
      2              9.800      0         0.000
SOLUTION OPTIMUM
ERROR BELOW TOLERANCE    0.000000



  LP3



VARIABLE    ENTRIES    SOLUTION       UPPER          LOWER
            TYPE       ACTIVITY       BOUND          BOUND


X1        B*    4       2.801  *899000.000            0.000
Z         B*    0       9.800  *899000.000   *399000.000
Y1        LL    0       3.000  *899000.000            3.000

Y2        B*    0       5.801  *899000.000            2.000
Y3        LL    0       2.000  *899000.000            2.000
X2        LL    4       0.000  *899000.000            0.000

X3        LL    3       0.000  *899000.000            0.000
X4        B*    4       0.200  *899000.000            0.000
X5        LL    4       0.000  *899000.000            0.000
```

We see that two iterations were required to find the minimal value of z. The answer is $z = 9.8$, $x_1 = 2.801$, $x_2 = 0$, $x_3 = 0$, $x_4 = .200$, and $x_5 = 0$. Allowing for small round-off errors this agrees with our previous solution.

Example 4
(This is Problem 9, p. 337.)

Minimize $z = 2x_1 + x_2 + 3x_3$ subject to $x_1 \geq 0$, $x_2 \geq 0$, $x_3 \geq 0$, and

$$y_1 = x_1 + 2x_2 + x_3 \geq 3$$
$$y_2 = 4x_1 + 2x_2 + 6x_3 \geq 4$$

```
NAME              LP4


COLUMNS

       X1        Z            2.000000
       X1        Y1           1.000000
       X1        Y2           4.000000
       X2        Z            1.000000
       X2        Y1           2.000000
       X2        Y2           2.000000
       X3        Z            3.000000
```

```
              X3          Y1             1.000000
              X3          Y2             6.000000

          BOUNDS

           FR BOUND      Z
           LO BOUND      Y1             3.000000
           LO BOUND      Y2             4.000000

          ENDATA



          LPSOLUTION
          ITERATION     VALUE OF    NUMBER      SUM OF
            NUMBER       'Z      '  INFEAS      INFEAS


              0            0.000       2         3.568
              1            2.000       0         0.000
          SOLUTION OPTIMUM
          ERROR BELOW TOLERANCE      0.000000



          LP4



          VARIABLE    ENTRIES    SOLUTION       UPPER          LOWER
                     TYPE        ACTIVITY       BOUND          BOUND


          X1      LL      3      0.000  *899000.000          0.000
          Z       B*      0      2.000  *899000.000   *899000.000
          Y1      B*      0      4.000  *899000.000          3.000

          Y2      LL      0      4.000  *899000.000          4.000
          X2      B*      3      2.000  *899000.000          0.000
          X3      LL      3      0.000  *899000.000          0.000
```

LPS has given us the simplest solution but, as we know, this problem has many solutions.

Example 5   Maximize $z = 3.12x_1 + 6.8x_2 + 7.43x_3 + 4.01x_4$ subject to $x_1 \geq 0$, $x_2 \geq 0$, $x_3 \geq 0$, $x_4 \geq 0$, and

$$
\begin{aligned}
y_1 &= 2.91x_1 + 2.44x_2 + .87x_3 + 4.34x_4 \leq 8.29 \\
y_2 &= 1.78x_1 + 2.19x_2 + 4.00x_3 \leq 5.98 \\
y_3 &= 4.28x_2 + 2.60x_3 + 3.92x_4 \leq 9.84 \\
y_4 &= 4.60x_1 + 3.60x_2 + 2.10x_3 + 1.80x_4 \leq 9.20
\end{aligned}
$$

```
NAME           LP5

COLUMNS

     X1       Z             3.120000
     X1       Y1            2.910000
     X1       Y2            1.780000
     X1       Y4            4.000000
     X2       Z             6.800000
     X2       Y1            2.440000
     X2       Y2            2.190000
     X2       Y4            3.600000
     X2       Y3            4.280000
     X3       Z             7.430000
     X3       Y1            0.870000
     X3       Y2            4.000000
     X3       Y4            2.100000
     X3       Y3            2.600000
     X4       Z             4.010000
     X4       Y1            4.340000
     X4       Y4            1.800000
     X4       Y3            3.920000

BOUNDS

  FR BOUND    Z
  UP BOUND    Y1            8.290000
  UP BOUND    Y2            5.980000
  UP BOUND    Y4            9.200000
  UP BOUND    Y3            9.840000

ENDATA



LPSOLUTION
ITERATION    VALUE OF   NUMBER      SUM OF
  NUMBER      *Z      * INFEAS      INFEAS


     0           0.000    0          0.000
     4          17.448    0          0.000
SOLUTION OPTIMUM
ERROR BELOW TOLERANCE    0.000000


  LP5



VARIABLE    ENTRIES   SOLUTION      UPPER        LOWER
             TYPE      ACTIVITY      BOUND        BOUND


  X1        B*   4      0.475  *899000.000        0.000
  Z         B*   0     17.448  *899000.000 *899000.000
  Y1        UL   0      8.290       8.290         0.000
```

Note that some of this data is out of order. The computer doesn't mind.

Y2	UL	0	5.980	5.980	0.000
Y4	UL	0	9.200	9.200	0.000
X2	B*	5	1.174	*899000.000	0.000
Y3	UL	0	9.840	9.840	0.000
X3	B*	5	0.642	*899000.000	0.000
X4	B*	4	0.804	*899000.000	0.000

This problem is a slight modification of Example 2. Notice that the answer is also only slightly different. Do not think that what happened in Example 5 always happens, however. Consider the following problems.

Problem 1   Maximize $y = 5.001x_1 + 100.01x_2$ subject to $x_1 \geq 0$, $x_2 \geq 0$, and $x_1 + 20x_2 \leq 20$.

Solution   The region of feasibility is a triangle with corner points $\begin{bmatrix} 0 \\ 0 \end{bmatrix}$, $\begin{bmatrix} 0 \\ 1 \end{bmatrix}$, and $\begin{bmatrix} 20 \\ 0 \end{bmatrix}$.
The answer is $x_1 = 20$, $x_2 = 0$, $y = 100.02$.

Problem 2   Maximize $y = 5.001x_1 + 100.03x_2$ subject to $x_1 \geq 0$, $x_2 \geq 0$, and $x_1 + 20x_2 \leq 20$.

Solution   The region of feasibility is the same as in Problem 1. The answer is $x_1 = 0$, $x_2 = 1$, $y = 100.03$.

The only difference between these problems is a change of 0.01 in one of the objective coefficients, but values of x_1 and x_2 in the answers have changed by 20 and 1, respectively. One important aspect of linear programming is called *sensitivity analysis*. It is concerned with estimating the effect of slight modifications of a problem on the answer. We shall not further discuss this aspect here.

Example 6   We now turn our attention to more practical matters. The game Morra is a game that is actually played by adults. While it is a simple game, it is not easy to play because it is played very fast —usually in a noisy bar. To simplify matters, we will discuss and solve a version of three-finger Morra.

In this game, each player shouts a number and extends one, two, or three fingers. The players shout and extend fingers simultaneously. If exactly one player happens to shout a number equal to the total of fingers extended by both players, then he or she wins an amount equal to that number. If both or neither guess correctly, the round is a draw. For example, if A extends 1 finger and shouts 2 while B extends 1 finger and shouts 3, A wins \$2. Clearly, it is silly to extend 3 fingers and shout 2 or 3. When we eliminate all bad moves, there are nine strategies left. We will indicate a strategy by two numbers, x and y. The x indicates the number of fingers extended and the y indicates the number shouted. It is easily checked that the payoff matrix for this game is:

$$
A \quad
\begin{array}{c}
 \\
1\ 2 \\
1\ 3 \\
1\ 4 \\
2\ 3 \\
2\ 4 \\
2\ 5 \\
3\ 4 \\
3\ 5 \\
3\ 6
\end{array}
\begin{array}{c}
\hspace{2em} B \\
\begin{array}{ccccccccc}
1\ 2 & 1\ 3 & 1\ 4 & 2\ 3 & 2\ 4 & 2\ 5 & 3\ 4 & 3\ 5 & 3\ 6 \\
\end{array} \\
\left[
\begin{array}{rrrrrrrrr}
0 & 2 & 2 & -3 & 0 & 0 & -4 & 0 & 0 \\
-2 & 0 & 0 & 0 & 3 & 3 & -4 & 0 & 0 \\
-2 & 0 & 0 & -3 & 0 & 0 & 0 & 4 & 4 \\
3 & 0 & 3 & 0 & -4 & 0 & 0 & -5 & 0 \\
0 & -3 & 0 & 4 & 0 & 4 & 0 & -5 & 0 \\
0 & -3 & 0 & 0 & -4 & 0 & 5 & 0 & 5 \\
4 & 4 & 0 & 0 & 0 & -5 & 0 & 0 & -6 \\
0 & 0 & -4 & 5 & 5 & 0 & 0 & 0 & -6 \\
0 & 0 & -4 & 0 & 0 & -5 & 6 & 6 & 0
\end{array}
\right]
\end{array}
$$

From the description we see that the game is symmetric; i.e., neither player has an advantage. Thus A and B will have the same mixed strategies. We proceed to solve this game by applying the simplex method to the associated maximization problem. Since there are nine variables and nine constraints, we elect to use a computer!

See Section 6-1

When the matrix is not positive, the simplex method may fail to work. Just in case, we decide to add 6 to each entry to get a positive matrix. While the failure of the simplex method would have been no disaster, it would have meant repunching the data. Here is the result:

See the Flow Chart, p. 284.

```
NAME            MORRA


COLUMNS

        X1      Z           1.000000
        X1      Y1          6.000000
        X1      Y2          4.000000
        X1      Y3          4.000000
        X1      Y4          9.000000
        X1      Y5          6.000000
```

X1	Y6	6.000000
X1	Y7	10.000000
X1	Y8	6.000000
X1	Y9	6.000000
X2	Z	1.000000
X2	Y1	8.000000
X2	Y2	6.000000
X2	Y3	6.000000
X2	Y4	6.000000
X2	Y5	3.000000
X2	Y6	3.000000
X2	Y7	10.000000
X2	Y8	6.000000
X2	Y9	6.000000
X3	Z	1.000000
X3	Y1	8.000000
X3	Y2	6.000000
X3	Y3	6.000000
X3	Y4	9.000000
X3	Y5	6.000000
X3	Y6	6.000000
X3	Y7	6.000000
X3	Y8	2.000000
X3	Y9	2.000000
X4	Z	1.000000
X4	Y1	3.000000
X4	Y2	6.000000
X4	Y3	3.000000
X4	Y4	6.000000
X4	Y5	10.000000
X4	Y6	6.000000
X4	Y7	6.000000
X4	Y8	11.000000
X4	Y9	6.000000
X5	Z	1.000000
X5	Y1	6.000000
X5	Y2	9.000000
X5	Y3	6.000000
X5	Y4	2.000000
X5	Y5	6.000000
X5	Y6	2.000000
X5	Y7	6.000000
X5	Y8	11.000000
X5	Y9	6.000000
X6	Z	1.000000
X6	Y1	6.000000
X6	Y2	9.000000
X6	Y3	6.000000
X6	Y4	6.000000
X6	Y5	10.000000
X6	Y6	6.000000
X6	Y7	1.000000
X6	Y8	6.000000
X6	Y9	1.000000
X7	Z	1.000000
X7	Y1	2.000000

```
        X7       Y2          2.000000
        X7       Y3          6.000000
        X7       Y4          6.000000
        X7       Y5          6.000000
        X7       Y6         11.000000
        X7       Y7          6.000000
        X7       Y8          6.000000
        X7       Y9         12.000000
        X8       Z           1.000000
        X8       Y1          6.000000
        X8       Y2          6.000000
        X8       Y3         10.000000
        X8       Y4          1.000000
        X8       Y5          1.000000
        X8       Y6          6.000000
        X8       Y7          6.000000
        X8       Y8          6.000000
        X8       Y9         12.000000
        X9       Z           1.000000
        X9       Y1          6.000000
        X9       Y2          6.000000
        X9       Y3         10.000000
        X9       Y4          6.000000
        X9       Y5          6.000000
        X9       Y6         11.000000
        X9       Y7          0.000000
        X9       Y8          0.000000
        X9       Y9          6.000000

    BOUNDS

    FR BOUND     Z
    UP BOUND     Y1          1.000000
    UP BOUND     Y2          1.000000
    UP BOUND     Y3          1.000000
    UP BOUND     Y4          1.000000
    UP BOUND     Y5          1.000000
    UP BOUND     Y6          1.000000
    UP BOUND     Y7          1.000000
    UP BOUND     Y8          1.000000
    UP BOUND     Y9          1.000000

    ENDATA



LPSOLUTION
ITERATION   VALUE OF   NUMBER    SUM OF
  NUMBER     'Z     '  INFEAS    INFEAS


     0          0.000    0         0.000
     6          0.167    C         0.000
 SOLUTION OPTIMUM
 ERROR BELOW TOLERANCE     0.000000
```

MORRA

VARIABLE	ENTRIES TYPE		SOLUTION ACTIVITY	UPPER BOUND	LOWER BOUND
X1	LL	10	0.000	*899000.000	0.000
Z	B*	0	0.167	*899000.000	*899000.000
Y1	B*	0	0.973	1.000	0.000
Y2	UL	0	1.000	1.000	0.000
Y3	UL	0	1.000	1.000	0.000
Y4	B*	0	0.987	1.000	0.000
Y5	UL	0	1.000	1.000	0.000
Y6	B*	0	0.987	1.000	0.000
Y7	UL	0	1.000	1.000	0.000
Y8	UL	0	1.000	1.000	0.000
Y9	B*	0	0.973	1.000	0.000
X2	LL	10	0.000	*899000.000	0.000
X3	B*	10	0.070	*899000.000	0.000
X4	LL	10	0.000	*899000.000	0.000
X5	B*	10	0.056	*899000.000	0.000
X6	B*	10	0.000	*899000.000	0.000
X7	B*	10	0.042	*899000.000	0.000
X8	B*	10	0.000	*899000.000	0.000
X9	LL	10	0.000	*899000.000	0.000

We read from this the answer

$$z = .167 \doteq \tfrac{1}{6}$$
$$x_3 = .070$$
$$x_5 = .056$$
$$x_7 = .042$$

with all other variables zero.

Since we added 6 to the matrix, the value of the game is the reciprocal of $z \doteq \tfrac{1}{6}$ minus 6, that is, about zero. This was to be expected since neither player had any advantage. The best mixed strategy is, then

$$1 \ 4 : \tfrac{5}{12}$$
$$2 \ 4 : \tfrac{1}{3}$$
$$3 \ 4 : \tfrac{1}{4}$$

all others 0.

Example 7   When my friend told the board of directors of the Vertigo Knolls Mountaineering Club that I had just completed a course in linear programming with this book, they invited me to their next board meeting. The chairperson asked me in and then explained their problem as follows:

"Each year we've been having an outing for executives from the big three sporting-goods manufacturers, Appalachian Creations, Adirondack Knapsacks, and Pocono Productions. We've hoped this would induce them to donate equipment. However, in the past the executives haven't been very happy. Here are the most frequent complaints:

(1) the tents were old;
(2) some executives weren't provided with a tent manufactured by their own company;
(3) there were too many off-brand tents on the site.
(4) Our camp manager even complained that there wasn't enough room in his big truck to take all the tents to the site in one trip.

"The board has decided to buy all new tents. Here is a list of the best tents available (Table 16).

Table 16

Tent number	Manufacturer	Size		Cost ($)
		No. of persons	Cubic in. (packed)	
1	AC	2	804	70
2	AC	2	616	105
3	AC	2	565	100
4	AC	2	565	80
5	AK	2	565	70
6	AK	2	650	98
7	AK	3	285	150
8	AK	2	678	125
9	AK	1	275	35
10	PP	2	427	110
11	PP	2	553	150
12	PP	3	554	142
13	PP	6	1428	195
14	PP	2	396	100
15	PP	2	565	125
16	MA	2	565	165
17	TA	2	285	43
18	CB	2	906	142
19	DB	3	1011	159

"The latest figures on attendance indicate that there will be 85 people, of whom 17 are from AC, 15 from AK, 13 from PP, and 40 from the club. There are 25,920 cubic inches (— 15 cubic feet) of truck space.

"How can we meet our requirements at minimal cost?"

I realized at once that this was an LP minimization problem. I let x_1 be the number of number 1 tents to be purchased, and so forth. The total cost to be minimized was then

$$70x_1 + 105x_2 + \ldots (\text{etc.}) \ldots + 142x_{18} + 159x_{19}$$

I isolated six conditions:

(1) There must be enough tent space for all 85 people. Thus,

$$y_1 = 2x_1 + \ldots + 3x_{19} \geq 85$$

(2) There must be room for at least 17 AC executives in AC tents:

$$y_2 = 2x_1 + 2x_2 + 2x_3 + 2x_4 \geq 17$$

(3) There must be room for at least 15 AK executives in AK tents:

$$y_3 = 2x_5 + 2x_6 + 3x_7 + 2x_8 + x_9 \geq 15$$

(4) Similarly, for the 13 PP executives:

$$y_4 = 2x_{10} + 2x_{11} + 3x_{12} + 6x_{13} + 2x_{14} + 2x_{15} \geq 13$$

(5) The total number of cubic inches must not exceed 25,920:

$$y_5 = 804x_1 + \ldots + 1011x_{19} \leq 25,920$$

(6) Finally, the total number of AC, AK, and PP tents must exceed all others:

$$x_1 + \ldots + x_{15} \geq x_{16} + x_{17} + x_{18} + x_{19},$$

that is,

$$y_6 = x_1 + \ldots + x_{15} - x_{16} - \ldots - x_{19} \geq 0$$

I punched the cards, submitted them, and obtained the following output:

```
NAME            VKMCLUB


COLUMNS

       X1        Z            70.000000
       X1        Y1            2.000000
       X1        Y2            2.000000
```

X1	Y5	804.000000
X1	Y6	1.000000
X2	Z	105.000000
X2	Y1	2.000000
X2	Y2	2.000000
X2	Y5	616.000000
X2	Y6	1.000000
X3	Z	100.000000
X3	Y1	2.000000
X3	Y2	2.000000
X3	Y5	565.000000
X3	Y6	1.000000
X4	Z	80.000000
X4	Y1	2.000000
X4	Y2	2.000000
X4	Y5	565.000000
X4	Y6	1.000000
X5	Z	70.000000
X5	Y1	2.000000
X5	Y5	565.000000
X5	Y6	1.000000
X5	Y3	2.000000
X6	Z	98.000000
X6	Y1	2.000000
X6	Y5	650.000000
X6	Y6	1.000000
X6	Y3	2.000000
X7	Z	150.000000
X7	Y1	3.000000
X7	Y5	285.000000
X7	Y6	1.000000
X7	Y3	3.000000
X8	Z	125.000000
X8	Y1	2.000000
X8	Y5	678.000000
X8	Y6	1.000000
X8	Y3	2.000000
X9	Z	35.000000
X9	Y1	1.000000
X9	Y5	275.000000
X9	Y6	1.000000
X9	Y3	1.000000
X10	Z	110.000000
X10	Y1	2.000000
X10	Y5	427.000000
X10	Y6	1.000000
X10	Y4	2.000000
X11	Z	150.000000
X11	Y1	2.000000
X11	Y5	553.000000
X11	Y6	1.000000
X11	Y4	2.000000
X12	Z	142.000000
X12	Y1	3.000000
X12	Y5	554.000000
X12	Y6	1.000000

X12	Y4	3.000000
X13	Z	195.000000
X13	Y1	6.000000
X13	Y5	1428.000000
X13	Y6	1.000000
X13	Y4	6.000000
X14	Z	100.000000
X14	Y1	2.000000
X14	Y5	396.000000
X14	Y6	1.000000
X14	Y4	2.000000
X15	Z	125.000000
X15	Y1	2.000000
X15	Y5	565.000000
X15	Y6	1.000000
X15	Y4	2.000000
X16	Z	165.000000
X16	Y1	2.000000
X16	Y5	565.000000
X16	Y6	-1.000000
X17	Z	43.000000
X17	Y1	2.000000
X17	Y5	285.000000
X17	Y6	-1.000000
X18	Z	142.000000
X18	Y1	2.000000
X18	Y5	906.000000
X18	Y6	-1.000000
X19	Z	159.000000
X19	Y1	3.000000
X19	Y5	1011.000000
X19	Y6	-1.000000

BOUNDS

FR	BOUND	Z	
LO	BOUND	Y1	85.000000
LO	BOUND	Y2	17.000000
UP	BOUND	Y5	25920.000000
LO	BOUND	Y6	0.000000
LO	BOUND	Y3	15.000000
LO	BOUND	Y4	13.000000

ENDATA

LPSOLUTION

ITERATION NUMBER	VALUE OF *Z *	NUMBER INFEAS	SUM OF INFEAS
0	0.000	4	60.003
6	2402.500	0	0.000

SOLUTION OPTIMUM
ERROR BELOW TOLERANCE   0.000000

```
VKMCLUB


VARIABLE    ENTRIES   SOLUTION      UPPER         LOWER
         TYPE         ACTIVITY      BOUND         BOUND


X1       B*    5         8.500  *899000.000        0.000
Z        B*    0      2402.500  *899000.000  *899000.000
Y1       LL    0        85.000  *899000.000       85.000

Y2       LL    0        17.000  *899000.000       17.000
Y5       B*    0     19837.999    25920.000        0.000
Y6       LL    0         0.000  *899000.000        0.000

X2       LL    5         0.000  *899000.000        0.000
X3       LL    5         0.000  *899000.000        0.000
X4       LL    5         0.000  *899000.000        0.000

X5       B*    5         5.667  *899000.000        0.000
Y3       LL    0        15.000  *899000.000       15.000
X6       LL    5         0.000  *899000.000        0.000

X7       LL    5         0.000  *899000.000        0.000
X8       LL    5         0.000  *899000.000        0.000
X9       B*    5         3.667  *899000.000        0.000

X10      LL    5         0.000  *899000.000        0.000
Y4       LL    0        13.000  *899000.000       13.000
X11      LL    5         0.000  *899000.000        0.000

X12      LL    5         0.000  *899000.000        0.000
X13      B*    5         2.167  *899000.000        0.000
X14      LL    5         0.000  *899000.000        0.000

X15      LL    5         0.000  *899000.000        0.000
X16      LL    4         0.000  *899000.000        0.000
X17      B*    4        20.001  *899000.000        0.000

X18      LL    4         0.000  *899000.000        0.000
X19      LL    4         0.000  *899000.000        0.000
```

The answer is $z = \$2402.50$, and

$$x_1 = 8.500$$
$$x_5 = 5.667$$
$$x_9 = 3.667$$
$$x_{13} = 2.167$$
$$x_{17} = 20.001$$

with all other variables zero.

When I returned to the club and showed them the answer, they were not impressed. "How can we possibly buy $8\frac{1}{2}$ tents?

$8^1/_2$ is a movie or a shoe size; it's not a whole number. Take it back and do it right."

Luckily for me, George, our mathematical programming expert, was in the computer room when I returned. I explained my problem, showed him this book, and the following dialogue ensued.

GEORGE: *This book is only an introduction. It has failed to explain several things that I feel could have been included, but perhaps the authors felt this would be too much for a beginning student.*

ME: *But what has this to do with tents?*

GEORGE: *Nothing. I thought I should warn you that you can't expect to learn everything from a one-semester introductory course. What's more, they've barely mentioned sensitivity analysis, branch and bound for integer programming, and several other important topics.*

ME: *But what should I do now?*

GEORGE: *Based on your experience with this course, what do you think?*

ME: *Well, I guess I'll have to round off. I'll try rounding some up, some down. I don't know. Maybe we can find more room in the truck or convince somebody not to go.*

GEORGE: *Mmm. Your problem is that you've gotten yourself into an* integer programming *situation. You realize, of course, that the transportation problem is an integer programming problem. However, it happens that the simplex method, when applied to transportation or assignment problems, always yields integer answers.*

ME: *That's lucky. But my problem is not a transportation problem.*

GEORGE: *Several techniques have been developed for more general situations; there are even computer programs available.*

ME: *Could we try some of them on this problem?*

GEORGE: *Unfortunately, this particular computing facility only purchased LPS. The integer programs were felt to be too specialized and not entirely satisfactory.*

ME: *Will rounding off work?*

GEORGE: *Generally not. Rounding off might destroy feasibility. It is unfortunate that the optimal integer feasible solution is not simply the rounded off noninteger optimum. It can happen that restricting the problem to integers radically alters the answer. The integer optimum can be very different indeed.*

ME: *What about in my case?*

GEORGE: *Oh, you're probably OK if you round off. Especially since your constraints can surely be stretched a little, and it appears that most of the tents are absolutely ruled out on account of size or price.*

ME: *OK. I'll take $x_1 = 9$, $x_5 = 6$, $x_9 = 4$, $x_{13} = 2$, $x_{17} = 20$, and refigure z by hand... \$2440. That should be acceptable.*

GEORGE: *Good. You know, you really should consider studying more LP. With your background from this course, it shouldn't be hard to get into more subtle aspects.*

ME: *I haven't got the time now. I just wanted to solve this one problem for the club. Oh, by the way, did you notice that this conversation we're having right now is recorded word for word in the last section of the book I used for the course?*

GEORGE: *No!*

Exercises 6.6   1. PHE store number 4173 in Ft. Scott, Kansas, has materials on hand as listed in Table 17. From these materials they make food items, which are sold for the profits listed in Table 18. Some of these items are assembled into special package meals, though they are also sold separately. The components and profits for the package meals are shown in Table 19. The store must sell at least 70 dogfeasts, exactly 22 fishfeasts, and at least 20 lbs. of potatoes; and they must use up all their hamburger patties. Maximize profit.

Table 17

Materials	Quantity
Hamburger patties	342
Cheese slices	194
Potatoes	35 lbs. (16 oz. in a lb.)
Cola syrup	26 qts. (32 oz. in a qt.)
Short wieners	244
Long wieners	110
Hamburger buns	284
Superburger buns	178
Short hot-dog buns	244
Long hot-dog buns	130
Fish patties	180

Table 18

Item	Materials	Profit (cents)
Hamburger	1 hamburger, 1 hamburger bun	7
Cheeseburger	1 ham., 1 cheese slice, 1 ham. bun	10
Superburger	2 ham., 1 cheese sl., 1 sup. bun	17
Small fries	3 oz. potato	6
Large fries	5 oz. potato	8
Hot dog	1 short wiener, 1 short h.d. bun	9
Foot-long hot dog	1 long wiener, 1 long h.d. bun	11
Fishburger	1 fish patty, 1 sup. bun	14
Small cola	0.4 oz. cola syrup	10
Medium cola	0.6 oz. cola syrup	15
Large cola	0.8 oz. cola syrup	20

Table 19

Package meal	Components	Profit (cents)
Burger snack	hamburger, small cola	19
Burger meal	cheeseburger, sm. fries, med. cola	35
Burgerfeast	superburger, lg. fries, lg. cola	52
Dogfeast	ft.-long hot dog, lg. fries, lg. cola	35
Fishfeast	fishburger, lg. fries, lg. cola	42

Supplementary Exercises

1. Solve the game $\begin{bmatrix} 1 & 2 & 3 \\ 4 & 5 & 6 \end{bmatrix}$.

A 2. Solve the game $\begin{bmatrix} 1 & -1 & 0 \\ -1 & 0 & 1 \\ 1 & 1 & -1 \end{bmatrix}$.

3. A and B play tic-tac-toe on a 3×3 matrix. A move like A(1,3) means A places X at address (1,3); move B(2,1) means B places O at address (2,1). Assuming the first two moves are A(2,2), B(1,2), write the complete winning strategy for A.

A4. Find the value of the game with payoff matrix $\begin{bmatrix} 1 & -1 & -2 \\ -1 & 1 & 1 \\ 2 & -1 & x \end{bmatrix}$

where $x \geq -1$.

5. Solve the matrix game $\begin{bmatrix} 4 & 2 \\ 1 & 3 \end{bmatrix}$.

^6. A and B each write the number 1, 2, 3, or 4 on a slip of paper. If the absolute value of the difference of the numbers written is less than 2, then B pays A $10. Otherwise, A pays B $5. How should each player proceed? Should one of the players pay something to the other to make the game fair? How much?

7. At the beginning, we had 10,944 customers and they had 11,520. During each of the first three weeks, $\frac{1}{2}$ of our customers switched to them and $\frac{1}{4}$ of theirs switched to us. During each of the next two weeks, $\frac{1}{3}$ of our customers switched to them and $\frac{2}{3}$ of theirs switched to us. At the end of the fifth week, how many customers do we have and how many do they have?

^8. Right now Shoddy Mfg. and Shiny Imports each have 1000 customers. Each month, $\frac{4}{5}$ of Shoddy's customers switch to Shiny and $\frac{1}{10}$ of Shiny's switch to Shoddy. Shoddy must have at least 225 customers to survive. How long do they have?

9. Given the Markov chain with transition matrix

$$\begin{bmatrix} \frac{3}{4} & \frac{1}{8} & 0 & \frac{1}{8} & 0 & 0 & 0 & 0 \\ 0 & 0 & 0 & \frac{2}{3} & 0 & \frac{1}{3} & 0 & 0 \\ 0 & 0 & 1 & 0 & 0 & 0 & 0 & 0 \\ 0 & 0 & 0 & \frac{1}{4} & \frac{1}{2} & \frac{1}{4} & 0 & 0 \\ 0 & \frac{1}{5} & \frac{1}{5} & \frac{2}{5} & \frac{1}{5} & 0 & 0 & 0 \\ 0 & \frac{2}{7} & 0 & 0 & 0 & \frac{4}{7} & 0 & \frac{1}{7} \\ 0 & 0 & \frac{1}{9} & \frac{2}{9} & \frac{1}{3} & 0 & 0 & \frac{1}{3} \end{bmatrix}$$

(a) Which states are absorbing states?
(b) Is this an absorbing chain?

^10. "Our campaign is fantastic," said the ad-agency director to the president of company A. "Look at the percentage of switches in one month between your company and companies B and C" (Table 20).

Table 20
Percentage of Switches

From	To		
	A	B	C
A	87.5	12.5	0
B	0	$33\frac{1}{3}$	$66\frac{2}{3}$
C	25	50	25

"Why, if this continues for a few years you'll wipe out your competition." What will in fact happen?

11. What fraction of a given distribution will finally enter each state of the Markov chain with the following transition matrix?

$$\begin{bmatrix} \frac{2}{5} & \frac{1}{5} & \frac{1}{5} & \frac{1}{5} & 0 \\ \frac{1}{4} & \frac{1}{4} & \frac{1}{4} & \frac{1}{4} & 0 \\ \frac{1}{6} & \frac{1}{6} & \frac{1}{3} & \frac{1}{3} & 0 \\ 0 & 0 & 0 & 1 & 0 \\ 0 & 0 & 0 & 0 & 1 \end{bmatrix}$$

A12. What will be the final ("limit") distribution of the initial distribution $[63 \quad 84 \quad 87 \quad 48 \quad 71]$ in the Markov chain with the following transition matrix?

$$\begin{bmatrix} 1 & 0 & 0 & 0 & 0 \\ 1/3 & 1/6 & 0 & 1/6 & 1/3 \\ 0 & 0 & 1 & 0 & 0 \\ 1/5 & 2/5 & 0 & 1/5 & 1/5 \\ 0 & 0 & 0 & 0 & 1 \end{bmatrix}$$

13. Find the maximum and minimum values of $y = -2x_1 + 14x_2$ subject to $x_1 + x_2 \leq 2$, $x_1 - 7x_2 \leq 18$, $3x_1 - x_2 \geq -6$.

C 14. A parfumier has two liquids. Liquid A contains 3 drams attar of roses and 15 grains ambergris per oz; liquid B contains 2 drams attar of roses and 4 grains ambergris per oz. The liquids cost, respectively, $120 and $80 per oz. How can he most economically prepare a perfume containing at least 15 drams attar of roses and 15 grains ambergris? He would like to use, as nearly as he can, the same amount of each liquid.

15. A(a) Transportation problems can also be solved by the simplex method. However, as we mentioned in Section 5-1, the problem is so special that in this case the simplex method can be completely recast and presented in the guise of patterns of change (see Chapter 4). When solving transportation problems using methods of this chapter, the number of variables is very large and the constraints take the form of equations instead of inequalities. Set up, but do not solve, the problem shown in Chart 1. Introduce the variables $x_1, x_2, \ldots, x_9$ as shown. For example, x_8 is the amount shipped from III to B. Write the objective variable and the constraint equations.

Chart 1

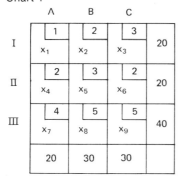

A(b) Solve the above problem by using the technique of Chapter 4. Express the result in terms of the variables in part a.

(c) Use the techniques of Section 6-4 to solve this problem. (Better use a computer.)

A 16. Maximize $y = 6x_1 + x_2$ subject to $x_1 \geq 0$, $x_2 \geq 0$, $x_3 \geq 0$, $2x_1 + x_2 \leq 2$, and $x_1 + x_2 + x_3 \leq 4$.

17. Let A be a square matrix with 1's on the "antidiagonal" (the diagonal from the lower left corner to the upper right) and 0's everywhere else. Find det A.

18. A(a) Let B be a 6 ×6 matrix whose determinant is $1/243$. Find det (3B).
   *(b) If D is a square matrix and r is a number, what is the relation between det (rD) and det D?

19. (a) Let $A = \begin{bmatrix} 1 & -2 \\ 3 & 2 \end{bmatrix}$ and $B = \begin{bmatrix} 3 & 2 \\ -1 & 4 \end{bmatrix}$. Calculate det A, det B, and det (AB).

(b) If C and D are square matrices of the same size, what do you think is the relation between det C, det D, and det (CD)?

A20. PHE is building a giant version of its corporate symbol, a tetrahedron. The coordinates of the corners of the tetrahedron, measured in feet from the statue of the PHE founder in the plaza of the headquarters building, are $\begin{bmatrix} 200 \\ 150 \\ 0 \end{bmatrix}, \begin{bmatrix} 350 \\ 400 \\ 0 \end{bmatrix}, \begin{bmatrix} 500 \\ 280 \\ 0 \end{bmatrix}, \begin{bmatrix} 400 \\ 300 \\ 650 \end{bmatrix}$. What is the volume of this monument to hamburger gastronomy?

21. Here is a little game, of no particular practical value. There are many different 2×2 matrices that can be made using the numbers 1, 2, 3, 4: $\begin{bmatrix} 1 & 2 \\ 3 & 4 \end{bmatrix}, \begin{bmatrix} 4 & 2 \\ 1 & 3 \end{bmatrix}$, etc. Of all such matrices, which ones have the largest determinant? One way to play this game is to try all possible matrices, but the real object of the game is to use the facts about determinants given in Section 6-5 to shorten your work. If you like this game, and if you are a little crazy, you might then play the game with 3×3 matrices made using 1, 2, 3, 4, 5, 6, 7, 8, 9. (We are told that the maximum is 412.)

APPENDIX A
Review of
Arithmetic
and Algebra

This appendix should not be considered as a general review of these subjects, since it covers only those topics required in this book. The appendix concludes with exercises, which can be used as a test of your preparedness.

**Section 1
Use of Parentheses**

An expression such as $13 - 4 \times 2$ is interpreted as $13 - (4 \times 2) = 13 - 8 = 5$; *not* as $(13 - 4) \times 2 = 9 \times 2 = 18$. If we had meant the latter expression we would have had to employ parentheses: $(13 - 4) \times 2$. The former expression, $13 - 4 \times 2$, will always be interpreted as $13 - (4 \times 2)$, whether or not the parentheses were inserted. In general in an expression in arithmetic, we always do the multiplications and divisions before the additions and subtractions unless there are parentheses to indicate otherwise. A few examples:

$$4 \div 2 + 3 \times 3 - 2 = 2 + 9 - 2 = 9;$$
$$6 \div (2 + 1) + (4 - 2) \div 2 - 6 \div 2 + 1 + 4 \div 2 - 2$$
$$= 6 \div 3 + 2 \div 2 - 3 + 1 + 2 - 2 = 2 + 1 - 3 + 1 + 2 - 2 = 1;$$
$$(1 + 2 + 3 - 4) \div (5 + 6 - 10) = 2 \div 1 = 2$$

**Section 2
Other Notations**

The product 3×4 is also written $3 \cdot 4$ and $(3)(4)$. Thus, $3 \times 4 = 3 \cdot 4 = (3)(4) = 12$. If one or more of the numbers in the product is denoted by a letter, such as $3 \times a$, the notation $3a$ is used. Thus, $3 \times a = 3 \cdot a = (3)(a) = 3a$; $7 \times x_1 = 7x_1$.

The quotient $6 \div 2$ is also written $6/2$, $\dfrac{6}{2}$, and $^6/_2$. Thus,

$$6 \div 2 = \frac{6}{2} = 6/2 = {}^6/_2 = 3.$$

A special notation for addition (mixed fraction notation) will be explained below.

**Section 3
Zero**

The laws for combining zero with any number a are:

$$a + 0 = a$$
$$a - 0 = a$$
$$a0 = 0$$
$$a \div 0 \text{ and } a/0 \text{ are undefined}$$
$$0 + a = a$$
$$0 - a = -a$$
$$0a = 0$$
$$0 \div a = 0/a = 0$$

Examples:

$$3 + 0 = 3$$
$$4 - 0 = 4$$
$$4 \cdot 0 = 0$$
$$7 \div 0 \text{ and } 7/0 \text{ are undefined}$$
$$0 + 8 = 8$$
$$0 - 8 = -8$$
$$(0)(8) = 0$$
$$^0/_{10} = 0$$

Section 4
Fractions

In the fraction $^3/_4$ the number 3 is called the *numerator* and the number 4 is called the *denominator*. From Section 3 we see that a fraction must not have denominator zero. Any whole number may be written as a fraction with denominator one: $7 = ^7/_1,\ 3 = ^3/_1,\ 0 = ^0/_1$.

4-1 Cancellation

Fractions can often be simplified by dividing their numerator and denominator by the same number. For example: $6/9 = (\not{3} \cdot 2)/(\not{3} \cdot 3) = 2/3$, $77/154 = \not{11} \cdot 7/\not{11} \cdot 14 = \not{7} \cdot 1/\not{7} \cdot 2 = 1/2$. It is usually a good idea to cancel, when possible, to make the denominator of a fraction as small as possible. It is not serious if you fail to notice a tricky cancellation, such as $3813/1763 = \not{41} \cdot 93/\not{41} \cdot 43 = 93/43$.

4-2 Mixed Fraction Form

In some cases fractions with a larger numerator than denominator, such as $^{18}/_7$, are better written in mixed form: $^{18}/_7 = 2^4/_7$. Here it must be kept in mind that $2^4/_7$ means $2 + ^4/_7$ and *not* $2 \times ^4/_7$. Another example, involving cancellation: $^{18}/_{12} = 1^6/_{12} = 1^1/_2$; another solution: $^{18}/_{12} = ^3/_2 = 1^1/_2$.

4-3 Multiplication of Fractions

To multiply fractions, multiply numerators and denominators, respectively, after cancellation where possible. Examples:

$$\frac{2}{3} \cdot \frac{5}{4} = \frac{\overset{1}{\not{2}}}{3} \cdot \frac{5}{\underset{2}{\not{4}}} = \frac{5}{6}; \quad \frac{4}{9} \cdot \frac{15}{14} = \frac{\overset{2}{\not{4}}}{\underset{3}{\not{9}}} \cdot \frac{\overset{5}{\not{15}}}{\underset{7}{\not{14}}} = \frac{10}{21};$$

$$\frac{9}{4} \cdot \frac{14}{15} = \frac{\overset{3}{\not{9}}}{\underset{2}{\not{4}}} \cdot \frac{\overset{7}{\not{14}}}{\underset{5}{\not{15}}} = \frac{21}{10} = 2^1/_{10}; \quad \frac{1}{4} \cdot \frac{4}{3} \cdot \frac{6}{8} = \frac{1}{\underset{1}{\not{4}}} \cdot \frac{\overset{1}{\not{4}}}{\underset{1}{\not{3}}} \cdot \frac{\overset{2}{\not{6}}}{\underset{4}{\not{8}}} = \frac{\overset{1}{\not{2}}}{\underset{4}{\not{8}}} = \frac{1}{4}$$

4-4 Division of Fractions

To divide one fraction by another, invert the divisor and multiply, as in 4-3. Examples:

$$\frac{2/3}{4/3} = \frac{2}{3} \cdot \frac{3}{4} = \frac{1}{2}; \quad \frac{4}{2/5} = \frac{4}{1} \cdot \frac{5}{2} = 10;$$

$$2\frac{3}{4} \div 3\frac{1}{2} = (\frac{11}{4})/(\frac{7}{2}) = (\frac{11}{4})(\frac{2}{7}) = \frac{11}{14};$$

$$3\frac{1}{2} \div 2\frac{3}{4} = (\frac{7}{2})/(\frac{11}{4}) = (\frac{7}{2})(\frac{4}{11}) = \frac{14}{11} = 1\frac{3}{11};$$

$$\frac{8/7}{6} = \frac{8}{7} \cdot \frac{1}{6} = \frac{4}{21}$$

See Section 5 for treatment of minus sign.

The *reciprocal* of a nonzero number is defined to be 1 divided by that number. For example, the reciprocal of 3 is $\frac{1}{3}$; the reciprocal of -3 is $1/-3 = -\frac{1}{3}$; the reciprocal of $\frac{4}{5}$ is $1/(\frac{4}{5}) = \frac{5}{4}$; the reciprocal of $-\frac{2}{3}$ is $1/(-\frac{2}{3}) = \frac{3}{-2} = -\frac{3}{2} = -1\frac{1}{2}$. The number 0 has no reciprocal since, if we had $\frac{1}{0} = x$, we would have $1 = 0 \cdot x = 0$, which is absurd. Thus, "$\frac{1}{0}$" is left undefined. (Note however that $\frac{0}{1}$ *is* defined, and in fact $\frac{0}{1} = 0$.)

The reciprocal of a nonzero number is also called the multiplicative inverse of that number. The reciprocal of 3 is also written 3^{-1}, read "3 inverse." Thus, $3^{-1} = \frac{1}{3}$, $(-3)^{-1} = -\frac{1}{3}$, $(\frac{4}{5})^{-1} = \frac{5}{4}$, $(-\frac{2}{3})^{-1} = -\frac{3}{2}$. Since 0 has no reciprocal, 0^{-1} is undefined.

4-5 Common Denominators

For purposes of addition, subtraction, and comparison it is convenient to write fractions with the same (common) denominators. For example, $\frac{1}{4}$ and $\frac{1}{6}$ can be written with common denominator 24: $\frac{1}{4} = \frac{6}{24}$, $\frac{1}{6} = \frac{4}{24}$. Actually, these fractions could be written with common denominator 12: $\frac{1}{4} = \frac{3}{12}$, $\frac{1}{6} = \frac{2}{12}$. Here, the number 12 is the smallest number that can serve as a common denominator; it is the *least common denominator* for $\frac{1}{4}$ and $\frac{1}{6}$. It is easy to find a common denominator for a collection of fractions: multiply together all of the different denominators. Usually the common denominator found in this way is not the least one. Finding that takes a bit more practice. Examples: the least common denominator of $\frac{2}{9}$, $\frac{4}{15}$, $\frac{5}{3}$ is 45: $\frac{2}{9} = \frac{10}{45}$, $\frac{4}{15} = \frac{12}{45}$, $\frac{5}{3} = \frac{75}{45}$; the least common denominator of $\frac{8}{5}$, $\frac{7}{4}$, $\frac{11}{6}$, $\frac{5}{2}$ is 60: $\frac{8}{5} = \frac{96}{60}$, $\frac{7}{4} = \frac{105}{60}$, $\frac{11}{6} = \frac{110}{60}$, $\frac{5}{2} = \frac{150}{60}$.

4-6 Addition and Subtraction of Fractions

To add or subtract fractions: (1) Find a common denominator for the fractions; (2) add or subtract, as required, the numerators of the fractions written with the common denominator; (3) put the result of step 2 over the common denominator; (4) cancel if possible. Examples: $\frac{1}{2} + \frac{2}{3} - \frac{3}{14} = \frac{6}{12} + \frac{8}{12} - \frac{9}{12} = (6 + 8 - 9)/12 = \frac{5}{12}$; $\frac{4}{15} + \frac{3}{5} - \frac{1}{15} = \frac{4}{15} + \frac{9}{15} - \frac{1}{15} = \frac{12}{15} = \frac{4}{5}$.

4-7 Applications (a) How many minutes are in $^7/_8$ of an hour? Answer: $(^7/_8)(60) =$

$$\frac{7}{\overset{1}{\underset{2}{8}}} \cdot \frac{\overset{15}{\cancel{60}}}{1} = \frac{105}{2} = 52^1/_2 \text{ minutes.}$$

(b) What fraction of an hour is 52 minutes? Answer: $^{52}/_{60} = {}^{13}/_{15}$ of an hour.

(c) A factory can make $^4/_5$ of one computer in a 40-hour work week. If the work week is increased to 50 hours, how many computers can they make? Answer: $(^4/_5)(^{50}/_{40}) = 1$ computer.

Section 5
Plus and Minus

The $+$ sign is omitted whenever possible. For example, we write $+ 18$ simply as 18, and $+ 8 + 9$ as $8 + 9$. The rules connecting plus and minus signs are as follows, with examples in brackets:

$+ -a = -a$	$[+ -6 = -6]$
$- + a = -a$	$[- + 5 = -5]$
$- - a - + a$	$[- - 4 = 4]$
$(-a)(b) = -(ab)$	$[(-^2/_3)(^3/_4) = -((^2/_3)(^3/_4)) = -^1/_2]$
$a(-b) - -(ab)$	$[4(\ ^5/_6) = -((^4/_1)(^5/_6)) = -^{10}/_3]$
$(-a)(-b) = ab$	$[(-^2/_3)(-^3/_4) = (^2/_3)(^3/_4) = {}^1/_2]$
$(-a)/b = -a/b$	$[(-4)/6 = -(^4/_6) = -^2/_3]$
$a/(-b) = -a/b$	$[(^2/_5)/(-^3/_4) = -((^2/_5)/(^3/_4)) = -((^2/_5)(^4/_3)) = -^8/_{15}]$
$(-a)/(-b) = a/b$	$[(-^4/_5)/(-^2/_{15}) = (^4/_5)/(^2/_{15}) = (^4/_5)(^{15}/_2) = 6]$
$a + -b = a - b$	$[6 + -4 = 6 - 4 = 2]$
$a - + b = a - b$	$[^2/_3 - +^1/_2 = {}^2/_3 - {}^1/_2 = {}^4/_6 - {}^3/_6 = {}^1/_6]$
$-a + b = b - a$	$[-7 + 12 = 12 - 7 = 5]$
$-a + b = -(a - b)$	$[-9 + 4 = -(9 - 4) = -5]$
$a - b = -(b - a)$	$[4 - 7 = -(7 - 4) = -3]$
$-a - b = -(a + b)$	$[-^1/_2 - {}^1/_3 = -(^1/_2 + {}^1/_3) = -^5/_6]$
$-a - -b = -a + b$	$[-7 - -8 = -7 + 8 = 1]$

Section 6
The Distributive Law

This law can sometimes make calculations easier. It says: if a, b, c are any numbers, then $ab + ac = a(b + c)$. For example: it looks hard to calculate $(^3/_4)(^4/_7) - (^6/_7)(^3/_4)$, but: $(^3/_4)(^4/_7 - {}^6/_7) = (^3/_4)(-^2/_7) = -^3/_{14}$.

Section 7
Decimals and Percent

One-place decimals are tenths, two-place decimals are hundredths, three-place decimals are thousandths, etc. For example: $0.3 = \frac{3}{10}$, $0.4 = \frac{4}{10} = \frac{2}{5}$; $6.3 = \frac{63}{10} = 6\frac{3}{10}$; $-7.2 = -\frac{72}{10} = -7\frac{2}{10} = -7\frac{1}{5}$; $-0.07 = -\frac{7}{100}$; $0.25 = \frac{25}{100} = \frac{1}{4}$; $-3.62 = -\frac{362}{100} = -3\frac{31}{50}$; $0.215 = \frac{215}{1000} = \frac{43}{200}$. To convert a fraction to a decimal, use long division to divide the denominator into the numerator. For example: $\frac{7}{10} = 0.7$; $\frac{4}{5} = \frac{8}{10} = 0.8$; $9\frac{1}{2} = 9\frac{5}{10} = 9.5$; $7\frac{6}{25} = 7\frac{24}{100} = 7.24$; $-\frac{151}{250} = -0.604$.

The long division in the last example above is:

$$
\begin{array}{r}
0.604 \\
250\overline{)151.000} \\
150.0 \\
\overline{1.000} \\
1.000 \\
\hline
\end{array}
$$

For some fractions, conversion to decimals can only be done approximately. For example: $\frac{1}{3} \doteq 0.333$, $\frac{2}{3} \doteq 0.667$; $\frac{1}{7} \doteq 0.143$. (The symbol $\doteq$ indicates that each result is approximate.)

To convert a fraction of a whole to a percent, write it as a decimal and multiply by 100, that is, move the decimal point two places to the right. For example: $\frac{7}{20} = 0.35 = 35\%$; $\frac{71}{200} = 0.355 = 35.5\%$.

Problem 1

If an 8-hour day is cut by 3 hours, what percent of the day is left?

Solution

$(8 - 3)/8 = \frac{5}{8} = 0.625 = 62.5\%$

Section 8
Less Than and Greater Than
Table 1

Table 1 summarizes these symbols.

Symbol	Read	Examples
$<$	is less than	$3 < 5$, $-2 < 0$, $-5 < -3$
$>$	is greater than	$7 > 2$, $6 > -2$, $-8 > -9$
$\leq$	is less than or equal to	$6 \leq 7$, $6 \leq 6$, $-3 \leq -3$
$\geq$	is greater than or equal to	$9 \geq 8$, $-7 \geq -7$, $-7 \geq -8$

Note that, for two numbers a and b, to say that $a \leq b$ is the same thing as saying $b \geq a$. For example the fact that $7 \leq \frac{21}{3}$ can also be written $\frac{21}{3} \geq 7$; the statement $-9 \geq -11$ can be expressed as $-11 \leq -9$.

A number x is said to be *positive* if $x > 0$; x is *negative* if $x < 0$; x is *non-negative* if $x \geq 0$. The number 0 is neither positive nor negative, but it is non-negative.

If two fractions are close together in value, it may be hard to tell which is larger. To do so, write them with a common denominator.

Problem 2 Which is larger, $^{12}/_{29}$ or $^{5}/_{12}$?

Solution $^{12}/_{29} = 144/12 \cdot 29$, $^{5}/_{12} = 145/12 \cdot 29$; hence $^{5}/_{12}$ is larger. (Note: there is no need to multiply out $12 \cdot 29$.)

**Section 9
Absolute Value**

The *absolute value* of a number is the distance from the number to 0. This distance is always non-negative. For example, the absolute value of -6 is 6, since -6 is six units from the origin, 0, on the coordinate line (Chapter 2, Section 1). The absolute value of 6 is also 6. Thus, 6 and -6 have the same absolute value.

The absolute value of a number x is written $|x|$. Thus, the above results may be written

$$|-6| = |6| = 6$$

The absolute value of zero is zero; the absolute value of any positive number is positive; and the absolute value of any negative number is also positive. We express this as follows:

$$|0| = 0$$

If $a \neq 0$, then $|a| > 0$

**Section 10
Working with Inequalities**

The same number may be added to both sides of an inequality. Suppose we know that $x_2 - {}^{2}/_3 \geq 3{}^{1}/_3$. Then we may add ${}^{2}/_3$ to both sides to obtain $x_2 - {}^{2}/_3 + {}^{2}/_3 \geq 3{}^{1}/_3 + {}^{2}/_3$, that is, $x_2 \geq 4$.

The same number may be subtracted from both sides of an inequality. Suppose we are given that $x_3 + 6 \leq 14$. Then we may conclude that $x_3 + 6 - 6 \leq 14 - 6$, that is, $x_3 \leq 8$.

If both sides of an inequality are multiplied by the same *positive* number, then the inequality sign stays the same. Suppose ${}^{2}/_3 x_1 \geq -{}^{7}/_3$. Then we may multiply both sides by the positive number 3 to obtain $3 \cdot ({}^{2}/_3)x_1 \geq 3 \cdot (-{}^{7}/_3)$, that is, $2x_1 \geq -7$.

If both sides of an inequality are multiplied by the same *negative* number, then the inequality sign is reversed, that is, $\leq$ becomes $\geq$, and $\geq$ becomes $\leq$. Suppose we multiply both sides of the inequality $\frac{2}{3}x_1 \geq -\frac{7}{3}$ by the negative number -3. Then we must change $\geq$ to $\leq$: $-3 \cdot (\frac{2}{3})x_1 \leq -3 \cdot (-\frac{7}{3})$, that is, $-2x_1 \leq 7$.

This last law, that multiplication by a negative number reverses inequalities, is most frequently used for the negative number -1. Multiplication by -1 changes all $+$ signs to $-$ signs, all $-$ signs to $+$ signs, the $\leq$ sign to $\geq$, and the $\geq$ sign to $\leq$.

Problem 3 Rewrite the inequality $3x_1 - x_2 + 4 \leq 2x_2 - x_3 + 1$ as a standard format constraint—that is, with all the x's on the left and a number on the right.

First Solution Subtract $2x_2$ from both sides: $3x_1 - x_2 + 4 - 2x_2 \leq 2x_2 - x_3 + 1 - 2x_2$, that is, $3x_1 - 3x_2 + 4 \leq -x_3 + 1$. Now add x_3 to both sides: $3x_1 - 3x_2 + 4 + x_3 \leq -x_3 + 1 + x_3$, that is, $3x_1 - 3x_2 + x_3 + 4 \leq 1$. Finally, subtract 4 from both sides: $3x_1 - 3x_2 + x_3 \leq -3$.

Fancy Solution Add $-2x_2 + x_3 - 4$ to both sides: $3x_1 - 3x_2 + x_3 \leq -3$.

Problem 4 Write the constraint of Problem 3 as a constraint in standard format with a positive constraint bound, that is, with the number on the right positive.

Solution If we multiply the final form of Problem 3 by -1 it becomes $-(3x_1 - 3x_2 + x_3) \geq -(-3)$, that is, $-3x_1 + 3x_2 - x_3 \geq 3$. Now the constraint bound is 3, which is positive.

Section 11
Finding Unknown Numbers

We only need to be able to solve rather simple problems of this type, such as the following.

Problem 5 What number should be added to $\frac{7}{5}$ to yield $\frac{3}{4}$?

Solution Call the unknown number x. We want to have $\frac{7}{5} + x = \frac{3}{4}$. Then $x = \frac{3}{4} - \frac{7}{5} = -(\frac{7}{5} - \frac{3}{4}) = -(\frac{28}{20} - \frac{15}{20}) = -\frac{13}{20}$. Check: $\frac{7}{5} + -\frac{13}{20} = \frac{28}{20} - \frac{13}{20} = \frac{15}{20} = \frac{3}{4}$ ✔.

Problem 6 What number should be multiplied by $-\frac{3}{2}$ to yield $\frac{4}{5}$?

Solution We want $x(-\frac{3}{2}) = \frac{4}{5}$. Then $x = (\frac{4}{5}) \div (-\frac{3}{2}) = -(\frac{4}{5})(\frac{2}{3}) = -\frac{8}{15}$.

Check: $(-\frac{8}{15})(-\frac{3}{2}) = \dfrac{\overset{4}{\cancel{8}}}{\underset{5}{\cancel{15}}} \cdot \dfrac{\overset{1}{\cancel{3}}}{\underset{1}{\cancel{2}}} = \frac{4}{5}$ ✔.

Problem 7   What number should be multiplied by $^7/_8$ so that when the result is added to $-^3/_2$ the sum is zero?

Solution   We want $-^3/_2 + x(^7/_8) = 0$. Then $x(^7/_8) = ^3/_2$. Then $x = (^3/_2) \div (^7/_8) = (^3/_2)(^8/_7) = ^{12}/_7 = 1^5/_7$. Check: $(^{12}/_7)(^7/_8) + -^3/_2 = ^{12}/_8 - ^3/_2 = ^3/_2 - ^3/_2 = 0$ ✔.

Exercises

(All answers are at the end of Appendix B.)

A1. Write each of the following in the form $^m/_n$, where m and n are integers, $n > 0$, and n is as small as possible. (Write $^m/_1$ simply as m.)

(a) 2×2
(b) $(-2)(-3)$
(c) $6 \div -3 + 2(4 - 3)$
(d) $(-^2/_3)(^9/_{20})$
(e) $(-^2/_3) \div (-^9/_{20})$
(f) $-^3/_4 + ^5/_{12}$
(g) $-(^{17}/_8 - ^2/_{12})$

A2. Write the answers to parts e and g in Exercise 1 in mixed fraction form.

A3. Find:

(a) 6^{-1} (e) $(^3/_2)^{-1}$
(b) $(-8)^{-1}$ (f) $(-^5/_6)^{-1}$
(c) -8^{-1} (g) $((^3/_4)^{-1})^{-1}$
(d) $(^2/_3)^{-1}$

A4. Find the least common denominator for $^5/_6$ and $-^7/_9$.

A5. Convert to decimals:

(a) $-7^4/_{25}$ (b) $-(-3.21 - ^{15}/_4)$

A6. By what percent do you increase a 40-hour work week when you change it to a 45-hour week?

A7. (a) Which is larger, $^{13}/_{24}$ or $^7/_{13}$?
(b) Which is smaller (that is, most negative), $-^{13}/_{24}$ or $-^7/_{13}$?

A8. Find $|-^3/_4|$, $|-70|$, $|8 - 12|$, $|8 - 8|$, $|--3|$, and $|12 - 8|$.

A9. Rewrite the inequality $-4x_1 + 3 - 2x_2 + x_4 \geq -5x_1 - 2x_2 - 2x_3 - 4 + 3x_4$ as a standard format constraint with a positive bound, that is, as an inequality with all the x's on the left and a positive number on the right.

A10. What number should be multiplied by $-^7/_2$ so that when the result is added to $^6/_5$ the final result is zero?

APPENDIX B
Solutions and Answers to Selected Exercises

Section 1-1   1. (a) 3×3 (d) 5×1 (g) 4×3

2. (b) There is no such entry in the given matrix.
   (d) 2

3. (a) $(1,1)$ entry   (c) $(1,3)$ entry

4. (a) $\begin{bmatrix} 7/_4 & 7/_4 & 64 \\ 64 & -3/_2 & 0 \end{bmatrix}$

5. (a) $\begin{bmatrix} 1 & 2 & 3 & 4 & 5 \\ 2 & 2 & 3 & 4 & 5 \\ 3 & 3 & 3 & 4 & 5 \\ 4 & 4 & 4 & 4 & 5 \end{bmatrix}$

(e) $\begin{bmatrix} 1 & 1 & 1 & 1 & 1 \\ 2 & 2 & 2 & 2 & 2 \\ 3 & 3 & 3 & 3 & 3 \\ 4 & 4 & 4 & 4 & 4 \end{bmatrix}$

(c) $\begin{bmatrix} 1 & 2 & 3 & 4 & 5 \\ 2 & 4 & 6 & 8 & 10 \\ 3 & 6 & 9 & 12 & 15 \\ 4 & 8 & 12 & 16 & 20 \end{bmatrix}$

(g) $\begin{bmatrix} 1 & 1\frac{1}{2} & 2 & 2\frac{1}{2} & 3 \\ 1\frac{1}{2} & 2 & 2\frac{1}{2} & 3 & 3\frac{1}{2} \\ 2 & 2\frac{1}{2} & 3 & 3\frac{1}{2} & 4 \\ 2\frac{1}{2} & 3 & 3\frac{1}{2} & 4 & 4\frac{1}{2} \end{bmatrix}$

6. $\begin{bmatrix} -3 & 4 \\ a & b \end{bmatrix}$

$$a + b = 1$$
$$-3 + a = 4 + b$$
$$a - b = 7$$
$$2a = 8$$
$$a = 4$$
$$b = -3$$
$$A = \begin{bmatrix} -3 & 4 \\ 4 & -3 \end{bmatrix}$$

7. The associated matrix is $\begin{bmatrix} 2520 & 2000 & 1950 & 800 \\ 3240 & 1780 & 1530 & 1360 \\ 760 & 370 & 250 & 125 \end{bmatrix}$.

8. (a) The matrix could be of size 35×5724 but this would be much harder to print out and read than size 5724×35, since a computer printout has a set width but unlimited length. Thus, in this and all succeeding PHE problems we will assume the production matrix has size 5724×35.

(b) The total number of entries is 5724 times 35, that is, 200,340 entries.

Section 1-2   1. (a) $R_2 = \begin{bmatrix} 2 & 2 & 3 & 1 \end{bmatrix}$

(b) $C_3 = \begin{bmatrix} -1 \\ 3 \\ 2 \end{bmatrix}$

(c) There is no such row in the matrix.

2. (a) $\begin{bmatrix} 2 & 3 & 7 \\ 4 & 6 & 8 \\ 6 & x & -1 \end{bmatrix}$; here, the x indicates that we are free to choose any number whatsoever for the $(3,2)$ entry since nothing in the problem specifies it. If we wish to give a

particular solution, we might choose x = 0 and obtain $\begin{bmatrix} 2 & 3 & 7 \\ 4 & 6 & 8 \\ 6 & 0 & -1 \end{bmatrix}$.

3. The numbers in R_2 represent the production of heavy sleeping bags. The numbers in C_3 represent the production of the Rutland, Vermont, factory.

5. 12

6.
(a) $\begin{bmatrix} 1 & 1 & 1 \\ 2 & 2 & 2 \\ 3 & 3 & 3 \end{bmatrix}$, for example.

(b) All of the entries are the same.

7. $\begin{bmatrix} 1 & 2 & 3 \\ 4 & 5 & 6 \\ 7 & 8 & 9 \end{bmatrix}$ $R_1 \leftrightarrow R_2$ $\Longrightarrow$ $\begin{bmatrix} 4 & 5 & 6 \\ 1 & 2 & 3 \\ 7 & 8 & 9 \end{bmatrix}$ $C_2 \leftrightarrow C_3$ $\Longrightarrow$ $\begin{bmatrix} 4 & 6 & 5 \\ 1 & 3 & 2 \\ 7 & 9 & 8 \end{bmatrix}$ $R_2 \leftrightarrow R_3$ $\Longrightarrow$

$\begin{bmatrix} 4 & 6 & 5 \\ 7 & 9 & 8 \\ 1 & 3 & 2 \end{bmatrix}$ (This is one way to do it.)

8. (d) The total production of foot-long hot dogs in the week of April 21 by PHE.
(e) The total number of menu items produced by the Tuscaloosa store in the week of April 21.

Section 1-3

1. (a) $\begin{bmatrix} -2 & 0 & 6 \\ 4 & 8 & 14 \\ 6 & 16 & 10 \end{bmatrix}$ (d) $\begin{bmatrix} 2 \\ -4 \\ 0 \end{bmatrix}$ (f) $[-4 \ \ 6 \ \ 10 \ \ 20]$ (j) $\begin{bmatrix} -3 & -9/2 & 15/2 \\ -9 & -21/2 & -6 \end{bmatrix}$

2. (a) $\frac{1}{2}\begin{bmatrix} 3 \\ 5 \\ -1 \end{bmatrix}$ (c) $\frac{1}{4}\begin{bmatrix} 2 & -3 \\ 1 & 0 \\ 4 & 3 \end{bmatrix}$ (e) $\frac{1}{12}\begin{bmatrix} 3 & 4 \\ 8 & 9 \end{bmatrix}$ (h) $\frac{3}{8}\begin{bmatrix} -16 & -12 & -1 \\ 4 & -5 & 0 \\ -20 & 6 & -8 \end{bmatrix}$

(i) $\frac{1}{120}\begin{bmatrix} -168 & 160 & -240 \\ 50 & -18 & 84 \\ 0 & 135 & 16 \end{bmatrix}$

3. (a) 2 (b) No number will work, since multiplication by a number does not alter the size of a matrix and the second matrix is a different size than the first. (d) No number will work here either, since to get the (1,1) entries correct requires multiplication by 2 but to get the (2,2) entries correct requires multiplication by 1. (f) $-\frac{12}{7}$

4. (a) $3\begin{bmatrix} 3 & -4 \\ 1 & 0 \end{bmatrix}$ (c) $6\begin{bmatrix} -2 & 1 \\ 1/2 & -1/2 \end{bmatrix}$ (f) $-\frac{3}{4}\begin{bmatrix} 3 & -4/5 \\ 16/21 & -20/3 \end{bmatrix}$ (See Appendix A, Section 11, if necessary.)

7. $\frac{800}{1000} \times \frac{15}{10} = \frac{8}{10} \times \frac{15}{10} = 4 \times \frac{15}{50} = \frac{60}{50} = \frac{6}{5}$

8. (b) 366V (leap year!) (c) 3653V (3 leap years)

Section 1-4

Available Output
Table for
November

2.
Item	Factory		
	Danbury	Springfield	Rutland
Light sleeping bags	90	90	72
Heavy sleeping bags	18	18	0
Down jackets	36	36	0
Two-person tents	126	198	0
Four-person tents	0	288	72

3. (a) $\begin{bmatrix} 3 & 4 \\ 4 & 5 \end{bmatrix}$ 　(c) They cannot be added. 　(f) $\begin{bmatrix} -3 & 0 & 0 \\ -1 & -2 & -1 \end{bmatrix}$

(h) $\begin{bmatrix} 1 & 1 \\ 1 & 3 \end{bmatrix}$ 　(l) $\begin{bmatrix} 4/5 \\ 9/5 \\ 1 \end{bmatrix}$ 　(p) $\begin{bmatrix} 5/6 \\ -1/6 \end{bmatrix}$ 　(r) $\begin{bmatrix} 2/3 & 1/8 \\ 3/4 & 1/2 \end{bmatrix}$ 　(s) $\begin{bmatrix} 0 \\ 0 \end{bmatrix}$

(t) $\begin{bmatrix} 1/2 & -7/4 & 1 & -4/7 \\ 7/3 & 14/5 & -5/4 & -1/4 \\ -5/3 & 5/8 & -5/2 & 9/4 \\ 2/9 & 1/8 & 7/15 & -5/6 \end{bmatrix}$

4. (a) $\begin{bmatrix} 11 & -13 & 24 \\ 13 & 5 & 26 \\ 12 & 22 & 8 \end{bmatrix}$ 　(c) $\begin{bmatrix} -3 \\ 3 \\ 1 \end{bmatrix}$

(d) The operation cannot be performed, since the matrices involved are different
sizes.

(f) $\begin{bmatrix} 2/5 \\ -33/100 \end{bmatrix}$ 　(g) $\begin{bmatrix} 2 & 0 & 0 \\ 0 & 2 & 0 \\ 0 & 0 & 2 \end{bmatrix}$

(i) While the first two matrices can be combined, the last matrix cannot be combined with the result since it has a different size from the other two. Thus,
the problem has no solution.

(j) $\begin{bmatrix} 363/40 & -59/20 \\ 61/40 & 23/4 \end{bmatrix} = 1/40 \begin{bmatrix} 363 & -118 \\ 61 & 230 \end{bmatrix}$

5. (a) $\begin{bmatrix} 5 & -1 & 1 \\ 4 & 3 & -2 \end{bmatrix}$ (c) $\begin{bmatrix} 0 & 0 & 0 \\ 0 & 0 & 0 \end{bmatrix}$ A matrix consisting entirely of zeros is called a
zero matrix. In this case we have the 2×3 zero matrix.

(d) Since they have different sizes, there is no matrix which when added to the first
yields the second.

(e) $[1/2 \quad -5/12 \quad -5/2]$

6. (a) $\begin{bmatrix} 4 \\ 0 \\ 1 \end{bmatrix}$ (c) No solution; the matrices are not both the same size.

(d) $\begin{bmatrix} -\frac{1}{3} & \frac{3}{4} & \frac{4}{5} \\ -\frac{41}{60} & \frac{7}{4} & -\frac{1}{7} \end{bmatrix}$

7. (c) $\mathbf{W} = \mathbf{F} - 2\mathbf{U}$

Section 1-5
1. (a) 47   (c) 6
2. (a) 3×5. Notice that if you multiply a matrix by a square matrix the answer is the same size as the original matrix. (c) They cannot be multiplied. (d) 6×5
3. Not enough information is given in the problem. They could have been a 5×2 and a 2×7 or a 5×4 and a 4×7 or infinitely many other sizes.

4. (a) $[11]$ (e) $\begin{bmatrix} 7 & 10 & 9 \\ 10 & 8 & 5 \\ 2 & 4 & 4 \end{bmatrix}$

(f) and (g) Notice that with even rather simple matrices, the order of multiplication matters. (h) and (i) Notice that sometimes (very rarely) the order of multiplication does not matter.

(j) $\begin{bmatrix} 3 & -1 & 5 \\ 2 & 4 & -2 \end{bmatrix}$ (k) They cannot be multiplied. (n), (o), (p), and (q) Here you are

verifying that $\left(\begin{bmatrix} -1 & 1 & 2 \\ 2 & -2 & 1 \end{bmatrix} \begin{bmatrix} 3 & 4 & 0 & -2 \\ 1 & -2 & 2 & -1 \\ 1 & -1 & 3 & 2 \end{bmatrix} \right) \begin{bmatrix} 2 & 3 \\ 1 & 2 \\ -1 & 1 \\ -2 & 1 \end{bmatrix}$ is the same as

$\begin{bmatrix} -1 & 1 & 2 \\ 2 & -2 & 1 \end{bmatrix} \left(\begin{bmatrix} 3 & 4 & 0 & -2 \\ 1 & -2 & 2 & -1 \\ 1 & -1 & 3 & 2 \end{bmatrix} \begin{bmatrix} 2 & 3 \\ 1 & 2 \\ -1 & 1 \\ -2 & 1 \end{bmatrix} \right)$. If there are three matrices to be

multiplied, it does not matter which multiplication is done first.

(r) $[0]$ (s) $\begin{bmatrix} 6 & 3 & -3 & 9 \\ -2 & -1 & 1 & -3 \\ 4 & 2 & -2 & 6 \\ -2 & -1 & 1 & -3 \end{bmatrix}$

5. (d) $\begin{bmatrix} 1 & 0 & 0 \\ 0 & 1 & 0 \\ 0 & 0 & 1 \end{bmatrix}$ (e), (f), (g), and (h) $\begin{bmatrix} 2 & 1 & 3 \\ 3 & 2 & 1 \\ 1 & 2 & 3 \end{bmatrix}$

(i) $\begin{bmatrix} 0 & 0 & 0 \\ 0 & 0 & 0 \\ 0 & 0 & 0 \end{bmatrix}$ (l) and (m) $\begin{bmatrix} 0 & 0 & 0 \\ 0 & 0 & 0 \\ 0 & 0 & 0 \end{bmatrix}$ (n) and (o) Notice how multiplying by

$\begin{bmatrix} 1 & 0 & 0 \\ 0 & 1 & 0 \\ 0 & 0 & 1 \end{bmatrix}$ does not change the other matrix. Compare what happens when

you multiply a number by the number 1.

6. (b) $\begin{bmatrix} 18 & 1 \\ 17 & 6 \end{bmatrix}$ (d) $\begin{bmatrix} ^2/_3 & 0 \\ ^{14}/_{27} & -^1/_4 \end{bmatrix}$

8. The answer matrix is $\begin{bmatrix} -1 & 7 & 13 \\ 11 & 19 & 35 \\ 3 & 7 & 12 \end{bmatrix}$.

9. (c) Since PQ is defined, Q must have 7 rows. Since Q has only one column, Q is 7×1. Then QP is 7×7 and thus has 49 entries.

10. (a) $\begin{bmatrix} 0 & 0 \\ 0 & 0 \end{bmatrix}$ (b) $A = \begin{bmatrix} 1 & 0 \\ 0 & 1 \end{bmatrix}$

    (c) There is no such matrix because a 2×2 matrix times any matrix must, if it is defined, be a matrix with just two rows.

    (d) $\begin{bmatrix} 1 & 0 & 1 \\ 0 & 1 & 1 \end{bmatrix}$

Section 1-6

1. $\begin{bmatrix} 3 & 3 & 0 \\ 1 & 5 & 3 \\ 2 & 4 & 2 \end{bmatrix} \left(^1/_6 \begin{bmatrix} -2 & -6 & 9 \\ 4 & 6 & -9 \\ -6 & -6 & 12 \end{bmatrix} \right) = ^1/_6 \begin{bmatrix} 3 & 3 & 0 \\ 1 & 5 & 3 \\ 2 & 4 & 2 \end{bmatrix} \begin{bmatrix} -2 & -6 & 9 \\ 4 & 6 & -9 \\ -6 & -6 & 12 \end{bmatrix}$

$= ^1/_6 \begin{bmatrix} 6 & 0 & 0 \\ 0 & 6 & 0 \\ 0 & 0 & 6 \end{bmatrix} = \begin{bmatrix} 1 & 0 & 0 \\ 0 & 1 & 0 \\ 0 & 0 & 1 \end{bmatrix}$

2. $3 \begin{bmatrix} 1 & 2 & 4 \\ 7 & 1 & 5 \end{bmatrix} - 2 \begin{bmatrix} 1 & 2 & 4 \\ 7 & 1 & 5 \end{bmatrix} = (3 - 2) \begin{bmatrix} 1 & 2 & 4 \\ 7 & 1 & 5 \end{bmatrix} = \begin{bmatrix} 1 & 2 & 4 \\ 7 & 1 & 5 \end{bmatrix}$

8. (i) $A^2 = \begin{bmatrix} 1 & -2 \\ 2 & 1 \end{bmatrix} \begin{bmatrix} 1 & -2 \\ 2 & 1 \end{bmatrix} = \begin{bmatrix} -3 & -4 \\ 4 & -3 \end{bmatrix}$

    (ii) $A^3 = A^2A = \begin{bmatrix} -3 & -4 \\ 4 & -3 \end{bmatrix} \begin{bmatrix} 1 & -2 \\ 2 & 1 \end{bmatrix} = \begin{bmatrix} -11 & 2 \\ -2 & -11 \end{bmatrix}$

    (iii) $A^5 = A^3A^2 = \begin{bmatrix} -11 & 2 \\ -2 & -11 \end{bmatrix} \begin{bmatrix} -3 & -4 \\ 4 & -3 \end{bmatrix} = \begin{bmatrix} 41 & 38 \\ -38 & 41 \end{bmatrix}$

    (iv) $A^{10} = A^5A^5 = \begin{bmatrix} 41 & 38 \\ -38 & 41 \end{bmatrix} \begin{bmatrix} 41 & 38 \\ -38 & 41 \end{bmatrix} = \begin{bmatrix} 237 & 3116 \\ -3116 & 237 \end{bmatrix}$

    (There are other ways to get the answer in four steps.)

9. $-2D$

Review Exercises Chapter 1

1. (a) $\begin{bmatrix} 40 & 20 & 30 \\ 10 & 50 & 30 \end{bmatrix}$ (b) R_2, C_3.   (c) 2×3 (d) (2,1), (1,2)

2. (a) $\begin{bmatrix} 60 & 30 & 45 \\ 15 & 75 & 45 \end{bmatrix}$ (b) $^3/_2 \begin{bmatrix} 40 & 20 & 30 \\ 10 & 50 & 30 \end{bmatrix}$

3. (a) 25　　(b) $\begin{bmatrix} 100 & 50 & 75 \\ 25 & 125 & 75 \end{bmatrix}$

4. (a) $\begin{bmatrix} 5 & 5 & 0 \\ 0 & 15 & 5 \end{bmatrix}$　　(b) $\begin{bmatrix} 35 & 15 & 30 \\ 10 & 35 & 25 \end{bmatrix}$

   (c) $\begin{bmatrix} 40 & 20 & 30 \\ 10 & 50 & 30 \end{bmatrix} - \begin{bmatrix} 5 & 5 & 0 \\ 0 & 15 & 5 \end{bmatrix}$

5. (a) $\begin{bmatrix} 20 \\ 30 \\ 15 \end{bmatrix}, \begin{bmatrix} 200 \\ 300 \\ 150 \end{bmatrix}$　　(b) $10 \times 200 + 50 \times 300 + 30 \times 150 = \$21,500$

   (c) $\begin{bmatrix} 10 & 50 & 30 \end{bmatrix} \begin{bmatrix} 200 \\ 300 \\ 150 \end{bmatrix}$　　(d) $\begin{bmatrix} 40 & 20 & 30 \end{bmatrix} \begin{bmatrix} 200 \\ 300 \\ 150 \end{bmatrix} = \$18,500$

   (e) $\begin{bmatrix} 40 & 20 & 30 \\ 10 & 50 & 30 \end{bmatrix} \begin{bmatrix} 200 \\ 300 \\ 150 \end{bmatrix} = \begin{bmatrix} 18500 \\ 21500 \end{bmatrix}$

6. (a) $\begin{bmatrix} -2 & {}^{19}\!/_{12} & -{}^{1}\!/_{5} \end{bmatrix}$　　(b) not defined　　(c) $\begin{bmatrix} -2 & {}^{4}\!/_{3} \\ 4 & -{}^{1}\!/_{6} \end{bmatrix}$

   (d) $\begin{bmatrix} -{}^{73}\!/_{60} \\ {}^{107}\!/_{60} \end{bmatrix}$　　(e) $\begin{bmatrix} 1 & -2 & 3 \\ 4 & 5 & -6 \\ 7 & -8 & 9 \end{bmatrix}$

7. (a) $\begin{bmatrix} 0 & 0 \\ 0 & 0 \end{bmatrix}$　　(b) $\begin{bmatrix} -4 & -12 \\ 12 & 4 \end{bmatrix}$　　(c) $\begin{bmatrix} 1 & 2 \\ 3 & 4 \end{bmatrix}$　　(d) not defined

   (e) $\begin{bmatrix} 0 & -11 \\ -3 & 7 \\ 16 & -6 \\ -15 & 6 \end{bmatrix}$

8. (a) $\frac{1}{2}\begin{bmatrix} 1 & 3 \\ -5 & 7 \end{bmatrix}$　　(b) $\frac{1}{6}\begin{bmatrix} 3 & 9 \\ -15 & 8 \end{bmatrix}$

   (c) $\frac{1}{840}\begin{bmatrix} -560 & -630 & 672 \\ 700 & -720 & 735 \end{bmatrix}$　　(d) $\frac{3}{5}\begin{bmatrix} 15 & -2 \\ -5 & 20 \end{bmatrix}$

Supplementary
Exercises
Chapter 1

1. 4×7

3. $x = {}^{2}\!/_{3}$

5. (a) $\begin{bmatrix} 3 & 0 \\ 0 & 1 \end{bmatrix}$　(c) $\begin{bmatrix} 1 & 0 \\ 0 & {}^{1}\!/_{2} \end{bmatrix}$　(e) $\begin{bmatrix} 7 & 0 \\ 0 & {}^{1}\!/_{4} \end{bmatrix}$

7. $5x = 10$, so $x = 2$. Then $3x + 2y = 2x - y$ becomes $6 + 2y = 4 - y$; so $3y = -2$, and $y = -{}^{2}\!/_{3}$.

8. (e) Its first and second rows are interchanged.
   (f) It becomes a matrix with C_1 the same as C_2 in the original matrix, C_2 the same as C_3, C_3 the same as C_4, and C_4 the same as $2C_1$.
9. 25×162
11. (a) The 1×25 matrix with all entries equal to 1.
   (b) The 162×1 matrix with all entries equal to 1.
12. (b) The amount motel 4 would have made on June 15 if it had had the same rates as motel 5.

Section 2-1

1. (a)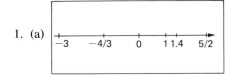

2.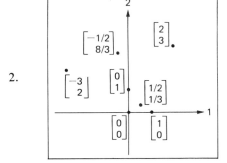

4. $\begin{bmatrix} -45\sqrt{2} \\ 45\sqrt{2} \end{bmatrix}$ or $\begin{bmatrix} 45\sqrt{2} \\ -45\sqrt{2} \end{bmatrix}$ according to whether second base is on the positive or negative 2-axis. (This result requires use of the Pythagorean Theorem.)

5. (b) 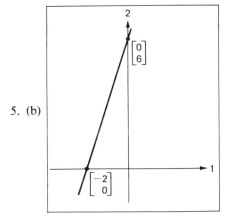   (d)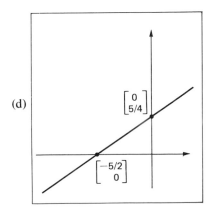

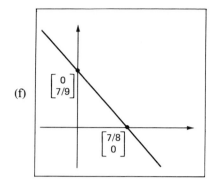

(f) $\begin{bmatrix} 0 \\ 7/9 \end{bmatrix}$ $\begin{bmatrix} 7/8 \\ 0 \end{bmatrix}$

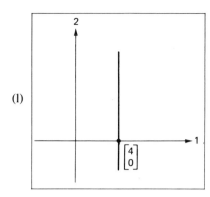

(l) $\begin{bmatrix} 4 \\ 0 \end{bmatrix}$

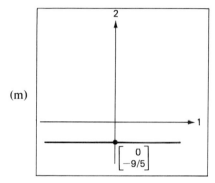

(m) $\begin{bmatrix} 0 \\ -9/5 \end{bmatrix}$

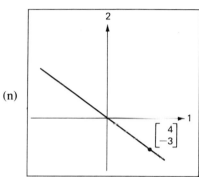

(n) $\begin{bmatrix} 4 \\ -3 \end{bmatrix}$

(p) 2-axis, (q) 1-axis

6. (c) For $x_1 = 80$, $x_2 = .5$. However, such a value for x_1 is not included in the graph and it is not proper to so extrapolate. In fact, essentially no insulation is required for sleeping at 80°F, so that the answer $x_2 = .5$ is incorrect.

   (d) Similarly, for $x_1 = -80$, the formula predicts that $x_2 = 4^1/_2$. However, again we have no reason to suppose that this answer is correct.

7. (a) $x_2 = -2$ (c) $x_1 = {}^{20}/_3$

9. (a) $x_2 = 6.2$ (b) $x_1 = 11$ (e) $3x_1 - 2x_2 = 0$
   (h) $5x_1 - 4x_2 = -20$ (i) $-x_1 + x_2 = 1$

10. (a) $[0 \quad 1] \begin{bmatrix} x_1 \\ x_2 \end{bmatrix} = 6.2$ (b) $[1 \quad 0] \begin{bmatrix} x_1 \\ x_2 \end{bmatrix} = 11$

    (e) $[3 \quad -2] \begin{bmatrix} x_1 \\ x_2 \end{bmatrix} = 0$ (h) $[5 \quad -4] \begin{bmatrix} x_1 \\ x_2 \end{bmatrix} = -20$

    (i) $[-1 \quad 1] \begin{bmatrix} x_1 \\ x_2 \end{bmatrix} = 1$

11. (a) $\begin{bmatrix} x_1 \\ -x_2 \end{bmatrix}$ (b) $\begin{bmatrix} -x_1 \\ x_2 \end{bmatrix}$ (c) $\begin{bmatrix} -x_1 \\ -x_2 \end{bmatrix}$

    (d) same as (c)   (e) $\begin{bmatrix} x_2 \\ x_1 \end{bmatrix}$

12. Sale of x_1 hamburgers gives a profit of $15x_1$ cents $= 0.15x_1$ dollars; hence a net profit of $0.15x_1 - 105$ dollars. Thus, the relation is $x_2 = 0.15x_1 - 105$, which we rewrite as $0.15x_1 - x_2 = 105$. Using the flow chart on p. 67, we see that the graph of this linear equation is the line through the points $\begin{bmatrix} 105/0.15 \\ 0 \end{bmatrix}$ and $\begin{bmatrix} 0 \\ 105/-1 \end{bmatrix}$, that is, the points $\begin{bmatrix} 700 \\ 0 \end{bmatrix}$ and $\begin{bmatrix} 0 \\ -105 \end{bmatrix}$. It will be convenient to mark off the 1-axis and 2-axis in units of 100 (Figure 1). This graph is mathematically correct but quite unsatisfactory in a sales presentation for the following reasons: (1) the negative values of x_1 are meaningless, (2) the terms x_1, x_2, 1-axis, 2-axis, will not be understood, (3) the negative values of x_2 may not be understood by the managers, (4) the graph rises too slowly to excite the managers about selling burgers. We therefore make the following changes in the presentation, resulting in Figure 2: (1) eliminate that part of the graph to the left of the 2-axis; (2) call the axes "burgers sold" and "profit"; (3) call the negative 2-axis "loss"; (4) mark the 1-axis in units of 1000.

Figure 1

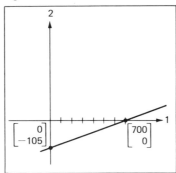

Figure 2

YOUR PROFIT DEPENDS A LOT ON YOUR DAILY HAMBURGER SALES!

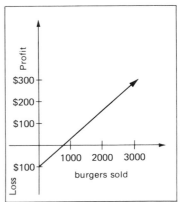

Section 2-2

1. $\begin{bmatrix} 1 & 2 & 3 \\ 3 & 8 & 13 \end{bmatrix}$ $R_2 - 3R_1$ ⟶ $\begin{bmatrix} 1 & 2 & 3 \\ 0 & 2 & 4 \end{bmatrix}$ $R_1 - R_2$ ⟶ $\begin{bmatrix} 1 & 0 & -1 \\ 0 & 2 & 4 \end{bmatrix}$

$\frac{1}{2}R_2$ ⟶ $\begin{bmatrix} 1 & 0 & -1 \\ 0 & 1 & 2 \end{bmatrix}$; answer: $\begin{bmatrix} -1 \\ 2 \end{bmatrix}$

3. $\begin{bmatrix} 0 & 3 & 16 \\ 1 & -6 & 0 \end{bmatrix}$ ⤬ $\begin{bmatrix} 1 & -6 & 0 \\ 0 & 3 & 16 \end{bmatrix}$ $R_1 + 2R_2$ ⟶ $\begin{bmatrix} 1 & 0 & 32 \\ 0 & 3 & 16 \end{bmatrix}$

$\frac{1}{3}R_2$ ⟶ $\begin{bmatrix} 1 & 0 & 32 \\ 0 & 1 & {}^{16}/_3 \end{bmatrix}$; answer: $\begin{bmatrix} 32 \\ {}^{16}/_3 \end{bmatrix}$

5. $\begin{bmatrix} 4 & -1 & 2 \\ -20 & 5 & -1 \end{bmatrix}$ $R_2 + 5R_1$ $\begin{bmatrix} 4 & -1 & 2 \\ 0 & 0 & 9 \end{bmatrix}$

$\frac{1}{4}R_1$ $\begin{bmatrix} 1 & -\frac{1}{4} & \frac{1}{2} \\ 0 & 0 & 9 \end{bmatrix}$; answer: no solution, the lines are parallel.

7. $\begin{bmatrix} -2 & 1 & 3 \\ 4 & -2 & -6 \end{bmatrix}$ $R_2 + 2R_1$ $\begin{bmatrix} -2 & 1 & 3 \\ 0 & 0 & 0 \end{bmatrix}$

$-\frac{1}{2}R_1$ $\begin{bmatrix} 1 & -\frac{1}{2} & -\frac{3}{2} \\ 0 & 0 & 0 \end{bmatrix}$; answer: $\begin{bmatrix} -\frac{3}{2} + \frac{1}{2}x_2 \\ x_2 \end{bmatrix}$, where x_2 may be any number.

The lines are identical.

9. $\begin{bmatrix} \frac{22}{5} \\ \frac{48}{5} \end{bmatrix}$

13. no solution, the lines are parallel

17. $\begin{bmatrix} 0 \\ x_2 \end{bmatrix}$ where x_2 may be any number. The lines are identical.

21.

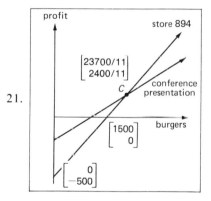

The "crossover point" was found by solving $\begin{bmatrix} 0.15x_1 - x_2 = 105 \\ x_1 - 3x_2 = 1500 \end{bmatrix}$, where the first equation came from the solution of Exercise 12, Section 2-1. The summary presentation to the 894 managers is: "As long as you sell at least 2155 burgers a day you will be beating the average Midwestern store, even if they sell as many burgers as you do."

Section 2-3   1. (a) Operate Springfield for one day and Waterbury for 2 days.
(b) Do not operate Springfield at all and operate Waterbury for 3 days.
(c) Operate Springfield for $\frac{2}{3}$ of one day and Waterbury for 4 days.
(d) Impossible, since the value for x_1 in the solution of the system is negative.

2. Dearborn	4	3	2	1	0
Oshkosh	0	2	4	6	8

Oshkosh	8	7	6	5	4	3	2	1	0
Dearborn	0	0.5	1	1.5	2	2.5	3	3.5	4

3. Let x_1, x_2, x_3, respectively, be the number of packets of type 1, 2, 3 used to make an expedition packet. Then

$$\begin{bmatrix} 3x_1 + 4x_2 + 2x_3 = 86 \\ 2x_1 + x_2 + 5x_3 = 88 \\ 3x_1 + 3x_2 + 3x_3 = 90 \end{bmatrix}$$

4. Let x_1, x_2, x_3, x_4, be, respectively, the number of hamburgers, superburgers, fish-feasts, and root beers in an order. We are asked to find all solutions (at least all "meaningful" solutions) of the system $\begin{bmatrix} 30x_1 + 40x_2 + 100x_3 + 25x_4 = 350 \\ 40x_1 + 60x_2 + 140x_3 + 30x_4 = 480 \\ 50x_1 + 70x_2 + 170x_3 + 35x_4 = 570 \end{bmatrix}$.

Section 2-4

1. (b) $\begin{bmatrix} -1 \\ 2 \\ -2 \end{bmatrix}$ (d) $\begin{bmatrix} 132 \\ 460 \\ -716 \end{bmatrix}$ (f) $\begin{bmatrix} -1 \\ 3 \\ 1 \end{bmatrix}$ (h) $\begin{bmatrix} 0 \\ 0 \\ 0 \end{bmatrix}$ (j) $\begin{bmatrix} 1 \\ 2 \\ 3 \end{bmatrix}$ (l) $\begin{bmatrix} -5 \\ 0 \\ 15 \end{bmatrix}$ (n) $\begin{bmatrix} -14 \\ -7 \\ 43 \end{bmatrix}$

(p) $\begin{bmatrix} 52/58 \\ 117/58 \\ -16/58 \end{bmatrix}$ (r) $\begin{bmatrix} 604/477 \\ 110/477 \\ 546/477 \end{bmatrix}$ (t) $\begin{bmatrix} x_3 \\ x_3 \\ x_3 \end{bmatrix}$ (v) $\begin{bmatrix} 53/16 \\ 41/16 \\ 3/16 \end{bmatrix}$ (x) $\begin{bmatrix} (91 - 25x_4)/31 \\ (22 - 35x_4)/31 \\ (39 + 7x_4)/31 \\ x_4 \end{bmatrix}$

(z) $\begin{bmatrix} 5 \\ -3 \\ 4 \\ 5 \\ 4 \end{bmatrix}$

3. (a) Run Springfield 6 hours, 40 minutes; close Danbury; run Mystic 5 hours.
   (b) Close Springfield; run Danbury and Mystic 5 hours each.
4. No. The system has no solution.
6. The mistake occurs in the second matrix, which should read

$$\begin{bmatrix} 1 & 4/3 & 0 & -1/3 \\ 0 & 2 & 3 & 1 \\ 0 & 8/3 & 4 & 4/3 \end{bmatrix} \xrightarrow{\begin{array}{c} R_1 - 2/3 R_2 \\ R_3 - 4/3 R_2 \end{array}} \begin{bmatrix} 1 & 0 & -2 & -1 \\ 0 & 2 & 3 & 1 \\ 0 & 0 & 0 & 0 \end{bmatrix}$$

The correct answer is $x_1 = 2x_3 - 1$, $x_2 = 1/2(1 - 3x_3)$. I found the particular solution corresponding to $x_3 = 2/3$. While such mistakes are possible, they are rare. Unfortunately, these kinds of errors are impossible to detect since the answer one obtains does, in fact, check. I was lucky enough to have given this problem on an exam so the students caught the error.

7. None

9. Let $x_1 = $ Lefty's take, $x_2 = $ Spike's take, $x_3 = $ Jimmy's take and $a = $ the total stolen. Then $x_1 + x_2 + x_3 = a$, $x_3 = 2(x_1 + x_2)$, and $x_1 = x_3 - x_2 - 10{,}000$. The system to solve is

$$\begin{bmatrix} x_1 + & x_2 + x_3 = & a \\ 2x_1 + 2x_2 - x_3 = & 0 \\ -x_1 - & x_2 + x_3 = & 10{,}000 \end{bmatrix}$$

Trying to solve this, we obtain

$$\begin{bmatrix} 1 & 1 & 1 & a \\ 2 & 2 & -1 & 0 \\ -1 & -1 & 1 & 10{,}000 \end{bmatrix} \Rightarrow \begin{bmatrix} 1 & 1 & 1 & a \\ 0 & 0 & -3 & -2a \\ 0 & 0 & 2 & 10{,}000 + a \end{bmatrix} \Rightarrow$$

$$\begin{bmatrix} 1 & 1 & 1 & a \\ 0 & 0 & 1 & 2a/3 \\ 0 & 0 & 1 & (10{,}000 + a)/2 \end{bmatrix}$$

This system will be inconsistent unless the last two rows give the same value to x_3. That is, we must have $x_3 = 2a/3 = (10{,}000 + a)/2$ or $4a = 30{,}000 + 3a$ or $a = 30{,}000$. Thus, $x_3 = 20{,}000$ and $x_1 + x_2 = 10{,}000$. Since Lefty got $\$1000$ more than Spike, $x_1 = \$5500$, $x_2 = \$4500$.

10. (a) no solution

(c) $\begin{bmatrix} 2 \\ x_2 \\ 3 \\ 2 \end{bmatrix}$ (e) $\begin{bmatrix} 2 - 3x_3 \\ 3 - 2x_5 \\ x_3 \\ 4 \\ x_5 \end{bmatrix}$ (g) $\begin{bmatrix} 2 - x_3 \\ \frac{1}{3}(-6 + 2x_3) \\ x_3 \\ 3 \end{bmatrix}$ (i) $\begin{bmatrix} 1 - 2x_3 \\ 2 - x_3 \\ x_3 \\ -2 \\ 2 \end{bmatrix}$

Section 2-5 4. They should ship 3 of packet I, 6 of packet II, 2 of packet III, 3 of packet IV, and 1 packet V.

5. The system is inconsistent. No solution.

Section 2-6 2. $\begin{bmatrix} 3 + \frac{3}{4}x_2 - \frac{3}{2}x_3 \\ x_2 \\ x_3 \end{bmatrix}$; the planes are identical.

4. $\begin{bmatrix} -\frac{1}{3} + x_3 \\ \frac{2}{3} - 2x_3 \\ x_3 \end{bmatrix}$; the planes meet in a line.

6. No solution—the planes have no common point.

8. $\begin{bmatrix} \frac{1}{3} - \frac{2}{3}x_3 \\ -\frac{2}{3} + \frac{1}{3}x_3 \\ x_3 \end{bmatrix}$; the planes meet in a line.

9. The equations of Section 2-6 represent planes in space—that is, 2-dimensional spaces in a 3-dimensional space. By analogy, the system of Exercise 4 in Section 2-3 represents three 3-dimensional spaces in a 4-dimensional "space"—whatever that is. Since the solution to this system contains one free variable, it must represent a single line in 4-dimensional space. So the three 3-dimensional spaces intersect in a 1-dimensional "space," namely a line. This is analogous to Figure 25, in which three 2-dimensional "spaces" (planes) intersect in a 0-dimensional "space" (a point), but in the present case all of the dimensions are one higher, which makes things more difficult to visualize.

Review Exercises
Chapter 2

1. (a) There are four possible work schedules (see Table 1)

Table 1
(in days)

	I	II	III	IV
Me	7	7	7	7
Jane	0	1	2	3
Bill	6	6	6	6
Fred	3	2	1	0

(b) There is only one possibility (Table 2).

Table 2
(in days)

	I
Me	6
Jane	4
Bill	6
Fred	2

(c) Again, there are four possibilities (Table 3).

Table 3
(in days)

	I	II	III	IV
Me	5	5	5	5
Jane	4	5	6	7
Bill	3	3	3	3
Fred	3	2	1	0

(d) Now there are three possible schedules (Table 4).

Table 4
(in days)

	I	II	III
Me	$3\frac{1}{2}$	$3\frac{1}{2}$	$3\frac{1}{2}$
Jane	0	1	2
Bill	$3\frac{1}{4}$	$3\frac{1}{4}$	$3\frac{1}{4}$
Fred	$2\frac{1}{4}$	$1\frac{1}{4}$	$\frac{1}{4}$

2.

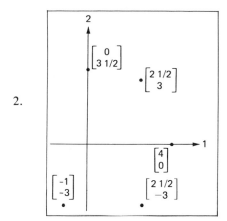

3. (a)

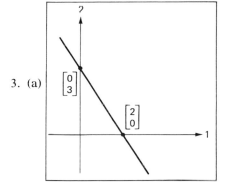

(b)

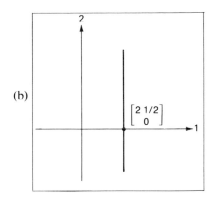

(c)

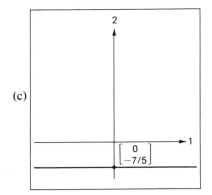

(d)

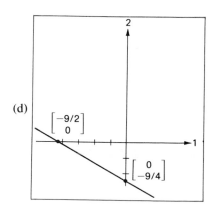

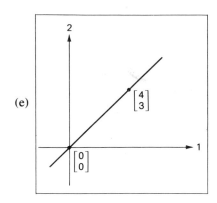

(e)

4. (a) $\begin{bmatrix} 2 \\ 3 \end{bmatrix}$ (b) No solution.   (c) $\begin{bmatrix} 7/2 - 1/2 x_2 \\ x_2 \end{bmatrix}$ (d) $\begin{bmatrix} 590/77 \\ 192/77 \end{bmatrix}$

(e) No solution   (f) $\begin{bmatrix} 2 - x_2 \\ x_2 \\ 1 \end{bmatrix}$ (g) $\begin{bmatrix} 12/5 \\ -3/5 \\ -1/5 \end{bmatrix}$ (h) $\begin{bmatrix} -2 \\ 1 \\ 3 \end{bmatrix}$

(i) $\begin{bmatrix} 3/2 - 3/2 x_4 \\ 1/4 + 3/4 x_4 \\ 5/2 + 3/2 x_4 \\ x_4 \end{bmatrix}$ (j) $\begin{bmatrix} 38 \\ 29 \\ 83 \\ 47 \end{bmatrix}$

Supplementary
Exercises
Chapter 2

2.

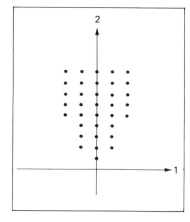

3. (b)

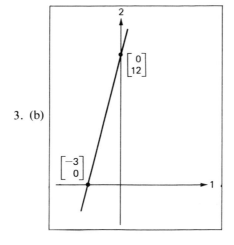

(d)

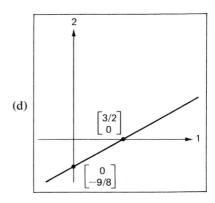

(f)

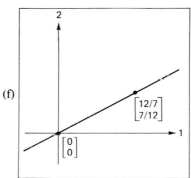

(h)

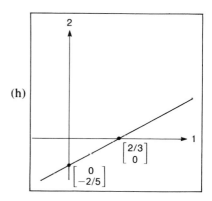

4. (b) $x_2 - -3$ (d) $3x_1 + 2x_2 = -6$

5. (b) The lines are parallel.   (d) $\begin{bmatrix} 4/5 \\ 5/6 \end{bmatrix}$

6. (b) $\begin{bmatrix} 7 \\ 3 \end{bmatrix}$ (d) $\begin{bmatrix} 0 \\ 5 \end{bmatrix}$ (f) $\begin{bmatrix} 1 \\ -1 \\ 2 \end{bmatrix}$ (h) $\begin{bmatrix} 1 \\ 2 \\ 1 \end{bmatrix}$ (j) $\begin{bmatrix} -2 \\ 3 \\ 1 \end{bmatrix}$ (l) $\begin{bmatrix} -4 + 2x_3 \\ -1 + x_3 \\ x_3 \end{bmatrix}$

(n) $\begin{bmatrix} 2 - x_2 \\ x_2 \\ 1 \end{bmatrix}$ (p) $\begin{bmatrix} -2 \\ 1 \\ 3 \end{bmatrix}$ (r) $\begin{bmatrix} 2 \\ 1 \\ 2 \end{bmatrix}$ (t) $\begin{bmatrix} -14/3 \\ 47/6 \\ 1/3 \end{bmatrix}$ (v) $\begin{bmatrix} 2 \\ 3 \end{bmatrix}$

7. (b) $\begin{bmatrix} 0 \\ 2 \\ -1 \\ 1 \end{bmatrix}$ (d) $\begin{bmatrix} 6 - 2x_4 \\ 3/2 \\ -2 + 2x_4 \\ x_4 \end{bmatrix}$ (f) $\begin{bmatrix} 3/2 - 5x_4 \\ 4 + 8x_4 \\ 5/2 - 20x_4 \\ x_4 \end{bmatrix}$ (h) No solution

(j) $\begin{bmatrix} -1 \\ 0 \\ 4 \\ -1/3 \end{bmatrix}$ (l) $\begin{bmatrix} 3/2 - 3/2 x_4 \\ 1/4 + 3/4 x_4 \\ 5/2 + 3/2 x_4 \\ x_4 \end{bmatrix}$

8. Let x_1, x_2, x_3 be the number of senior execs, juniors, and workers, respectively. Then $x_2 = 5x_1$, $x_3 = 50x_2$, $1000x_1 + 500x_2 + 200x_3 = 107,000$; so we solve

$$\begin{bmatrix} 10x_1 + & 5x_2 + 2x_3 = 1070 \\ 5x_1 - & x_2 & = & 0 \\ & 50x_2 - & x_3 = & 0 \end{bmatrix} \text{ to obtain } \begin{bmatrix} 2 \\ 10 \\ 500 \end{bmatrix}$$

10. 2 A's, 3 B's, 10 C's, 9 D's, 8 F's. (Despite the tremendous thought that went into my grading policy, the college fired me for incompetency.)

12.

$$\begin{bmatrix} 0 & 1 & 2 \\ 3 & 1 & 0 \\ 1 & 2 & 3 \end{bmatrix} \xrightarrow{R_3 + R_1} \begin{bmatrix} 0 & 1 & 2 \\ 3 & 1 & 0 \\ 1 & 3 & 5 \end{bmatrix} \xrightarrow{R_1 - R_3} \begin{bmatrix} -1 & -2 & -3 \\ 3 & 1 & 0 \\ 1 & 3 & 5 \end{bmatrix}$$

$$\xrightarrow{R_3 + R_1} \begin{bmatrix} -1 & -2 & -3 \\ 3 & 1 & 0 \\ 0 & 1 & 2 \end{bmatrix} \xrightarrow{-R_1} \begin{bmatrix} 1 & 2 & 3 \\ 3 & 1 & 0 \\ 0 & 1 & 2 \end{bmatrix}$$

13. (a) $-x_1 + x_2 = 1$ (c) $3x_1 + 4x_2 = 0$ (f) $-3x_1 + 2x_2 = 12$

Section 3-1

1. (a) $\begin{bmatrix} 0 & 1 & -1 & 1 & 3 \\ 2 & 1 & 0 & -3 & 2 \\ -1 & -1 & 3 & -1 & 0 \end{bmatrix}$ (b) $\begin{bmatrix} 0 & 1 & -1 & 1 \\ 2 & 1 & 0 & -3 \\ -1 & -1 & 3 & -1 \end{bmatrix}$ (c) $\begin{bmatrix} 3 \\ 2 \\ 0 \end{bmatrix}$

(d) $\begin{bmatrix} 0 & 1 & -1 & 1 \\ 2 & 1 & 0 & -3 \\ -1 & -1 & 3 & -1 \end{bmatrix} \begin{bmatrix} x_1 \\ x_2 \\ x_3 \\ x_4 \end{bmatrix} = \begin{bmatrix} 3 \\ 2 \\ 0 \end{bmatrix}$

3. (a) $\begin{bmatrix} 1 & 0 & 0 & -8/5 \\ 0 & 1 & 0 & 1/5 \\ 0 & 0 & 1 & -4/5 \end{bmatrix}$

(b) The system does not have a unique solution.

(c) $\begin{bmatrix} -1 + 8/5 x_4 \\ 4 - 1/5 x_4 \\ 1 + 4/5 x_4 \\ x_4 \end{bmatrix}$, where x_4 may be any number.

6. (a) It will. (c) It won't. (f) It will.

7. (b) The system has a unique solution. (d) The system does not have a unique solution.

2. (a) 100 (b) 9900

4. (a) $\begin{bmatrix} 2 & 3 & 4 \\ 5 & 6 & 7 \\ 8 & 9 & 10 \end{bmatrix}$ (c) $\begin{bmatrix} 2 & 3 & 0 \\ 5 & 6 & 0 \\ 8 & 9 & 0 \end{bmatrix}$ (e) $\begin{bmatrix} 2 & 3 & 4 \\ 0 & 0 & 0 \\ 0 & 0 & 0 \end{bmatrix}$ (g) $\begin{bmatrix} 8 & 9 & 10 \\ 5 & 6 & 7 \\ 2 & 3 & 4 \end{bmatrix}$

5. (a) $\begin{bmatrix} -18 \\ -25 \end{bmatrix}$ (d) $\begin{bmatrix} -31/3 \\ 301/3 \\ 209/3 \\ -20 \end{bmatrix}$

7. (b) $\frac{1}{5}\begin{bmatrix} 3 \\ 2 \\ -6 \\ 18 \end{bmatrix} = \begin{bmatrix} 3/5 \\ 2/5 \\ -6/5 \\ 18/5 \end{bmatrix}$

8. No, their product is $\begin{bmatrix} 1 & 0 & 0 \\ 1 & 1 & 3 \\ 0 & 0 & 1 \end{bmatrix}$, not the 3×3 identity matrix.

11. (a) $x + 2z$, $3x + 6z$ (b) $\begin{bmatrix} x + 2z = 1 \\ 3x + 6z = 0 \end{bmatrix}$ (c) No solution

(d) The matrix does not have an inverse.

12. $-\frac{4}{9}\begin{bmatrix} 13 \\ 6 \\ 4 \end{bmatrix}$ (First divide R_1 by 2.)

1. $\begin{bmatrix} 1/2 & 0 \\ 0 & 1/3 \end{bmatrix}$

3. $\begin{bmatrix} 1/2 & -2/3 \\ 0 & 1/3 \end{bmatrix}$

5. $\begin{bmatrix} 3/14 & -2/7 \\ 5/14 & -1/7 \end{bmatrix}$

7. $\begin{bmatrix} -105/217 & 105/124 \\ -28/31 & -70/93 \end{bmatrix}$

9. No inverse
11. No inverse

13. $\begin{bmatrix} -1 & 22 & -8 \\ 0 & -5 & 2 \\ 1 & -19 & 7 \end{bmatrix}$

18. $\begin{bmatrix} -27/32 & 9/32 & -3/16 \\ -3/4 & 1/4 & -3/2 \\ 69/64 & 9/64 & -3/32 \end{bmatrix}$

19. $\dfrac{1}{226}\begin{bmatrix} -14 & 28 & 49 & 15 \\ -34 & 68 & 6 & -12 \\ 64 & 98 & 2 & -4 \\ -4 & 8 & 14 & -28 \end{bmatrix}$

20. $\dfrac{1}{5}\begin{bmatrix} -115 & 145 & -64 & -18 \\ 50 & -60 & 26 & 7 \\ 5 & -10 & 6 & 2 \\ 10 & -10 & 3 & 1 \end{bmatrix}$

21. $X = \begin{bmatrix} 8 & -22 \\ -2 & 14 \end{bmatrix}$

22. $(\dfrac{1}{76})\begin{bmatrix} 69 & -135 & -21 & 73 & 126 \\ -20 & 92 & 16 & -52 & -96 \\ -26 & 74 & -2 & -22 & -64 \\ 3 & -29 & 9 & 23 & 22 \\ 24 & -80 & -4 & 32 & 100 \end{bmatrix}$

Section 3-4

1. Order 1: $\begin{bmatrix} 0 \\ 3 \end{bmatrix}$; Order 2: $\begin{bmatrix} 1 \\ 1 \end{bmatrix}$; Order 3: $\begin{bmatrix} 18 \\ 12 \end{bmatrix}$; Order 4: cannot be done; Order 5: $\begin{bmatrix} 3 \\ 61 \end{bmatrix}$; Order 6: cannot be done; Order 7: $\begin{bmatrix} 1087 \\ 2190 \end{bmatrix}$; Order 8: $\begin{bmatrix} 1 \\ 0 \end{bmatrix}$; Order 9: $\begin{bmatrix} 0 \\ 1 \end{bmatrix}$.

3. (c) $\begin{bmatrix} 2 & -11 \\ -1 & 6 \end{bmatrix}\begin{bmatrix} 10 \\ 9 \end{bmatrix} = \begin{bmatrix} -79 \\ 44 \end{bmatrix} = \begin{bmatrix} x_1 \\ x_2 \end{bmatrix}$; (we first switched the equations).

   (e) $\begin{bmatrix} 2 & -11 \\ -1 & 6 \end{bmatrix}\begin{bmatrix} 2 \\ 3 \end{bmatrix} = \begin{bmatrix} -29 \\ 16 \end{bmatrix} = \begin{bmatrix} x_2 \\ x_1 \end{bmatrix}$; (we had to interchange x_1 and x_2).

5. Multiply orders by $\begin{bmatrix} -6 & -6 & 7 \\ {}^{11}/_5 & 2 & -{}^{12}/_5 \\ {}^{24}/_5 & 5 & -{}^{28}/_5 \end{bmatrix}$.

7. (a) $\begin{bmatrix} 4 \\ 11 \\ 0 \end{bmatrix}$ (b) $\begin{bmatrix} 0 \\ 0 \\ 50 \end{bmatrix}$ (c) $\begin{bmatrix} 16 \\ 48 \\ 0 \end{bmatrix}$ (d) $\begin{bmatrix} 72 \\ 0 \\ 0 \end{bmatrix}$

Section 3-5

1. $\dfrac{1}{15}\begin{bmatrix} 1 & 0 & {}^1/_2 \\ 0 & {}^1/_3 & 2 \\ -2 & -1 & {}^1/_2 \end{bmatrix}$

2. $\begin{bmatrix} 12 & -18 & -8 \\ -16 & 24 & 10 \\ 14 & -22 & -8 \end{bmatrix}$

5. $\begin{bmatrix} -{}^{89}/_{90} & {}^{23}/_{30} & -{}^{97}/_{90} \\ -{}^{49}/_{90} & {}^{13}/_{30} & -{}^{47}/_{90} \\ -{}^{20}/_{90} & {}^1/_6 & -{}^{20}/_{45} \end{bmatrix}$

6. $\begin{bmatrix} {}^1/_{225} & 0 & 0 \\ 0 & {}^1/_{225} & 0 \\ 0 & 0 & {}^1/_{225} \end{bmatrix}$

7. (a) False. The matrix $\begin{bmatrix} 1 & 2 \\ 3 & 6 \end{bmatrix}$ has no inverse, so the left side of the equation is meaningless.

(b) True. (c) False. (d) False. (e) True.

8. $(ABC)^{-1} = (BC)^{-1}A^{-1} = C^{-1}B^{-1}A^{-1}$

9. $\frac{1}{4}B^{-1}C^{-1} = (4CB)^{-1}$

10. $X = A^{-1}C^{-1}E$

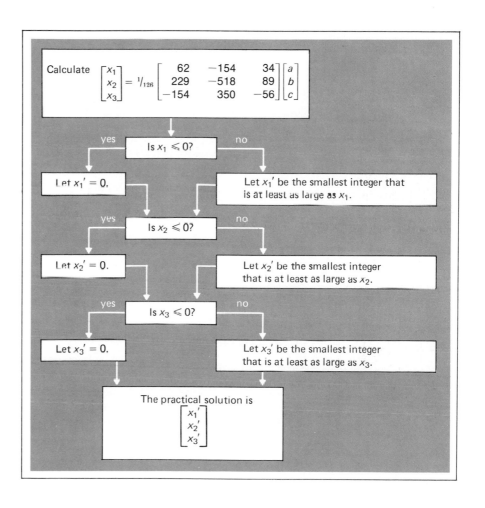

Section 3-6

1. (a) $\begin{bmatrix} -2 & 1 \\ 3/2 & -1/2 \end{bmatrix}$ (b) no inverse (c) $\begin{bmatrix} 2 & 1 \\ 3/2 & 1/2 \end{bmatrix}$ (d) no inverse

(e) $\begin{bmatrix} -117/283 & -528/283 \\ -442/283 & -1456/849 \end{bmatrix}$ (f) $\begin{bmatrix} 0.08 & -0.11 \\ 0.15 & 0.24 \end{bmatrix}$ (approx.)

2. (a)

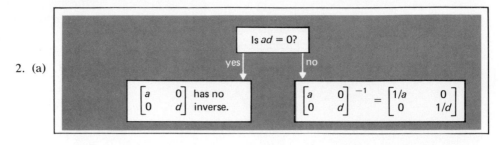

Is $ad = 0$?

yes — $\begin{bmatrix} a & 0 \\ 0 & d \end{bmatrix}$ has no inverse.

no — $\begin{bmatrix} a & 0 \\ 0 & d \end{bmatrix}^{-1} = \begin{bmatrix} 1/a & 0 \\ 0 & 1/d \end{bmatrix}$

(c)

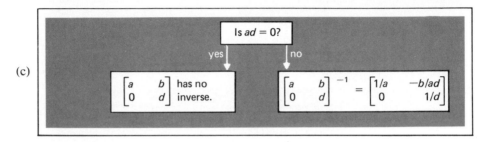

Is $ad = 0$?

yes — $\begin{bmatrix} a & b \\ 0 & d \end{bmatrix}$ has no inverse.

no — $\begin{bmatrix} a & b \\ 0 & d \end{bmatrix}^{-1} = \begin{bmatrix} 1/a & -b/ad \\ 0 & 1/d \end{bmatrix}$

4. $\begin{bmatrix} a & b & 1 & 0 \\ 0 & d & 0 & 1 \end{bmatrix} \Rightarrow \begin{bmatrix} 1 & {}^b/_a & {}^1/_a & 0 \\ 0 & d & 0 & 1 \end{bmatrix}$

$\Rightarrow \begin{bmatrix} 1 & 0 & {}^1/_a & -{}^b/_{ad} \\ 0 & d & 0 & 1 \end{bmatrix} \Rightarrow \begin{bmatrix} 1 & 0 & {}^1/_a & -{}^b/_{ad} \\ 0 & 1 & 0 & {}^1/_d \end{bmatrix}$

5. $\begin{bmatrix} a & b & 1 & 0 \\ c & d & 0 & 1 \end{bmatrix} \Rightarrow \begin{bmatrix} 1 & {}^b/_a & {}^1/_a & 0 \\ 0 & {}^{(ad-bc)}/_a & -{}^c/_a & 1 \end{bmatrix}$

$\Rightarrow \begin{bmatrix} 1 & 0 & {}^d/_{(ad-bc)} & -{}^b/_{(ad-bc)} \\ 0 & 1 & -{}^c/_{(ad-bc)} & {}^a/_{(ad-bc)} \end{bmatrix}$

6. $\begin{bmatrix} k & 0 & 0 \\ 0 & a & b \\ 0 & c & d \end{bmatrix}^{-1} = \begin{bmatrix} {}^1/_k & 0 & 0 \\ 0 & {}^d/_{(ad-bc)} & -{}^b/_{(ad-bc)} \\ 0 & -{}^c/_{(ad-bc)} & {}^a/_{(ad-bc)} \end{bmatrix}$

Review Exercises
Chapter 3

1. Note that $\begin{bmatrix} 3 & 4 & 5 & 2 \\ 1 & 5 & 6 & 0 \\ 5 & 5 & 6 & 5 \\ 1 & 1 & 1 & 1 \end{bmatrix}^{-1} = {}^1/_3 \begin{bmatrix} 5 & -2 & -3 & 5 \\ -1 & 1 & -3 & 17 \\ 0 & 0 & 3 & -15 \\ -4 & 1 & 3 & -4 \end{bmatrix}$.

	une 1	June 8
	15	0
	15	35
	15	0
	15	63

e. (c) $\begin{bmatrix} 0 & 1 & 0 \\ 0 & 0 & 1 \\ 1 & 0 & 0 \end{bmatrix}$

$\begin{bmatrix} 3 & -1 \\ -1 & -1 \\ -1 & 3 \end{bmatrix}$

en $A = A'I = A'(AA') =$

$\begin{bmatrix} 0 \\ -1 \end{bmatrix}$.

$\begin{bmatrix} 1 \\ 64 \end{bmatrix}$ (f) $\begin{bmatrix} 5 & -6 & -13 \\ -4 & 5 & 11 \\ -2 & 2 & 5 \end{bmatrix}$

$\begin{bmatrix} 1 & 4 \\ 1 & 0 \\ -5 & -8 \end{bmatrix}$ (n) $^{1}/_{102} \begin{bmatrix} 58 & -15 & 25 \\ -32 & 3 & -5 \\ 8 & 12 & 14 \end{bmatrix}$

$\begin{bmatrix} 81 & -77 & 48 \\ -60 & 70 & -30 \\ 18 & -6 & -6 \end{bmatrix}$

of X. We must have $-x_1 + x_2 + 3 = 3$ and x_1 x_2

This system has no solution; hence there is no such

d, it will be a matrix with only two columns. Hence this

r.

Exercise 1(i).

$\begin{bmatrix} 23 & 24 & -17 \\ -19 & -18 & 13 \\ -5 & -6 & 5 \end{bmatrix} = \begin{bmatrix} ^{100}/_3 & 36 & -^{76}/_3 \\ -^{1}/_3 & 0 & ^{1}/_3 \\ -^{103}/_6 & -18 & ^{79}/_6 \end{bmatrix}$

4. (b) $\begin{bmatrix} 4/_5 \\ 3/_5 \\ 3/_5 \\ -2/_{15} \\ 1/_3 \end{bmatrix}$

7. $17x_1 + 18x_2 + 19x_3 + 20x_4 = 23$

12. $\begin{bmatrix} 4 & 1 & 3 & -6 & -1 \\ 5 & 2 & 1 & -4 & 2 \\ -1 & 0 & 3 & -4 & -4 \end{bmatrix}$

14. $-1/_2 \begin{bmatrix} k+4 & -k-2 \\ -k-3 & k+1 \end{bmatrix}$

15. $\begin{bmatrix} 4 & -3 & -4 \\ -3/_2 & 5/_4 & 5/_4 \\ -3 & 2 & 3 \end{bmatrix}$

16. If A has an inverse A' and if A'' is also an inverse of A, th
$(A'A)A'' = IA'' = A''$, i.e., $A' = A'' = A^{-1}$.

17. For example, $A = \begin{bmatrix} -1 & 0 \\ 0 & 1 \end{bmatrix}$, $B = \begin{bmatrix} 1 & 0 \\ 0 & -1 \end{bmatrix}$, $C = \begin{bmatrix} -1 \\ 0 \end{bmatrix}$

18. $\begin{bmatrix} 3x_1 - 9/_2 x_2 + 6x_3 = 3/_2 \\ 4/_3 x_1 - 2x_2 + 8/_3 x_3 = 2/_3 \\ -16/_5 x_1 + 24/_5 x_2 - 32/_5 x_3 = -8/_5 \end{bmatrix}$

19. False. For example let $A = [1 \quad 0]$, $B = \begin{bmatrix} 1 \\ 0 \end{bmatrix}$.

20. $X_1 = \begin{bmatrix} 1 & 2 \\ 2 & 3 \end{bmatrix}$ $X_2 = \begin{bmatrix} 1 & -1 \\ 2 & 0 \end{bmatrix}$

Section 4-1

Shipping Costs per Pound (in cents)

	Gallup	Silver City	Carlsbad	Raton
CLAMS				
Los Angeles	142.5	150	207.5	202.5
Houston	225	200	145	185
SHRIMP				
Los Angeles	154.5	162	219.5	214.5
Houston	225	200	145	185
FILLETS				
Los Angeles	142.5	0.5	207.5	202.5
Houston	234	209	154	194